DEPARTING EARTH FOREVER

BOOK ONE
WARNING AND PROMISE

DEPARTING EARTH FOREVER

BOOK ONE

WARNING
AND
PROMISE
DENNIS
CHAMBERLAND

QED
Quantum Editions

Library of Congress Cataloging-in-Publication Data

Chamberland, Dennis, 1951-

Departing earth forever - book one – warning and promise / by Dennis Chamberland.

p. cm.

ISBN 978-1-889422-33-6 (pbk. : alk. paper)

I. Title.
GC66.5.C43 2007
910.9162--dc22

Printed in the United States of America

Photos and Illustrations licensed from Shutterstock.com
or are in the public domain unless otherwise attributed.
The likeness of Jamestown Colonist "Jane" is used with the
gracious permission of the Jamestown Rediscovery Foundation.

Departing Earth Forever

REVIEWS

"A spellbinding read... game changing views, superbly written and illustrated, and a bold charge to civilization's spacefaring future!"

– Alan Stern

- Principal Investigator for the New Horizons Mission to Pluto and former NASA Associate Administrator – Science Mission Directorate

---

"Captivating... provides a convincing portrayal of the future rewards of space travel for humankind."

– Dr. David W. Miller

Adjunct Assistant Professor, University of Illinois, Bionucleonics
North American Technical Center Director, ISOE,
International Atomic Energy Agency

"A chilling warning and potentially fatal end game to humanity's reach for the stars. A siren song on steroids."

– Jim Bacquet

Nuclear Health Physicist (Ret.)

"We should carefully consider these facts and arguments as if our future as a species depends on it. Because it does!"

- Mike Hall

Major General, United States Air Force (Ret.)

"A hyperintelligent lesson on the dangers we face as our explorers prepare to embark on a journey of uncertainty."

– James F. Reed III

Brigadier General, United States Army (Ret.)

"A brilliant expose... the Bible for human space exploration beyond our moon."

– LtCol Robert S. Miller

United States Air Force (Ret.)

"Exciting... engrossing... one of the most courageous books of our time."

– Barbara Cox

Retired NASA Management and Program Analyst, John F. Kennedy Space Center

---

"Dr. Chamberland invites the reader into a thought-provoking conversation that will alter perspectives regarding exploration beyond our planet."

– Donna J. Corley, Ph.D.

"*Departing Earth Forever* sounds the boots-on-the-ground, experienced voice of safety for future space exploration, delivering workable solutions and solid guidelines...a must read."

– Dr. Ken Babington

"*Departing Earth Forever* represents a deep and profound vision of a new frontier offering solutions presented in ways that a non-scientific person can grasp."

– Dr. Glenn Galtere

"Space is a relentless, unforgiving toxic environment (as so eloquently described by Dr. Chamberland) and until Spacecraft engineering mitigations are in place (which they are not currently), no informed/sane Agency or entity should fly unprotected humans into deep space. A superb read!"

– Charles Smallwood, MD, MBA

Former Medical Director, Kennedy Space Center Occupational Medicine and Environmental Health Services

---

"...a gripping, truthful, insightful, and spellbinding composition."

– Dr. Dennis Steindler

Former Executive Director of the McKnight Brain Institute

"It's a rare author who underemphasizes the magnitude of his work, but Dennis Chamberland has done just that. The work could easily be titled Chamberland's Rules for Interstellar Exploration, and it should go down in history alongside such notable works as Asimov's thoughts on robotics. When a Zefram Cochrane finally arrives on the scene, this book should find a spot on his bookshelf as well as that of every aspiring astronaut and explorer who preceded him."

– Dr. Milind Ramesh Limaye

Physician and Commercial Airline Pilot

"Dr. Chamberland writes with a passion and a vision for the future that is seldom seen in such a scholarly undertaking, and I urge readers to approach these volumes with an open mind. A new path lies before us all, if we have the wit to take it."

– Joseph M. Bishop

Former DOD Nuclear Engineer - Aquanaut Engineer

"Dennis Chamberland sounds a stunning and unambiguous warning while offering clearly articulated and viable solutions as humanity races toward sure calamity in human space exploration."

– The Lifeboat Foundation

"...unabashedly identifies the flaws in NASA's current deep space human exploration paradigm that, unless altered, will result ultimately in failed missions and loss of life."

– Tim Novak

Explorer

"Dr. Chamberland provides the missing manual on NASA's development culture, outlining the problems in the engineering organization that need to be addressed for safe travel and rapid development in space. A must-read for anyone considering the trip themselves, and a should-read for anyone wanting to understand the numerous challenges with long term extra-orbital space travel."

– J Marshall Presnell
Systems Architect

"*Departing Earth Forever* brings out strategic points that need to be taken to heart for humanity's survival. Dr. Chamberland lays out a shocking surprise – 'We don't know as much as we thought.'"

– COL Christine V. O'Donnell, RN
United States Army (Ret.)

"Dennis Chamberland draws on his wide-ranging scientific background to write the primer for deep space travel that is readable by the layman – yet technical enough to provide a cautionary tale for the experienced space scientist. This is the dawn of an era when we will see much of the content of the book become reality as man prepares to expand his quest for new horizons."

- Buddy Oakes

"LET THERE BE NO MISUNDERSTANDING, NO FAULTY assumptions, and no ambiguity. Humankind will move off this planet and will build powerful civilizations on Mars, on Luna, on Ceres, and in great solar orbiting cities around the Sun. We shall do so by using our intellect and will not fall victim to our own preconceived prejudices or mindless politics. We are going to accomplish this most monumental of all human achievements by leveraging our philosophical prowess with the tools of our science. And, by doing so, literally nothing can stop our advance outward. We can only lose this quest if we fail to learn from our past and intelligently correct our course on our way to the stars."

– Dennis Chamberland

To
Dr. Corinne Johnson Rutzke
the one who inspired the need for this book
by her passion for exploration and
her love for the students of today
who will become the explorers of tomorrow

And...

To
Elon Musk
the once in a thousand-year genius -
the game changer who is
brilliantly guiding humanity
outward to the new empires

–

fortes fortuna juvat

The First Principle of Human Exploration

"In every exploration system, we must require the systems we build to adapt to the human standard rather than expect the human to adapt to the machine or the environment – and in every design activity, we will protect the human as a primary objective."

– Dennis Chamberland

From *Undersea Colonies* – 2007

DEPARTING EARTH FOREVER

BOOK ONE

NASA
The View
From Inside
NASA

"Take me out, to the black
Tell 'em I ain't coming back
Burn the land and boil the sea
You can't take the sky from me."

Ballad of Serenity - Firefly

NO ONE DRIVES TO THE SAME PLACE FOR THREE decades interfacing with the same people year after year without soon waking up and realizing that they have invariably become as close as flesh and blood family members. That became the reality in my mind and heart after less than a year at the John F. Kennedy Space Center and only deepened in the 30 years I was privileged and blessed to occupy several positions there. I have long considered this in the underlying reason, tone, and message of this book. Each morning at my desk, writing over a period of three years, I have kept this one idea at the top of the stack: I would never sell out my family for a buck on the writing and speaking circuit – never. However, just like any other family, NASA occasionally needs the adult in the room with the courage and deep respect to stand up and open some corrupted wounds so that they can heal. And then to correct the family course that is invariably headed out over the rails to disaster, heartache, and historic loss of credibility in an Agency that has fully earned its historic, world class reputation over more than half a century.

Yet while the perspective I bring to this work is *from* one as a NASA family member, this book is written specifically *to* a far larger human space exploration family that is rapidly expanding quite broadly outside the NASA family walls.

Departing Earth Forever: Warning and Promise – Book One has a key audience:

(1) Those of you who are going to depart Earth forever and permanently colonize Mars and our Moon.

(2) Those of you who are *not* getting on those one-way outbound ships but are quite fascinated in following in detail the epically historic migration of humanity as a multiplanetary species.

As someone who has lived for decades and has intimate inside knowledge about the NASA sausage making factory, I am morally obligated to tell this story so that those who are going to finally harvest the incredible work and accomplishments of NASA will have the tools to make it in one piece to those distant worlds. Thus, these pages were carefully mapped and extensively researched, and a coherent strategy was laid out before the first words were written. In the big picture of your life, future Mars colonist, you definitely need this detailed, expanded picture from an inside view of where we are today, what is about to blow up in our faces, and what can be done to fix it all before you climb on the interplanetary ship on its way out to deep space.

You can trust me on one thing – we *are* going! There is literally nothing that can stop us now, thanks in whole part to Elon Musk's genius. Unfortunately, the sketchy course on which we are embarking is truly in a monumentally desperate need of fixing in everyone's rocket factories. Fortunately, the course adjustment is easy to understand, very straightforward, and just a single decision away – the decision to fix it *now* rather than after the body bag phase. That detailed discussion is what this pair of books is all about.

Much has been made over the endless fun of living and working in microgravity. But few astronauts ever publicly speak of the toxic effects of months spent trying to actually live in it or the fact that astronaut crews often hide the epic misery of trying to live and work in it every single day. Fortunately, there is also an easy fix that is described in detail inside this book. Further, as a professional nuclear engineer, I am very aware that no expert with any brain just begins a detailed warning about the dangers of terrestrial ionizing radiation without first defining some basic terms and limits. And the same is even more true for the far more powerful and massive species of galactic cosmic radiation. The problem is that in today's world of internet educated desk jockey scientists and engineers, the knowledge base is truly in a wickedly disastrous shambles. Further, NASA's press releases are catastrophically skewed toward their own self-interests and have not educated the public but have made the problem infinitely worse just to protect their catastrophic human exploration model. But here, hopefully out of reach from NASA's muzzle and censure, I can give you the big picture, having lived the life of a NASA space life sciences engineer. Together, fellow space traveler, we may establish a common language of the history and science of cosmic radiation, microgravity, interplanetary travel, deep space, and the wild, wild deserts of Mars and Luna. And all of this is told exclusively from the insider's view written specifically to those of you who need it most – the near future crewmembers who require that viewpoint to stay safe and alive and make healthy Mars babies as you grow humanity on distant worlds.

As you turn these pages, you will find that I speak of the people of NASA (the family I was blessed to join) in the most reverent of terms. That NASA family is truly the best and the brightest; the most committed, ethical, and innovative of any civil servants. And NASA employees have consistently voted NASA as the best agency in government year after year for good reason. Yet, NASA is also an arm of a body consisting of many moving parts collectively called the U.S. Government. It is a massive juggernaut that is tasked with making all those parts function in the interests of the nation. It consists of a complex quilt-work of constituencies, powerful interest groups, and corporations that often do not align well with the interests of human space exploration – or even fundamental ethical considerations.

And thus, in a careful reading of this pair of books, notice I am pointedly addressing the areas of the NASA-government interface that represent the small part of NASA that interfaces with Congress – one that frequently capitulates to, or even participates in, government's unfettered permissive power (which is, pointedly, on loan exclusively to itself). In this work, I have no other descriptive choice but to address them as "the Agency." But let me be clear that the most fundamental intent of these books certainly does not include any criticism of the vast majority of NASA – the rank and file who come to work every day as the *crème de la crème* of the American people. We have been, and still are, more grieved than any other element of Americans at the choices made for the NASA exploration vision by our government masters that are quite out of our control. And yet we show up to work each day and dutifully perform our responsibilities, still expressing the same vision, the identical innovation, and the brilliance of our forefathers... when and where we can.

If anyone needs proof of that bold statement, let me answer it in just three words: Mars Exploration Rovers. This program alone (even though matched in genius by the brilliance of New Horizons Pluto serial missions into the Oort Cloud, the James Webb Space Telescope, and many other dozens of equally amazing uncrewed ventures into the deepest parts of space) is the perfect example of a NASA innovative genius that has not lost any of its brilliance since the glory days of Apollo. The startling, steady state, historic stream of astonishing originality of NASA over its vast national infrastructure has become so commonplace that it is expected and emulated by other nations and programs all over this planet. And these incredible people and teams have certainly earned this historically significant reputation fully and completely.

Finally, please understand that NASA is already being wholly and inevitably eclipsed in the future of human exploration and settlement of deep space by the clear frontrunner, SpaceX. Yet, adding to SpaceX is a growing list of serious entities determined to join in the near-term human exploration of the regions beyond low-Earth orbit that include China National Space Administration, Roscosmos State Corporation for Space Activities, Japan Aerospace Exploration Agency, United Arab Emirates Space Agency, the Blue Origin-led National Team, Dynetics, and other nations and organizations that have publicized their deep space exploration goals. Very soon now, NASA's congressional rulers will have deliberately maneuvered the Agency to become a minority player in this venture because they have literally priced us out of the game by protecting an obsolete human exploration system, as you will discover in the next chapters. Again, that pathway was never decided by NASA, but it is one contrived by their swarm of congressional masters for their own political purposes and advantages. It certainly does not arise from a strategy that emerges from a coherent, seamless, human exploration strategy that transcends mere political administrations.

For these reasons, it is essential to understand this fundamental idea even before the book begins:

Not one nation or private enterprise on Earth is going to be spared from the wholesale, deadly collapse of the current Apollo-era model of human space exploration. No amount of taxpayer or state money on the part of any nation can continue to prop it up by applying more countermeasures in deep space. Each nation or entity on the planet that is building on the Apollo model success is unwittingly overwriting their individual mission exploration philosophies. Every current design, and those on the drawing boards from every nation and corporation, are perpetuating the fatally flawed Apollo model. NASA is used as the prime illustrative example in this book because I spent my career in the Agency, I have standing in the discussion, and I am an eyewitness to its perpetuation. But substitute any name for NASA in this book – SpaceX, China National Space Administration, or Roscosmos State Corporation for Space Activities – and the tragic conclusion will *inevitably be the same.*

It really makes no difference how advanced the rocket is, how reusable it is, or how reduced its cost-per-pound to fly if it does not protect its crew, and by extension its mission, as its primary design objective. It is still primitive and will inevitably lead to catastrophic mission failure on every extended deep space mission.

But it does not have to be this way…

We still have time if we immediately recognize the dead-end down which we are charging ahead at flank speed, stop, and redesign. Then we may conquer all of deep space without any limits at all! That is why *Departing Earth Forever* had to be written – and from a view inside the Agency that created the current collision with humans and the immutable natural laws that occupy and rule over everything in deep space above our infinitesimally small cocoon of safety. These books are not written to slow the inevitable migration of our species outward. On the contrary, the fundamental purpose of these books is to encourage – *even demand* – that we press ahead to that historic migration with all possible speed. These books suggest investing in a quintessential, philosophically driven strategy that drives success: exchange the time required for the pause to redesign and reap the ultimate reward of an accelerated and tragedy free exploration of the solar systems after. It will take far less time to redesign than it will to depart on a system that will inevitably fail, and then have to be investigated, redesigned, and rebuilt after the tragic collapse.

It would be just great if we could, for once, skip the knuckle dragging body bag phase on our one-way trip outbound. It is true that all exploration has its inevitable risks. But it is *not* true that it is ethically and humanely acceptable to deliberately engineer in the risks that will always injure and ultimately claim the lives of some fraction of its crews, and then solemnly pontificate at the funerals about the "inevitable cost of human exploration." Such is certainly not the price of exploration. It is instead the unavoidable cost of intransigent human stupidity maladroitly veiled by eloquent eulogies. It is the going-in assumption of these books that humanity is better than that. We do not have to go there this time, regardless of our repetitively tedious exploration history. If the human community has evolved even just a little, we can stop the awful inertia dragging us out into unprotected space and fix the problems before the clock ticks down to zero. If we are capable as a species of stopping and correcting our course, then very soon after we do, we shall depart Earth forever to successfully conquer space all the way to the end of the solar system and beyond. It has always been our destiny.

"In questions of science, the authority of a thousand is not worth the humble reasoning of a single individual."

— Galileo

Public Confessions of a Space Jockey

I have known in my heart of hearts from boyhood that I was a member of the NASA family. I was born into the NASA clan at 12 years of age the day I received the first package of materials I requested from the Kennedy Space Center. I somehow fortuitously ended up on a permanent mailing list and began receiving my much-coveted envelopes monthly from NASA Kennedy Space Center. When I was 15, a glossy photo of the brand new Vehicle Assembly Building arrived, which went up on my wall and accompanied me to OSU where it was proudly displayed until May 1972 when a popular poster of model Barbie Benton displaced it (I wasn't a total geek!). Curiously, I was thus literally imprinted as a NASA family member at a very vulnerable age, and from those early years, I began my long quest to someday walk as an employee through the front door of the Kennedy Space Center (that I knew with all certainty was destined to be my family home).

When I departed active duty as a U.S. Naval officer in 1980, I entered an engineering graduate program at my alma mater, Oklahoma State University. I immediately called the Kennedy Space Center. I was in search of a NASA scientist who would answer my most essential career question: what I should study in my engineering curriculum so that I could come to work at KSC in the field of advanced life support; specifically in development of a Controlled Ecological Life Support System, or CELSS. I was routed to Dr. William Knott III, and when I explained my desires, he sent a letter that detailed the academic path that would best meet that goal. From that day, I followed the academic pathway that would lead to my job interview seven years later in the transfer aisle of the same Vehicle Assembly Building of my dreams. Just before graduating, Dr. Knott put me in touch with NASA Civil Servant Perry Williams, who, with extraordinary patience, took my calls every Friday and directed me weekly for four years post graduation through a Civil Service path until I was finally hired by NASA. I have often said that I was hired just to get me off their phones!

My first boss at NASA, Bob Woods, asked me at the end of my interview in the VAB under the shadow of one of the shuttles, "So, Dennis, do you want to come and work here?" to which I replied, "Sir, I would pay you to come and work here!" And I was not joking. At that moment, I realized that the long road was over. I was finally home!

30 years later at my retirement coffee in the Neil Armstrong Operations and Checkout Building's Mission Briefing Room, I confessed to my NASA family that "Every day for the last 30 years I have driven out here each morning, parked my car, and prayed most sincerely, 'Thank You, God, for this job!'" There was never a single day that I did not mean every word of that prayer of deepest sincerity.

Whether I was working as an engineer or in the role of a Principal Investigator on a science investigation, I never once in three decades ever felt a single moment of regret, nor did I ever feel like I could have had a better job elsewhere. NASA was my family then and remains so today. And just like any other family, I love them even though the family sometimes drives me crazy. And even though they sometimes make choices that astonish me, they are still family – and will be until the end.

Every single day I was privileged to drive through the gates of the John F. Kennedy Space Center, I was humbled that I was doing my part by participating in planning for a new human civilization on other worlds. No matter how trivial the task, or how annoying the bureaucratic nonsense, I never once lost sight of our astonishing mission.

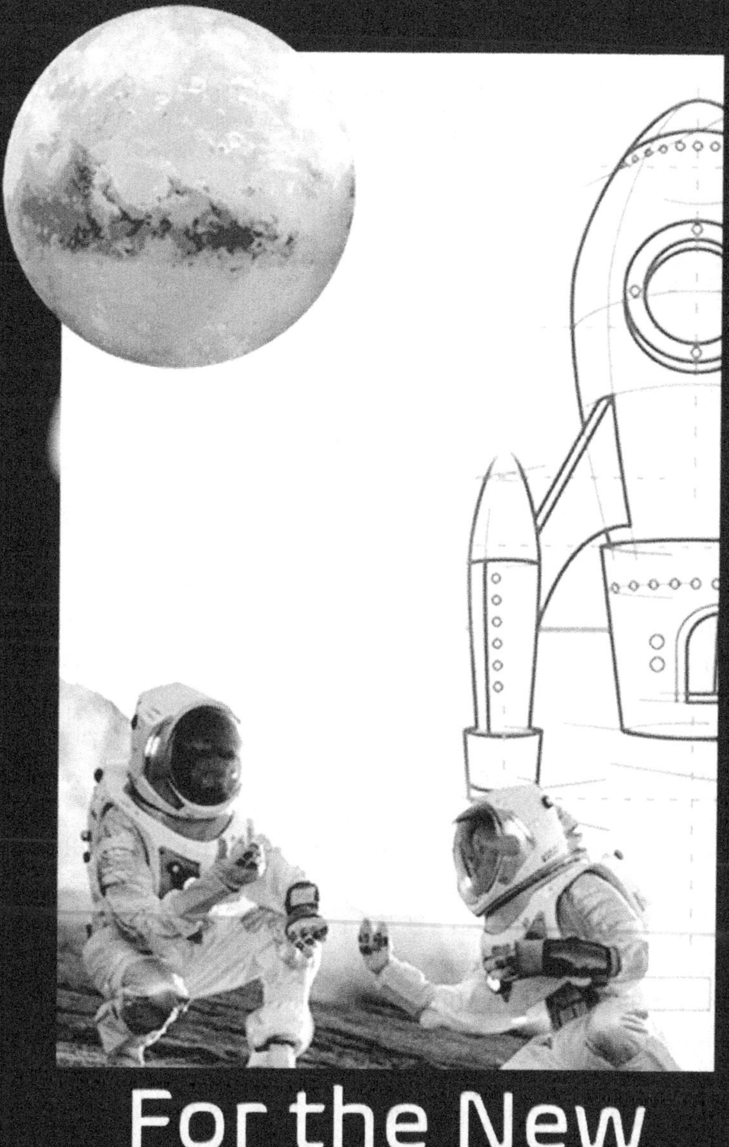

For the New Citizens of Other Worlds

"All civilizations become either spacefaring or extinct."

— Carl Sagan

THIS IS THE VERY FIRST MOMENT IN HUMAN HISTORY that this book could have been created. It is specifically written by a career NASA space life sciences engineer, conceived for you who are destined to soon become Mars and Lunar settlers. As a Bioenvironmental Engineer, it has been one of my tasks to communicate biology to engineers and engineering to biologists. And as a professional writer, it is my task to clearly communicate biology and engineering in the form of advanced space life support systems to the 21st-century astronaut crews who will depart Earth forever to settle new worlds.

And why "biology"? As we shall discuss in a future chapter titled, "It's the Biology Stupid," you will soon discover (one way or another) that human rated rocket science is all about biology first!

Each of you who make up these future astronaut crews will require the information contained inside these pages to successfully reach and then survive in good health on a distant world. It was never written to intentionally evoke controversy, although it surely will (of which I reserve the right to wholly ignore). I only care about the welfare of those of you who will actually be departing Earth forever. Therefore, let the critics chained to their keyboards get a real life somewhere outside their front door. Authentic explorers have much better things to do with our precious time than to act as a critic of someone who gets up every day and actually works toward building a new human empire.

Many of you future interplanetary colonists who are reading these words already know exactly who you are and what your destiny will be – and it's not on this planet. You have successfully ignored all the nay-sayers and clueless noisemakers and have packed your bags for the one-way trip into deep interplanetary space. Just a couple of decades ago, most of you did not even exist. But I know very well who each of you naked apes are because... I am one of you (with just a few more cycles around the nearest star under my belt). And when I teach, speak, and write about Mars, I know what I am talking about. While many reading these words were still in diapers, I was living out and participating in the creation of the professional science, engineering, and technology required for colonizing a distant world. And all the while, I was also collecting the materials for this book on your behalf. And because these books were written specifically for *your* eyes....

you should listen.

This is the certified, steely-eyed missile man's no bullshit zone. If your eyes are scanning these words, you are sitting in my classroom, as so many hundreds of students have done in various college science classes before you. And when you sit in my classroom, colonist, I want you to know all of it – the good, the bad, and most especially, very specifically, *the very ugly*.

Not only does your life depend on it, but anyone who lives and works beside you on an alien world depends on your level of knowledge and expertise. These are not going to be tourist class flights. Everyone's lives and fates are inextricably linked together by circumstance, sound engineering, good, authentic science, and a shared knowledge of the universe the way it really is. Out there, we all count on one another – what we know, what we respect, and what we fear, love, and are ready to handle. Your journey and my journey as described between these covers are hard-wired exclusively to the very best authentic peer reviewed science, competent engineering, and a management structure that actually knows what the hell it is doing – one that is ethically unwilling to exchange my life or yours for a politically driven mission goal. (You will read about REID later in this book.)

On Mars, when we step off that pad onto the sands, the very first thing every new colonist will confront is the truth as it really is. Not a single word here is science fiction. It is as real as the cold, drifting sands of Mars. A voyage to Mars is not a video game or a movie set. The planet is not there to impress or entertain. Mars demands respect. Mars will kill quickly, silently, and without pause – or with lots of noise and violence – if the new colonist from Earth does not truly understand her.

If you think the COVID-19 Pandemic back on Earth was bad, you absolutely *do not* want to be a part of a mass human die-off isolated on another world, caused not by a viral speck of violent nucleic acids but by a tsunami of human ignorance perpetrated either by any clueless colonist or by the designers of our equipment, our life support shields, or a clueless mission plan.

I want there to be no surprises for anyone. We will go there to go to work, not stand around pissed off that the real Mars sucks. Can you imagine the effect of a single new arrival that shows up appalled at what they find on this new, alien planet – whining to go back home from day one? Laying out what seems at first like a harsh truth in detail and in advance is the purpose of this book. It is my objective to fully prepare you and the rest of your peers who are destined to be the first generation of Mars colonists. If you buckle up your favorite rear into one of those ships solely informed by the Internet, pop-science media, Sci-Fi novels, Hollywood, and NASA, you are pretty much screwed before T minus 0 – trust me.

So, buckle in, fellow interplanetary explorer and settler of new worlds. Pull your seat restraints tight. You're going to need them!

Imminent Catastrophic Collapse

"Civilization is in a race between education and catastrophe. Let us learn the truth and spread it as far and wide as our circumstances allow. For the truth is the greatest weapon we have."

— H. G. Wells

"Beware of false knowledge; it is more dangerous than ignorance."

— George Bernard Shaw

THE SUBTITLE OF THE ONLINE REVIEW OF THIS BOOK which may have intrigued you enough to convince you to part with your hard-earned money, is worth restating up front. Book One is about:

"THE TRUE STORY
of the imminent catastrophic collapse
of the old order of human space exploration
and the astonishing technical revolution
that will take its place."

Although bluntly declaring the warning of that impending catastrophic collapse is vitally crucial and eminently necessary, this book is *far and away profoundly more.* These pages are all about the promise that will inevitably be harvested by all the successes that will follow after the correction of the human space exploration model is in place.

Why

Humans are, unfortunately, quite often exhaustively and boorishly self-aware of their alleged intelligence. Pointedly, that alone represents the core of most of humanity's problems since the dawn of civilization and, indeed, leads directly to the necessity for this book. Said Elon Musk, one of the individuals to whom this book is dedicated, "The biggest mistake I see people making is to assume they're smart." In proper context, he was referring to the contrast between human intelligence and synthetic intelligence (AI). And yet, Musk put his finger directly on the problem that is identified in the many following pages. A superficial self-awareness of personal intelligence leads, in most cases, directly to hubris – especially in a cultural context when humans completely overestimate the aggregate functional intelligence of their civilizations.

While the alternate purposes of these books are certainly not debating sociological eccentricities, nor were they written for species bashing, it is very useful to start this conversation with this deliberate statement of fact: "Human beings are not as functionally intelligent as they would suppose, particularly when organized as functional units."

Rarely do human organizational components build on and leverage the constellation of the assembled intelligencia. The reverse almost always ultimately evidences itself in entrenched absurdity, such as has infected the human spaceflight model today and directly threatens the lives of the crews and the success of their soon-forthcoming extended deep space missions for prolonged periods.

Unfortunately for these crews now enthusiastically lining up to become the auspiciously willing subjects of a cosmic cookout, entrenched absurdity without exception inevitably leads to mass death and the utter ruinous collapse of the best laid plans of the cultures that launch them. The models for these historically cyclic activities provide us, humans, with the requisite wherewithal by which we typically go about intentionally harming ourselves with all the ancillary celebratory bravado common to our species of naked, often quite challenged, bipedal apes with an attitude. As of this writing, our kind is closing in on these events quickly and without evident pause.

As history has repeatedly demonstrated throughout the broad expense of the human experience, epochal change is more often than not heralded by a catastrophic collapse of old orders. We are about to witness just such a catastrophic collapse in human space exploration caused by the imminent, inevitable breakdown of the fatally flawed exploration model itself. This wholly defective model is produced by the deeply substandard and hopelessly naive exploration philosophical foundation upon which the entire house of cards was erected more than half a century ago and then thrust into the 21st century on steroids.

The primary purpose of Book One in this two-book series is to carefully identify the elements of this collapse. We also examine in detail why these impending epic serial disasters are developing as they are. But herein lies our more important purpose – a clear delineation of how, by the implementation of specific engineering methodologies (most notably not more useless layered band-aid "countermeasures"), a seismic, historic disaster can be forestalled with intelligent forethought by every spacefaring nation on Earth, or individually by non-governmental entities.

As will be discussed in exhaustive detail in just a few pages in the next chapter, "The Skies are Clear," the dire warning of this book is certainly *not* intended for *any* relatively brief (less than six months) tourist-class missions into suborbital space or low-Earth orbit, or even short-term Apollo-length cislunar space jaunts.

As is stated again in the pages ahead of you, the first nation or private enterprise that understands and actually heeds this message for long duration deep space exploration and human settlement and makes the necessary corrections in engineering their human space exploration paradigm, will singlehandedly rule the space frontier all the way to the edges of the Solar System and beyond. For, you see, this work is not one of projected doom. It is to herald and fully celebrate the promise of the new human exploration that will inevitably follow – *but only after* the aggregate errors have been corrected by the brighter few of those who will ultimately become known as history's truly significant explorers. Darwin Award explorers that kill themselves by their lack of intellect and proper expedition planning rarely rate very high on history's plaques of honor.

As for those entities who do not heed these clear warnings – they will need to stockpile body bags before the rush. And then they will have to arrange for any impaired survivors to return to Earth, pay the lawyers, settle with the victims' families, and then start it all over again. Unfortunately for these first casualties out of the gate, the more intelligent competition will soon come behind them flying First

Principle ships, and will pass their graves as they successfully and permanently stake out every strategic interplanetary prize deep into the Oort Cloud and ultimately beyond. As we shall soon discuss, this is not at all a new human quandary, but merely the latest in our next intellectually challenged steps of serial species debasement in human expansion: learning by fire and packing plenty of matches.

The cause of this imminent catastrophic collapse is very well known. The entire edifice of the future of all long-term human space exploration above 36,000 miles is built on a fictitious reality that is now being engineered into "human rated" deep spaceships emblazoned with the fake Hollywood slogan of "Failure is Not an Option" when, in truth, failure for them is the only possible outcome by virtue of their already obsolete designs. These fantasy driven expectations are ultimately powered more by hubris than intellect because they certainly do not pass even the most rudimentary of logical assessments of already accumulated space based data. The tragic die has already been cast. Now we must destroy the die and intelligently fabricate a new one or face the disastrously nasty clean-up afterward.

Unless there are radical, immediate changes, their integrated advanced space life support systems will inevitably fail and either seriously impair or permanently injure the crews sent out into the great void. Government and private entities are now planning to launch crews from the home planet with inadequate protection from the assaults of multimorbidity provoked by the unified bio-physiological attacks of galactic cosmic radiation, microgravity, and a wholesale ignorance, misunderstanding, or egregious underestimation of interplanetary risks of long-term exposure to space by the hopelessly inadequate 1960s Apollo era spaceflight model.

Still, tens of thousands of internet-meme-educated space jockeys will queue up in very long lines to throw away their bodies and long-term health to very sketchy pop-science expectations because the ancient hormonally driven lure of the "Gentleman and Gentlewoman Adventurer" is far and away more powerful than their actual IQ. When asked the Facebook meme generated question, "If given a chance, would you sacrifice your life for an experimental human mission to Mars knowing that it may end your life?" with great enthusiasm, one Facebook space pioneer wannabe replied with gusto, "With every part of my being, would do anything to have that opportunity. Would go NOW today!!!! (sic)"

The good news is that it does not have to be this way at all, and, as this book clearly describes, there is a better, actually workable plan for going forward. Not one single explorer astronaut must depart Earth with the expectation of a high probability of death – strapping in and ticking off the laundry list of "ways we are going to die" at the one-way altar of human absurdity. Indeed, there is, in fact, another *far better* way!

The secondary objective of Book One is to clearly define and describe a new, powerful foundational philosophy that displaces the wholly inadequate Apollo model that cannot cope with the challenges of deep space and interplanetary class human rated missions. That philosophy is called the First Principle of Human Exploration. From that foundation, a new human space exploration model is defined in detail and very strongly advocated for immediate implementation. By this plan, the current pace of exploration is only paused, purposefully, with a clear and radical improvement to the design so that human space exploration is not crushed and stalled potentially for decades by tragedy with the direct cost of precious human lives and the long term health of many.

Sooner than later, the other worlds of the Solar System will certainly end up populated and claimed by expansive colonies of the more intelligent explorers who understand the clear distinction between foolish and acceptable risk and who will, for the most part, all arrive alive, healthy, and ready to claim all these new worlds where they and their descendants will live for subsequent generations. Not because they have failed to accept a courageous challenge, but because they are knowledgeable and individually intelligent enough to precisely enumerate risk based on current peer reviewed research and are profoundly and conscientiously selective about the foundational philosophy of their energizing explorational methodology. In the very near future, no truly intelligent human being will ever purposefully board an outbound ship for deep space that is not a certified First Principle vessel.

Redesign now, or redesign after the serial tragedies. We who will depart Earth forever (including this author) get to choose. That is one key element of *Departing Earth Forever*. It is about the warning and about the promise. It is about the collapsing old-school exploration paradigm and the emerging new one that will take its place. It is about the explorers who will not survive the journey well or not at all, just as it is also about the explorers who will not only survive but who will thrive and create whole new robust and self-sufficient civilizations on new worlds far, far away from our home planet.

Book Two of this pair of publications is about specific ways to rebuild our models and crewed hardware using the First Principle of Human Exploration and how the First Principle is being used today. It also suggests deep space exploration models, vehicles, and strategies that guarantee success and the continuation of our species as citizens of these new worlds we dare to permanently occupy. Book Two also details the latest discoveries about the alien nature of these worlds and the engineering and advanced life support strategies we have available today that will enable these planets to be safely occupied from day one.

There will *certainly be imminent change* in the deep space human exploration model, whether it is brought about by what is now developing into an inevitable mass catastrophe or rather by intelligent foresight and correction before the disaster – a deliberate process that begins now instead of after the mass burials on another world. For humans, this would represent a rare, historically unprecedented reflexive bout of true situational and personal self-awareness – but it is certainly not impossible. Sometimes, just sometimes, the naked apes do actually manage to respond to clear and fair warnings and act just in the nick of time to prevent disaster.

Departing Earth Forever: Books One and Two represent that clear and fair warning.

The response to these chapters going forward will mean the difference between success and failure. It will determine which nation or entity will actually be the first to occupy the strategic high grounds with living, healthy human crews who arrive for the long haul – departing Earth forever. Those who are committed to the old resolve – to get there fastest and first despite the toll – will inevitably only end up populating the first tombs of humans on other distant worlds and will eventually be displaced by their conquering intellectual and genetic superiors.

Fellow deep space explorer, wait for the inevitable fix – then fly! You and your descendants will never regret your foresight, intellect, and very close attention to the actual details of human spaceflight in deep space *before* you strap yourself into the seat that you have purchased at so great a cost. Let us just do all that we can do together to assure that the cost is measured only in dollars and not your life.

Skies are Clear!

"That so few now dare to be eccentric marks the chief danger of our time."

— John Stuart Mill

IT IS THE FIRST PURPOSE OF THIS BOOK TO SOUND A warning to the class of long duration human crewed space missions that are functionally unprotected from galactic cosmic radiation and altered gravity. That group is comprised of missions with a duration greater than six months in low-Earth orbit or 30 days in deep space beyond the Van Allen radiation shields above 36,000 miles in altitude.

Below the shields, in low-Earth orbit, let me offer the very first advice to those who intend to travel there:

Beneath the shields, in low Earth orbit, the skies are clear, astronaut! Stop worrying so much and have fun!

Upfront and foremost, I want to separate the issue of traveling as an astronaut in suborbital or low-Earth orbital space for relatively short missions as compared to traveling on extended missions in deep space above 36,000 miles. If you very seriously intend to fly in space in suborbital or low-Earth orbit, as many people are (including this author), the message of warning contained in this book is NOT intended for you.

Where and What is Space?

This pair of books titled *Departing Earth Forever* is all about space and the worlds embedded in space, including our home planet, Earth. Generally speaking, and for all practical purposes, the vast, fundamentally limitless void we call space is anywhere outside the denser lower levels of the atmosphere of Earth. But since the atmosphere does not just suddenly stop at a certain altitude, the question has become not only complicated but controversial. As it turns out, "space" has more than just one different formal and widely accepted definition.

Ask any scientist who owns their own spacesuit, and they will tell you that space is not at all defined as where the atmosphere ends – because, as we shall soon discuss, the International Space Station actually orbits in our atmosphere. Rather, space is defined very concisely as the approximate point where orbital dynamic forces become more influential on the spaceship you are riding in than the aerodynamic forces of what remains of the atmosphere at high altitudes. Fundamentally, that is the point where the atmosphere is insufficient to support the aerodynamically dependent control surfaces of a vessel at suborbital, or even orbital, velocities.

The non-scientist should then ask, "Fine, but *exactly where is that*?"

Much of NASA, the Federal Aviation Administration (FAA), the United States Space Force, and the National Oceanic and Atmospheric Administration (NOAA) – basically, the U.S. government – all agree that those conditions begin at around 50 miles altitude, or just over 80 kilometers above sea level. But the headstrong people in NASA Mission Control disagree with their own Agency and the U.S. Government, fixing the boundary of space at 76 miles, or 122 kilometers, according to Emily Nightingale of the Science and Technology Policy Institute in 2014.[1]

Nearly all the rest of the world disagrees with both of these assessments, fixing the boundary of space at 100 kilometers, or 62 miles, above sea level. This specific height is called the Kármán Line, established by Hungarian American physicist and engineer Theodore von Kármán (1881–1963) in the early 20th century as the elevation where the atmosphere becomes too weak to support aerodynamic flight. Ironically, and mostly lost to 21st-century history, von Kármán himself originally established by his calculations that the boundary of space began at 52.08 statute miles (83.81 kilometers or 275,000 feet) above mean sea level. He determined that at this altitude, the aircraft must be faster than the orbital velocity to derive sufficient lift to support itself.[2]

Even American lawyer Andrew Gallagher Haley (known as the world's first practitioner of space law) coined the now familiar term "von Kármán Line" in reference to the original 275,000-foot figure.[3] Thomas Gangale notes that the 52-mile line was *arbitrarily redefined* to

62 miles (100 km) as the result of an agreement in 1961 between the U.S. and U.S.S.R. delegates to the Fédération aéronautique internationale (FAI/IAF) – an international record-keeping body for aeronautics. Gangale states in his 2017 work *The Non Kármán Line: An Urban Legend of the Space Age* that "it is not clear that the von Kármán calculation and the IAF number (100 km) **ever had anything to do with each other**."[4] Again, the argument was and continues to be not where the Earth's atmosphere actually ends, since the International Space Station flies in the near-vacuum-like extraordinary thin uppermost layers of the planet's atmospheric thermosphere region at 254 miles altitude above Earth. After all is said and done, space is where (again) orbital dynamic forces become more influential on the spaceship you are riding in than aerodynamic force.

In reality, the Earth's "detectable" atmosphere extends out to about 6,214 miles – some 25 times higher than the altitude of the ISS as it orbits the planet.

So, why is there so much controversy over a mere 63,360 feet difference (essentially two Mount Everests) between the estimates of "legal space"? Remember that an orbital bound spacecraft will cover this distance in a blink of an eye – roughly less than two seconds. Unless you are a rocket scientist, lawyer, or government regulator, apparently the only reason anyone else cares is for them to be able to return from these fringe regions in the far reaches of the upper atmosphere and qualify to claim the title of "astronaut."

What is an Astronaut?

Astronaut is a title given to any individual who has flown into space.

That's it. Notice there is nothing added to the requirement. Astronaut is a title fairly and legitimately earned by any individual who has flown into space. Period.

NASA (and everyone else) borrowed the term "astronaut" from the 1930 sci-fi story by Neil R. Jones, "*The Death's Head Meteor*."

Without acknowledging its original use by Jones, NASA's website states, "The term 'astronaut' derives from the Greek words meaning 'space sailor,' and refers to all who have been launched as crew members aboard NASA spacecraft bound for orbit and beyond. The term 'astronaut' has been maintained as the title for those selected to join the NASA corps of astronauts who make 'space sailing' their career profession."[5] Cosmonaut and taikonaut are the Russian and Chinese terms, respectively, for the same Greek derivations.

However, in one of two possible readings of the definition from the NASA website, some people feel that NASA apparently is promoting an odd possessive claim that "astronaut" is a term that applies exclusively to their space faring crews, period, as though the Agency contrived the term. Yet even if true, NASA is not alone in making title-limiting claims and placing restrictions on the designation of astronaut. Adding to the distinction of simply being an individual who has flown into space is the additional claim that the "legitimate" astronaut title comes with government restrictions as one who must also be a designated "professional astronaut" or a person with "special astronaut training."

Why are these arguments and title limitations suddenly appearing out of the inky vacuum with such fervor? Because of the very real specter of astronaut elitism. The term astronaut has long been reserved for a subclass of heroes and heroines in our culture and is now suddenly under threat by a whole new class of astronaut – the tourist space flyer. At least a few professional astronauts (and many astronaut fanboys) do not like this imminent adulteration of their hero status at all.

11 July 2021 – The Floodgates Open

Reality suddenly reared its ugly head in earnest on 11 July 2021 when the very first crop of tourist class astronaut crew riding the newly designated spacecraft models departed Earth on board Virgin Galactic's SpaceShipTwo VSS Unity. Along with two seasoned military turned commercial jet pilots at the helm were an individual designated "Astronaut 001" (Sir Richard Branson, the billionaire owner of Virgin Galactic) and three other non-professional astronaut crew. It was the first entirely tourist designated spaceflight in history, and everyone onboard earned their astronaut wings because the definition of an astronaut is "any individual who has flown into space." Branson, the pilots, and crew rode Unity up to 53 miles (85 kilometers) above the Earth.

Just nine days later, another billionaire (Amazon owner and, on that day, the wealthiest man on Earth), Jeff Bezos, departed Earth with three other tourist class space travelers – all returning safely to terra firma as bonafide astronauts. Bezos and crew flew to 66 miles (106 kilometers) above Earth on board the Blue Origin New Shephard crewed spacecraft, eclipsing Branson by approximately 69,000 feet in altitude. One of the crew members on board was 19-year-old Oliver Damien, the youngest human at that time to be launched into space.

And as if to spectacularly overshadow this historic pair of suborbital excursions into the fringes of near-Earth space, less than two months later, on 16 September 2021, the next all "civilian" crew of four astronauts departed Earth on a SpaceX Falcon 9 rocket, riding a SpaceX Crew Dragon spacecraft into low earth orbital space. Called the Inspiration4 mission, the Dragon was placed into an orbit of 364 miles in altitude and concluded two days, 23 hours, and 3 minutes later with the first crewed splashdown in the Atlantic Ocean since Apollo 9 in 1969. The orbiting SpaceX astronauts experienced a whopping 1,440 times the comparatively brief experience of the suborbital astronaut adventurers riding on either Virgin Galactic's SpaceShipTwo VSS Unity or Blue Origin's New Shephard, as well as flying higher – at 364 miles – than the International Space Station, and even the Hubble Space Telescope. This flight was the brainchild of billionaire Jared Isaacman who also wrote the check for the mission. As of the summer of 2021, anyone – *literally anyone* with the cash – suddenly had the ability to design their own space mission, rent a rocket and independently fly spacecraft to orbital space. Nothing would ever be the same again!

A few weeks later, on 13 October 2021, *Star Trek's* famous Captain James Tiberius Kirk's human alter ego (a 90-year-old actor named William Shatner) launched into suborbital space onboard a Blue Origin Capsule along with three other astronauts and returned safely to Earth. Shatner became the oldest human to be launched into space, quite literally going where no nonagenarian has gone before…

Abruptly, in the days and weeks following 11 July 2021, the group of humans wearing the title of "astronaut" expanded by 25 individuals flown by Virgin Galactic, Blue Origin, SpaceX, and Space Adventures by the end of the first year. In the near future, that number will expand exponentially, thus – as many complain loudly – watering down the elitist astronaut "class." The gates to space have now been opened to everyone from teens to astronaut crews in their 90s. The only watershed change remaining ahead is the reduction in ticket prices by a magnitude, and then the wide-open gates to the cosmos will be officially bum-rushed by the masses.

So Who Gets To Wear The Wings?

After seeing the Tower of Babel handwriting on the spacecraft walls, exhibiting extraordinary and quite atypical restraint for a government agency, the Federal Aviation Administration quickly sidestepped the epic social argument and got out of hanging commercial astronaut wings on anyone after the first day of 2022. After covering the late 2021 launches, they shuttered their Astro-wings office forever and turned out the lights.

For a government agency, it was a brilliant move in every respect. They never had any standing in a predominantly pointless social argument and no budget for the distribution of the thousands of wings per year that will quickly follow.

The FAA's prescient vision is underscored by the case of the Virgin Galactic flyers. Their astronaut titles were immediately attacked by those who declare space begins at the currently recognized Kármán Line. Branson's crew flew above the NASA 50-mile boundary, but Besos' crew flew above the arbitrarily established 62-mile/100 km 1961 version of the Kármán Line, their 20 July flight topping out at 2.4 Everest's higher.

Peevishly, a new elitist class suddenly crops up in the tourist astronaut ranks – are you a 50-mile astronaut wannabe, or are you a Kármán Line "genuine astronaut"?

All this snobbish elitism stupidity is absolute rubbish, of course. Said veteran NASA Astronaut Mike Massimino about the vast numbers of people about to become astronauts on tourist class flights, "I think it's great! The more the merrier!"[6]

NASA has a set of wings that it issues to non-military crew who fly on NASA spacecraft. And each military service branch issues its own slightly different astronaut wings to its NASA flown astronauts.

Publicly, the private companies have brilliantly ignored the absurd dispute by designing and handing out their own astronaut wing designs, shown here in order of private flights:

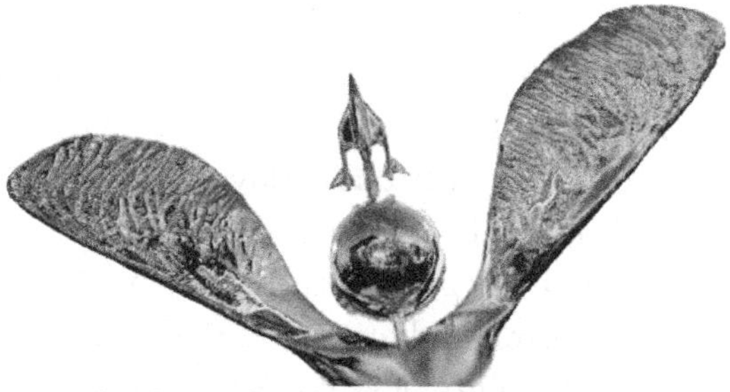

Virgin Galactic Astronaut Wings

Blue Origin Astronaut Wings

SpaceX Astronaut Wings

I liken this entire discussion to "aeronaut" elitism. Anyone who has ever lifted off terra firma strapped in the seat of a heavier-than-air flying machine (airplane or balloon) fairly earns the title of "aeronaut." Why have you likely never seen anyone flashing around a business card with aeronaut emblazoned as their title on the front? Have you ever seen an adult wearing the plastic wings handed out to the kids by the pilot on the way out of the door? The answer should be very clear – such a card or plastic wings would never pass the so-what test. The days of aeronaut elitism have long since passed, just as the days of astronaut elitism are about to become a thing of a stuffy, bygone era ready to fade into the recesses of history. And that, fellow space explorer – is a very good thing!

What Definitely NOT to Worry About!

In over nearly a half-century of living and working in the microgravity of space in low-Earth orbit (LEO), we have discovered that if someone journeys there for only a few days – or even a few weeks – microgravity induced anomalies return to normal relatively quickly once the astronaut's feet hit the green grass of Earth. The same is true of the cumulative dose of cosmic radiation you will receive on such short missions in LEO. The probability of any worry is never altogether zero, but it is *statistically not quantifiable* under these short-term conditions. (See the chapter "The Hiroshima Unit" in Module Two of this book.)

The mild aggravation of the effects of microgravity notwithstanding, from an excursion in length of up to around six months or so (it is highly variable from one astronaut to another), there does not seem to be an alarming reason to believe that the general misery an individual will suffer should lead to any perceived level of long-term health concerns. Thus, if you have your tickets or plan to fly suborbital or orbital on Virgin Galactic, Blue Origin, Axiom, Sierra Nevada, Roscosmos, SpaceX, Orbital Reef, Axiom Space, or any other commercial vendor – relax, have fun, and live it up in the adventure of a lifetime! And please keep a sharp lookout for Dennis and Claudia Chamberland because you will most definitely see us frolicking and vomiting up there along with everyone else. We are expectantly counting the days until we can fasten in and ride up.

It is one thing to understand, fully respect, and avoid the real hazards of poorly designed, archaic, unprotected ships *in deep space*; but it is also altogether absurd to classify all Apollo-era modeled missions and hardware flying *in low-Earth orbit* as physiologically dangerous – because they are flown and managed under the quite different conditions of relatively short term, low-Earth orbital missions. We have a *complete set of peer reviewed data* to declare that these skies are cleared for extravagant, unlimited fun and safety!

However, fellow space traveler, if you also have plans for deep space excursions above the Van Allen shields, desire to travel to cislunar space between Earth and the Moon, orbit the Moon, land on the surface of the Moon, or venture out to Mars or celestial bodies beyond Luna, this book was written specifically for you. In these regions, for prolonged missions, the skies are *not* clear.

(The statistically "safe" period of exposure to the full energy and spectrum of Galactic cosmic radiation on a Lunar voyage is on the order of no more than about 55 days, even though the Apollo crews' health effects on much shorter missions there is in significant question, as is discussed in detail later in the upcoming chapter, "A Pretense of Knowledge.")

The Warning, The Promise, and the Direction

So, pull up a comfortable chair, pour yourself a favorite beverage, and read on. This is a book of **warning** for you – written by a guy who was knee-deep in human spaceflight and space life sciences for three decades. But this book is also a book of **promise** and certainty for you, and for me, and for our families. Our **direction** is that we are headed out together with you on First Principle Solar System class ships and habitats, and we are going there with zero plans to return to Earth. This book is the place where you will get the real story about your future travel plans – about what ship to choose to fly on and which ones you should definitely *not* fly on! This story is all about the new generation ships and habitation structures that will come about because, in this book, we are expressing our full faith that humanity is too intelligent to fail in this. The space vehicles and surface habitats represented between the covers of this book are advanced, next-generation engineering wonders built from a new, energizing philosophy, and they specifically do not resemble the current deep space versions of the "unsinkable" *Titanic*!

You are about to read a book divided into two parts with these core ideas represented between the pair of volumes:

(1) A detailed, compelling, unambiguous, and explicit **warning**. I promise that I will not hold *anything* back from you. What I learned in three decades in space life sciences and engineering at NASA, I will tell you in detail.

(2) The **promise** of human empires beyond Earth, which is specifically expressed in the detailed engineering and clear process solutions to the warning.

(3) A comprehensive look at humanity's **direction**. A unique, no BS look at Luna (our Moon) and Mars – the first pair of new worlds of the new

human empire. If you are planning to go there, someone needs to let you know the details about where you intend to build your new life (the really good, the really bad, and the really stunningly ugly) and what you can do to make it more like a place you would want to call home.

Let us also face one nasty truth before we even get started. Just because you fell off the top of the launch tower and are headed down, riding on gravity's elevator at exponential velocity toward the pad, out of all the NASA career space geeks, I am not space exploration's version of Sheldon Cooper pointing out the inertial calculations of your fate as it relates to gravity's final solution before impact. I will be happy to point out the ugly truth to you – and I certainly intend to do so. But we are going to have a good time together even as we examine this cosmic hot dog roast about to get rolling in earnest!

With that picture in your mind, let me share another image with you before you turn to the next page. This author is *not* the guy at the launch pad elevator waving "The sky-is-falling" sign and screaming, "Everyone's going to die!" while shooting you the bird as you board the spacecraft (although something similar actually happened to me as I drove through an anti-nuclear protest staged at the KSC front gate against the launch of the Galileo Jupiter probe– signs, yelling about Armageddon, extended middle finger and all). Get real. I am your fellow traveler into deep space, dutifully dressed in typical NASA casual white shirt and thin black tie under my spacesuit.

It is important that you understand that there are three very different personalities involved in this discussion taking place right now across the world. It does not take much of a genius to understand who they are.

One is the total pessimist. That individual uses their data sets to prove that humankind can *never* build a civilization on another world because of cosmic radiation and other hazards of deep space travel. I am not one of these losers, trust me.

Then there is the opposite end of this spectrum, and these individuals are actually just as hopeless. These are the people who claim that all we need to do is to build an inflatable spaceship connected to a thin aluminum can and strap rockets on its end. Game over. Now we endlessly ride about the solar system in these Apollo-era contraptions, adding countermeasure band-aid after countermeasure band-aid and dreaming about changing our very genes until we somehow morph into true Buzz Lightyear human hybrids. They are certain that the human body will endlessly adapt if we can just tack on enough of these nearly always useless coverups for inadequate engineering designs and a complete lack of any coherent exploration philosophy. While I used to work for this company, I *definitely* do not want to be counted in this crowd, either.

Then there are those of us who favor the authentic spaceflight engineer and genuine scientist's approach. Using lessons learned and current data streams, armed with a driving and energizing philosophy that everyone understands, we design and build advanced systems which are constructed from the ground up to protect crews first and, in so doing, safeguard mission integrity. It is just fundamental. Responsible science applied to good engineering practices without the need for very sketchy safety waivers and idiotic countermeasure Band-Aids will guarantee a permanent human presence in the space frontier. That describes your author (a career space engineer and scientist) as well as the direction of the remainder of this book.

This is not a book about why we cannot succeed. At the same time, it is also certainly not a book about continuing to patch up an obsolete human space exploration paradigm that has long since expired. **This book is about having the vision and courage to correct our course to ensure success *today*.** And, oh, by the way, *I'm going with you!* All the way. To Mars. Beyond. Move over...

This is a book about how we shall fly safely and confidently in deep space as far out as we can dream. It is likewise a book on how definitely *not* to fly in deep space – which, unfortunately, also happens to be the pathway down which we are now headed with all engines burning bright.

Finally, dear fellow space traveler, I have "been there and done that" – a certified, card-carrying NASA geek for life – and literally have a ton of NASA t-shirts, pins, patches, and a 30-year certificate to show for my experience. So, if you think for one second that I am going to allow you, my friend, to get on one of those old-school, Apollo paradigm inspired deathtraps headed out to the inky black, you are seriously mistaken. Now, go ahead and turn the page and find out why. Before the book is over, you will also know what kind of ship we *will* be flying out together on our one-way mission to new worlds of our making. We are, after all, the first of a new generation of Martians who will fly outbound *only* on the intelligently designed new generation spacecraft built by the authentic steely eyed missile crews of the 21st century.

Essential Note to Readers

You will encounter a pair of specific phrases throughout this volume and the next: *"Functionally Unshielded"* and *"Functionally Unprotected."*

Every space vehicle is minimally "shielded" by its pressure vessel walls and other spacecraft equipment, such as a re-entry heat shield. For example, the Curiosity Mars Rover was minimally shielded during its interplanetary cruise phase by its aeroshell and the equipment stacked above and below it, but it still received a dose of 1.8 millisieverts per day (the unit "**mSv**" is defined in detail on the next page) of cosmic radiation despite its minimal shielding. But for the purposes of this book, the term "functionally unshielded" has a very specific meaning. It refers *only* to human rated spacecraft on long term deep space missions (not tourist class missions). A functionally unshielded human spacecraft exposes its crews to cosmic radiation levels greater than the current United States limit for commercial nuclear power workers, which is 50 mSv/year (or 0.14 mSv/day). Although this upper limit is rarely attained by any nuclear power worker, even if it were, it is still some 12.8 times less than the average daily cosmic radiation exposure received by the Curiosity rover over its 253-day outbound mission to Mars.

Likewise, when the term "functionally unprotected" is used in these volumes, it refers *only* to human rated spacecraft on long term deep space missions (not tourist class missions). Long term and long duration missions are those that are on the order of no less than 55 days in deep space above 36,000 miles which exposes crews to statistically significant risks to their long-term health based on exhaustive, peer reviewed research. A functionally unprotected human spacecraft exposes its crews to the long duration toxic mix of cosmic radiation over the federal limit for radiation workers and the synergistically toxic effects of microgravity, as is discussed exhaustively in this book.

The Most Important Unit

You will see the unit millisievert (mSv) more than a hundred times in the coming pages. The Sievert is a unit that describes the effect of one joule of gamma radiation energy deposited in a kilogram of human tissue. In other terms, the Sievert considers the relative biological effectiveness of different forms of ionizing radiation in humans. In this book, I refer to the biological international unit millisievert (mSv) which is one-thousandth of one Sievert. The U.S. equivalent of this unit is expressed as the rem (roentgen equivalent man), and one Sievert is equivalent to 100 rem.[7]

Everyone is exposed to about 2-3 mSv per year of natural, terrestrial ionizing radiation. Likewise, everyone is exposed to about 0.33 mSv per year of galactic cosmic radiation (GCR) that somehow makes it to the Earth's surface. One chest x-ray is about 0.02 mSv. A whole-body CT scan is rated at around 12 mSv. An astronaut in Low Earth Orbit or on the surface of Luna or Mars will be exposed to about 0.66 mSv/day on average. In deep space, the burn rate is a hefty 1.8 mSv per day. About half of the people exposed to an acute dose of around 5,000 mSv would die from that exposure. Documented probability of ionizing radiation exposures that can cause cancers later in life begins with acute exposures of ~20 mSv. In 2021, the National Academy of Sciences recommended to NASA that the lifetime limit for exposure of astronaut crews be established at no more than 600 mSv during their careers. A typical functionally unshielded Mars voyage exposes crews to around 1,000 mSv.

The Sievert was named for the Swedish medical physicist Rolf Maximilian Sievert (1896-1966) who pioneered the documentation of the effects of ionizing radiation on the human body.

Why We Will Depart Earth Forever

Reason 1

"Don't tell me that man doesn't belong out there. Man belongs wherever he wants to go - and he'll do plenty well when he gets there."

— Wernher von Braun

THE PRIMARY PURPOSE OF THIS BOOK IS TO LAY OUT A clear warning that ideally will result in a relatively simple pair of changes to our exploration model before we actually commit humans to deep space. The initial thrust of this work is devoted to that discussion. The follow-up to that warning is a solid, coherent plan to address and fix the problems. And finally, these pair of books exist to clearly articulate a promise that is already built into our shared genome. Driving that collective message and promise is an implicit, overriding, very legitimate question that is featured prominently throughout each of the two companion books:

***Why* will we depart Earth forever?**

Why would we take such risks? Why would we leave all that we know to face so much uncertainty in a universe that we realize is alien in every conceivable way?

In these companion books, *Departing Earth Forever: Books One and Two,* I will answer that question with six different responses. At the end of each section, you will see this query addressed in expanded detail, enumerated as powerful, compelling reasons why we shall indeed depart this Earth forever – and very soon!

I have written extensively and spoken publicly to thousands of people on space and undersea exploration and permanent colonization for more than three decades. But more than just speaking about human exploration, I freely admit to overt advocacy. I addressed each group – small or large – as a permanent space and ocean settlement activist.

Invariably, when the Q&A session rolled around, one of the most repeated questions that nearly every gathering asked was, "Why?" At the end of this chapter, I will give you my most frequent answer. It may surprise you!

One June afternoon in 1987, I stepped onto Space Shuttle Pad 39A at the John F. Kennedy Space Center and slowly walked up the steep incline to the enormous rotating service structure at the top of the pad's sloping surface. As I made my way up, I was overwhelmed with the sense of just where I was walking. It was from this very concrete covered mound that 24 individual humans flew to the Moon between 1968 and 1972. And, as of that day, it was from where the first 24 Space Shuttle orbiters also had launched. It was truly breathtaking to me. I viewed everything with the hungry eyes of a child, sweeping it all inside, ravenously soaking in the vista of mankind's most important achievement as I could feel the potency beneath my feet. It was a moment of reverence. The ground on which I walked may as well have been sacred in so many ways.

I ended my trek at the pad's steel girded base. 247 feet of girders and crossed steel members towered above me. I stopped and peered down into the cavernous flame trench scorched from the intense heat of immense rockets igniting above it, searing silica sand particles into sparkling specks of glass. Then I stepped over to the elevator, entered, pushed the button for the 195-foot level, and rode the car attached to the pad's fixed service structure upward. The door opened, and I turned and faced the 65-foot swing arm that led to the White Room – the small vestibule where humans take their final step off the Earth and into their spaceships. Startled by the expansive view of the sprawling Space Center and the endless, vacant beach some 20 stories below, I slowly traveled across the open walkway, marveling at the stunning 360-degree panorama in the clear Florida sunlight.

Finally, I moved inside the relatively small White Room and was seized by what I saw. Facing me at the opposite end of the space was an open web of nylon straps – the single, flimsy, laced barrier between me and the ground 20 stories below. Through the thin bands, I could see the beach and the ocean stretching out for miles to the horizon. I approached the void slowly and carefully looked out. On launch day, the open door of a crewed spacecraft filled that area. And then it occurred to me. The last humans to cross that threshold on the rare, bitterly cold Florida morning of January 28, 1986, were seven astronauts, four men and three women: Francis R. Scobee, Commander; Michael J. Smith, Pilot; Ronald McNair, Mission Specialist; Ellison Onizuka, Mission Specialist; Judith Resnik, Mission Specialist; Gregory Jarvis, Payload Specialist; Christa McAuliffe, Payload Specialist, and teacher. These seven individuals had entered the Space Shuttle Challenger at this very point just 17 months earlier – and they never returned.

Why? Why did they do it? Why did they take this elevator and walk across this swing arm and then step through this threshold? They knew the risks. And yet they willingly chose to fly anyway. They knew – yet they dared to ride out of the atmosphere into space and take their chances. But those seven never made it to space. Instead, high over the North Atlantic Ocean on that frigid January morning, in the words of President Ronald Reagan, they touched the face of God.

These individuals were anything but stupid. They were well qualified, trained, and enthusiastic. And they wanted to explore all the way to the innermost edge of deep space. Why? Why would anyone take such a risk? They climbed on board a class of spacecraft that had only flown 24 times before. It was a vehicle that traveled up to 23.2 times the speed of sound to achieve low-Earth orbit. Nothing good can happen to a human body at Mach 1 – or Mach 9 or Mach 23.2 – if the slightest thing goes wrong. But they individually decided privately, silently, intimately, to themselves and to their families, that it was to them somehow worth all the risk.

And how do I, as the author of this book, verbalize that intimate discussion that each of them conducted within themselves? I know it very well because I have had it with myself – for real. No, I have never flown into space. But I am fully prepared to do so at a moment's notice. And I have likewise decided unequivocally that I am ready to participate, plan, or lead a permanent human undersea colony, or even space colony, as the opportunity allows. I have had and settled the discussion within my own heart and soul long ago. For decades I have been fully prepared and packed to lead or follow on a one-way trip out.

I looked carefully at that row of crossed nylon straps at the very threshold to other worlds as it waved gently in the Atlantic sea breeze that June afternoon. I knew even then that I was ready, even though the last seven explorers who had crossed over that threshold that faced me never made it back home. It is something wired into the genetic

Ultimately, that genetic wiring ensures our survival as a species. Fortunately for us, these behaviors are already a built-in part of who we are – or at least a few of us. So, am I saying that the genetically hardwired survival activities of eating and breeding are instinctive equivalents to exploration and expansion of the species? I believe that it is quite obvious that they are indeed, as even a cursory examination of human history illustrates. In my mind, asking me why I want to depart Earth forever and live in an undersea colony or in a colony on Luna or Mars is exactly the same as asking me why I want to have sex. My reflexive, private, internal response is always, "Are you joking? Is this really a question?" But professionally, I bite my tongue and answer the question as civilly as I can. Sometimes with the canned answers – but almost always now with the unvarnished truth.

At the front of each section in the two volumes of *Departing Earth Forever*, I will continue to answer the question "Why We Will Depart Earth Forever." There may be a few black-tie answers here and there, but I am determined to just get real with you. In the end, those of us who decide to go – those who pay the fare, the ones who, with trembling hands, pack their bags and climb up to the White Room that separates this world from the others – will each depart this planet of our birth forever for a very good, overriding, most important reason. It is the reason that supersedes and makes all the less significant reasons possible.

We will depart Earth forever *because we want to.*

Of course, there are other less important answers, as well. But, after everything is said and done, they all relate to survival – just as they all will bring our species great joy and fulfillment, with the full expectation of perpetual expansion. And at the end of everyone's day, eating, breeding, and exploring are so fundamentally human that they really need no explanation at all. Whether a black tie or cut-off shorts and t-shirt answer, these are things that, at the most distant boundary of our capacity to reason, render mere words absolutely irrelevant. It is indeed the driving and most powerful justification for becoming a multiplanetary species – the ultimate assurance of our long term survival.

Module
Zero
Ch.01
Ch.02
Ch.03
Ch.04
Ch.05
Ch.06
Ch.07
Ch.08
Ch.09
Ch.10
Ch.11
Ch.12
FOUNDATION
Let the Revolution
Begin

Arbiter of Empires

"It is never too late to give up your prejudices."

— Henry David Thoreau

"The most damaging cognitive bias is overconfidence - illusory superiority..."

— Paul Gibbons

BROADLY INSCRIBED ACROSS THE EMERGING STORY OF the birth of the first human interplanetary empire is the astonishing conclusion that the 21st century's crewed expeditions to Mars were born on 26 April 1607. It was the same moment that awakened the British Empire as a world power. While these two exploration models may seem wholly unrelated, separated one from another by four centuries, still, each of them was birthed by a single momentous act of 17th-century exploration. Thus, the first indirectly begat the latter by virtue of the historically momentous result of that lone 17th-century voyage.

Unfortunately, and specifically because of these empires' shared inherited ancestry, even though separated by a mere four hundred cycles around Sol (which counts as not even a single tick of epochal time), each empire also thus shares the identical genius, as well as the unbroken curse, of their genetically endowed natures. Daring human enterprise always seems inevitably plagued by its genomic heritage emblazoned with simultaneous measures of sheer brilliance, hopeful confidence, and a sharp intellect that is often likewise corrupted by the negative energies of stubborn ignorance and arrogance. Humanity frequently shows itself to be in the same moment both ingenious and likewise infected with knuckle-dragging stupidity. Unfortunately, this genetically driven package is also quite dangerous when it regularly morphs into an emboldened and wholly irrelevant sense of entitlement that invariably leads to self-deception that consistently folds into disaster. It is the cyclic curse of the true brilliance of humans habitually entangled in their inexorable sense of misplaced awesomeness.

Early in the 17th century, three ships departed England and sailed due west. Aboard the *Susan Constant*, the *Discovery*, and the *Godspeed* were 104 optimistic colonists and, as well, the hopes and dreams of a new empire. They left England convinced they knew what they were about to experience on the *terra incognito* they called Virginia.

They actually had no clue what they were about to sail into or, for that matter, what they were doing. Had they known the truth in advance, no one in their right minds would have boarded the ill-fated trio of ships in England.

Thus, they were not just uninformed – they were tragically so. And from a sense of Old-World entitlement, in their blinding overconfidence, most of them died together from disease, predation, and starvation – which ultimately led to cannibalism. (See the chapter just ahead titled "Jamestown, Virginia.")

Four centuries later, their direct descendants are about to embark on a voyage of striking similarities. From the world empire that arose from their ancestors, a nascent, new, interplanetary empire is already being envisioned and vigorously engineered. It is indeed the direct offspring of the acts of empire building of our English forefathers. Thus, the former directly begets the future that is now morphing into reality before our eyes at a truly uncanny speed.

And so it is that exactly as before, we 21st-century explorers are equally convinced that we know precisely what we are about to experience and what lies just ahead of us in the vast *spatium incognito*. Many of these same descendant people also actually believe that they know what they are doing and what they will encounter on the far distant shores. They are faultlessly confident that they already know these planets to be uncannily like Earth, and they will shape them to become the new world empires that they have fixed so resolutely in their dreams.

Sadly, most of these dreamers have no clue whatsoever of what lies out there waiting to invisibly, silently, slowly, and insidiously stalk and kill them. But it is not the dreamers who are entirely at fault. It is the responsibility of those of us who do know this void and these worlds very well – those of us who have invested our lives uncovering their secrets. Those of us who are obligated not just to warn the dreamers but to give them a far better plan knitted together from countless peer reviewed investigations. This new plan will actually give them a chance. It will allow each of them to weave their individual, precious, and unique dreams of empire into a reality and, in the end, live to tell their triumphant stories to their descendants.

But among the vast landscape of the misinformed are, astonishingly, many who also happen to be the ones in charge. They are not just uninformed; they are tragically so – and they will, unfortunately, play no small part in determining the lives, health, and wellbeing of many.

Thus, considering all this, we stand on the threshold of a historic moment of human expansion that our species has derived from an authentic sense of innate, natural human privilege born of a genetically driven vision of manifest destiny. Yet, unless we immediately stop, take a moment to consider what really lies just ahead, and make the necessary course corrections now, we press on toward inevitable calamity of identical historic import as that which beset our equally clueless forefathers.

As described in some detail in these volumes, the tragedy, if not averted, will be the result of leveraging a very human characteristic that Nobel Prize winner Friedrich von Hayek designated as "a pretense of knowledge" onto the very exacting tapestry of space and the unrelenting laws by which it unforgivingly binds everything together.

Deep space cares not one whit about human entitlement, privilege, or anyone's sense of destiny, and is swift to assume the role of the unmerciful mediator of arrogance, any form of charade, or, in particular, any deeply held belief. Nature extends well beyond the tiny cocoon of Earth's atmosphere, encompassing the whole of intergalactic space, and is always, invariably and inevitably, the final arbiter of everyone's empire. She is also the quite unmerciful umpire of dreams. Nature has zero tolerance for a pretense of anything. Far more than needing to come to terms with and understanding ourselves at this pivotal moment in human history, we must primarily come to terms with the empty blackness and the abjectly alien natures and characters of each of the new worlds to which we aspire.

Yet things do not have to end in tragedy. We have a golden, historic opportunity now, at this moment, to literally stop the madness and redirect our intellectual energy into a better, nearly failsafe approach to interplanetary exploration and colonization. But we must seize the occasion with an urgent immediacy before the combined inertia of the world's already obsolete Apollo-era human space exploration paradigm is just too big to slow and redirect without inevitable tragedy.

Make no mistake – we will soon escape the bonds of Earth forever. Just as the early deck-watch onboard the sailing vessel *Discovery* gazed through the mists on 13 May 1607 and saw that the landfall of a new empire lay just ahead – very soon now, human eyes will likewise witness Marsfall as it unfolds when the first human-crewed ships enter the Martian atmosphere. It is only a question of how many lives we will permit to be wasted in the process of getting it right.

We get to choose that number very precisely.

The catastrophe can be avoided, but only if we use our intellect and lessons already learned as we also fastidiously reject the ignorance, arrogance, and self-deception at which we as humans are so astoundingly adept.

This chapter of human empire does not have to be decided by the capriciousness of our own most unfortunate attributes, nor by the insanely disfigured nature and structure of our absurdly organized bureaucracies. We have the clear opportunity – right now – to decide the success of the opening years of the new empire by the sound application of science and engineering based exclusively on, and only by, the hard data *we already hold in our hands*. We get to choose which path we take on the way to our actual manifest destiny. Lessons of the 21st century do not have to be paved with the corpses representing stubbornness and human stupidity, as mankind has managed to accomplish recurrently in our history. As a race, we have hopefully learned the lessons of our past. And, perhaps now, finally, after all these millennia, we have the intellect to examine, refine, and command our own nature… before we set out to design and build the tools to rule our Solar System.

Every story always has a starting place – whether it is on a dark and stormy night or waking up on a magnificent sunny morning lying beside a beautiful princess. This story has a pair of starting places. One is from the end of a wooden plank in 17th-century Virginia, and the other is from a seat aboard a NASA jet at the end of the Space Shuttle landing strip at the John F. Kennedy Space Center in the 21st century.

Ready and Prepared

"Space travel can't ease the pressure on a planet grown too crowded not even with today's ships and probably not with any future ships – because stupid people won't leave the slopes of their home volcano even when it starts to smoke and rumble. What space travel does do is drain off the best brains: those smart enough to see a catastrophe before it happens, and with the guts to pay the price – abandon home, wealth, friends, relatives, everything – and go. That's a tiny fraction of one percent. But that's enough."

— Robert A. Heinlein

IN THE SPRING OF 2006, OUR NASA TEAM TOUCHED DOWN ON the long Space Shuttle Landing Strip at the John F. Kennedy Space Center (KSC) onboard a NASA business jet. We were returning after a day trip north from KSC to the Brookhaven National Laboratory's Relativistic Heavy Ion Collider in Upton, New York. There, our team had finalized plans to consider one of the first investigations into a simulated Mars journey involving an acute cosmic radiation exposure to a vertebrate animal. Just a few days before that trip, I had been handed a check by former Space Shuttle Astronaut Dr. Sam Durrance from the Florida Space Research Institute for our combined government and private team to fully fund the investigation.

I was honored to join the impressive team comprised of an amazing assembly of some of the most capable minds on the planet: Dennis Steindler, the Director of the McKnight Brain Institute at the University of Florida; Harry Nick, also of the McKnight Brain Institute; Grigori Enikolopov and Juan M. Encinas of Cold Spring Harbor Laboratory; Marcelo Vazquez of Brookhaven National Laboratory; Howard Levine and Philip Scarpa from the John F. Kennedy Space Center; and Robert Switzer of Neurosciences Associates Corporation.

The project began one morning about a year earlier when a smiling and most sincere gentleman unexpectedly walked into my office at the KSC Space Life Sciences Laboratories. He stood at my door and began to ask pointed, detailed questions about live animal experiments. He had found the right guy. At that time, I was the longest serving Committee Chairman of any previous Agency Institutional Animal Care and Use Committee (IACUC), responsible for assuring the enforcement of the Animal Welfare Act of 1966 for all vertebrate animals designated for spaceflight out of KSC. He identified himself as Dr. Duane Graveline, former Physician Astronaut and Director of Medical Operations at KSC for the first four shuttle missions. "Doc" Graveline was named in June 1965 as one of six members of NASA's fourth group of astronauts. The scientist-astronauts were the Agency's first trainees recruited for their academic backgrounds rather than flight experience.

As we sat together in my office, I was quite impressed by not only his passion and insight but also by his deep knowledge of the subject matter. It was to be the very first day of years of friendship between us until his death a decade later. According to Dr. Graveline, NASA appeared to be either overlooking or, at the least, minimizing the potentially devastating effects of cosmic radiation on flight crews – and in a vacuum of any meaningful data at all. As a physician astronaut, he seized my attention immediately. He was right on then – and, even after his untimely passing, he is still correct today. Our team was soon assembled as a direct result of Dr. Graveline's enthusiasm, energy, and persistent acumen, and the work was accomplished per his seminal vision.

The results of our investigation were eye-opening, to say the least. Before that time, except for the brilliant insights of Dr. Dennis Steindler, few even suspected that the harsh, relentless exposure of the brain to cosmic radiation could cause irreparable damage to the most sensitive layers of adult stem cell sites of neural regeneration in the core of the brain. Such chronic damage could have devastating effects on a crew's ability to cognitively function during an extended deep-space mission. Because of the injury, they could even lapse into symptoms of dementia during flight. Since our research findings, several other investigations have all supported the same concluding effects, as mentioned in detail later in this book.

And what we have discovered is that the data now being considered for the effects of human exposure to radiation during long term exploration in deep space, in interplanetary space, or on the surface of other worlds, is from World War II terrestrial ionizing radiation models developed from atomic bomb survivors – *not* data from the far different and epically more powerful and massive species of galactic cosmic radiation that we will actually have to deal with everywhere beyond our planet. Further, we have used this very sketchy data model to establish limits for our crews' exposures based solely on long term cancer risks while wholly ignoring what may be far more injurious and dangerous in both the short and long term – the effects of this radiation on the sensitive, fast growing epithelial and endothelial tissues found throughout our bodies, in the brain, cardiovascular system, and even in the digestive organs. (This is discussed in detail in the future chapter titled "Hitting the Physiological Wall.")

Why have we not revised our models to account for the actual environment? Why do we not also have sensitive tissue limits as well as the stochastic cancer limits in place to protect our crews? Because we do not have the data on vertebrate animal models in deep space and have apparently, preposterously, deemed such research an unnecessary waste of our resources.

And yet, while some of the data is slowly being gathered (at least to a limited extent) in fundamentally inadequate ground-based experiments, the space exploration agencies of the planet are literally primed and becoming more and more prepared than ever to conduct long term human Lunar and Mars expeditions in the unfiltered environment of deep space as soon as possible. To be concise, they are planning to allow astronaut crews and civilians to exit the relatively safe confines of low-Earth orbit and enter deep space beyond 36,000 miles, to the surface of the Moon, and in cislunar space (cislunar space is defined as that volume within the Moon's orbit, or a sphere formed by rotating that orbit) as well as on very long human flights to Mars, in the full glare of unfiltered, minimally shielded cosmic radiation in extended microgravity. In reality, they will thus be unethically using these astronauts as vertebrate animal experimental subjects.

It is as though there is a persistent disconnect in NASA and elsewhere between the emotionally driven psychological desire to hurry up and go explore these distant and dangerous places and the responsibility to actually acquire a far more sober and complex technical understanding of what they intend to do before actually departing Earth. Therefore, the essential question becomes: are we as scientifically and technically ready as we are emotionally prepared? If the two do not blend seamlessly into both emotive readiness and technical capability, then certain disaster inevitably follows, as history clearly and repeatedly demonstrates.

The answer to that question, as evidenced by our antiquated expedition design paradigms, is that we are clearly *not* prepared or ready for the expeditionary sequel. Indeed, we are now gearing up to use humans as test subjects in deep space instead of laboratory animal models as though that were standard operating procedure. And while NASA is moving with all due momentum to normalize this scientifically and morally disreputable approach, it is most assuredly not the way ethical science does business!

Human expansion outward and away from our home planet is inevitable. The human species *will* travel to Mars and *will* settle permanently there. This work is by no means about why we cannot. Instead, it is all about how we must – about *how we will* – correct our current path and make it successfully happen with minimal loss of life or health of our crews; hopefully, driven far more by intelligent foresight than simply by our adolescent passions. The purpose of this pair of volumes is to correct our system before (not after) we start stacking up the bodies of those whose deaths we have carefully and methodically engineered as mere data points. The objective here is to fix things *before* they break, as they most certainly will many millions of miles away from home with no easy way back except to wait for the cosmic realignment of the planets.

The imperative questions that overlay this certainty and this expectation, however, are monumentally sober ones. How do we achieve this correction considering where we are today? How will we execute the fixes based on our lessons learned and correct our course as we set about to build the machines to make this expansion a reality? How many hopeful colonists will die in the quest to reach and settle these other worlds – not because of their courage and willingness to sacrifice, but because of the shortsightedness, ignorance, and stubbornness of we who lead them to the edge of space and to the very boundaries of our knowledge? How many career scientists and engineers know the details but are afraid to speak openly and publicly because of the inertia and billions of dollars already invested in an obsolete, dangerous human spaceflight system? These are the core questions of this book and the reason it had to be written.

Make no mistake about what you are about to read. The message of this book is that ***we are going*** outside our atmosphere to permanently live and create new human civilizations. As President John F. Kennedy said so presciently, "Not *because* it is easy – but *because* it is hard!" The overarching message in this work is that we have reached a tipping point in our explorations, and now ***we are obligated*** to take stock of what we have learned. And, using that information and experience, ***we now must change our course,*** or many explorers will pay the price – with their lives or with crippling, life altering injuries.

Thus, this book's message is both about certainty and intellectual resolve. We are certain to depart our home planet just as we are certain to be ultimately successful – but only if we resolve to first sequentially understand the hard requirement for a new core exploration philosophy and then actually employ it in our quest.

Human beings are creatures of pattern and design. Generationally we invariably repeat the adolescent errors of our parents. In the course of multiple generations, we consistently duplicate our own history, routinely replicate the failures of our culture's past, and walk along false paths in blind alleys down which our forefathers have strode before us.

The late Dr. Duane Graveline understood that to be ready and prepared, we have to first have the data in our hands, connect the dots, and ***then*** design the space exploration architecture. We have unfortunately been busy building the machines based on antiquated technological expectations with grossly insufficient data and internet generated fake facts, fully intending to countermeasure the hell out of the mess later and make it fit by forcing humans to adapt to the difference.

It is the intent of these volumes to try and straighten that mess out and define what is actually "ready and prepared" on a more coherent, logical, and scientifically responsible path.

But before we cast our eyes out toward deep space, let us take a short glance backward and understand the roots of our seminal tale of the far-flung empire we are planning to launch across our Solar System. The deepest origins of humanity's historic migration to Mars actually begin in Jamestown, Virginia, on 4 May 1607…

Jamestown Virginia

-

1607

The multitudes remained plunged in ignorance... and their leaders... did not dare to undeceive them."

— Winston Churchill

IT WOULD BE GROSSLY IRRESPONSIBLE TO DISCUSS THE human exploration and colonization of Mars without first considering the British settlement of Jamestown, Virginia, at the turn of the 17th century. Humans are creatures of great potential, and we have made astonishing technological leaps. And yet, in so many other ways, we have made no advances at all. As the centuries and millennia have passed, we have retained a core tendency to be creatures of illogical habit, of a maddeningly short-sighted, irrational stubbornness bordering on intellectual dementia with the inability to retain even short-term historical recall. In so doing, we often relegate ourselves to creatures of quasi-intellectual capacities often described by such aphorisms as,

"*Those who fail to remember history are doomed to repeat it.*"

But it does not have to be this way. And, to ensure that it does not, it always pays huge dividends to take stock not only of our greatest strengths but also our greatest weaknesses and failures before setting off on any epic adventure, such as the first human migration and settlement of other distant worlds.

While there are countless other examples of colonization in our history besides Jamestown, Virginia, few offer such profound lessons in human nature and the breakdown of leadership – all because of glaring, if not altogether obvious, failures in the effort's fundamental philosophical underpinnings (which we are repeating today in hi-tech spades). This is particularly appropriate since, in so doing, we are also examining our own historical antecedents with all of their powerful lessons that must not be lost to time. As we forensically reexamine this painful chapter in the advancement of western civilization's manifest destiny in the following perilously succinct summary, it is instructive to frame it in the singular person of a 14-year-old female colonist who was posthumously named Jane.

The first commercial vessels managed by the Virginia Company of London arrived in the New World on 4 May 1607. The initial British foothold in the New World was a business venture and was deemed so historically noteworthy that historian William Kelso declares that its founding marked the beginning of the British Empire itself.[1] And yet, in all of its momentous significance, its formative years were marked with horror and tragedy – and that due principally to a breathtaking cultural arrogance fused with an organizational incompetence that was inevitably measured in precious human lives.

The following example is given with a purpose: so that we do not repeat another Jamestown disaster on Mars or any other distant world – a disaster born of our own 21st-century arrogance, scientific incompetence, and historical amnesia.

The briefest analysis of the serial tragedies of Jamestown is that English colonists departed the Old World fully expecting to find England waiting for them when they arrived in the Americas. What they found instead was an alien world that they were wholly incapable of dealing with. They left England with seeds for crops but were forced to eat their seed when the indigenous natives prevented them from leaving their tiny fort to plant enough for a barely adequate harvest. Yet even before that, their life support plan abruptly failed in that the two social classes of settlers were incompatible with the system they exported to a new world. It was a case of a single blended old world cultural system that was wholly irreconcilable to a new world paradigm.

Farming was beneath the dignity of the gentleman class, and the indigent class did not know how to farm. After the famine had already begun, only then did the colony's leader, Captain John Smith, finally disallow the gentleman class free run of the food stores and implement the Biblical injunction that if any man, regardless of their class, did not work, he did not eat.[2] But this urgently enacted, enforced adjustment in their social structure came far too late. The belligerence of the natives was also not expected, and by the third winter in Jamestown, due to crop failures and drought, famine set in with a vengeance.

At the end of the third winter, only 60 of the 500 colonists who arrived on the first ships had survived – a mere 12 percent of their population.[3]

They had come from England by the hundreds and expected to find another England – or to at least remake the New World in its image. Instead, they found themselves in a completely alien land with unique weather patterns, soil, water, and belligerent natives they could not control. And, if that were not enough peril, they arrived with an Old World philosophical system that would not, and could not, work in this new land.

They set sail for another England. Instead, they arrived in an alien world. Most of them died utterly unprepared in this strange new land that demanded something they simply did not have – knowledge of and preparation for an alien place called Virginia that was definitely not England. One of the victims of this unfolding tragedy was a 14-year-old girl who was named Jane (her likeness shown here) by the archeologists and anthropologists who discovered her remains in 2012. Her skull was unearthed among the skeletal remains of animal bones in a garbage pile. Etched on her cranium were the clear marks from where her fellow colonists had removed the tissue from her face for cannibalistic consumption. From her remains, forensic anthropologists have recreated what they believe Jane looked like in life.[4] It is an emotionally powerful moment to see her restored as she was in the bloom of youth.

The image of this young girl strikes a profound emotional chord deep within me. Nothing is known of her story: how she ended up in Jamestown as a colonist, who she was in England before her voyage, or what her hopes were for a new life in the Americas. Nor do we know how she died or what madness drove her fellow colonists to cannibalism and to then discard her remains in the garbage. But we do know precisely who and what was ultimately to blame for this catastrophe and the serial failures that led to her historically tragic and untimely death. The loss of her life and the lives of so many others in Jamestown was the direct responsibility of those who created the opportunity for her to board the ship, those who encouraged her to sail over the horizon to her demise – those who thought they knew all there was to know about the new world but who actually did not know enough to keep those under their purview alive. They enticed her and her fellow travelers with promises of a new life. Instead, they dumped them onto an alien shore that engulfed them all in a tragedy of horrific and historic import.

But what of those who sent her? Were they prosecuted for their crimes? There was indeed eventually official retribution against the financial entity itself. A petition was presented to King James I to formally investigate the tragedies at Jamestown. This led to a trial before the King's Bench in May 1624. The Virginia Company was found culpable and was thus dissolved, leaving the territory of Virginia to be declared a royal colony.[5]

What we discover in the 21st century is a lesson waiting to be learned – for our time and for our culture. The question we must come to terms with is this: *how* will we learn it? Will we learn it by repeating the tragedy because of human shortsightedness and ingrained organizational stubbornness? Will we repeat it because the interplanetary systems we have built are now too expensive to tear down and start over? Has our systemic inertia already exceeded our intelligence? Are our habitually defended cultural practices greater than our capacity to allow the raw data already in hand and meticulously analyzed by the scientific community to correct our models? Or will we actually intelligently heed the warning and do whatever we have to do to avoid tragedy and ultimately repeating the repulsive horrors of our forefathers?

Now, in the 21st century, as I stand soberly at Jamestown just inches from Jane and look for long minutes through the glass case that holds her etched and scarred remains, I am powerfully reminded that I have been entrusted with at least a small part of her ultimate legacy. I have a full moral obligation to clearly communicate to my fellow citizens of this new empire that arose from the one from which she came, that all the Janes who are alive today and looking outward to new worlds –

who have the same hopes and dreams of a new life, just as she did – do not have to suffer her fate. We can learn from Jamestown, and we can refuse to repeat their failures of cultural and philosophical ignorance, and the gross incompetence of leaders rooted in a glorious past that is absolutely incompatible with the future. We can learn from a new philosophy they never had and employ it like the powerful tool that it is so that there will be no Jamestown examples in the future of space exploration and settlement of Luna, Mars, and other distant worlds.

We certainly know what to do. We also know how to do it. We know and understand redesign. We realize the ethical imperatives. The question is: do we likewise have the will? Do we have any intention at all to recognize our arrogant blindness and human tendency to forge ahead with but a pretense of knowledge?

Just as the Jamestown colonists eagerly awaited landfall on the New World, as we expectantly look outward toward an approaching Marsfall, let us respectfully pin Jane's photograph on our spacecraft walls. Let us carry her likeness with us to remind us that even as the centuries have passed and our technology has expanded exponentially, *we humans have not changed at all*. Let her unthinkably horrific demise become a permanent reminder to each of us of who we really are – certainly not representatives of a contemporary British-esque cast system, and likewise, certainly not our self-inflated Hollywood personas. Let us renew ourselves by what we have learned. And, based on those lessons, let us build an intelligent and hopeful new civilization on these distant worlds – places where humans are not merely convenient organisms for experiments or forced adaptation of any kind.

Let us learn from the serial errors of Jamestown and, with eyes wide open, deliberately sidestep their fate by redesigning our engineering systems to match reality, not our antique human spaceflight system designs leftover from another era. Let us *never* depart Earth imagining that we shall encounter another Earth in the vast reaches of deep interplanetary space. Let us instead sail out fully prepared on that broad expanse of the sea of interplanetary space that poses its own challenges, and then arrive safely and healthy on an alien and unique planet that we completely understand and are equipped for at the moment of touchdown on her dusty plains.

There, we will *not* represent our home nations – nor will we even represent Earth! There we will represent humanity – the very best and the very brightest our species can offer. Let us arrive to live and thrive, not to plant the flag among footprints in the dust, and there die from our own arrogance and intransigent refusal to change when the time for the sanity check and necessary modifications came *before* the interplanetary engines were ignited.

We can do this. But not unless we learn as much from our legacies of victory as we do from our equally important legacies of failure – our Jamestowns, our Janes, our Apollos, Challengers, and Columbias, as well as all of our cumulative brilliant successes in the space frontier – and then boldly correct our deeply flawed exploration model while there is still time.

It definitely bears repeating:

We certainly know what to do. We also know how to do it. We know and understand redesign. We understand the ethical imperatives. Thus, the questions are these:

Do we likewise have the will and intellect to pull it all together before it is too late?

Or will we be victims of our own denial and stubborn human intransigence for which we are so renowned?

Let The Revolution Begin

"Again and again the imaginary plan on which one attempts to build up that order breaks down, and then we must try another. This imaginative vision and faith in the ultimate success are indispensable.
The pure rationalist has no place here."

— Max Planck

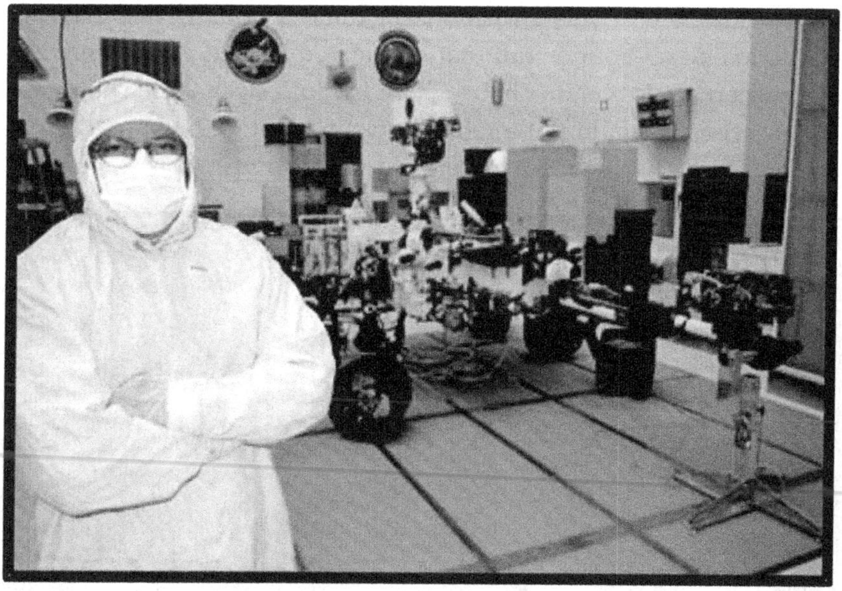

IN THE FALL OF 2011 (AS SEEN IN THE PHOTO ABOVE), I was dressed in the required cleanroom garments and standing mere inches away from the Mars Science Laboratory Curiosity Rover as it was being prepared for its 253-day flight to Mars. I was privileged to be part of the Radiological Team that was about to implant the nuclear source in the rover's Alpha Particle X-Ray Spectrometer (APXS) instrument. But delay followed delay that morning at the Kennedy Space Center, and I remained altogether enraptured at our station near one of the most astonishing pieces of hardware ever constructed. After a very long time, I somehow found the will to force my eyes from the rover, paced a dozen steps away, and paused directly in front of the aeroshell envelope that would protect and encase the $2.5 billion Curiosity on its way through 352 million miles of deep space.

For years I had personally seen and stood by space hardware sent to Venus, Jupiter, and even to Saturn and its giant moon, Titan. But on this day, I was completely in awe of what rested before me. I looked into the empty aeroshell that within days would hold the carefully folded lander-rover that was the size of a small pickup truck. With the uncorrupted imagination of my teen years, I stared intently for a long time, peering through the open square hatch inside this cocoon that was soon to fly away to the Red Planet. I could not resist wondering what it would be like to ride in this Mars-bound shell to the planet itself as its sole occupant. It was truly an astounding quarter-hour for me, standing so close to a Mars-bound craft that I imagined I could almost sniff the thin Martian air. I thought about where I would sit in there riding out to Mars, where I would put my backpack and laptop. I would ask that they replace the hatch with a glass window, of course…

Mars Science Laboratory Curiosity Rover Aeroshell Pre-flight

But as I stared into that empty shell, I also soberly remembered what my NASA research team had discovered just five years earlier. As one of the Agency Principal Investigators on the project, our team[1] was examining the effect of high energy cosmic radiation on the most sensitive parts of the adult brain, as was mentioned just a few pages ago. This region of the brain, known as the hippocampus, is essentially the nursery for the birth of new brain cells in a process called neurogenesis. The cosmic ray modeling for our investigation was conducted at the Brookhaven National Laboratory's Collider Accelerator, where we "created" cosmic radiation from elemental iron and focused the beam of highly energetic particles traveling at near light speed on genetically engineered mice models.

We discovered that the effect on the process of neurogenesis was devastating. So devastating, in fact, that we titled our paper, "Quiescent adult neural stem cells are exceptionally sensitive to cosmic radiation." As I remembered what we had discovered while I stared into the inside of Curiosity's spacecraft to Mars, I began to rethink whether or not I really wanted to travel as a passenger to the Red Planet *in this particular vehicle* or any other that had *not been specifically created to protect the occupants as a priority design philosophy.*

Ultimately, my wife, Claudia, and I watched the Curiosity rover launch from the Cape Canaveral Air Force Station on 26 November 2011. After it successfully landed on 5 August 2012, the rover began to send back data that it had collected and stored while in flight to the Red Planet. It reported that it had recorded a level of cosmic radiation that had streamed into the spacecraft during its eight-and-a-half-month journey that was equivalent to 35.5 full-body CT scans! It is also significant to note that Curiosity flew during a relatively quiet solar period, and recorded that only three percent of its dose was from sun-generated particles during its voyage.[2]

This annual dose rate was some 273 times the average background radiation exposure everyone experiences on Earth. Astonishingly, it was also 13.5 times higher than the dose permitted for professional nuclear power workers. And all of that high energy cosmic radiation was recorded on only the outbound journey to Mars while *inside* the walls of the spacecraft. Clearly, a flight of living humans to Mars would require engineering some radical differences into a crewed vehicle than the craft designed for a fleshless robot before such an endeavor could in any way be deemed safe. And none of those considerations even dealt with what would happen to a colonist's body after exposure to microgravity for such an extended time, or how the cosmic radiation exposure and microgravity would interact together inside the human body, as is discussed in detail later in this book.

My date with genetically modified mice and the massive Brookhaven collider had actually begun at the Kennedy Space Center over two decades before. I had come to the Space Center in the summer of 1987 as a Flight Vehicle Safety Specialist in the first wave of NASA hires after the Challenger disaster. With my degree as a Bioenvironmental Engineer and experience as a Navy Civilian Nuclear Engineer, I was soon assigned to the Biomedical and Life Sciences division as a Bioengineer of Macro Systems in the laboratories that were designing advanced life support systems for Moon and Mars bases.

Our task was to design and build bioregenerative life support systems that could be transported to the Moon or Mars and provide food, oxygen, and water for the crew while scrubbing their carbon dioxide (CO_2) and recycling their "waste." It was a complex pioneering endeavor wrapped in an engineering paradigm called CELSS – an acronym for Controlled Ecological Life Support System. This was the penultimate dream of any bioengineer of macrosystems: designing and building an engineering system that leveraged life as its core components. It was an amazing time in my life – having the opportunity to go to the office every day, dreaming, engineering, and building the essential advanced components for permanent human cultures on other worlds.

I was given the opportunity to lead the CELSS component that on Earth would have been dubbed "waste processing," but I coined the more appropriate phrase "resource recovery," which was later adopted Agency-wide. After my boss, the late brilliant NASA scientist Dr. William Knott III, Chief of the KSC Advanced Life Support Laboratories, handed me my assignment in waste processing, I went off and explored the task ahead of me by interviewing the CELSS team. I thereupon returned and explained to Dr. Knott that I did not actually have a job in waste processing because in these space systems, by definition, there can be no waste. Everything, no matter how unpleasant, is something the crew will need tomorrow to survive. Hence, everything in the system is tomorrow's resource that must be recovered. It is a lesson future Mars and Lunar colonists will need to realize before they depart Earth in order to ensure their long-term survival. And it is a lesson every citizen of Earth must soon learn for the same reason. Dr. Knott agreed and relabeled my responsibility as leading the Resource Recovery team. (For more on Dr. Bill Knott, see the final chapter of Book Two titled "Postscript.")

In 1994, Bill gave me the opportunity to test our advanced life support systems components in a remote and extreme environment on Earth. We created a portable hydroponic plant growth chamber and moved it to the Marine Lab undersea laboratory in Key Largo. There, in the OCEAN Project (Ocean CELSS Experimental Analog), I was privileged to plant and harvest one of the first full agricultural crops of any kind ever grown in a crewed undersea habitat – an effort first accomplished by my friend and long-term aquanaut record holder Rick Presley who planted and harvested various herbs on his 1990 undersea mission. Our crop of choice was common loose-leaf lettuce, *Lactuca sativa,* because of its short growth cycle (and ultimately, therefore, its impact on the final cost of our mission.) It was in that environment that I personally discovered that all environments off the surface of Earth that were not "one-Earth-normal" were subject to unexpected surprises; because living systems, however complex and however robust, are forced into an adaptation mode when taken out of the one-Earth-normal environment.

Our crop was grown in the undersea laboratory with a bottom depth of 30 feet under the ocean (or fsw – feet of saltwater). The hatch depth was at 21 fsw. That depth was established because it is the absolute limit threshold before human occupants have to undergo decompression to return to the surface following stays of greater than nine hours. The Marine Lab laboratory was frequently used for saturation dives of one to many days. Claudia and I were certified as aquanauts there and had lived and worked in it on several previous saturation dives as a part of our privately planned and funded expeditions. But plants were not humans, of course, and how they would respond was anyone's guess since this crop of extensively studied lettuce was the first to be grown in the environment undersea at a pressure of 1.57 atmospheres.

Lettuce has a circulatory system, to be sure, but it is obviously quite different than humans. Lettuce also actively exchanges gasses and water, just like humans, but its living systems, again, are quite different. There was one certainty, and that was that the environment was significantly different than one-Earth-normal. The question we were eager to understand was what the adaptation limits of the plant under those specific conditions were. We knew very well that we were planning to take these systems to other kinds of extreme environments. At least this analog experiment would give us a look at the plant's response in one kind of extreme environment.

We discovered that in comparison with our control plants grown in ideal conditions in the laboratory at KSC, our subject plants were decidedly stunted. Their biomass production was down more than half. It was an interesting process of investigating the systems, and even crew behavior, in the "postmortem" to determine relevant factors and ascertain why we had such results.

Several possibilities were considered. The Marine Lab habitat was supplied with a constant stream of air pumped through a compressor on the surface and down to the habitat. There was some indication that an inline filter may have been altering the natural carbon dioxide level, leaving the plants without adequate carbon nutrients since carbon dioxide is essential to plant growth. Further, when the duty aquanaut/researcher, Christopher Olstad, made his rounds several times per day, the volume of the Marine Lab was small enough for his presence to cause the carbon dioxide to spike in the habitat to levels far above what would be considered the normal range for lettuce.

What we observed was simply a failure of the organisms to thrive. They adapted. They survived. But they certainly did not thrive. Was it caused by too little CO_2? Was it caused by too much CO_2? Was it caused by wildly vacillating CO_2? Was it a pressure induced effect? One thing was certain – the plants were challenged to the point that they did not adapt normally and even exhibited what is labeled in an agricultural setting as a pathogenic response. Yet, in the same environment, the human aquanauts thrived with no noticeable ill effects.

These were all incredibly valuable lessons learned for future space voyagers who intend to grow all, or at least part, of their own food. In addition to the environmental challenge of alien environments, we can also add significant challenges induced by equipment, and even radical periodic changes provoked just by the very presence of the crew. For example, the average CO_2 level in the International Space Station (ISS) is 5,250 parts per million (ppm) day-to-day.[3] On Earth, we are exposed to a level of 400 ppm.[4] The ISS levels are some 13 times higher than on Earth due to the limit of the ISS equipment to remove the CO_2 to a lower level. In most greenhouses, the maximum CO_2 level for healthy plants is limited to around 1,000 ppm, after which the CO_2 becomes deleterious to prime plant growth potential. It ultimately becomes toxic to plants at levels above 5000 ppm[5] – which just happens to be the ISS CO_2 level.

Finally, we also learned that humans are like plants in one essential way: both are living organisms. If you change their environment to one that is different than one-Earth-normal, any living organism is forced to adapt. The questions that we must understand are ones that define the *limits* of adaptation – and when do those off one-Earth-normal conditions begin to induce systemic injury and a failure to thrive, just as we witnessed with lettuce plants in a human-rated undersea habitat environment?

After a while, NASA, in its often unfathomable behemoth-like judgment, shut down the Advanced Life Support System work throughout the Agency in order to focus resources on a flight program that soon after was canceled. When that happened, I was moved to a different but equally exciting task in Space Life Sciences. I was appointed as Chairman of the Kennedy Space Center's Institutional Animal Care and Use Committee (IACUC). For the next 14 years, I led the approving council for all the Life Sciences payloads flying out of the Kennedy Space Center onboard the Space Shuttles, as well as payloads sent to the Russian Space Station Mir and to the International Space Station.

From this incredibly interesting position, I was able to hear firsthand the details of every single space biology payload (from both international and U.S. researchers) having anything to do with Space Life Sciences involving vertebrate animal models. It was from this vantage point that I gained a keen insight into the questions involving biological spaceflight in microgravity as well as issues of cosmic radiation. What I learned was both equally exciting and discouraging.

From this Chair, I was also inevitably privy to many of the internal politics of Space Life Sciences, grants, and institutional biases, as well as ongoing ethics concerns. Thus, unfortunately, I have no other choice here but to digress into the gutter for just a moment, and then we will return to the more refreshing science and engineering discussions. I am quite certain that the next few paragraphs will get all of my future invites to NASA Christmas coffees permanently revoked. However, in order to adequately understand the rest of this book, it is a depressing tale that must be told. If you wish to understand the real NASA – read on.

As U.S. human spaceflight evolved beginning on 5 May 1961, we rapidly migrated from a very brief foray in human space exploration in deep space (Apollo Lunar exploration) back to a fixed focus in low-Earth orbit (Skylab, Space Shuttle, and International Space Station). In fact, I had one astronaut confide to me personally on a Kennedy Space Center launchpad that she felt that the Space Shuttle was nothing more than a very high altitude aircraft and that, as an astronaut, she was very much looking forward to the day when the United States departed the limiting confines of low-Earth orbit and returned to deep space. Such a sentiment, I discovered throughout the years, was very common in the Astronaut Corps. Yet there is a well defined political reason why we were stuck in low-Earth orbit for so many decades. But please understand that, more often than not, "well defined" political reasons and rational reasons are *rarely* the same.

It never came as a surprise to me that the politics in NASA were identical to politics in any other field everywhere else on the planet. People are people, and politics are politics. In the end, it is merely a simple process of following the money and not making the foolish mistake of equating a money trail to any connection with competent engineering. The money and the "rational plan" rarely meet in the human sphere.

The trail for following NASA money always begins in the White House. The NASA Administrator serves at the pleasure of the President. NASA budgets are built with the leadership of the administration, coordinated with the Office of the NASA Administrator, and sent to Congress for funding. But Congress insists, of course, that the funding must be somehow rationally linked to the mission as *they* see it. The NASA mission itself is ultimately decided based on the White House's vision of that mission, ideally with NASA input all along the way. Hence, the mission and its concomitant money flow are interrupted and fundamentally reset with each new administration while still requiring the blessing of the funding arm of the government – Congress (and its vision). It is the ugly sausage making of DC.

Because it seems each new President seeks their very own historically unique JFK "Apollo moment" within a few months of entering the Presidency, the new Chief Executive (with a couple of exceptions) invariably scraps their predecessor's vision for human space exploration and announces a replacement for it in a grand ceremony before the cameras. It is all about the President's legacy, it seems. Bizarrely, it has nothing whatsoever to do with a coherent, consistent, intelligent legacy of human space exploration that benefits everyone. In the end, this sickening, incomprehensible waste of opportunity and money is chalked up to Presidential ego by Chief Executives on both sides of the political spectrum. But, of course, no one in government can complain or they will lose their jobs.

For example, when George H.W. Bush entered office in 1989, he announced he was leading the United States to explore Mars. He made his grand declaration, and, as NASA employees, we were issued buttons (shown above) that were labeled "MARS or BUST." NASA tooled up for the new Mars mission and headed forthrightly in that direction, spending billions to make it happen and marshaling every imaginable resource to pull it all together.

Four years later, President Bill Clinton was sworn into office and the mission immediately and radically changed to "Mission to Planet Earth." We were literally told to throw our "MARS or BUST" buttons in the trash. It was no longer politically correct to wear them. But some of us hid them in our desks for good reason. Sure enough, eight years later, President George W. Bush was elected, and we were headed back to Mars once more! For those of us who had enough foresight to keep them, out came the "MARS or BUST" buttons. The NASA Administrator, working with the President, ended Shuttle because we were headed back to deep space after being stuck in low-Earth orbit for decades. The NASA Administrator stated in a meeting that I was attending that he was committed to "break Shuttle to the extent that it could not be fixed later." In its place, the Constellation program was conceived, designed to fly astronauts into deep space once more, and was a key component of future human Mars mission planning.

Eight years later, President Barack Obama was elected. He almost immediately cancelled the Constellation program, and even the discussion of crewed Mars missions was adamantly discouraged. The recurrently shameful "MARS or BUST" buttons went back into hiding. I clearly remember an interesting conversation I had one-on-one with a Center Director at that time. I could not hide my disappointment that

Mars was once again out of favor. He looked about an empty room considering his words carefully, and then said to me in a quiet tone of voice, "Dennis, we're going to Mars... *we're going to Mars*."

In December of 2017, President Donald Trump announced that it was back to the Moon and Mars... again. The well-worn "MARS or BUST" buttons were suddenly welcome once more on all NASA campuses.

With a refreshing and noteworthy departure from the norm, when President Joe Biden entered the White House in January of 2021, his administration (early on) continued NASA's Lunar-Mars pathway initiated by the Trump Administration nearly unchanged. Yet, prior to President Biden, during all of this never ending and even catastrophic upheaval (to any semblance of coherent human space exploration direction), countless billions of dollars were squandered and wasted on ethereal Presidential legacies. Along with the invariable toll on human capital was the unconscionable waste of historic exploration opportunities.

One fact is clear: were it not for insufferable Presidential egos and Congressional sausage making, there would already be permanent human colonies on both Luna and Mars today. But because of their egregious abuse of taxpayer funds as the cost of manufacturing their legacies, the United States' future in space was disgracefully squandered by unfettered, myopic egocentricity.

But corporate NASA does not resemble its Administration counterparts, waking up like a goose in a new world every morning. The Agency's people – unequivocally the finest brain trust in the world – were all committed and working in concert to make world class, historic missions to space happen regardless of the unchecked, ongoing state of serial administration "political psychosis" (a pair of words that are inextricably linked).

Even though their fundamental pathways kept shifting, the life scientists, rocket scientists, engineers, physicists, mathematicians, administrators, and office staff continued to work on assigned projects as smoothly and as seamlessly as was possible without losing the corporate memory of how to actually go explore space. Each mission directive announcement by a new President usually trickled down to the NASA Centers as massive reorganizations a year or so later. Thankfully, the politically generated disruptions to the employees and contractors were kept as palatable as possible by the savvy leadership of NASA, who valued their people and NASA's ultimate missions off planet.

While all of this vacillation on Mars exploration was ongoing, the Space Shuttle program was still fully operational until 2011, flying as many space missions as the budget and the infrastructure could handle. But with the ever-shifting priorities in space, what was the Shuttle actually doing up there flying its requisite billions of miles round and round the planet? In a pair of words: microgravity research.

I repeat, *microgravity research.* That fact is key to understanding how NASA managed to screw up the two-car parade in low-Earth orbit and exactly why it never actually reached its ultimate destination.

The goals of the Space Agency were radically changing and eternally shifting with, much of the time, few employees even knowing where the target was during their careers, much less being able to articulate it day to day. But one thing never changed since the first low-Earth orbit mission post-Apollo – and that was microgravity research. It was the steady, reliable cash cow that justified and funded a fat majority of all of Shuttle's scientific packages. Hence, it became the *de facto fundamental reason* for human space exploration. All the rest was really background noise. Of all the payloads that have flown on crewed space missions since the Apollo program, a very large cross section of them has been funded to do microgravity research, and as of this writing, nothing has changed.

As far as space life science is concerned, a very clear preponderance of all the protocols that came through my IACUC committee was involved in microgravity research. It sometimes appeared that the very idea of microgravity research became the bottom-line justification alone for all human space exploration and space biology, even overriding what seemed to me to be the validation for the proposed science. Indeed, such a sentiment is still very powerful in the Agency, as is evidenced by this rather candid, revealing quote from the journal *Nature* (2015): "Microgravity research is significantly enabled by space, and in this case, it is not the 'gravy,' but rather, the 'goal.' It has its own value."[6]

Such a blatantly bizarre statement can be taken in more than one way. But from the vantage point of a NASA insider to the process, I know the unfortunate reality of this comment to be entirely accurate in its most negative sense.

NASA has invested what may be as much as $100 billion in the International Space Station.[7] Its primary scientific purpose? For use as a platform for microgravity research, of course. Hence, microgravity research became *the* cottage industry in NASA. The great moving star in the heavens is all about microgravity. And what is the product of that industry? Arguably, thus far, it has principally been scientific papers and PhDs.

By the mid-1990s, planning for the ISS was hitting a fever pitch as the Space Station was undergoing serial billion-dollar re-designs due to seemingly incessantly shifting design requirements and chronic mission creep (driven principally by successive administrations and Congress). Within the life science community at NASA, there were various scientific and engineering camps and factions typically entrenched amid the multiple Centers around the country. Among them were the bioregenerative life support adherents who believed that producing food on Luna and Mars would eventually break even in transport costs and allow some level of security and self-sufficiency off-Earth. The physiochemical camp (using only chemical and mechanical processes to recover air and water while shipping all food from Earth) believed that their relatively smaller, and in some ways simpler, life support system would always offer the advantage over a bioregenerative system that they viewed as unstable and unreliable.

The inevitable, quick-witted, albeit one sided, response from the life scientists was that the mechanical and chemical engineers did not have the background, knowledge, or capacity to keep a house plant alive, much less a complex advanced life support system such as a CELSS. I personally always assumed (and impishly reminded many of them) that they had bought a house plant in college and, after keeping it on the windowsill without watering or caring for it for six months, piled over with their dirty underwear, that the plant eventfully gave up and died. So, from that moment on, they forever considered plants to be fickle, unstable, and unreliable.

In sober retrospect, after 14 years of reviewing every individual Space Life Sciences protocol for animal subjects, I believe that the most significant of all the scientific findings from the hundreds of papers we appraised was a continuous validation of the harmful effects of exposure to long term microgravity on any biological system. I recently conducted a sweeping metadata study of all major findings in microgravity research from the onset of the Space Shuttle program to date in search of a comprehensive list of significant scientific findings and/or benefits of microgravity research. What have we discovered from all these decades of research into microgravity that we can bring back home to Earth and use in our lives? After throwing away one of the most important of all lessons to be learned in the microgravity environment (see the future chapter titled "The Partial Gravity Scandal"), after all the billions have been spent, now, finally, what did we actually get for our money?

An intriguing piece appeared in the academic online journal, *The Conversation,* which concisely summarized the results in an article titled "Five Key Findings from 15 Years of the International Space Station."[8] The summary of the five key findings from the ISS in 15 years includes the conclusion that microgravity is harmful to human physiology, and four hopeful suggestions as to how microgravity research may *someday* lead to some important basic scientific findings or new technologies on Earth. Apparently, in 15 years on the ISS and all the decades before, except for the injuries, no vital finding has actually been uncovered or made it to the marketplace. At least Tang was available on the supermarket shelves in the 1960s and '70s.

Our primary and most important findings concerning microgravity are that microgravity injures human beings over long periods of exposure (*not short-term spaceflights*). Some of those injuries are permanent. For example:

In 2015, awards were given to the preeminent scientific discoveries from the ISS.[9] Of the four awards, one was given for the discovery that there might be some hope that we can develop drugs that may counter permanent bone loss in microgravity. The second was awarded for the discovery that microgravity-induced bone loss mimics the pathological bone loss process called osteoporosis in Earthbound humans. The third award was given for a study of the behavior and propagation of flames in microgravity. And the fourth award was given for a study of the suppression of the human immune system in microgravity. The predominant importance of decades of study of microgravity is that it is hazardous to human health.

After my 30 years in the Space Life Sciences business, perhaps the most astounding conclusion I have reached from all of it is this: If we can somehow end the insanity of forcing every member of the entire space exploration community to live in and thereby sustain the inevitable injuries induced by exposure to the microgravity investigator's laboratory 24/7, then the rest of humanity can get on with safely exploring space in good health and stop the hemorrhage of scarce research dollars designed simply to keep NASA's human hardware in orbit. (Note to Congress – the Emperor is really naked. Get over it and move on.)

Thus, if I may be permitted what is considered an obscene blasphemy in the Agency: there is more to study in space than microgravity and its endless, mostly ineffectual workarounds. Someone needs to mercifully terminate the endless study. We have already reached the single, profound conclusion of that research: long term exposure to microgravity is toxic and harmful to human beings.

Let us definitively cut our losses. Give the microgravity community the ISS to continue their microgravity studies while the rest of us who are committed to human exploration beyond low-Earth orbit move on to build an artificial gravity-enabled and adequately shielded platform for the safe human exploration of space. It is time to stop the endless, abjectly ridiculous countermeasure madness, injuring our crews while we force them to live in toxic environments in order to document, once again, that it is indeed *toxic* on long term missions, just to ensure that taxpayer dollars flow to orbit.

For some bizarre reason, NASA has not yet figured this out. It is not lost on me that simply by making these inherently revolutionary and countercultural statements, I stopped following the money and have insolently disregarded five decades of political bargaining. I choose to take my stand on the side of ethical, sound, and responsible science and engineering, because there is, in fact, another way.

But then, that is exactly what this book is all about – the other way.

In human space exploration, now is the time for the long overdue revolution to begin.

Are We There Yet?

"The bad news is time flies.
The good news is you're the pilot."

– Michael Altshuler

THIS BOOK IS ALL ABOUT HUMANS TRAVELING TO LUNA AND Mars and living there permanently. But it likewise applies to the voyager who wishes to travel between many other worlds in the Solar System spending multiple months or cumulative years in deep space between planets. It is impossible to separate human cosmic radiation exposure and the conjoined toxic effects of microgravity from the interwoven hostile environment of deep space during the transits to and from any distant world. Therefore, it is helpful to know exactly how long it actually takes to get there and the duration of our stay. We must also determine the time required for a round trip, if one is desired.

Even to those people who are very knowledgeable and interested in participating in human space exploration or colonization, transit times to Luna or Mars in deep space are something of a puzzle. This is so because each transit time is somewhat different. Since every target destination in space is always moving – just as the departure point on Earth is also simultaneously in motion due to the rotation and solar orbit of the planet – no two distances between points are ever quite the same, as you will see in the examples below. For instance, in the case of Mars, the relative approach or departure speeds between the two planets may be as low as 3.6 miles per second or as high as 31.6 miles per second (almost 10 times faster).

And, as if to make the question even more complex, no individual rocket systems or propelled masses are ever exactly the same, even from mission to mission. Therefore, the time required to travel to Luna or Mars will vary slightly – or even significantly – for every mission. In the following discussion, I have provided you with the average transit times based on a half-century of crewed and uncrewed expeditions to Luna and all uncrewed missions to Mars. That should help dispel at least some of the mystery and give you a single average number to lock into your memory. But please keep in mind that since the launch pads on Earth and the target landing zones on these other celestial bodies are always in relative motion between one another, an average is just an average.

Luna – Earth's Moon

In the case of Luna, there is not so much ambiguity because the proximity is significantly different than the expanse between Earth and other planets such as Mars. The six Apollo landing missions demonstrated that the trip out to the Moon required an average of 3 days, 8 hours. That time was calculated from liftoff at Kennedy Space Center's Pad 39A to the moment the rockets were fired to enter Lunar orbit. Yet the total average mission time for the Apollo 8 and 13 circumlunar voyages was 6 days, 1 hour. (Even with the catastrophic failure on the Apollo 13 craft and loss of most power, their complete mission duration was only 5 days, 22 hours, 54 minutes, 41 seconds.) The round trip for the six landing missions (including a day or so on the surface and the trip back) required an average of 10 days, 7 hours. But the average time for all Apollo missions together was 9 days, 5 hours.

Mars

The voyage to Mars requires coordination with the cycles of the planets around the Sun and employment of the most energy efficient trajectory outbound. The most energy efficient orbital trajectory to and from any planet in the solar system is called a Hohmann Transfer Orbit. One cannot just "fly to Mars" directly. The voyages consist of months-long parabolic orbits to rendezvous with the planet at a certain predetermined point in its orbit – much like a quarterback tosses a pass downfield at a receiver who is running to catch up with the ball in flight. Further, because of this cosmic dance, the "window" of time (or launch opportunity) that a spacecraft can be launched toward Mars and return on these most efficient orbital trajectories is limited to approximately one month every 26 months. And this 26-month launch window cycle repeats itself every 14.95 years.

Counting a representative sample of 28 Mars missions that successfully made it to the planet (although not all of them successfully landed) from all nations between 1964 to the 2021 Chinese Tianwen orbiter with its Zhurong Rover and the United States Mars Perseverance Rover with its Ingenuity helicopters, these voyages required an average of 243 days travel time in deep space from Earth to Mars. Remember that number. It will be the average transit time to Mars that we will use throughout *Book One* and *Book Two* of this work – 243 days or 0.67 years.

Once the rockets are fired to depart Earth orbit, there is no way to turn around and return to Earth before the mission is complete due to the unforgiving realities of relative planetary motions and alignments around the Sun – or, by its proper name, Sol. To those accustomed to traveling about on the fixed surface of Earth, this concept can be somewhat difficult to grasp. On Earth, there is no such thing as a permanent No-U-Turn sign. If an aircraft gets into trouble, it can turn around and head back to base. If you forgot your toothbrush, you turn around and go back home to get it. If a passenger ship loses a main generator, it returns to port. If Pooh Bear is left at the hotel in Dahlonega, Georgia, Dad stops the car, turns around, and makes the recovery. (Been there; done that…)

But in space, none of this is possible with our current knowledge and technologies. Once the rockets are fired to enter the transfer orbit, every ship is thus required to travel all the way to Mars, land successfully, and wait for the next window of opportunity to return to Earth. The laws of space travel are quite unforgiving. Following them very precisely is the only way to successfully fly out and fly back. Thus, the typical round-trip Mars mission is just shy of 1000 days, with *no chance* of a return to Earth before then. Obviously, in the future, advanced rockets (such as nuclear and ion drives – or an Astrophage powered spin drive like the one in Andy Weir's *Project Hail Mary*) will change that picture to a significant degree.

As of this writing, the range of days required to fly out one way to Mars (shortest to longest) was from **131 to 334 days, with an average time of 243 days.** Mariner 7 required 131 days in 1970. The Mars Polar Lander voyage required 334 days to reach Mars in 1999 – and was subsequently lost on arrival.

NASA is currently advertising 180-day human Design Reference Mission (DRM) to the Red Planet. But, again, the average time required to fly to Mars since 1964 has been 243 days, which is two months longer than NASA's DRMs. As two classic and well-known examples, the Curiosity Rover spacecraft launched on 26 November 2011 and landed on the surface of Mars 253 days later on 2 August 2012. And the NASA

Rover Perseverance launched on 30 July 2020 and landed 204 days later on 18 February 2021. Together, they represent an average of 229 days between missions – just 14 days short of the total historical average. Between the two nearly identical spacecraft (in mass and configuration), Curiosity required 10 days longer than the average, and Perseverance arrived 39 days faster than average. Such common differences represent the realities of the cosmic dance.

Therefore, throughout the remainder of this book, I will use the average time established over half a century of spaceflight and those transit times reflected in the Hohmann Transfer schedule catalogue (243 days one way and 1000-day round trip missions) as a point of reference. The time difference between NASA's advertised DRM requirement of 180 days and the historical average of 243 days is significant for reasons that will become apparent in this book. That 63-day dissimilarity is very important to the biology onboard any Mars bound crewed spacecraft, given the primitive human rated spacecraft planned for use today.

Future Missions

The Hohmann Transfer orbit interplanetary transit technique is certainly not the final word on interplanetary travel. New rocket propulsion technologies such as ion drives and more powerful boosters will change transit times in the future. But for the technology that is available as this book goes to print, 243 days is the historical average transit time to Mars.

Why We Will Depart Earth Forever

Reason 2

"My definition of a free society is a society where it is safe to be unpopular."

— Adlai Stevenson

"I'll tell you what freedom is to me – no fear."

— Nina Simone

IN THE SPRING OF 2018, CLAUDIA AND I ROUNDED A SHORT flight of steps in a colonial home in Virginia and stood in the personal study of the fourth President of the United States, James Madison. The study was restored not only to its original furnishings, but the actual personal implements and individual items of Madison's stationery were in place on his desk. On his bookshelves were copies of his books (the originals were lost) that he used as his crucial references for a historic purpose – the drafting of a majority of what would become known as the Constitution of the United States.

Madison locked himself in his study with the collective philosophic wisdom of the Old World. Using the lessons learned from the failed remnants of governments and the wreckage of individual freedoms under centuries of European monarchies, he drafted an instrument of government that was nothing less than a brilliant distillation of all the best and a rejection of the worst. In the end, Madison's document elevated the democratic principles of majority rule while at the same time, specifically protecting the personal freedoms of every individual citizen, and *especially* the rights of the minority – those who were not counted in the majority. In so doing, James Madison created a society in which everyone's niche was protected from harassment from every other in the same geographical footprint. No nation of people had ever lived in such an advanced democratic republic protected against the excesses of both the subjugation of the governing elite and the tyranny of the majority. The nation it was created to govern was to become known as a representative republic whose citizens were specifically shielded from the oppression of the mainstream zeitgeist.

As I lingered there in that sacred place, I knew very well the full significance and historical consequence of where I was standing. When the crowd left, led away by the docent, I hung behind to be the last one out of that upper room at Montpelier. I closed my eyes and felt the presence of genius that still lingers in that place for all those who have eyes to see and ears to hear its timeless message to every other human government that will follow – even on future distant worlds.

Madison's message is elementary and yet as profound as any ever spoken through the ingenious thoughts of humankind: freedom is not relegated to one person or one government or one majority – but it is safeguarded for every individual in the culture, whether they are of the majority or not. Each citizen is granted a profound and revered liberty that is sealed with the words penned 13 years earlier in the Declaration of Independence that founded the nation for which this Constitution was written. Each citizen was considered by the most fundamental precepts of the law to be endowed by their Creator with certain unalienable Rights – and among these are Life, Liberty, and the pursuit of Happiness. Based on these essential prescient principles that had already been voiced in the Constitution's reason for being (the Declaration of Independence by Madison's friend Thomas Jefferson), Madison thereupon created his work of extraordinary brilliance and enduring power.

Whether in a cage full of rapidly breeding gerbils or on a planet full of rapidly breeding higher organisms, there always comes a time in both habitats that the individual socio-niches become more and more compressed to the point that they are forced to merge. That is where the trouble always begins. Regardless of the prevailing propaganda to the contrary, not all niches are compatible, leading to very unhappy organisms perpetually struggling against one another's concepts of rights, definitions of social allowances, and that set of doctrines and faith to which each individually holds as consecrated, absolute, and inviolable.

On Earth, for example, the times of greatest social upheaval have rarely ever been precipitated by a true lack of geography (no matter the perception) but always by a lack of niche compatibility. The expansion of Western civilization across the Atlantic Ocean is a perfect example.

In Europe, nonnegotiable religious constructs collided between separate niches in the same geographic space. Protestantism collided with Catholicism, and both collided with aligned states, each bound to their inflexible doctrines. Each nation-state had chosen a side, and their citizens who had opposite or different beliefs were marginalized, persecuted, jailed, and routinely executed. And although "colonization waves" to the Americas from Europe is a hotly contested subject, I consider the British colonies at Roanoke Island (1585) and Jamestown settlement (1607) the "first wave" of colonization (the Spanish settlement at St. Augustine, Florida, established 42 years earlier, was strictly a military garrison not intended for colonization). In the Americas, the "second wave" of colonists crossed the ocean on the Mayflower 13 years after the founding of Jamestown to open up a new niche that was fully compatible with their own immutable religious beliefs. It was brilliantly successful, and soon, many more followed, escaping the niche conflicts that had formerly bound up countless lives in hopelessness, misery, and fear.

In the 21st century, as we look outward to settling new worlds beyond Earth, the second reason to depart the Earth forever is the same as in previous centuries:

Freedom

The Constitution of the United States is unquestionably the most brilliant governing document of any culture of people in history. It was designed specifically to *prevent niche incompatibility* and limit the power of government, thereby preserving and protecting individual freedom. Learning from the failures of the Old World governments, the United States Constitution created a nation in which the legal construct of the society was specifically designed to sort out niche conflicts with authentic, evenly applied justice, and provide opportunities for everyone to express their individual freedoms.

But, as is discussed in greater detail in the upcoming chapter found in *Book Two* of this series, "Weaving the Microcultural Fabric," the new colonies (those which are not founded as national outposts) will start their governments from zero on the first day. The questions are, "What kind of freedom will the citizens of these new worlds find?" and "What will motivate them to even start a new colony to begin with?" As we look back to the previous centuries, it is evident that our forefathers in the United States wanted to be unshackled from state religions and from the sociocultural limits imposed by European class stratifications. But what about the 21st century?

It is very likely that purely religious communities will be established as theocracies by those who can fully afford to send the equipment and the colonists, such as from the Middle Eastern Muslim nations flush with resources paid from vast petroleum enterprises. Obviously, there will be a loud outcry from the areligious and from other religions that the new words are being immediately "corrupted" with the same old problems that have vexed humanity – religious strife resulting from select exclusive niches. On the other side of that argument is this consideration: it is, after all, humanity that is being transplanted across worlds, not robots, remember? Humanity always travels with all its problems and deepest convictions intact and is certainly not going to be "culturally cleansed" to fit any given ideology of what cultural cleansing may even mean from one ideology to another. If cultural cleansing actually worked, there would be no reason to escape a world filled with submissive human robots.

But Muslims are not at all shy about their wish to dominate their own cultures and absolutely do not tolerate any competing ideology. That statement is not insulting to Islam. It is a factual and accurate representation of their sincere religious devotion dedicated to unquestioning adherence to maintaining the uncompromising purity of their doctrinal precepts. The same is true of nearly every major religion that wishes to preserve the unfettered integrity of its individual ideologies. But it is the Islamic peoples of the oil-rich nations that have the financial wherewithal to successfully establish off-world colonies early on. And, frankly, it would be surprising if they do not.

I am fully aware of the impact of this statement on the community of space colonization advocates. I am relatively certain that very, very few of them have ever considered that the first independent colonies on the new worlds may very well be religious theocracies. Please take notice: the reasons for establishing 21st-century off-world colonies have not changed *at all* from those of the original American colonies. The very first waves of colonists to the New World came almost exclusively to establish religious theocracies.

Others came to escape. Some came to dominate. Humans always, always travel with their precious inherited ideological baggage. They never leave home without it, and if they did – *they would not be human*. Our nature has not changed at all in the past 400 years. Therefore, the crystal ball of the future lies in clearly understanding what happened behind us before we prognosticate about what may lay ahead. Yet of all the key, irreplaceable understandings of historical and future human colonization is the central fact that human cultures will always be made up of humans – who always travel with their ideologies and the numerous, shifting narratives that energize and drive them.

It is incredibly interesting that this discussion alone will immediately result in expanded decisions on ways to *prevent* such a thing from ever happening on Mars. The fascinating part about that very human reflex is how humans always seem to immediately plot to limit the freedoms of other individuals to dictate their "freedom" to dominate the ideological landscape with *their* ideologies and narratives. Yet, that is precisely why those others want to depart in the first place – to find freedom from these incessant plots which limit their ever-shrinking niches here, even if they have to travel 300 million miles to an alien shore to get away from them. Over the centuries, not much has changed at all.

Of course, the same discussion also applies not only to Muslim and Christian communities but to Hindu, Buddhist, Jewish, and Wiccan, as well as all variations of anti-theist communities. Each will want their pure philosophical niches preserved in the new worlds. And then there are the new-world-scientific-colony ideas, liberal communities, conservative communities, progressive communities, and literally dozens of others framed around everything from a genuine democracy to a clone of the U.S. constitutional government to true libertarianism. Many early founding colonies may be completely socialistic or communistic, reflecting the very powerful governments of their origin.

If anyone wants to get a snapshot of what the new colonies on other worlds will look like, I can let you in on a very accurate view of the future. Ride out to cislunar space, to Lagrange point 5 (L5), turn around, and take a photo of Earth. Wait 12 hours and take another. Place these images side by side. The cultural civilizations that exist on Earth are exactly what the human colonies will look like on distant worlds – all the good, all the bad, and all the ugly!

Finally, this is such a powerful quote by Adlai Stevenson that it bears repeating as a very concise summary of which colony I personally would prefer to live in:

"My definition of a free society is a society where it is safe to be unpopular."

This sentiment is repeated in the quote from Nina Simone:

"I'll tell you what freedom is to me – no fear."

It is just such a society that our U.S. Constitution was originally designed to guarantee for its citizens – a society that allows everyone in it to be truly free from the oppressive opinions, narratives, and strong-armed tactics of a majority of the others. It is to this place, this kind of freedom on other worlds, that people and power will ultimately travel across the trackless millions of miles of deep space to find and settle. Because, for human power and creativity to actually thrive, people must be able to live and produce their individual visions protected by the shield of true liberty under the wings of absolute freedom and not shackled by the overshadowing fear of the iron fisted and inevitably oppressive majority.

Module
One
Ch.01
Ch.02
Ch.03
Ch.04
Ch.05
Ch.06
Ch.07
Ch.08
Ch.09
Ch.10
Ch.11
Ch.12
Blind Alleys of
Human Space
Exploration

The Limits of Human Adaptation

"There are three kinds of men.
The one that learns by reading.
The few who learn by observation.
The rest of them have to pee
on the electric fence for themselves."

- Will Rogers

Humans have a fixed adaptation capacity.

ACKNOWLEDGING THIS RULE IS ESSENTIAL BECAUSE ON its inviolate life-or-death truth stands the inevitable success or failure of any human space mission outside the Van Allen Belts in deep space. This maxim is so critical that it bears repeating:

Human beings have a fixed adaptation capacity.

It is likewise essential because so much denial and bloody history are wrapped around the stubborn human defiance of its incontrovertible truth. This is evidenced in today's crewed space exploration business where the Band-Aid codicil is:

"Humans have a fixed adaptation capacity… **but**… we can add countermeasures."

That fairy tale continues with this false promise:

"With enough countermeasures, humans have no adaptation limit."

But the saga's final summary is:

"Humans are creatures whose brains write checks their bodies can't cash."

Remember the previous comment about "denial and bloody history"? The endless countermeasure fantasy is where it all began.

Driving any human organism beyond its limits of adaptation results in physiological injury. Every. Single. Time.

While that limit is variable among individuals, this trait is especially endemic to explorers, particularly astronauts.

But let me be clear: rational human adaptation is *never* characterized by pushing the human body to adaptation by the healing of a forced injury! Healing from injury is an adaptation response, but it is by no means an adaptation process that would actually be included as a part of any sane or successful exploration endeavor. Planning to injure crews as a part of the expedition strategy "because they'll get over it" is truly over the top barbaric and irrational. And yet, as we look forward to the future of human spaceflight in deep space, such an evident lack of sanity is already underway. Today, before the carnage begins, that is actually the going in plan for exploring deep space.

The unambiguous statement that all humans have a finite adaptation capacity should be common sense. But, by a combination of a slate of unfortunate social pressures (including a persistent Hollywood mythos and cultural idol worship), common sense has been shoved to the sideline. The end result is that otherwise intelligent people offer up others and themselves as human sacrifices to a 21st-century hero worship even as we set our sights on Mars. It is very reminiscent of the Virginia Company's flyers designed to attract Jamestown recruits that represented Virginia as paradise. As was ultimately revealed, it was more accurately corporate advertising to recruit flesh and blood specimens to be used as human test fodder. In the end, the test failed, and hundreds died. The flyers were pretentious ego-oriented crap. But the stupid humans always seem eager to take the bait as they did – and as they certainly will in the near future.

While somewhat variable within a relatively small window, human adaptation capacity is by no means unlimited, nor can it with *any* conceivable level of accuracy even be described as *highly variable.* The incredibly complex human organism is designed to constantly seek a biochemical balance with its ongoing internal processes and its external environment in a process called homeostasis. This predominantly hormone driven process defines the human's ongoing adaptation process with itself and the variable conditions outside itself to achieve a relatively stable physiological equilibrium. Homeostasis is a natural but quite limited function. Beyond its limits, the failure of the body to successfully achieve homeostasis results in injury. If not corrected, the human organism will transition from repairable injury to permanent injury and, ultimately, death.

For example, the absolute safe external temperature range for humans is in a restricted band between 50 and 85 degrees Fahrenheit. Many factors determine the ultimate outcome, and (even in this very narrow 35-degree band) permanent injuries can occur if the body fails to adapt in a nominal way by wearing inadequate clothing or through other effects such as wind, rain, or humidity.[1]

The nominal internal body temperature range is much narrower. It is a 6-degree band from 95° to 101° Fahrenheit.[2] It is exactly the same story for exposure to low pressure, low partial pressures of oxygen, transitions from high pressures to low pressures, exposures to ionizing radiation, and even prolonged exposures to microgravity. The human body is extraordinarily resilient and adaptable – but – *only within its narrow design bands*. Outside those bands, it is *decidedly not*. Outside the range, nominal adaptation ends, homeostasis fails, and injury begins. It will be either acute injury (from which the subject recovers in a given period of time), chronic, permanent injury, or even death.

Consider the following example of a failure to achieve nominal adaption. If the human body cannot cool itself down by sweating or finding relief from the heat, failure to achieve homeostasis will result in an injury called heatstroke. If not immediately treated, the brain, heart, kidneys, and muscles may be injured, leading to permanent damage.

Yet, within our narrow functionality range, human adaptation is ongoing every minute of every day. Our bodies automatically metabolize more when we need additional energy to maintain our core temperature. We will invariably sweat if we need to cool down. Our bodies frequently repair background radiation damage or damage from medical x-rays that injure our cells. We are literally engineered to be constant engines of adaptation. Adaptation is what we do 24/7. However, we are designed to adapt within set, pre-established ranges – NOT outside those limits!

The process of adaptation is constantly ongoing within the human physiology until it reaches its limit, which is defined as that point at which the body can no longer biochemically cope with the off-normal condition (of whatever challenge it is addressing), and homeostasis collapses. At that point, one or more of the body's systems begins to malfunction, tear, break down, and ultimately fail. I have defined these conjoined internal and intercellular processes as the Adaptation Breakdown Syndrome.

In 1936, Hans Selye developed a theory he called the General Adaptation Syndrome, which is a stress model that explains the body's physiological response to external psychological stress. Selye described these stress responses as divided into three stages: alarm, resistance, and exhaustion.[3] Many of these same process mechanisms are involved in the Adaptation Breakdown Syndrome but on a biochemical and physiological level.

The Adaptation Breakdown Syndrome is characterized by numerous, widely variable physical and psychological indicators that can be portrayed by what I have defined as a "failure to thrive" physically or emotionally as the injury begins to take its general, whole-body toll. Either individually or systemically, the symptomatology indicates that one or more internal systems have been so adversely affected by the off-normal, out of limits environmental stress or stresses that it has elicited a biological reaction that is more accurately categorized as a pathological response instead of one of nominal adaptation.

The three stages of Adaptation Breakdown Syndrome are:

(1) The body's immediate physiological rally to cope with any off-normal condition,

(2) The body's collective processes which ultimately reach their adaptation limits, and,

(3) Acute or chronic injury, which is the final stage when nominal adaptation is exceeded and the human organism expresses a pathological response.

Unfortunately, for a suite of reasons, in current human space exploration models, the philosophy of expecting ever expanding human adaptation limits has driven the exploration narrative. But we have finally exceeded the limits of this philosophical dead end. We are now seeing clear evidence of an emerging pattern of an evolving pathology of spaceflight which is unambiguously driven by those flawed and increasingly unethical philosophies using humans to discover the limits – not of nominal adaptation, but of pathology driven by the fatally flawed exploration narrative. As a government agency, NASA has pressed this limit to the absurd and corrupt by assigning a "death score" to their mission planning called the REID, which is discussed in detail in later chapters.

The perfect example of this 21st-century breakdown in ethical limits is the prohibition of the use of non-human primates in space medical experiments. It is now universally unacceptable to use non-human primates in spaceborne medical experiments because it is generally regarded as an unethical use of higher animal subjects, exceeding the legally allowed limits of pain and suffering. Therefore, it is the human primates who have picked up the pain and suffering responsibility, and all medical experiments on higher animal species are now conducted *on astronauts*! How do the world space agencies determine the ultimate effects of microgravity, space radiation, and the multimorbidity of exposure to long-term spaceflight? By using astronaut subjects – because non-human primates are totally off limits for *ethical* reasons.

From our human spaceflight experiments, we have encountered many of these biological pathology responses as individual systemic biochemical reactions to each stressor. But they do not act in isolation from one another. In space – as on Earth, Luna, or Mars – the human body represents an astonishingly complex synergy of interacting physiological systems. One physiological challenge is nearly always ultimately expressed as a whole system biochemical challenge. Further, when there are multiple pathologies (exposure to microgravity plus cosmic radiation) ongoing at the same time (multimorbidity), the Adaptation Breakdown Syndrome is accelerated, and failure of the human body to thrive begins in a non-linear progression of decline that may lead to death. This systemic process is discussed in detail in the chapter "Breaking Down – Multimorbidity."

Limited Variability

In the human population, every individual has a highly variable response to environmental challenges, whether it be exposure to microgravity, or the injuries caused by any exposure to cosmic radiation, atmospheric pressure, temperature extremes, and even off-nominal diurnal light patterns. Some of the reaction is based on genetic predisposition. Some is based on age and general state of health. And yet, the human organism as a discreet entity has ultimate limits that no individual in the species can adapt to under any circumstances. Somewhere in the middle of these limits is the range of nominal human adaptation. At either end of that scale is off-nominal human adaptation, which is defined here as a pathological process leading to injury. Collectively, these injurious processes are the genesis of the Adaptation Breakdown Syndrome.

How do the space agencies of the world handle this reality? By leveraging their antiquated human spaceflight philosophy left over from the 1960s Apollo legacy onto 21st-century flight crews. They have designed their exploration philosophies based on a previously highly effective NASA model with the expectation that the crews will continue to adapt to the space environment if successfully augmented with a suite of "countermeasures." That philosophy was founded on only a handful of very brief treks to the Moon through deep space followed by years of orbiting Earth below the Van Allen belts that are totally dissimilar to the future plans of extended time and exposure in the deep space environment. Thus, with the advent of longer spaceflights, even in low-Earth orbit, adaptation has reached the limit of its viability, and the Apollo-era philosophy is now unambiguously breaking down at the cost of injury to astronaut flight crews. This is so specifically because we have reached the far end of the variability spectrum in which *all astronauts* exposed to long-term spaceflights are exhibiting the various pathologies of the Adaptation Breakdown Syndrome.

The Error Driven by a Faulty Philosophy

The current human space exploration philosophy is protecting its advertised rationale for existence and cannot change without a revolution in exploration purpose which leads to another, more powerful replacement ideology. Its reason for being, again, is to conduct microgravity experiments, which is an exceedingly narrow discipline with very little to show to date for hundreds of billions of dollars of expenditure. And yet, once again, it is so politically positioned and powerful that nearly all human space exploration since Apollo has been sponsored by this science that constrains all others. Thus, the only countermeasure necessary (induced gravity) to protect a crew from pathological exposure to microgravity has been judged as inappropriate – and even branded "impossible" – to protect the status quo. In this case and many others, human intellect has shown itself capable of compartmentalizing logic to the point of an irrationality that can only result in disaster, which is then typically finally recognized and documented during the inevitable post catastrophe investigation. Ultimately, even following tragedy, systemic reorganization, and high moral pontification, the humans invariably continue with their previous activity after the news cycle shifts to some other crisis.

In this monumentally absurd scheme, a flight crew is required to live 24/7 in a hazardous work environment in order to sustain their experimental designs – which are, ironically, predominantly doing experiments that are investigating the reasons why that specific working environment is hazardous. It is a ridiculous but highly lucrative endeavor hidden by an avalanche of scientific verbiage quite beyond the average taxpayer's and lawyer-politician's vocabulary. Thus, it is indefinitely self-sustaining – which has always worked quite brilliantly in NASA and still does to this day.

Further, because this exploration philosophy also protects the faulty notion of broad human adaptation to every other space hazard, the mass related energy cost for adequate shielding to protect astronauts from the inevitable, immediate pathology of cosmic radiation is deemed unnecessary. Astronauts and colonists are thus expected to "adapt" to high energy cosmic radiation effects. Because of a gross failure to understand the true nature and long-term physiological effects of this galactic cosmic radiation species, this disaster in progress has been allowed to continue virtually unchallenged.

Only a radical change of exploration philosophy can transform the current irrationality of the status quo. Unfortunately, even as I write, the near-future exploration architecture for deep space exploration beyond the Van Allen Belts is already being developed based wholly on the current fatally flawed exploration paradigm. Unless this model is changed soon, a very heavy cost will be extracted from the lives and health of astronaut crews, the ultimate success of their missions, and the future of human exploration and settlement beyond Earth.

The "Superman disorder" is common to humanity. We consistently think of our physical and mental abilities as far more advanced than what they really are. This leads to immediate trouble when we try leaping off tall buildings, stopping speeding locomotives, or lifting our middle finger to the high energy, high mass radiation particulates expelled by a suite of black holes, supernovae, and PeVatrons (described in the chapter "A Primer on Cosmic Radiation"). Unfortunately, all of these space environment realities will have the same inevitable effect on soft human organisms and their complex and exceptionally sensitive cellular biology as the tall building leap and speeding locomotive impact. When the biochemical checks finally hit the bank, they will inevitably be stamped "*INSUFFICIENT*."

Discovering the Human Adaptation Zone

Claudia and I have homes in Florida and Tennessee. In January, the conditions between the two are radically different, to say the least. And yet, whether I go outside my home in 60-degree Fahrenheit sunshine or a minus five-degree ice storm, I can still function. I can function in Tennessee because of appropriate countermeasures such as wearing gloves, long johns, a warm scarf and hat, and a thick coat. Or I can function in Florida with far fewer countermeasures. But if I left Florida for Everest Base Camp in Nepal, the necessary countermeasures would be vastly different. And if I made that transition in a single day, there would be a greatly elevated risk of me suffering from high altitude sickness, and possibly even death. Some humans could handle this alteration with no ill effects, while others would die of high-altitude cerebral and pulmonary edema just by making the sudden transition. The difference between the three environments is that by the time I arrive at Everest Base Camp, I have either reached the far end of – or exceeded – the human adaptation limit at the extreme of the zone.

Under current NASA philosophies, the response would be to pile on the countermeasures immediately: supplemental oxygen, UV snow glasses, Primaloft gloves and ensemble, stainless crampons, and so forth. Here – at or beyond the very limit of the human adaptability zone – endless countermeasures are equivalent to dressing the body in an expensive suit for the funeral. Yes, it is true that with enough luck, preconditioning, and training, I may be able to make the trek up to the summit of Everest, take my photo at 29,029 feet, and perhaps even successfully descend.

But I will have to step around more than *200 bodies* that litter the mountain on the way to the top. And those bodies represent only a fraction of those who returned home without hands, fingers, ears, and even noses.

The conditions on Everest are so extreme that it is unsafe to even try to bring the corpses down to their families for burial! Yes, the countermeasures worked – for some. But for many others, the numerous amputations and cases of permanent cerebral damage prove that at the limit of the human adaptability zone, countermeasures are often like pasting a Band-Aid over the abdomen of a patient suffering from pancreatic cancer.

If things go terribly wrong at my Tennessee estate, I can just as easily die of exposure there if I exceed the human adaptation limit at the end of the zone. But this is far less likely to occur for an important reason: I am not nearly as close to the far end of the zone in the Tennessee Appalachians as I would be in the Nepalese Himalayas. In Tennessee, the countermeasures will work perfectly as intended if I intelligently apply them, and I will not wind up permanently injured. But on Everest, no matter how many countermeasures I stack up against the zone, permanent injury in many forms becomes a measurable high probability.

This is a discovery of limits – and the limit is called the Human Adaptation Zone.

As evidence of clear malpractice by the world's space agencies, there is yet no defined Human Adaptation Zone in crewed space exploration because that idea is detrimental to the Apollo era exploration philosophy that has always driven the gravy train. Thus, the faulty idea of nearly unlimited human adaptation held up by endless countermeasures still thrives. The exact same error in exploration philosophy ended the push to permanently settle the undersea regions in the 1970s, leading Jacques Cousteau himself to declare that such a goal was "unrealistic."[4,5]

Cousteau had adopted the philosophy of Dr. George F. Bond, the U.S. Navy physician who became the father of deep saturation diving and believed that human adaptation to pressure and depth had virtually no limits.[6] But, once the limit had been exceeded and resulted in a pile of evidence that saturation at depth led to inevitable injury, the quest to settle the undersea regions of Earth ended with Cousteau and the United States Navy declaring it impractical. Had they recognized the blind alley they were trapped in and merely modified their exploration philosophy, the result would have been far different.[7]

The tragic premature pause in undersea colonization was caused by a failure of exploration philosophy related to human adaptation outside the Human Adaptation Zone. And today, crewed space exploration is headed down the same false path that will have identical, inevitably tragic results – both for the participants and for the future of permanent human settlement of the cosmos itself.

To clearly understand the concept of a Human Adaptation Zone, it is essential to go back to the beginning and review the enabling philosophy of all human discovery:

The First Principle of Human Exploration

"In every exploration system, we must require the systems we build to adapt to the human standard rather than expect the human to adapt to the machine or the environment – and in every design activity we will protect the human as a primary objective."[8]

The enabling philosophy, again, is to protect the human as a *primary objective* and not demand or expect adaptation to the environment, which, at the end of the Human Adaptation Zone, leads to inevitable injury. Note that the First Principle of Human Exploration requires an engineering system as the initial shell of protection. That same system also provides the safe cocoon to supply an approximated one-Earth-normal environment as much as is possible, keeping the human explorer as close to the safe end of the Human Adaptation Zone as achievable.

Thus, the Human Adaptation Zone is defined as the zone where the human explorer is capable of functioning at near 100% physiological and mental capacities without risking permanent injury by indefinite exposures to multiple toxic off-nominal one-Earth-normal environmental conditions. If any exploration system were constructed such that it was energized and enabled by the First Principle of Human Exploration, and if the exploration system was designed around preserving the Human Adaptation Zone, then there would be very few limits to successful human exploration on or off Earth.

Let me give a perfect example of this unlimited exploration capacity in regions far outside the Human Adaptation Zone: the United States Navy nuclear submarine fleet. I could not possibly lend a more ideal illustration than this.

The United States Navy nuclear submarine fleet operates in the most extreme environmental conditions of any human endeavor in history – in many cases, far worse than any space environment. The phenomenal craft operate in a void of total darkness, in freezing water in which a human could live for mere minutes, at extreme pressures, and submerged for months at a time in very close proximity to the most powerful compact nuclear reactors ever built. Plus, the crews of these extraordinary vehicles collectively number in the hundreds of thousands and have successfully carried out their undersea missions day after day, year after year, decade after decade. Unlike NASA, the United States Navy does not assign their crews "death scores" based on their occupational activities. After the completion of their expeditions, they return to port, depart from their submarines, and routinely come home to their families until the next assignment that puts them back in the sub and the extreme environment in which it operates. Then, after 20-30 years, they retire and live normal lives with no lasting medical sequelae.

No former nuclear submarine sailor is writing exposés upon returning home about injuries induced by the environment, or even of any periods of physical maladjustment or injuries. There is no fear of chronic exposure to radiation, even though they literally live and sleep next to a nuclear reactor with an exceptionally high concentration of super-enriched fuel. They have no pressure induced maladies even though they cruise at hundreds of feet of depth and may remain submerged for months. Their advanced nuclear-enabled life support systems create a shirt-sleeve environment that protects them in comfortable surroundings where they eat steaks and watch movies before turning into soft beds with clean, laundered sheets every night.

Comparing these sailors to United States astronauts is a far different story. The state of affairs for the astronaut is marked not only by uncertainty, but it is abundantly clear now that they have already crossed the boundary of safe exposure to the space environment – and the interplanetary voyages have not even begun! So how do NASA and the world's other private and governmental space agencies ultimately attain the pinnacle of safe exploration that the United States Navy submarine fleet has already achieved so brilliantly?

There will be a lot of whining about how the space environment is "radically different" than the ocean and how a redesign is "just impossible" for a slate of reasons that protect the Apollo-vintage old model expectations and have little or naught at all to do with physics. It is nothing but whining, nonetheless. The same degree of hand wringing was also levied against Elon Musk and SpaceX when he dared to design for Falcon 9 booster return and re-use – with NASA leading the pack of nay-sayers. But in the end, innovation and creativity ultimately and clearly revealed the sad, Luddite mentality they represented.

The difference is that the United States Navy has provided a safe, shielded cocoon for its crews that so effectively protects them that they remain well within the Human Adaptation Zone at almost one-Earth-normal every moment of their voyages. The Navy follows what I have identified and articulated as the First Principle of Human Exploration. Thus, the U.S. Navy requires the systems they build to adapt to the human standard rather than expect the human to adapt to the machine or the environment – and in every design activity, they protect the human as a primary objective. Why? For the simplest of all reasons – to safeguard the mission. Sick and injured crew cannot practically carry out mission objectives.

In the spirit of full disclosure, I admit that I am a former United States Naval Officer. But I am likewise a retired 30-year NASA engineer and space life scientist. I have lived and served in both worlds. Therefore, making this stark comparison is a position I have earned through the expertise I have gained via the investment of my life.

And thus, it is reasonable for me to say that NASA can do exactly the same for their crews as the Navy has accomplished. All they need to do is create the mission statement and rid themselves of the fatally flawed, Apollo-era "we-are-so-awesome" corporate persona. But before they can do so, they will have to have a radical moment of self-discovery, recognize the false path down which they have embarked, come to terms with the various byzantine political machinations that have brought them to this point, and then get on with a wholesale redesign of their space exploration paradigm. It will cost more money, to be sure. But it will cost a lot less than the periodic serial re-tool of the space architecture that accompanies each new administration in Washington. And it will be a one-time cost since all future exploration missions will be modeled after that redesign.

If they do not incorporate First Principle standards, NASA's human space exploration days are numbered. Death by anachronism would be a sad demise for the once great bastion of human innovation and exploration.

The pathway to the big-fix begins with understanding and actually implementing an enabling and empowering philosophy of the First Principle of Human Exploration, which in turn simply recognizes the limits imposed by a nonnegotiable Human Adaptation Zone. If the

Agency persists on the current outdated and destructive path and does not implement the overarching First Principle of Human Exploration, many will die or endure permanent injuries as the direct cost of cultural arrogance. Such results from institutional intransigence are described in depth and in detail in the Apollo 1, Space Shuttle Challenger, and Space Shuttle Columbia disaster investigations.

Things will certainly change in NASA, as they always do – but only as the Agency has repeatedly undergone after the body bags were counted.

Yet...

It does not have to be this way – *this time.*

First Principle Solar System Class Human Exploration Engineering Systems

Throughout Books One and Two of *Departing Earth Forever*, all human rated spacecraft flying in deep space and human habitats on the surface of other worlds that are engineered to protect the crew as a primary objective, are outfitted with artificial gravity, and have cosmic radiation shielding that guards them against exposure exceeding the United States OSHA legal standard for radiation workers are defined as **First Principle Solar System Class Human Exploration Engineering Systems.**

Elon Musk - SpaceX

A Brief Primer on Microgravity

"We can't solve problems by using the same kind of thinking we used when we created them.

- Albert Einstein

OF ALL THE REASONS TO FLY INTO SPACE, PERHAPS none is as exciting to novices as the idea of being weightless. But are astronauts *actually* weightless? Here are five typical misconceptions about the apparent absence of gravity that astronauts experience in space:

(1) There is no gravity because of the vacuum of space. (Incorrect)

(2) There is no gravity in space at all. (Incorrect)

(3) The astronauts have achieved a high enough altitude that the influence of gravity has disappeared. (Incorrect)

(4) There cannot be gravity outside the envelope of air around the planet. (Incorrect)

(5) You know you have entered space when gravity totally disappears. (Incorrect)

Remember, all of these statements are inaccurate. Only number three contains a very tiny seed of truth. It is correct that the further away from a celestial body you can get, the less its gravity acts upon a given mass (the astronaut), *declining in a non-linear way* according to the inverse square law, which describes gravity as reducing itself in inverse proportion to the square of the distance from the source. That is why things in Earth's orbit are held firmly by Earth's gravity. The Moon is captured quite securely by the gravity of our Earth, even though it orbits out at 250,000 miles! Thus, if Earth's gravity disappeared at the edge of our atmosphere, all satellites, including our Moon, would immediately drift away. Similarly, in the absence of any gravitational restraints, the Earth and every planet in our Solar System would drift off into interstellar space, just as every galaxy would likewise completely dissipate. At the altitude of the International Space Station, Earth's gravity is still exerting its influence at around 89% of that on our surface. And out at Luna, the Earth's gravity is still being expressed at 0.028% – more than enough to keep the Moon in tow.[1]

So, let us dispel the non-scientific rumors and begin to understand microgravity for what it really is.

Early in the manned space programs of the 1960s (and they were all "man-ned" except for the 16 June 1963 Vostok 6 mission when Cosmonaut Valentina Tereshkova soloed for 3 days in space), the terms "zero gravity" (or zero g's) and "weightlessness" were frequently used to express the character of freefall around the Earth in orbit. The reason an astronaut "felt like" they were in zero gravity was the fact that they were traveling at such a velocity and altitude that gravity was literally pulling them down toward the planet at the same velocity that they moved around it, giving the very *false* impression of having escaped the gravity of Earth. Instead, they were actually in a state of perpetual freefall *around* the planet while in orbit, still in the 89 percent grip of Earth's gravitational field that was, ironically, in fact, responsible for their perception of zero gravity! While they were always in the relentless grasp of Earth's gravitational field at 89 percent of one gravity of force, because of the flight dynamics of their orbit, they were actually experiencing the full physical reality of gravity pulling them endlessly around the planet. They were freefalling in space due to the constant expression of Earth's gravitational field – which describes their misperception of being in zero gravity.

Again, please do not miss the irony of this situation. The only reason that an astronaut "perceived" an absence of gravity was specifically because of the *effects* of gravity.

To completely understand this, it is important to add that any "freefall," whether from an airplane as a skydiver or around the planet as an astronaut, involves constant *acceleration* because gravitational force is always expressed as mass in acceleration. Therefore, the opposite of gravitational force – apparent zero gravity – must also be countered with the opposite vector of acceleration. Thus, a skydiver is accelerating downward at 9.8 meters per second squared, in "apparent zero gravity" freefall until their chute pops open – or they hit the ground. The astronaut is also in freefall around the planet. If their acceleration was lower than the expression of gravity at their altitude, they would lose orbit and fall to Earth. Likewise, if they traveled any faster, their altitude above Earth would increase, and, with enough energy, they would eventually depart Earth's orbit.

The really clever student would then ask, "If they are constantly accelerating, they should be gaining speed like the skydiver. So how can the astronaut be in a state of constant acceleration and yet not be gaining any speed in orbit?" This is a very astute question! The answer is found in the principles of circular motion. The gravity of Earth (or the Sun) lends a centripetal force requirement to allow for gravity to express a constant inward force of acceleration in a pattern of circular motion we call the orbit. Thus, the orbiting astronaut's tangential velocity remains the same, while the effects of constant acceleration are due entirely to the influence of centripetal force. (You may be relieved to know that this is as complicated as the discussion gets here, so please read on.)

Such a "physical reality" of being free of gravity specifically *because of* the physics of gravity and spaceflight is a real mind-bending science fact that always seems a bit contrary to common sense and logic. And yet, there it is – a fact that constantly expresses itself on humans and machines in space. In 21st-century parlance, "It is what it is!"

In practice, once on orbit, careful measurements were made of the actual state of apparent zero gravity. It "felt" like a faultless expression of zero gravity in orbit. But it was not. Even while still held in 89 percent of the full grip of gravity, it was discovered that astronauts orbiting around the home planet in spaceships experience an imperceptibly impure state of zero gravity for many reasons, often encountering micro-tugs of gravity from numerous sources. These include gravitational anomalies because of varying densities of Earth's crust that the astronauts fly over, and even micro-changes in the attitude of their spacecraft and tiny motions of their craft, which all induce very small, typically imperceptible alterations in their state of gravity. Hence, zero gravity was more accurately renamed *microgravity* to indicate the impurities in their *almost* perceived gravitation-less micro-world. Thus, microgravity more precisely defines the actual experience.

It always requires a bit of Physics 101 thinking to finally adjust to the idea of microgravity being expressed as a result of motion in – and because of – gravity. Just imagine the astronaut in perpetual freefall around the planet – somewhat like a skydiver who leaps out of a plane far above the globe at just the right angle so that they never strike the Earth or its atmosphere as they are continually falling around the Earth 24/7.

Yet, at some point in this discussion, it does become a tad more complicated. When the astronauts fire their rockets to escape the gravitational field of Earth when traveling to Mars, they also experience a state of microgravity the moment the engines stop firing and escape from Earth's gravitational capture is complete. So, if they are no longer in freefall around Earth and are moving fast enough now and with enough inertial energy to have escaped the grip of Earth's gravity (but not its force acting on them even as they fly away), then how do they experience the same state of microgravity if they are no longer falling around the curvature of the Earth?

The reason is quite simple. They have escaped the gravitational bonding of Earth's gravity and are in freefall around the Sun! Their motion is now commanded by the much greater influence of the star that orders the motion of Earth and the planets of our solar system. Their orbit is determined by the inertial energy they were just given by their rocket. If they lost all that energy, they would literally fall into the Sun – just as if they had lost their orbital energy around Earth, they would have fallen back to the Earth. But once they have escaped Earth's gravitational grip, they are still in freefall in space nonetheless, experiencing what appears to be an absence of gravity even while still in the grip of a full stellar gravitational field blended with the diminishing gravitational influences of the receding Earth. They would have to fly many times the distance of Pluto to escape the total influence of the Sun's enormous gravitational influence on their spacecraft, *even as* functionally they would remain in microgravity for the entire voyage.

The real light-bulb moment comes when you realize that there is actually nowhere in the universe that an astronaut can possibly escape gravity – *nowhere*! No matter where we fly in space, we will never be free of gravity's force acting on us. Because of the inverse square law by which gravity is expressed, even millions of light years beyond the nearest galaxy, the cluster of galaxies of which ours is a member will all express their influence on you and your path through space. So, if you are looking to be free from the bonds of gravity, you will have to look to another dimension. It cannot be found in this universe!

This is where many people become either lost or just lose faith in this theoretical construct. It is one thing to imagine perpetually falling around a planet that you can visualize. It is quite another to imagine this around a star. And it requires a further, rather grandiose leap of faith to imagine this effect functioning within the intergalactic sphere of an entire galaxy or galaxy cluster. How can we wrap our minds around this more easily? Dr. Albert Einstein already anticipated this issue in his

General Theory of Relativity in 1905-1907. He called it the Principle of Equivalence in one of his most well-known "thought experiments," an exercise which he defined as "the happiest thought of my life."[2]

Einstein postulated in his Principle of Equivalence that if there was an observer on board a rocket-propelled windowless room being accelerated at 9.8 meters per second squared – or if the observer was in an equivalent room sitting still on the surface of the Earth – that there was no experiment that could be conducted inside the closed room that would inform its occupant of which state they were in, whether they were sitting still on Earth or being propelled in space at 1 g of force.

The point of this experiment was to pin down one essential gravitational fact: that *gravity* and *mass in acceleration* are, in fact, *identical* expressions of reality to the observer embedded in the fabric of space and time. Hence, whether an astronaut *vis a vis* their mass in acceleration is freefalling in suborbital space, around a planet, around the Sun, or around the galaxy, they are, in fact, embedded in larger gravitational spheres.

Microgravity Exposure

When an astronaut is living and working in a gravitational field that is different than 1 g, their body is being exposed to an "off-nominal" gravitational field (i.e., anything less or greater than 1 g). Please take careful note of a word not normally used when talking about the state of apparent weightlessness: *microgravity exposure*. The astronaut is being "exposed" to a gravitational field that is different than one-Earth-normal, and therefore that exposure has an immediate effect on the human's physiology created to work nominally in one gravity. The moment the crewmember is exposed to microgravity, their physiology immediately biochemically responds with a homeostatic reaction to attempt to adapt to the condition and thereby keep the body working without malfunctioning on any level. After decades of careful study of microgravity, we have determined that the body is unable to respond fully and the results become negative, constantly degrading over a period of months until the effects reach the end of the Human Adaptation Zone and result in injury. These injury endpoints are highly variable between individual astronauts.

Hence, onboard the ISS, crew members are exposed to microgravity. On Mars, astronauts are exposed to altered gravity at

0.376 g. And on Luna, they are exposed to an altered gravity field measuring 0.17 g. Again, exposure to microgravity over merely days or, in many cases, even a few months, has a far different physiological effect than up to six months or a year. This is discussed in much greater detail in the upcoming chapter "Chronic Microgravity Exposure Etiology."

Exposure Limits – How Much is Too Much?

If someone has never tried their hand at living with microgravity day in and day out, most people think it would be an endless lark not having to fight the constant constraints of gravity pulling on them ceaselessly. But for those who have been there, many veteran astronauts openly define it as a "miserable" existence.[3,4]

At first, the misery is engendered by motion sickness, which strikes 60 to 80 percent of all astronauts.[5] Said U.S. astronaut/MIR cosmonaut Jerry Linenger, "While everyone was able to smile during press conferences, the newcomers were glad when the cameras were turned off so they could remain still and feel miserable alone."[6] He even went so far as to subtitle his *Off The Planet* book "Surviving Five Perilous Months Aboard The Space Station MIR."

The good news is that within a few days (very much like motion sickness at sea), the feeling of nausea begins to wear off. Several days later, phase two of a new, more prolonged misery begins. But instead of this being a temporary vestibular reflex disruption, the new symptoms are a sign of a creeping failure of the human body to adapt to microgravity. This failure to adapt is not serious at first. But after several months or more of spaceflight, more permanent changes begin, and injury is inevitable at the limit of the Human Adaptation Zone.

The human body is designed to function in a one-gravity field. For example, we are all equipped with check valves in our circulatory system to prevent gravity from pooling our blood in our lower extremities. And our capillary and glandular systems are designed specifically to pump against gravity's force. Our bones are constructed to maintain their health and normal state of rigidity and density as they are stressed against the constant gravitational forces we must counter to even move. Our whole-body musculature is intended to stay healthy by constantly straining against gravitational forces. And our heart maintains its health by all its work necessary to keep our blood circulating to meet the unending demands gravity places on our bodies. In the absence of gravity, these functions are effectively switched off. Blood does not circulate as it was designed, the lymphatic system is short circuited, unchallenged bones begin to lose their density, and muscles (including the heart muscle) begin to atrophy.

In the area of cosmic radiation, as you will discover in a few pages, there are limits of exposure that we can measure with our instruments and apply to our astronauts. But this is not the case with exposure to microgravity. After several months or more of exposure to microgravity in low-Earth orbit, longer term physiological, biochemical, and genetic changes begin to accumulate that – as the exposure time is extended – in some cases, become permanent injuries.[7] But there has still not been an established limit for exposure to microgravity.

As was discussed previously, in over a half-century of living and working in the microgravity of space, we have discovered that if you journey there for only a few days – or even a few weeks or a few months – everything returns to normal relatively quickly once the astronaut's feet hit the green grass of Earth. The same is true of the cumulative dose of cosmic radiation you will receive on short missions in low-Earth orbit. But the long-term exposure to these physiological challenges is another issue altogether.

In the next section, I will discuss details of the physiological changes that occur in human exposures to microgravity.

A Brief Primer on Cosmic Radiation

"No amount of standing on hilltops on dark nights and surveying the heavens could prepare a man for the actuality of space travel, because the earthbound observer saw only the stars, not what separated them. They glittered in his vision, filling his eyes, and he had no choice but to assign them a position of importance in the cosmic scheme. But The space traveler saw things differently. He was made aware that the universe consisted of emptiness, that the suns and nebulae were almost an irrelevancy, that the stars were nothing more than a whiff of gas diffusing into infinity. And sooner or later,
that knowledge began to hurt."

— Bob Shaw

"For those who want some proof that physicists are human, the proof is in the idiocy of all the different units they use for measuring energy."

— Richard Feynman

FOR THE FEW, ESPECIALLY FORTUNATE PEOPLE WHO have ever stood outdoors gazing upwards, far away from city lights on a dark, crystal clear night with even the most basic understanding of what they are witnessing, the experience is often life-altering and profoundly transformative. Staring deeply into the yawning, three-dimensional universe, a vast and endless ocean of infinity reveals itself in all of its naked splendor, splayed out in incomprehensible size and majesty. It is breathlessly self-illuminated by the spectacle of shimmering bands of a sextillion stars and, closer in, by individual points of light, some associated with stars that we can easily see from Earth that are a thousand or more times the size of our own

Sun. Each of these is accompanied by its own following retinue of planets and stellar bodies.

For humans like me, we see not just specks and bands of light, but far more. We imagine distant, whole new worlds with boundless variety that in the 21st century have just been detected – a vast majority of which lay still undiscovered before us. The very next leap of thought to us is quite reflexive, logical, and natural. In that unlimited void, we see places that we can travel to and explore. And some of us extend that thought yet another step. Perhaps we can go there to live – depart from Earth forever and make new human civilizations. A few of us even have no problem leaving our beautiful home planet with no intention of ever returning. I know very well that I am not alone in these thoughts – these wild and rebellious ideas that are consuming, sometimes vexing, and always, always, inevitably addictive. When we look outward at any clear night sky, we witness not only a natural spectacle that is quite beyond just stunning and magnificent – it is a living force that beckons us with a supernatural and irresistible intensity that some of us will not get over except to depart Earth forever.

The first time I saw the night sky for what it really is was when I was a young teen looking through my small telescope from my front lawn at a perpetually dark Oklahoma prairie sky. Typically, I would eventually grow tired of hunching over the telescope. I would spread a blanket down on the grass, and, in a more comfortable position, I would lie back and feast my eyes above me on the marvelous all-that-there-is, just as my Native American ancestors had done on this same flat and expansive landscape. I would spend these precious moments of my "solo star party" (except for my loyal dog Max beside me) identifying stars with my star wheel, combing the specks for planets. Finally, I would ultimately allow my boy-eyes to search diligently for any passing UFOs.

Although many decades have passed, it still seems like it was just last night. I intimately remember the feelings of wonder and of an awe that I could not adequately express then or now. But, since that time, I have also been witness to the extraordinary miracles of exploration and discovery up close and very personally. I even made my career out of the quest to explore that region that lies just above our atmosphere – day by day stepping onto sacred ground and reaching up to touch that which has always been just out of reach. In retrospect, if it were even possible, all of it has only increased my awe and my wonder, and the insatiable yearning to go there personally has grown relentlessly ever stronger.

As I sit back in my retirement chair on the other side of my life and remember all that I have seen, learned, and experienced, I recall that of all the problematic or difficult things we have discovered in the past 60 years of human spaceflight, nearly every single one has a relatively easy fix – or a few relatively simple countermeasures if you feel like incorporating some Space Agency jargon.

Yet even now, as I look up at that magnificent field of dreams just above me, one discovery stands alone in its scope and magnitude. This understanding is most surprising, uniquely startling, and worrisome. It is the realization of the extent, nature, and threat of cosmic radiation just beyond our terrestrial radiation shield. Even in science fiction, there has not been a more vexing or serious problem that stands between us and our cosmic dreams. It is one that is not only going to require that we step back and re-think everything we thought we knew about our own delicate physiology in the context of inserting it into the expansive cosmos...

...but it is one that will require that we *completely redesign* most of our human space exploration model from scratch.

If your eyes have scanned the last 20 words, I fairly warn you that phrase today is equivalent to high treason in some of the space exploration community. Yet, it was those 20 words that formed the foundational reason that this book had to be written by someone from *inside* the Agency. I have zero illusions about the fact that many will do all that they can to discredit me, even though the ideas expressed between the covers of both these volumes are framed by basic logic. But science is science, wishful thinking is wishful thinking, and playing political game semantics with the waste products of supernovae will sooner than later prove to be deadly not just for some but for all - *and there are no exceptions*. It is true – and you can trust me on this one – a *lot* of people absolutely do not want to hear any of this! If you do not believe me, just read the reviews of this book. Yet, here is my perspective on that very debate in the words of Matt Hooper from *Jaws*:

"I'm not going to waste my time arguing with a man who is lining up to be a hot lunch."

Indeed. But in the case of humankind at the dawn of the 21st century, when staring down in the gaping maw of the power generated in the heart of supernovae, arguing about the inevitable, ultimately pernicious effects of cosmic radiation on the human body is not just a waste of time, it also ultimately and inevitably leads to one single possible conclusion. Paraphrasing the words voiced so unforgettably by *Jaws* Amity Sheriff Martin Brody, we are not going to just need a *bigger* boat – we are also going to need *a very different boat!*

Invisible Energy

I am one of those perhaps unusual individuals who hold a post-graduate degree that is a hybrid between engineering and science: a Master of Science degree in Bioenvironmental Engineering – a graduate school degree that required a laboratory research thesis. In my career, I have worked both as an engineer and as a research scientist. I returned to my Oklahoma State University alma mater for post-graduate studies on the GI Bill after six years in the U.S. Navy and, due to my Native American heritage, was privileged to work as the College of Engineering, Architecture, and Technology Undergraduate Student Advisor for Minority Students while pursuing my degree.

My first job out of university was as a Nuclear Engineer, and, in 42 years of government service, I have never lost my professional connections in the nuclear industry, even as I was employed as an Aerospace Engineer by NASA, who liberally utilized my expertise as a Space Life Scientist for three decades. I have a very clear understanding of the design and operation of the various forms of a nuclear reactor and have had in-depth briefings from the most competent and informed leaders at the International Atomic Emergency Agency and other premier sources about what went terribly wrong at the Fukushima Daiichi and Chernobyl power plants. As a Space Life Scientist and engineer, I have a very clear understanding of what happens when a human stands in front of a nuclear source, and of the exhaustively researched safe limits to that exposure.

It is not difficult to miss the fact that there is as much misinformation about nuclear energy as there is scientific fact. The Internet is a vast, hopelessly tangled soup of some truth blended with mostly fiction on which countless people have, unfortunately, actually firmly established their world view on nuclear power and nuclear sources of all kinds. Unfortunately, the Internet is the place where many go to mine cherry picked "facts" that only support their beliefs and label everything else as "fake news."

For example, there is one particular meme floating about regarding the polluting of the North Pacific Ocean by radioisotopes escaping from the Fukushima Daiichi nuclear accident. The chart depicted on the graphic shows a rainbow-colored plume extending from Japan's eastern coast all the way across the North Pacific Ocean to the western shores of North America. The little story that accompanies this often-repeated image is this message: The pollution is quite obviously serious – just look and see for yourself. View the plume with your own eyes and conclude for yourself what this means for all life in the ocean. See how the entire ocean is hopelessly spoiled by nuclear radiation.

However, none of the colors are labeled concerning what they represent. There are no radioisotopes identified or the species, depth, or density of the purported contaminants. There is only this creepy picture laid bare, which, like a Halloween mask, carries emotional impact totally devoid of any information or accuracy whatsoever.

For a while, I felt duty bound to make my scientist-engineer comments on these idiotic Internet fabrications, but I have long since given up. The vast, deep ocean of ignorance is far too expansive to squander my precious time to take on the legions of true believers of the crisis de jour as published by the esteemed college of Google PhDs. Yet, this meme is also a perfect example of how little understanding of nuclear energy there is in the wider world. And nearly all these dreamscape fabrications concern merely terrestrial ionizing radiation – not even galactic cosmic radiation!

While this book is not intended to be a comprehensive treatise on nuclear energy and the far wider spectrum of nuclear radiation, if there is not the most basic understanding of any of the elementary principles, then the core idea of this book will be absolutely lost. So please fasten your seatbelt and hang on for the fast-track exposition on nuclear energy and radiation of several associated varieties. I promise to make it as easy as I can, and hopefully also somewhat entertaining along the way.

Allow me to begin with the words of one of the first scientists who discovered the very nature and physics of nuclear power and its attendant radiation qualities. Unfortunately, this study also cost her her life. She literally got too close to the invisible energy show for too long, and it killed her. I would like to allow the much esteemed and admired Madame Marie Curie (one of my own personal heroes of science) to establish the perfect tone for this discussion and, indeed, for the rest of this book. Madame Curie said most eloquently and profoundly,

"Nothing in life is to be feared, it is only to be understood. Now is the time to understand more, so that we may fear less."

Radiation 101

Our subject is cosmic radiation. But cosmic radiation is only one of many varied kinds of radiation in the universe and on Earth. This subject is framed most concisely in the following pair of very succinct sentences:

"Radiation is the emission of energy as electromagnetic waves or as moving subatomic particles, especially high-energy particles which cause ionization. Broadly speaking, radiation is a way in which *all forms of energy move* from one place to another."[1]

Therefore, radiation includes a very wide-ranging spectrum of many forms of energy, from light to heat to radio waves, and even the energy that comprises television signals, sound waves, and the barking of your favorite dog!

A few minutes ago, I finished my morning shower, bathed in the radiation of the visible spectrum emitted by the overhead light (whose energy I selected by the spectrum of my personal preference), sprinkled with the radiated heat of the water while I listened to the sound radiation emitted from my Bluetooth speaker. I could easily sense and feel all these very welcome and satisfying non-penetrating radiation sources and prefer not to ever imagine life without them.

But what I could not feel, or have any way at all of discerning with my physical senses, was the much more powerful and penetrating natural terrestrial ionizing radiation surrounding me that was being emitted by the radon gas seeping in at minuscule quantities from the concrete of the floor of my bathroom. As well, my body was impacted by one of the rare but occasional cosmic rays, perhaps created near the singularity of a black hole from a distant galaxy, that sometimes slip past Earth's Van Allen radiation shield and our atmosphere. It easily penetrated my roof and ceiling and instantaneously passed through my body with no sensation or any way of my knowing that it even happened.

After leaving my bathroom, I passed by my cats' litter box and was irradiated with a few stray gamma rays from the trace amounts of uranium and thorium embedded naturally in the bentonite clay of the litter. On the way to my study, I climbed the stairs under my smoke detector, precisely 24 inches above my head, which encases a speck of radioactive alpha emitter Americium-241, which never reached me, but which contains 98.4% of the alpha ionization energy of Plutonium-240, the material which was used in the atomic bomb at Nagasaki, Japan.

And finally, for breakfast, I ate a banana and very much enjoyed it, even though I knew that it contained an easily measurable amount of naturally concentrated potassium-40 radioisotope in the fruit. I never felt any of these radiation sources, nor did I have any conscious indication that they were passing through my flesh. Except for the smoke detector's Americium alpha ray (which cannot escape its plastic enclosure or even make it through the few inches of air above my head), all of the other sources most certainly penetrated my body.

The average amount of radiation (counting all natural terrestrial radiation sources and including medical x-rays) that a person is exposed to on Earth is about 1.5-3.5 millisieverts (mSv) a year.[2] As a reminder, a Sievert is a derived unit of ionizing radiation dose in the International System of Units (SI) and is a measure of the health effect of low levels of ionizing radiation on the human body.[3] A mSv is one-thousandth of a Sievert. The sievert represents the equivalent biological effect of the deposit of a joule of gamma-ray energy in a kilogram of human tissue.[4]

Penetrating and Ionizing Radiation

When I go to the dentist or the orthopedist, I occasionally receive an x-ray. An x-ray is a type of radiation that is similar to visible light but has a higher energy and shorter wavelength. In fact, it carries so much energy that it literally penetrates the flesh, bones, and teeth of my body at the speed of light. The final x-ray product displays an image depicting how much of the energy is absorbed by my body's variable structural density on its way to the film. Although x-rays also have naturally occurring sources such as lightning bolts, an x-ray may be artificially produced by striking a target with a stream of electrons, as found in many medical applications.

Another class of penetrating terrestrial radiation besides the x-ray is the gamma-ray. A gamma-ray is the result of the normal "settling" of a nuclear structural component in the core of an atom undergoing natural atomic decay. In the examples above, a gamma-ray is a high energy radiation that includes natural sources such as the decay products of radon gas, a partial byproduct of the potassium 40 in bananas, and the gamma emitting radioisotopes of radium, thorium, and potassium in cat litter. Think about it. As you were chewing your banana at breakfast, an atom of potassium 40 settled into a lower orbit and emitted a powerful gamma ray that passed right through your brain without your permission or notice!

Compare these penetrating radiations to one kind of non-penetrating radiation energy known as visible light. Light is comprised of quantum packets of energy called photons. At its highest energies, light radiation only penetrates a very thin layer of skin and does not create ionization events, so it is often referred to as "non-penetrating – non-ionizing radiation."

We care about high energy, penetrating, ionizing radiation sources because they can, and do, interact with our microscopic biological intercellular structures through ionization (the process by which a molecule or atom acquires a positive or negative charge by losing or gaining electrons). If the penetrating radiation carries sufficient energy to detach electrons from atoms or molecules as it passes through the cellular structures (such as x-rays or the broad spectrum of ubiquitous natural gamma rays), it thereby ionizes them, and thus it is called ionizing radiation.

So why is ionizing radiation dangerous? When atoms inside living cells become ionized, one of three things can happen: the cell dies, the cell repairs itself, or the cell's mitotic mechanism (the process of cellular division) is changed and can become mutated (biochemically altered) or even cancerous.

Not every cell is affected by ionizing radiation in the same way. All cells receive natural and/or medical radiation eventually, but how many of those interactions will result in cancer in your body? That process has been scientifically enumerated by probability tables developed in a mass study of Japanese atomic bomb survivors and from large studies of patients receiving medical radiation doses.[5]

Van Allen Radiation Shield

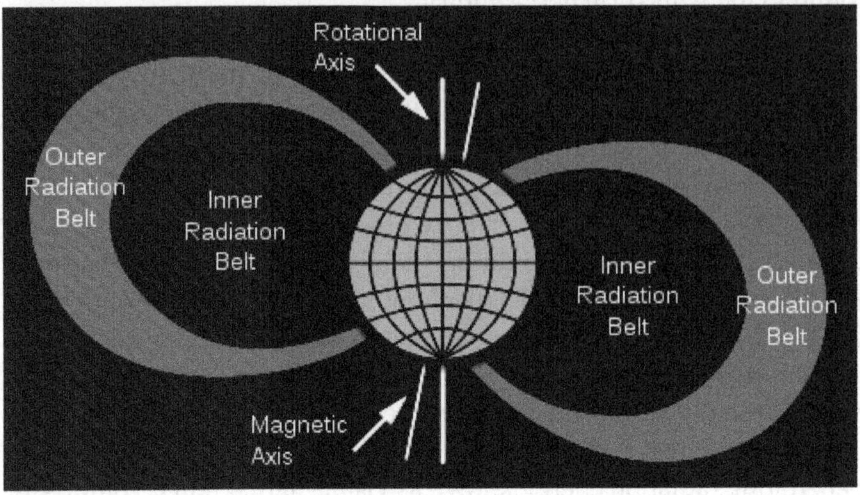

The Van Allen Belts (our natural cosmic radiation shield that traps and redirects high energy particles streaming in from galactic space) are located from 400 to 36,000 miles above Earth. The collective penetrating, high energy radiation located *beneath* this shield will be referred to throughout the remainder of this book as "terrestrial ionizing radiation." *Above* the belts at an altitude greater than about 36,000 miles, the extremely complex and very different collective high energy radiation species from many different solar, stellar, and galactic sources will be referred to as "cosmic radiation." And above our natural shields at greater than 36,000 miles, the combined sources of cosmic radiation – less the radiation from our Sun – are referred to as galactic cosmic radiation, or GCR.

In this book, the Van Allen shield itself also defines the boundary between low-Earth orbit and Deep Space, which begins at around 36,000 miles above the surface of Earth. The belts are located in symmetrical bands aligned with the Earth's magnetic axis – not necessarily matched perfectly with the planet's rotational axis. Therefore, the distance of the belts slightly fluctuates from point to point around the Earth's surface.

The difference between the radiation environment on Earth below the Van Allen shield and that above it is profound. While you may look up through a clear atmosphere on any exceptional viewing night and see nothing between you and the Moon, planets, stars, and even distant galaxies, your eyes, senses, and the perceived visible reality have collectively deceived you in ways that are difficult to imagine or appreciate on any scale. Without the Van Allen shield, most life as we know it on Earth (including humans and most animal life) could not survive. In a sense, Earth would probably resemble the sterile, lifeless deserts of Mars.

Please allow me to offer this simplistic contrast so that you can better catalogue this in your mind:

Van Allen Radiation Shield – Earth
No Van Allen Radiation Shield – Luna and Mars

Van Allen Radiation Shield = Abundant Lifeforms
No Van Allen Radiation Shield = Sterile Wasteland

Yet, humans have no senses available to us to even perceive the existence of penetrating ionizing radiation of any variety, whether it be mere inches from our bodies, on our breakfast plate, immediately surrounding us, or at a cosmic distance – period. Further, we also cannot visibly see the Van Allen radiation shield. On any given night, you stare straight through it as though it were not even present at all. Only the northern (aurora borealis) and southern (aurora australis) lights are exceptions that illuminate a small part of its structure under certain circumstances.

Finally, we have no unassisted way of visualizing radiation flux, densities, or strengths. Therefore, as we look outward into space, it is very easy to project that the environment we live in on Earth and surrounds us is universally the norm, even as we prepare to depart Earth and head out into deep space above the shields.

But it is not. It is fundamentally and radically different.

Indeed, the radiation environment we live in on Earth is not even close in spectra, quality, or physical manifestation of the radiation environment in deep space above 36,000 miles. It is at once foolish and dangerous to assume otherwise. And yet, like the silly Fukushima radiation contamination plume graphic, egregiously faulty assumptions have become the standard, just as our very limited human senses and linked emotionally driven feelings demand. But just above our heads, where we routinely stare, in places we can clearly see from the ground, the actual state of cosmic affairs is far different than we can perceive from the surface of our home planet.

Let me offer a clear illustration. I may stand comfortably in thin pajamas and bare feet in my Tennessee home and look outside my glass walled study onto a cold, snow blanketed mid-winter forest morning with an outside temperature of minus 5 degrees Fahrenheit. But if I fail to understand or remember the difference between the heat radiating around me from my fireplace and the outside temperature, and I exit my abode in my pajamas and bare feet without adequate shielding from the radiation differences, I could easily and quickly injure myself.

And yet, staring through the large glass windows, it is easy for me to forget the radical differences and dangers between the two very dissimilar radiative environments, separated by just a fraction of an inch of distance, as I am effectively shielded by the transparent pane of glass. The same is precisely true with each of us as we stare upward at the exceptionally thin barriers between us and deep space. What we can see out there and what we are experiencing on Earth are not at all the same.

Why Luna and Mars Have No Appreciable Radiation Shields

Earth has a potent magnetic field that is powered by the planet's fluid outer core (the intrinsic field) that actively generates more than 95% of the magnetic energy of the observed geomagnetic field at the surface that influences our protective Van Allen Radiation Shield magnetic barrier to GCR.[6]

It is created by the motion of convection currents comprised of molten nickel and iron in Earth's core. These very powerful currents create our formidable geomagnetic main field and the planet's geodynamo. NASA explains:

"At the heart of our planet lies a solid iron ball, about as hot as the surface of the sun. Researchers call it 'the inner core.' It's really a world within a world. The inner core is 70% as wide as the moon. It spins at its own rate, as much as 0.2° of longitude per year faster than the Earth above it, and it has its own ocean: a very deep layer of liquid iron known as 'the outer core.'

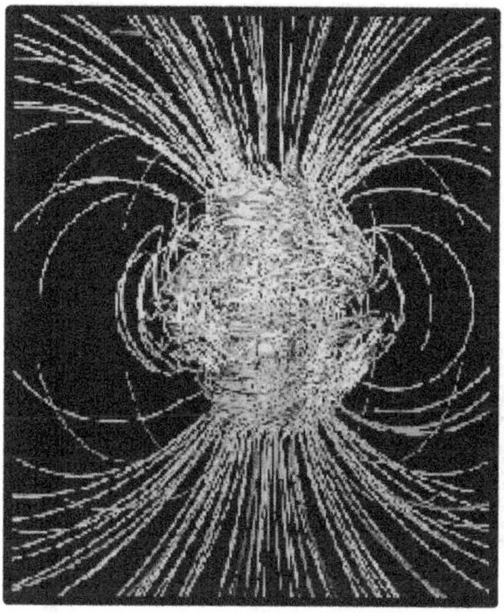

"Earth's magnetic field comes from this ocean of iron, which is an electrically conducting fluid in constant motion. Sitting atop the hot inner core, the liquid outer core seethes and roils like water in a pan on a hot stove. The outer core also has 'hurricanes' – whirlpools powered by the Coriolis forces of Earth's rotation. These complex motions generate our planet's magnetism through a process called the dynamo effect."[7]

Luna and Mars now have solid cores, which are often called "extinct cores." And yet, through other static natural morphological features, both worlds still have magnetic characteristics, although they are exceptionally weak. Lunar and Martian magnetic field features are depicted in the illustrations below. Each of these graphics enumerates the fields in nT (nanotesla).

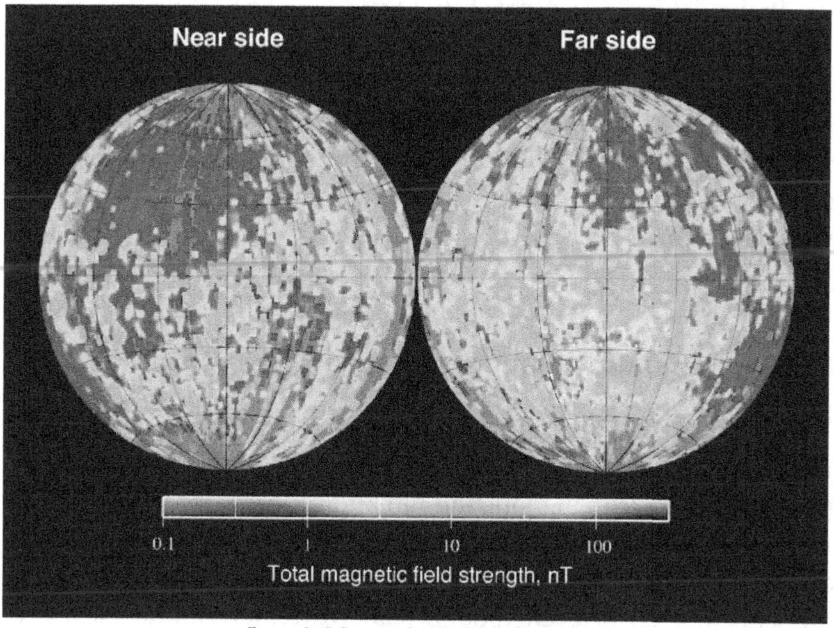

Luna's Magnetic Field Strength

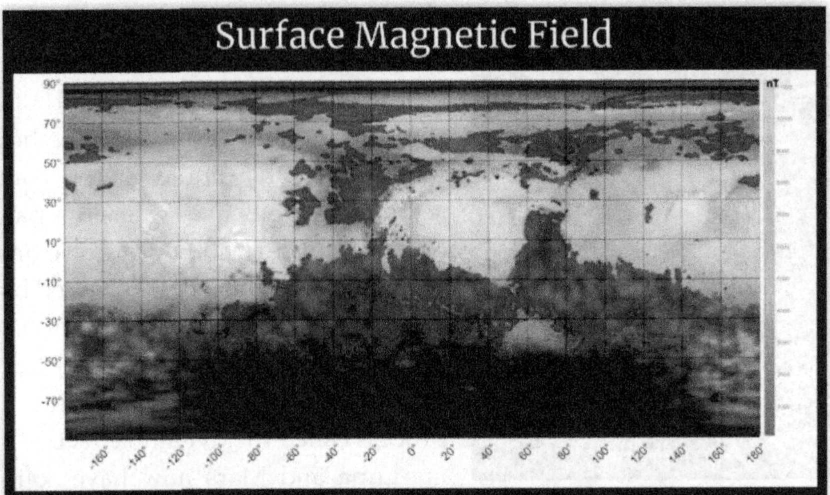

Surface magnetic field predictions at 4 pixels/degree calculated using Langlais et al. 2019, regions where the predicted field is below 250nT are excluded. Background is the global map map generated using EXI images.

Mars' Magnetic Field Strength

The Mars Surface Magnetic Field graphic is taken from the 2022 *A Mars Atlas,* assembled by the United Arab Emirates findings from their Mars Probe, "Hope" (Al-Amal مسبار الأمل).

Comparatively, the Earth's active magnetic field intensity is magnitudes greater and very uniform at roughly between 25,000 - 65,000 nT (0.25 – 0.65 gauss). The significance of these differences is that Luna and Mars, with solid cores and static fields, have almost no capacity at all to effectively filter any galactic cosmic radiation compared to Earth.

Cosmic Radiation – Solar Particle Events

In deep space, there are many kinds of penetrating particles, or "rays," that inhabit the cosmic environment outside Earth's boundaries. These naturally occurring ionizing radiation sources in deep space are collectively defined as cosmic radiation. Again, the environment in deep space is radically different than the radiation environment we experience on Earth, even though some extremely limited levels of cosmic ray activities do make it to the ground each year. As a matter of fact, we are exposed annually to extensively filtered galactic cosmic radiation on the surface of the Earth at a level some 1,990 times *less* than in deep space!

Further, there is also not just a single kind of cosmic radiation. If we travel above the Van Allen Belts and begin to journey about our solar system, we immediately start to measure two very different and distinct species of cosmic radiation around us, identified by their sources and relative energies. The first of these cosmic radiation sources is the nearest star, our own Sun.

When I was a younger man, I particularly enjoyed going out to the beach and exposing as much of my skin to the bright Hawaiian or Florida sunshine as I could. There was just something particularly physically and psychologically compelling about it (blame it on my Native American heritage). The feeling of the Sun's warmth on my entire body, the offsetting coolness of the moist ocean breeze, the wave after wave of negative ions produced by the surf, and the sheer beauty of it all made it an almost supernatural setting. There, in that time of my life and in those settings, the Sun was my best friend.

Decades later, now, the negative effects of my former BFF are with me forevermore. The flesh of my arms, the area most exposed to the solar radiation that reached Earth's surface, is not the same. The sunshine almost destroyed the skin on my arms' elasticity, and the high energy ultraviolet rays literally cooked the epidermal layers away so that the skin that was most exposed is now exceptionally thin, dry, and prone to injury from the slightest bump or scrape. And that does not even mention the induced cellular mutations that created basal and squamous cell carcinomas. The injury and damage from the Sun's collective radiation are totally irreversible. My BFF is no longer welcome!

Yet, out on the beach, I was actually quite fortunate to get by with as little injury as I did. But, in stark contrast, in deep space, without the benefits of 50 miles of dense atmosphere and the Van Allen shield, the Sun's threat is far more real and immediate. In my sci-fi novel trilogy *Apocalypse Moon,* I created an underground beach inside a colony beneath the surface of our Moon. But it would not be a good idea at all to lay out in the open with just your skin showing on any sandy location *anywhere else* on the Lunar surface! Not even considering the pressure and temperature issues, the unfiltered sunshine in just a few minutes would make my decades of skin damage appear not so bad by comparison. Truly, outside Earth's atmosphere and Van Allen shield, the Sun's radiation, both penetrating and non-penetrating, is a very different story.

The radiation my skin layers had to contend with on the Hawaiian and Florida beaches was infrared (heat) and, at the opposite end of the electromagnetic radiation scale, ultraviolet. The infrared was a radiation I could actually sense, picked up by the nerve endings in my epidermal layers, and it made me literally feel good all over. But I have no natural sensors for acknowledging the ultraviolet radiation. And, in the end, it was this far more biologically destructive radiation that permanently damaged my cells. In the direct sunshine, I was receiving a straight (yet well filtered) dose of non-ionizing but penetrating radiation from the Sun. Most was in the form of the higher-end energies of the ultraviolet spectrum that did all the damage. But it is a far different experience than had I been exposed to the Sun in deep space – a very unique environment. There, astronauts will have one particular thing in common with a sunbather on a Hawaiian or Florida beach. They will not be able to feel or otherwise physically detect all forms of cosmic radiation which *will* penetrate their body from our Sun (unless, of course, they are fully protected by adequate shielding that mimics the protection that we enjoy on the surface of the home planet). As of today, no astronauts are going to be that fortunate – until the first wave of First Principle ships are built and launched.

In deep space, beyond 36,000 miles above Earth, our Sun is a constant source of a kind of cosmic radiation we call Solar Cosmic Rays (SCRs). The Sun is never quiet and is in continuous flux, constantly emitting a steady background stream of energetic, penetrating particles of relatively low linear energy transfer (LET).[8] But on occasion, the Sun will experience a solar particle event (SPE) and eject massive plasma clouds into deep space consisting predominantly of protons and some heavy ions that, while almost entirely filtered from reaching Earth's surface, will cause a radiation surge in crewed systems in deep space.[9] On interplanetary class missions, depending on the 11-year solar cycle, astronaut crews may expect to be exposed to SPEs of highly variable energies. On the nearly nine-month cruise to Mars, the Mars Science Laboratory (MSL) spacecraft housing the Curiosity Rover recorded five distinct SPEs above background that impacted the spacecraft.[10]

SPEs will vary considerably in strength and their effects on astronaut crews. And, if not traveling in vehicles constructed according to the First Principle of Human Exploration, a large fraction of them will require astronauts to take refuge in a predesignated radiation shelter onboard their craft or orient the ship with its densest configuration relative to the Sun to enhance the protection offered by the spacecraft's structural shielding geometry.

SPEs have a unique dose distribution with respect to whole body irradiation. For example, SPE epidermal or skin doses are from five to ten times higher than those experienced by internal organs due to the epidermis' susceptibility to absorption in the relatively lower energy SPE spectra of protons and nuclei. The risk of astronaut injury or death is highly dependent on the severity, frequency, and cumulative duration of SPEs, but is widely considered low from SPEs alone.[11] Yet, this apparently "widely considered low probability" in the literature depends entirely on the focused energy of the solar event as has been witnessed by real-time events in deep space.

A substantial solar storm occurred on 4 August 1972. It was considered significant for a pair of reasons. It was rated in the "historically powerful" category of solar storm events that featured an intense to extreme solar flare, solar particle event, and geomagnetic storm components.[12] The storm was so intense that it detonated 20-25 U.S. laid underwater mines in North Vietnam's Haiphong Harbor.[13] The transit time for the exceptionally powerful Sun-ejected radiation particles was the fastest ever recorded, clocking in at 14.6 hours, which is almost 5 times faster than the average transit time of 72 hours.

The second and far more ominous importance of this event was that it is the single recorded storm to take place during the only active human program that has ever occurred in deep space. It took place between two Apollo manned missions to the Moon: four months after Apollo 16 and four months before Apollo 17, the final U.S. manned Moon expedition.

A lunar orbiting astronaut in either mission, however, would have fared far better. They would have been totally shielded from the storm on each orbit for the period that they were behind the Moon, placing the bulk of the Moon between themselves and the Sun, effectively halving their total dose from the stream of incoming particles. And, by orienting the Apollo Command and Service Module's (CSM) dense heat shield toward the storm when on the Sun-facing side of each orbit, it would have provided a much more significant barrier for its occupant than the functionally unshielded crew on the Lunar surface below. Thus, the lunar orbiting astronaut would have almost certainly survived with a heavy radiation dose but fully capable of returning to Earth – with or without his lunar exploring crewmembers.

The Apollo program literally cheated disaster by a mere roll of the cosmic dice. In the 48 months between December 1968 and December 1972, humankind conducted routine, systematic excursions into deep space with crews comprised of 24 unique individuals. Yet during that time, when there was a historically significant solar storm for which NASA was woefully unprepared, no astronauts were in space. Without the intervention of chance and a lot of luck, a disaster from cosmic radiation in space, or even on the lunar surface, was averted.

After December 1972, humanity returned to conduct space frontier operations in the far safer confines beneath Earth's planetary Van Allen Belts shield where prudence was not nearly as essential. Under these spaceflight conditions, escape from low-Earth orbit back to the complete safety of the home planet's surface was always only a little over an hour away, on average. There, during the ensuing nearly half-century of constantly shielded human spaceflight, NASA apparently learned little or nothing about the weakest link in their exploration paradigm. Alternatively, if they learned anything at all, they absolutely failed to turn those lessons learned into crucial changes in their exploration model and exploration hardware. Crews of today preparing for the Artemis Lunar missions are just as vulnerable as their Apollo counterparts were half a century ago. NASA still evidently has not learned any lessons regarding the flying anachronisms represented by their 21st century Apollos on steroids.

SPEs do not only pose a radiation hazard to human crews, as was surprisingly discovered on 4 February 2022. On 3 February, SpaceX launched 49 of their Starlink satellites from the John F. Kennedy Space Center. Six days previously, the Sun had erupted 92 million miles distant, sending a wave of charged particles Earthward. The energetic particles struck the Earth's upper atmosphere and the energy it deposited caused the atmosphere to warm and expand, increasing the density and drag in the thin upper regions of the thermosphere where the satellites had been deposited. In turn, 40 of the new satellites were dragged to their fiery doom by the thickened atmosphere whose density had increased by some 50%.[14]

Galactic Cosmic Radiation

As we now transition from radiation events originating from the nearest star (our Sun) to radiation events that originate from the cosmos itself, let us observe one relatively important distinction, both from a deep space spacecraft design perspective and a deep space operations perspective. Solar radiation events are fundamentally unpredictable, sporadic, and somewhat erratic. These events range from mostly a mere nuisance to occasionally deadly. While the 11-year solar activity cycle allows some low-level range of imprecise predictability, it is only "generally predictive" and cannot at all determine safe windows of deep space flight. Predicting a solar storm is very similar to predicting a terrestrial hurricane's appearance, track, or final strength. For the broad purpose of their predictive attributes, both are fundamentally chaotic systems.

Penetrating ionizing radiation that does not originate from the Sun is called galactic cosmic radiation (GCR – an acronym that is used widely in this book), or alternatively, galactic cosmic rays. Obviously, Earth is continuously exposed to the onslaught of GCR, but it is shielded by the Van Allen Belts and the atmosphere itself.

As we shall see, the flux, or rate of transfer, of GCR is much more constant than SPEs. While there is a comparatively long-term 11-year variability of incoming GCR particles from interstellar space linked to the solar cycle itself, there are no such things as "GCR storms."

For example, when the Sun's energy output is higher (meaning the Sun's radiant activity is higher than average, based on an 11-year cycle from high activity to low activity), the subsequent solar wind that eventually spreads outward toward the edges of the solar system deflects GCR away from the solar system to some extent. But in a lower solar cycle, the GCR encounters less solar wind resistance and impacts our solar system more. This effect is dependent not on just the cycle itself but also on the relative cycle's energy. For example, in the low point of Solar Cycle 24, which ended in 2019, the inner solar system (where Earth orbits) experienced some of the highest steady-state GCR influx ever recorded. This was because the solar minimum was the lowest observed since the beginning of the space age to that date.[15] (Sun cycle recordings began in 1755, which is known as Solar Cycle 1.)

"It's a bit counterintuitive," commented Nathan Schwadron, a space physicist at the University of New Hampshire. "Solar Minimum may actually be more dangerous than Solar Maximum."[16] Dr. Schwadron's comments describe the difference between the relative risks of lower energy SPEs and a potential solar storm event versus the risk of a steady state flow of a large amount of high energy GCR over time to astronaut crews.

The Solar System's Heliosphere Shield

Close in – a mere 36,000 miles off the surface of the Earth – we live under the safety of the Van Allen Shields and a thick atmosphere. At sea level, humans receive a fraction of the galactic cosmic radiation that exists above the Van Allen shields (around 2,000 times less). And we are also protected by a vastly larger shield that exists as a part of the Sun's highly active and energetic electromagnetic field (magnetosphere). Plus the Sun's atmosphere (astrosphere) is shaped like a bubble that extends out and surrounds our entire solar system. This massive bubble shaped envelope called the Heliosphere is continuously inflated by the energy of our Sun's "solar wind." At its boundary layer is the "interstellar plasma" that represents our galaxy's (the Milky Way) interstellar environment. The difference between the environment inside our Heliosphere and that in the unshielded interstellar void is substantial.

The solar system cruises at 515,000 miles per hour as we orbit the Milky Way Galaxy. In June 2021, it was reported in the *Astrophysical Journal* and Space.com that outside the relative shielding effects of the Heliosphere at the edge of interstellar space (where the solar winds cease to have any effect at a boundary called the "heliopause"), our interstellar probe Voyager 2 and our Interstellar Boundary Explorer satellite have discovered that the solar winds protect everything inside the solar system from 70% of the actual galactic cosmic radiation levels in interstellar space[17,18] Therefore, utilizing our interplanetary GCR levels measured by Curiosity at 1.8 mSv/day would imply that the interstellar GCR levels could actually rise to 6.0 mSv/day! Livescience.com reported this discovery from Voyager as it broke through into interstellar space: "Eventually, when the levels of cosmic radiation skyrocketed and stayed that way, it was clear the Voyager 2 had entered a new region of space, beyond our sun's domain."[19]

While interstellar travel is not the subject of this book, it is evident from what we have already discussed that the amount of shielding required by an interstellar crew headed toward another star system would be greater than that necessary to fully protect exploration within our solar system. For example, in a functionally unprotected interstellar ship, the crew would be exposed to the current U.S. legal radiation limit in just a little over one week, and in two weeks, their exposure would exceed one Hiroshima Unit (defined in the chapter "The Hiroshima Unit").

The Sun's radiation may be viewed as highly variable and potentially devastating over short bursts (high flux, relatively lower energy). But the galactic cosmic radiation streaming in from the cosmos can be viewed as steady and far more potentially injurious to human physiology over an extended period of time (low flux, very high energy). It is important to note that the energy of GCR is far greater per impact on the human body than a particle of solar radiation – and that difference is often measured as millions of times greater or even more.

The combined shields of the Van Allen Belts and our atmosphere are not perfect, and, inevitably, a small fraction of cosmic rays manage to make it all the way to the ground. Indeed, that is how cosmic rays were discovered in the mid-20th century using a relatively simple device called a "bubble" or "cloud" chamber. As a young teen encouraged by my Haskell High School science teacher, the late Mr. Claude Archer, I purchased my own bubble chamber as part of a youth's experiment set called the "Atomic Energy Lab" from Edmund Scientific Company ($8.95 in the 1960s). Using it, I witnessed firsthand the fine, gossamer trails and traces of cosmic rays that made it past the Van Allen Belts and through Earth's atmosphere zipping through my bedroom. The average annual dose or exposure on the surface of Earth from this cosmic radiation is 0.33 mSv (33 millirems), or 11% of a person's average yearly exposure due to all natural sources of radiation.[20]

In deep space, outside our twin terrestrial shields, how much exposure can we expect from both solar and galactic sources? Sensors placed onboard the Mars Science Laboratory encasing the Curiosity Rover measured the space radiation environment continuously during its nine-month transit to Mars from Earth, including solar and GCR. The combined dose of radiation from all *solar activity*, counting solar flares (including the strongest flare of the previous solar cycle), accounted for only about 5% of Curiosity's total dose. The remaining 95% resulted from galactic cosmic rays (GCRs). And, according to another study carried out onboard the Cosmic Ray Telescope for the Effects of Radiation (CRaTER) instrument, which has been circling the Moon onboard NASA's Lunar Reconnaissance Orbiter spacecraft since 2009, the cosmic radiation impinging on the solar system in the period of relatively quiet solar output during the years 2018-2019 was as high as any ever recorded, for the previously stated reasons.[21] Yet, it only accounted for 5% of Curiosity's total radiation dose.

In that regard, while in space, the solar component obviously streams in from the actual physical location of the Sun – considered a "point-source" in radiation control jargon. But the galactic cosmic radiation (GCR) component flares inward as an ultra-high energy ejecta of high energy particles from every direction simultaneously. While crews can afford themselves limited protection against predominantly proton SPEs by turning their ship's heaviest mass geometry toward the Sun during a solar storm, no such protection is available from the influx of GCR. Such is the lethal conundrum of the current wholly inadequate, obsolete, and potentially deadly Apollo era human spaceflight model.

Even though only 5% of the cosmic radiation on a vast majority of days in deep space is received from our Sun, when the SPE explodes from the Sun's surface, and if it is pointed toward any human crew in deep space, "adequate shelter" (defined specifically as effective shielding) is their only hope of survival or, at least, minimizing injury and depressive radiation exposure. Advantageously for all, a First Principle Solar System class spacecraft that is already designed with adequate, effective shielding protecting the crew from the everyday influx of GCR will likewise offer them unparalleled protection from the relatively rare but quite powerful solar storm. Any human rated deep space craft must be designed for all these eventualities, which I shall discuss in depth throughout both volumes of *Departing Earth Forever*.

The Engaged Calculus of the Astronaut Explorer

When any human decides to depart the shielded confines of Earth and travel into deep space above 36,000 miles, a rather radical shift in thinking is required. Humans who enter deep space are thrust into a region quite decidedly unlike that on the surface of the home planet or even in low-Earth orbit. Conditions are different here – significantly different. But to the First Principle Solar System class astronaut, these unique conditions are intimately and directly related to the ongoing interaction of their physiology to this environment! In stark terms, the antiquated, obsolete human space exploration equipment of the early 21st century is functionally unshielded, allowing a majority of the cosmic radiation dose unimpeded access to the bodies of astronauts. No near-future professional explorer would *ever* launch into deep space under these conditions.

Let me give an example of the engaged calculus of an aquanaut explorer undersea that is directly applicable to an astronaut in space. As an aquanaut, if I strap on my dive gear, enter the water of the ocean, and head downward, because my SCUBA gear is so well designed, I hardly detect any sensations of changing pressure on my body at all as I descend deeper. What I feel at one foot of depth is (by equipment design) the same as I feel at 30 feet. This is also true at 100 feet or even 300 feet. And yet, as an educated and trained diver, I know very well that this "feeling" I have is absolutely irrelevant to what is actually happening to my physiology as I tick off the fathoms and time of exposure heading down into the deep.

As a professional undersea explorer, I am constantly aware of what is going on inside my body as I descend from the results of numerous earlier studies by many divers before me. As I pass 21 feet of salt water (FSW), I understand that I have just passed my nine-hour saltwater saturation depth limit, and if I remain at or below this depth for more than that period, I will have to schedule a decompression interval coming back to the surface or injure myself coming home. Every foot below that depth reveals that there are new physiological changes and limits that I must keep track of by my engaged calculus. Beyond 100 feet, the nitrogen gas in my air supply will begin to have narcotic effects on my mind. Deeper than 200 feet, the oxygen in my air supply becomes toxic. But I do not "feel" any of these effects at all – until they have already significantly impacted my physiology! As a certified, professional aquanaut, I understand my environment, my body's exposure to it, and how they are actively interacting together – even though there are no sensations of any of these significant events as they are happening in real time. I clearly understand and plan for these things long before the dive even begins. In the mantra of all trained divers: "Plan your dive – dive your plan."

The same is true when an astronaut enters the space environment. Exposures to microgravity and galactic cosmic radiation have immediate effects on the body. And, just as in the undersea regions on Earth, those effects are dictated by time, altitude (the space equivalent of "depth"), and the equipment that may or may not shield the astronaut. The key is for every astronaut to have a clear and untainted understanding of those effects and what they are actively doing to their physiology as they increase their exposure, especially when they "feel" no sensation of it making changes to their body. It is the real, ongoing, ceaseless calculus of the astronaut explorer.

One essential insight of that relationship to an exposure and its physiological consequences is understanding the nature of the element that is impressing itself on your body every second. In this case, we are considering galactic cosmic radiation. As such, the very first integral in this calculus is the determination of how to even verbalize and describe GCR accurately within our conversations with one another.

In that respect, let me be blunt. The words typically used to describe GCR in deep space are inaccurate and quite inappropriate terms borrowed from the comparatively gentle environment of our incredibly safe home planet and its relatively very mild background terrestrial ionizing radiation. We have yet to even begin to apply accurate and appropriate terminology to GCR. For example, this is what galactic cosmic radiation *is not*:

Galactic cosmic radiation is not a "rain" of high energy particles.
Galactic cosmic radiation is not a "stream."
Galactic cosmic radiation is certainly not just hanging out in the "background."

Rather, galactic cosmic radiation is better described using terms that portray it for what it really is – extraordinarily energetic physical particles ejected from the most powerful known elemental forces in the universe. These submicroscopic, heavy cosmic primary particles with nearly indescribable energies comprise a significant part of the deep space environment. Therefore, this is what galactic cosmic radiation *is:*

Galactic cosmic radiation is most accurately expressed as omnipresent, omnidirectional, cosmic emissions.

Galactic cosmic radiation is the high mass, cosmic particulate ejecta of the most powerful known elemental forces in the universe – black holes, supernovae, and the stellar interactions of massive and epically energetic cosmic sources called PeVatrons.

However you choose to describe it, GCR is most certainly *not* hanging out in the background. It is actively penetrating and interacting with the interior of your body, your cells, and your DNA in ways that no terrestrial medical or natural radiation can even attempt to mirror.

A Pernicious Toxic Milieu of Cosmic Radiation in Deep Space

Again, it is of no small consequence that the radiation environment in deep space is so extraordinarily different than anything we experience on Earth. Indeed, as we shall discuss in the next paragraph and upcoming pages, the predominant cosmic radiation particles responsible for most of the deleterious effects on humans are up to 1,842 times more massive than any terrestrial radiation source we are familiar with, including the radioactive fallout of atomic weapons, and are accelerated at near light speed velocities by the most powerful energy sources in the universe, again, including supernovae and the external vortices of black holes.[22]

These individual particles possess energies that we currently cannot duplicate with our most advanced particle accelerators and are not even replicated by the most energetic forces of our own Sun! Many of these particles will pass through the thin skin of crewed spacecraft, eliciting secondary radiation particles, thereby provoking a significant dose of secondary radiation. One investigating team summarized their findings in their peer reviewed research results by describing the deep space radiation environment as a "pernicious toxic milieu of cosmic radiation."[23]

Quite unlike terrestrial ionizing radiation, which consists primarily of high energy massless photons, the incident penetrating particle species in deep space are comprised primarily of heavy atomic protons that are far more massive than mere electrons and are traveling at relativistic velocities (nearly the speed of light). Therefore, these galactic-source cosmic particles have the collective capacity to do far more damage to the human body than their much less energetic terrestrial ionizing radiation counterparts.

The GCR spectrum is predominantly comprised of approximately 87% hydrogen ions (protons) and 12% helium ions (alpha particles). The remaining one to two percent of the spectrum is comprised of high charge and extremely high energy nuclei traveling at relativistic speeds with exceptional energies measured in teravolts (TeV – tera, or trillion, electron-volts), which is equivalent to a million MeV (million electron-volts). This species is designated as HZE particles (HZE is an acronym derived from high [H], atomic number [Z], and energy [E]).[24]

Although HZE ions comprise only a small proportion of cosmic rays, their high charge plus high energies cause them to contribute significantly to the overall biological impact of cosmic rays, making them as significant as protons in regard to biological impact.[25] The most dangerous GCR are heavy ionized nuclei such as 26 Fe, an iron nucleus with the atomic number (Z) of 26, and nickel (Z=28) nuclei.[26] Such heavy particles are much more energetic (millions of MeV) than typical protons accelerated by solar flares (only tens to hundreds of MeV).[27] HZE ions can, therefore, penetrate through thick layers of shielding and body tissue, breaking the strands of DNA molecules, damaging genes, and killing cells.[28]

Without adequate research, it is not yet possible to describe GCR's exact effect on typically thin Apollo-era type shielding. But it may well be that the shower of secondary radiation induced by an impacting HZE particle can be even more damaging for an astronaut than the single particle itself. In the end, the only way to protect astronauts from the full spectrum of GCR, whether in deep space or on the surface of Luna or Mars, is through the use of adequate shielding.

Terrestrial Ionizing Radiation Sources Wave vs. Particle Catalogue

In this chapter, you have been given examples of different natural penetrating sources of ionizing radiation all around you. One of those was my smoke detector, 24 inches above my head in my stairwell, which encases a speck of radioactive Americium-241 – primarily an alpha emitter. Remember also the radon gas in my bathroom and the kitty litter. Those are primarily gamma emitters. Here is a short catalogue of the radiation sources that are literally surrounding us on the surface of the Earth.

Alpha Radiation

All ionizing radiation is dangerous to the human body in large enough quantities. Although a typical alpha emitter is the "weakest" form in nature (not even being able to penetrate a piece of paper or more than 1.2 inches of air), an alpha particle is mass-equivalent to the nucleus of a helium atom consisting of two protons and two neutrons.

Because of its very high mass and charge relative to all other forms of terrestrial ionizing radiation, an alpha particle's relatively lower inertial energy per particle mass limits its penetrating capacity. Nonetheless, extensive internal tissue damage can result if an alpha particle is inhaled. Too much alpha radiation on the outer skin is called an alpha burn. Yet, because of keratin in the epidermis, external alpha burns are usually limited to only mild reddening of the outermost layer of skin.

But the black-robed scythe toting granddaddy of all dangerous alpha emitters is Plutonium. When plutonium particles are inhaled, they become trapped in the tissues of the lungs. There, the alpha particles can easily kill pulmonary tissues, scarring the lungs, which results in additional lung disease and cancer. Plutonium then enters the body's circulatory system from the lungs and will migrate to the kidneys, exposing both blood and kidneys to alpha radiation. Ultimately it will concentrate in the bones, liver, and spleen. For these reasons, therefore, the lowly alpha emitter is one of the most dangerous radionuclides on the planet.

Beta Radiation

While relatively rare in nature when compared to the ubiquitous gamma, beta emitters are the big boy on the block inside our bodies and are responsible for most of the human's natural radioactivity. The lowly banana contains a beta *and* gamma emitter called potassium-40. But do not blame just bananas. This radionuclide is a part of all potassium rich foods.

If aliens captured you today in their flying saucer-UFO and started their *e-vil* medical experiments on you, the first thing they would discover is that you were naturally radioactive. Here is what they would quickly find out:

Inside your body, potassium-40 is undergoing two types of beta decay at a rate of 4,000 decay emissions per second. In about 90% of events, the potassium-40 decays to calcium-40 with the emission of a beta particle ($\beta-$, an electron) with a maximum energy of 1.31 MeV and an antineutrino, and then a 1.460 MeV gamma ray. In the other 10%, potassium-40 decays to argon-40. When will all this potassium-40 radioactivity finally be over after eating that banana? In about 125 billion years, based on the potassium-40 half-life! In other words, eons after your death, your ashes will still be quite active. And all this is due to a beta and gamma emitter inside your body.

A purely beta radiation decay event is the result of the release of far lighter particles than the massive alpha particles. The beta releases electrons or positrons which, because of their much lighter mass, are appreciably more penetrating than alpha particles and can span the distance of a few meters through the air.

But beta has an interesting characteristic. Under the right conditions, a beta particle can create a more penetrating gamma ray. Recall that a beta particle can penetrate a sheet of paper but can easily be stopped by a thin sheet of either a solid plastic barrier or aluminum. However, much denser materials such as lead or even steel do not make such good beta radiation shields. Beta radiation impacting these very dense metals materials come to a stop so suddenly that their energy is instantly converted into gamma radiation which then passes straight through.

Gamma Radiation

The third type of ionizing radiation is by far the most common on Earth – gamma. Gamma rays have the smallest wavelengths and the most energy of any wave in the electromagnetic spectrum. Gamma rays can easily penetrate spacecraft walls and human flesh – such as the gamma rays utilized in many medical therapies, especially for the treatment of cancer.

Gamma rays and x-rays (discussed previously) have the same fundamental commonalities but originate from different parts of the atom. X-rays are emitted from processes outside the nucleus, but gamma rays originate inside the nucleus. Thus, manmade x-rays are generally lower in energy and therefore less penetrating than predominantly naturally occurring sources of gamma rays.

Galactic Cosmic Radiation

Galactic cosmic radiation consists of subatomic particles moving at almost the speed of light that are 1,837 to 1,842 times as massive as the relatively tiny electrons emitted by beta radiation and the electron streams that generate the massless x-rays that most of us are familiar with and regularly interact with on Earth.[29] Hence, the potential injury to any given cellular structure is far greater than an x-ray or terrestrial gammas specifically because of the sheer inertial difference.

To comprehend this better, we must briefly discuss a bit of quantum mechanics and particle physics. Ready? Electrons are fermions (particles that follow Fermi-Dirac statistics) that rotate two full turns (720°) to return to their original configuration (a spin of ½).[30] X-ray photons are bosons (particles that follow Bose-Einstein statistics) and have a spin of 1 (360°). Electrons are charged particles with mass, while photons are uncharged, massless particles. (There, that wasn't so bad, was it?)

While it is impossible to compare the massless gamma ray or x-ray photons with cosmic particles, we can compare the electron (from beta radioactive decay, for example) with a cosmic particle to illustrate the mass and relative potential for inertial damage in any biological system.

In quantifying radiation effects, there is a term defined as Linear Energy Transfer (LET), which describes the amount of energy that a radiation particle transfers to the material it penetrates per unit distance.

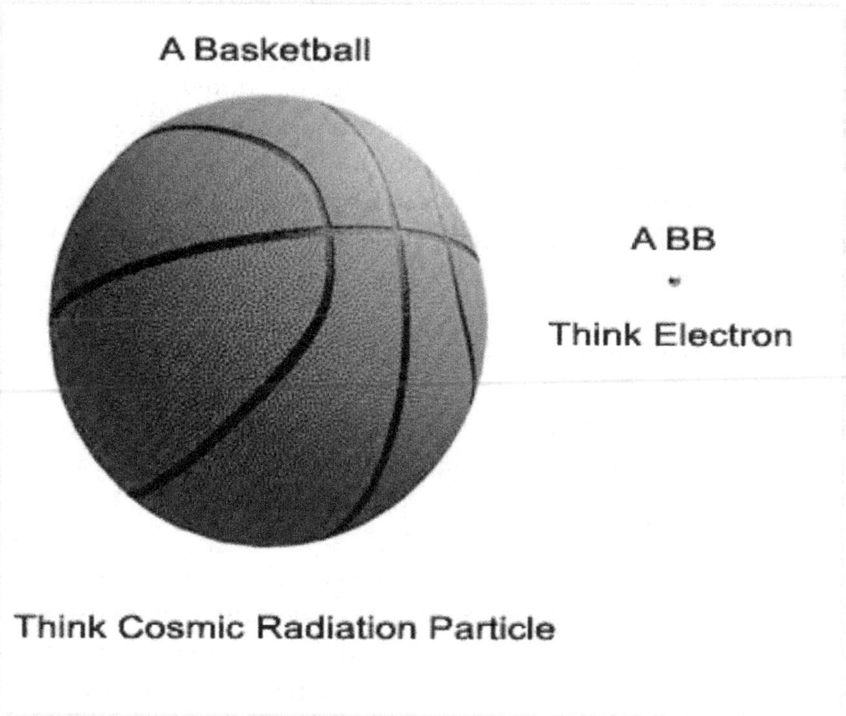

In the case of terrestrial and medical radiation sources compared to GCR, it is useful to compare the electron (which in many medical applications generates the massless x-ray photon) to the cosmic particle. To very roughly contrast the masses and the relative inertial energy difference between the two would be to relate the relative mass-in-motion of a BB (the terrestrial ionizing producer of an x-ray photon) to the relative mass-in-motion of a basketball (HZE cosmic radiation), as seen in the above image. The massless photon will impact the cell with a given amount of electromagnetic energy. But the cosmic particle will impact the cell with from 1,837 to 1,842 times the mass of a tiny electron at near light speed and with an inertial energy that the mere gamma or x-ray cannot duplicate! The physical difference to each cell interaction is not even in the same league. Think about it. If you are the hapless victim standing in the center of the dodgeball ring, would you rather be opposite Arnold Schwarzenegger ready to propel a basketball at your favorite head, or Pee-wee Herman swatting a BB in your direction? Get the picture?

Any cellular nucleus taking a direct hit from a single particle of cosmic radiation (basketball) will respond far differently than if impacted with a completely massless terrestrial ionizing medical source x-ray photon generated by the electron (a BB). That same cell, when struck with a single x-ray photon, will more than likely, in nearly every case, escape totally unaffected. But remember, when comparing the effects of the massless x-ray and gammy ray's potential to damage biology, their effects are energetic damage, not inertial damage from heavy cosmic particle particulate collisions within the cells. However, the cell's fate after taking a direct hit with an HZE particle is very uncertain, at best, and depends on whether the particle strikes the cell's nucleus directly or a series of vital organelles inside the cell. (The biology of GCR on cells is discussed in detail in the chapter "Chronic Cosmic Radiation Exposure Etiology.")

An astronaut in deep space will encounter enough cosmic radiation of all forms so that every cell in their body will be traversed by a hydrogen ion particle every few days. But the same astronaut will also encounter the probability of a transit in every cell of their body by far heavier HZE nuclei (^{16}Oxygen, ^{28}Silicon, and ^{56}Iron) about once a month with an exponential increase in individual particle energy.[31] Although their incidence is much lower, because of their mass and energy, these HZE nuclei constitute a significant biological threat *not encountered in low-Earth orbit*.[32] In other words, even though we have amassed man-decades worth of GCR data in LEO, we have almost no data on the effects of full dose-full spectrum GCR in deep space. Even so, what little information we have appears to demonstrate negative long-term consequences (given exposures of months or years).

Often, these discrete HZE particles are also sometimes referred to as cosmic primaries and, likewise, primary cosmic rays. Again, reemphasizing, HZE radiation consists of charged particles that have been accelerated to enormous relativistic energies by astrophysical sources of speculative origin. They have a range of energies (from 109 electron-volts [eV], which is equivalent to 1 Giga [billion] electron-volts [GeV], up to 1020 eV, which is 108 Tera [trillion] electron-volts [TeV]) that dwarf the best we can currently achieve in the most advanced accelerators on Earth (only about 7 TeV). Comparatively, the energy of our most powerful ionizing medical x-ray at the time of this writing is only 100 Kilo (thousand) electron-volts (KeV).

It is useful here to unwrap these abbreviations. A "powerful" medical x-ray clocks in with an energy of about 100,000 electron volts. The "weakest" HZE particle will clock in with ten thousand times more energy at 1,000,000,000 electron volts. So, a heavy primary with an energy of 1020 eV is around 1,000,000,000,000,000 *times more* energetic than the most powerful medical x-ray.

The truly astonishing energies possessed by these particles would require an equally astonishing seminal energy source to create them. Even our nearest star (the Sun) only contributes minimally to the cosmic ray flux – merely about 5%, and at comparatively low energies of up to a few hundred GeV max.[33] Again, Earth's most powerful atomic accelerators cannot even come close to generating these kinds of particle energies.

In a breakthrough study published in the journal *Nature* on 17 May 2021, it was reported that more than 530 ultra-high-energy (UHE) photons from gamma ray sources with energies greater than 0.1 quadrillion electron volts (>100 teraelectronvolts) were recorded over the initial 11 months of operation in 2020 of the Large High Altitude Air Shower Observatory (LHAASO) located on Haizi Mountain in China's Sichuan Province.[34] The highest UHE photons measured astonishingly at almost 1.4 quadrillion electron volts (peta electron volts or 1.4 PeV). Before May 2021, the most powerful cosmic ray ever documented was just under a quadrillion electron volts (1 PeV). Keep in mind that these cosmic energies were recorded *on the surface* of Earth. These rays made it through the Van Allen Belts and the thick atmosphere of our planet to the detectors on Mt. Haizi, 13,230 feet (4410 meters) above sea level.

Also of great significance, the same report from LHAASO announced the discovery of a dozen UHE cosmic natural particle accelerators within the Milky Way. In response to the article, the Chinese Academy of Sciences stated, "These findings overturn the traditional understanding of the Milky Way and open up an era of UHE gamma astronomy. These observations will prompt people to rethink the mechanism by which high-energy particles are generated and propagated in the Milky Way... The results show that the Milky Way is full of PeVatrons."[35] The cosmic natural particle accelerators are known as PeVatrons since they accelerate particles to the PeV range, while the Large Hadron Collider at Cern (the largest accelerator on Earth) merely accelerates particles to 0.01 PeV. The results from LHAASO force the renewed consideration of high energy cosmic ray acceleration and propagation in our galaxy and their potential effects on crewed deep space exploration. As of yet, we have no instruments in deep space designed and calibrated to measure these kinds of cosmic ray energies, their untainted full strengths, their flux, or their frequencies, nor how they will affect human crews.

The LHAASO technology, calibrated for Earth based equipment and tuned into these extraordinary energies, gives us but a first glimpse into the cosmic energies that will inevitably impact both spacecraft and crews when we move into deep space on long missions. But the question remains, if these detected particles already make it to the surface of Earth (where they were recorded), then why are they not adversely affecting our health here on the surface? That is because a vast majority of them are either shunted away by the Van Allen Shields, or they are nearly all absorbed by the atmosphere on the way down to the ground. But do the few that make it to the surface potentially kill cells? Of course they do! But the number of cells killed by these PeVatron byproducts is minuscule compared to the number of cells in your body that die off naturally every hour of every day. In other words, while there is always the remote probability that encountering one of these cosmic wrecking balls on Earth will trigger a cancer, it is so probabilistically low that we cannot even measure it. That is the good news about the newly discovered ultra-high energy cosmic particles that make it to the surface of our planet.

But the bad news above 36,000 miles is not nearly as positive. We already know that it is the full spectrum, unfiltered flux of ultra-high-energy (UHE) photons from gamma ray sources (including these PeVatrons) and high energy cosmic rays (HZEs) that cause the majority of the damage to human physiology in deep space. They are responsible for cognitive, cardiovascular, and whole system endothelial and epithelial damage. So, the bad news embedded in this May 2021 study is that now we have discovered a whole new range of energies that are another magnitude greater included in the already sparsely documented full spectrum of GCR. Again, quoting the article, "the Milky Way is full of PeVatrons." I emphatically repeat, if humanity does not responsibly and ethically study these hazards with laboratory animal cohorts *before* flying our crews into this toxic soup, then we will inevitably pay the tragic price measured in precious human lives for venturing beyond our reach at the unprincipled end of our broken exploration model.

Now, let us compare galactic cosmic radiation and terrestrial x-rays. An ion of GCR has a high relative biological effectiveness (RBE) value because its linear energy transfer (LET) is high and can produce dense, damaging ionizing events all along the particle track through human tissue. But the RBE and subsequent exponentially lower LET of terrestrial X-rays is relatively minimal, and X-rays produce far fewer ionizing events except at high energy and flux rates, as in some medical therapeutic applications.

Most studies have revealed that galactic cosmic radiation (high LET radiation) induces relatively complex DNA damage. This contains a unique class of DNA-focused structural damage that includes two or more individual lesions within one or two helical turns of the DNA molecule. Such injury is more difficult to repair than individual lesions, and in some instances, the damage is not reparable at all.[36] When this occurs in a given cell, the cell typically enters "programmed cell death" (apoptosis) and dies.

It is also essential to understand that regardless of their enormous energies, these interactions with soft human tissue all occur sub-microscopically (in the case of ionizing radiation) and microscopically (in the case of cosmic radiation). An astronaut can sit in a flux of cosmic or ionizing radiation, feel nothing at all, and not even be aware that anything is occurring. Yet - the injury is all taking place within the walls of the individual cells being exposed to the onslaught of energy. With one exception (mentioned below), there is no involvement of nerve cells in a GCR field. Likewise, in the case of a vast majority of medical x-rays, the patient would not even know they were being exposed at all unless they were specifically told. As well, in deep space, an astronaut will generally have no sensation of being impacted by cosmic particles, even though their energies are vastly greater and the cell damage far more significant.

The exception to this is what astronauts have described as light flashes they can "see" even with their eyes closed. The first formal report of these occurrences originated from Apollo Astronauts Neil Armstrong and Buzz Aldrin after they had departed the Van Allen Belts and flew into deep space. But they are not alone. Most astronauts on International Space Station, Space Shuttle, Apollo, Skylab, and MIR missions have reported the same phenomenon.

One U.S. astronaut on the Russian MIR space station, Jerry Linenger, was so vexed by this phenomenon that he could not sleep because of the incessant flashes of light in his brain. He finally stacked two rows of lead acid batteries and rested with his head shielded between them to cut back on the flashes that prevented him from sleeping during a solar storm. It is notable that he also reported that even these stacked lead shields did not stop all of the bright bursts.[37]

What are these flashes? There are three theories:

(1) Radiation penetrates the astronaut's skull and strikes individual neurons of the optic center of their brain, creating a flash.

(2) While traveling through the *vitreous humor* of the eye, the energetic particle emits light in the form of a nuclear phenomenon called Cherenkov radiation. It is the same effect that causes a nuclear power station's heat transfer pool to glow with a blue light.[38]

(3) Or (a favorite among some astronauts), there is the speculation that heavy cosmic particles enter the eye itself and strike the optic nerve causing the flash.

A majority of flight surgeons and neurophysiologists appear to agree that these flashes are caused by the heavy primaries of galactic cosmic radiation impacting the visual cortex of the astronaut's brain. Said one physician, "Each flash represents the death of a neuron after impact by a cosmic primary on the visual cortex."[39]

A single terrestrial x-ray particle passing through a cell may well induce genetic damage, yet it would not even be microscopically visible as its impact is submicroscopic. But the damage inflicted by an HZE particle as it passes through a living cell is so significant that these pathways of damage are routinely imaged by microscope, as demonstrated in the photos below. If that heavy particle (such as a proton of iron shown in Figure 1) passes through the cell's nucleus instead of other less vital parts of the cell, the cell will probably not survive.[40]

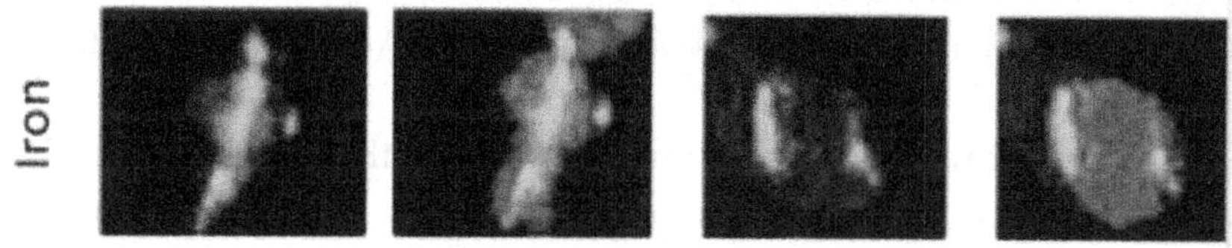

Figure 1

This fact was not lost early on in nuclear research activities. To distinguish the relative biological effects of many different radiation species (gamma, alpha, beta, and cosmic), nuclear scientists came up with a scale (formerly known as quality factor, which is now called radiation weighting factor) for the type and energy of radiation incident on the body. For example, the much smaller gamma ionizing radiation species that we are all more accustomed to encountering on the surface of Earth has a radiation weighting factor of 1.0, while galactic cosmic radiation is weighted from 5 to 20, depending on their energies. The BB vs. basketball analogy is fitting and was the very reason the radiation weighting factor was employed, as not all radiation is created equal.

As professional nuclear engineers, we were taught early on that there were three primary strategies to lessen our exposure injury to radiation sources: time, distance, and shielding. Time spent in a radiation field was to be absolutely minimized. The greater the distance away from the source, the less exposure you accumulated. And the greater the density of the shielding you could arrange in the vicinity of the source, the less exposure was accumulated.

In space travel, exposure time to cosmic radiation is minimized by increasing your speed to the destination – or spending less time in higher radiation fields. But there is absolutely nothing you can do in space to minimize your distance from the source since it "appears" that the source of cosmic radiation is the center of the universe itself. Lastly, there is shielding. Adequately shielding astronauts from space radiation while in transit to their destination and while they are on-site is absolutely essential.

Unfortunately, increased density of materials for shielding and viable space travel have been condemned in the past as antithetical concepts to deep space exploration. Dense materials are expensive to launch into orbit and require much more energy to accelerate and then decelerate in the conduct of any interplanetary voyage. That fact has rendered the discussion of adequate shielding moot. Until now...

Current options for shielding a human-rated deep space vehicle are covered in detail in the Book Two chapter titled "An Advanced Human Space Exploration System."

At this point in the decision-making process, the philosophy of the mission planners either results in a solution to the dilemma, or human exploration of space is finished, with low-Earth orbit being the final destination for humanity. While that reality is absolutely not accepted by many mission planners today, they will change their minds as their poorly conceived and inadequately designed vehicles inevitably return to Earth with corpses, brain damaged crew, and permanently maimed explorers. There are basically two choices to continue exploring and permanently settling deep space.

One mission design philosophy will require the resources and design strategies to safely protect the astronauts and will designate and allocate resources as a priority rather than risk the health or lives of the crew. Ultimately, after the serial deep space tragedies, the nation or private entity that launches these very successful First Principle Solar System Class expeditions will successfully dominate the space frontier with essentially zero competition from their more primitive competitors.

The other choice is represented by the ongoing but already obsolete design philosophy which demands the astronauts to "adapt" to the hazardous environment and thereby unavoidably personally assume the life risks and injuries associated with the realities of the deep space environment. However, while government astronauts may have life-long health coverage for their injuries paid for by the taxpayer, none of the private astronauts (a vast majority of future space travelers) will fly with such a luxury.

These functionally unprotected expeditions flying out into deep space on obsolete, primitive technology will all catastrophically fail without exception. How can I be so certain? Because the GCR environment and piggybacked microgravity exposure multimorbidity effects are not negotiable social constructs. The relentless forces of nature pay no attention to useless dreams of empire or absurd, stacked, and poorly conceived countermeasures. All those who are foolish enough to depart Earth unprepared will die *en masse* or return with permanent injuries and physical disabilities. This fact is not even up for rational debate. Those expeditions are truly comprised of ships of fools.

Once at a Lunar or Mars base, the issues become much easier to mitigate. Using a minimum of six feet of Martian or Lunar soil (regolith) available on site to cover crewed structures, this natural material can adequately shield the people from damaging GCR for longer term exploration activities on the planets.[41] If properly constructed, such shielded habitation shelters are as GCR safe as any living room on Earth.

Understanding Exposures

It is practically certain that exposure to natural radiation on Earth is absolutely inconsequential to most readers. In fact, if the experts who measure it had not informed us, we would not even know we were being exposed at all. Likewise, when we are exposed to medical and dental x-rays, we sit still for a few seconds, wait for the buzz and click, and then it is over. We do not feel a thing when it happens and, unless it is higher doses for cancer treatment, there are hardly ever any aftereffects. Further, the exposure is almost always limited to a relatively small part of our bodies.

In space, the exposure to cosmic radiation is to our entire body, and the streaming, massive, ultra-high energy rays passing through our bodies never stop unless and until we can retreat to habitation spaces with proper shielding. Further, the power of cosmic radiation is millions of times greater, and the subsequent probability of biological damage is significantly different. Here is the comparison:

On a <u>one-way trip</u> out to Mars, according to Curiosity data, we would be exposed to about 1.8 millisieverts/day of whole-body radiation, less the spacecraft's shielding effects (which vary quite widely). This reading was taken inside the spacecraft, where the rover was nestled beneath its rather thinly shielded walls. That is 273 times *per day* higher than the same annual dose on the surface of Earth, almost all of which is comprised of far lower energy terrestrial ionizing radiation.[42] That amount of radiation exposure is equivalent to a rate of 55 whole body CT scans per year, or 4.58 whole body CT scans per month, or about one every week!

Think about it. Any reputable physician would lose their license for even *suggesting* exposing *any* patient to even a fraction of that level of far lower energy medical radiation for *any* therapeutic reason. And yet, this amount of exposure is now actually being normalized by NASA and private agencies as the given expectation for deep space interplanetary astronaut crews. Go ahead and ask any physician if they would write you a prescription for a whole-body CT scan once a week for three years – the average length of a Mars voyage. They would probably immediately refer you to a psychiatrist for evaluation. And yet, that same scenario is the full expectation of current Mars voyage planning!

Standing unshielded on the surface of Mars, an astronaut's radiation exposure is cut to between one-half to one-third (0.67 mSv/day) due to the density of the planet that shields half the incoming radiation plus the thin atmosphere between the astronaut and the cosmic source of the radiation. This exposes each unshielded crew member on the surface to slightly less than two full body CT scans per month (the equivalent of 240 chest x-rays over the same period).[43]

Each astronaut crew member is driven consciously or not by the human space exploration philosophical model that has existed from the first days of crewed space mission planning and execution. It is the "Right Stuff" philosophy that demands physiological and intellectual adaptation by carefully selected humans that are specifically chosen for their rugged physical and mental capacity to take any kind of abuse they are ordered to endure without question or concern for their own wellbeing. That is why at the beginning of the space program, nearly all astronauts were military or ex-military candidates. While the candidate pool has significantly changed, the expectations apparently have not.

Nonetheless, no human who has ever been born or will be born genetically unaltered to the human genus can adapt to such absurd levels of physiological assault, deliberate, systematic abuse, and inevitable permanent injury. National space agencies and private companies who demand that their crews be exposed to such levels of cosmic radiation above federal standards for radiation professionals are participating in unequivocal violation of fundamental humane, ethical standards for employees, volunteers, and flight crews. Despite knowing this, they proceed because they have not only fixed their missions to adhere to the antiquated and obsolete human spaceflight paradigm, but, in the case of NASA, they have also deliberately exempted themselves from compliance to the federal laws governing radiation exposure, as we will discuss in detail later. Had they not taken this unethical and extraordinary step, they would have been forced to change their model or not fly into deep space at all. So they have consciously determined that instead of redesigning and changing their model they have elected to injure and kill their crews.

Galactic Cosmic Radiation in Low Earth Orbit and Deep Space

As stated previously, the boundary between low-Earth orbit (LEO) and deep space is specified by the extent of the Van Allen radiation shield that extends from 400 to 36,000 miles above Earth. As is described in this book, deep space is further identified as the region above the Van Allen shield exposed to the full, unfiltered space environment. Some define deep space as the region of space that is beyond the orbit of our Moon, but that definition is considered far too arbitrary for this work.

Based on averaging the data from all ISS missions and the Mars Science Laboratory flight carrying the Curiosity rover to the Red Planet, a crew member's GCR dose rate in deep space is magnified up to nearly three times (covered in detail in the next chapter, "Cosmic Radiation Measured in Deep Space"). And, as was discussed in detail earlier in this chapter, the individual species of that GCR radiation may display distinctly different characteristics, such as particle types and energies.

Visualizing Cosmic Radiation

If you have any interest in cosmic radiation in deep space, there is a most amazing GIF video that is believed to show the actual background flux of cosmic radiation in deep space. The reason this video is so instructive is that it gives a visual representation of the incidence of high energy particles present in the cosmos. They are so abundant that you can see evidence of their passing with your own eyes. Even the location where this little video clip originated is truly extraordinary. Unfortunately, books are difficult platforms for video content, so I have included a photographic single frame out-take as Figure 2. The GIF video is available at:

https://www.livescience.com/62394-comet-snow-rosetta-twitter.html

The video is also available on the Companion Website linked to this chapter at **http://DepartingEarthForever.com**

Figure 2

The video has an intriguing story. The 2.5-second clip was created by Twitter user landru79. It was assembled from 25 minutes of linked images broadcast from the European Space Agency's (ESA) Rosetta spacecraft on 1 June 2016. The camera was located on the craft as it orbited around the towering cliffs of comet 67P/Churyumov-Gerasimenko. Keep in mind that this clip was NOT sped up 600 times but was assembled from just a handful of images from the entire collection that was taken in a span of 25 minutes. What you see is what was happening in the split second the picture was taken in those few individual frames that were selected from the total shots gathered over the 25 minutes. Therefore the "rate" of the active GCR/HZE elements in each image over 25 minutes cannot be accurately determined.[44]

The image displays a backdrop that looks similar to falling snow. But note that there are several things actually going on simultaneously. As Mark McCaughrean, Senior Advisor for Science and Exploration at ESA, reports, "The driving force behind the snowy illusion is the dense backdrop of stars that appear to form a curtain of falling snow in the distance. But of course, they are not 'falling.' The motion of the stars is the result of a combination of Rosetta's changing position as it snaps each image and the comet's rotating motion."[45]

But in front of the floating "flakes," long steaks traveling in many different directions across the scene are clearly visible. These streaks are left behind by cosmic radiation. As the HZE particles strike the camera's sensors, they register as veins of light.

Now, let us focus a little more intensely on Figure 2. Please carefully observe that the apparent flakes, which are resolved as stars, are only visible in the dark fraction of the sky between the cliff-face and the right side of the image. The remainder of the image (including the cliff escarpment and the ground) is star free. The steaks in the image are presumably a mix of cosmic radiation and dust. There is no way of discerning the contribution of either – and, of course, the argument over that fraction is fierce. But, if you want to have some understanding of the reality of cosmic radiation in the deep regions of interplanetary space, this video clip will give you a true appreciation.

Figure 3 is another photo of the direct result of a cosmic ray strike on the camera of the Mars rover Curiosity, recorded 16 June 2019. JPL engineers clarified what they believe the bright light on the right of the photo really was. "It's not in the left-Navcam image taken at the exact same moment. It's a cosmic ray hit," Doug Ellison, visualization producer at NASA's Jet Propulsion Laboratory, explained.[46]

Figure 3

In another depiction of cosmic radiation from Mars, NASA released the Figure 4 image included in the 2008 *Space Faring: The Radiation Challenge Middle School Educator Guide* accompanied by this legend: "Figure 5 shows GCR falling onto the surface of Mars."[47]

Figure 5: GCR appear as dots in this image. Image credit: NASA.

Figure 4

Figure 5 is an illustration of the ambient radiation exposure in deep space recorded on a film that was worn on the ankle of Neil Armstrong during the Apollo 11 lunar landing mission in July 1969. Individual silver grains due to gamma rays and electrons form a background against which numerous proton tracks can be seen, along with one track of a heavy ion (HZE) with an estimated Z of 10-12.[48]

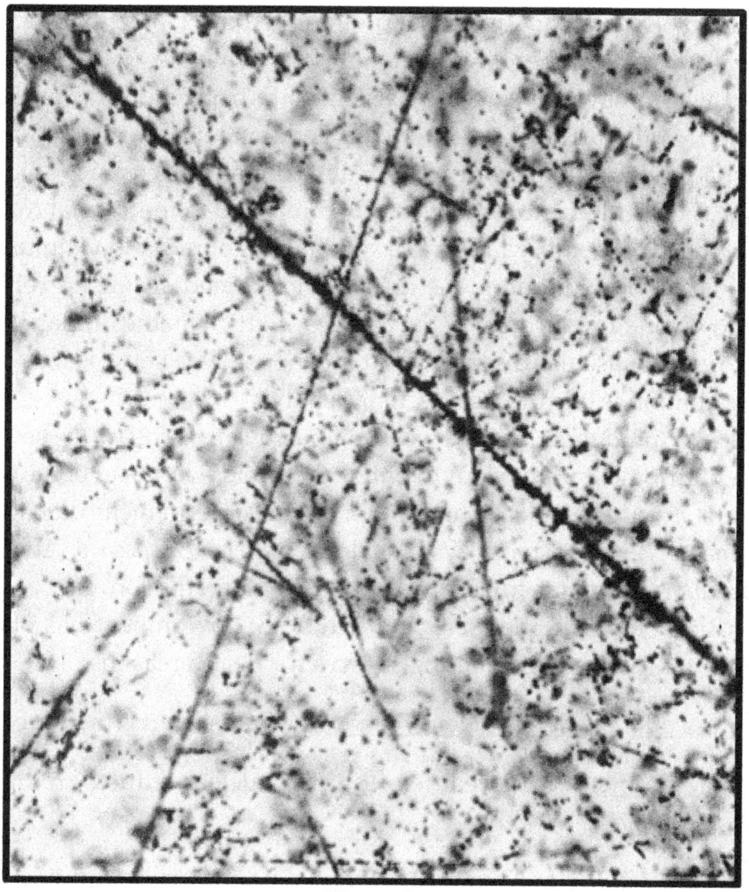

Figure 5

Exposure Limits – How Much is Too Much?

There are primarily two different reasons for establishing limits to exposure: acute, direct, immediate injury – and – inducing a kind of delayed genetic and tissue degeneration inside the astronaut, which is ultimately expressed by an increase in the probability of radiation-induced cancers and the long-term breakdown of cardiovascular tissues and other fast growing, sensitive endothelial and epithelial cells (effects that are genetically delayed over decades.) The chapter in this book titled "Hitting the Physiological Wall" offers a detailed discussion of this effect unique to *Departing Earth Forever* that has not appeared on anyone else's radar as of this writing. These radiation sources permanently alter the genetic material in some cells and degrade cellular repair mechanisms. As of today, there is no way to quantify this permanent damage or identify which human will ultimately become afflicted with cancer or disease, except by probabilities linked directly to the duration of exposure because there is not enough data available.

But we do have repeated, consistent data from the 24 individuals who did fly into deep space and studies conducted on Earth simulating deep space effects that confirm the reality of galactic cosmic radiation damage. Concluded one research study published in the journal *Nature*, "Researchers also exposed mice to a similar type of radiation (simulated GCR) and after six months, the mice demonstrated sustained cellular breakdown and impairment of the arteries—which, in the human body, leads to cardiovascular disease. What the mouse data show is that deep space radiation is harmful to vascular health."[49]

Keep in mind that this data was obtained by an exposure to the deep space environment equivalent to only 9 days and 5 hours average GCR exposure, yet it was sufficient enough to cause permanent damage to the subjects' cardiovascular systems *decades after the exposure*. And this time frame represents only 0.009 of a typical Mars mission.

Starting on Earth, there are limits the U.S. government has placed on exposure to ionizing radiating sources for health reasons:

(1) Professional Nuclear Power Workers: 50 millisieverts (mSv)/year

(2) General Public: 1 mSv/year (exposure to nuclear industry radiation sources)

But NASA (a federal agency) exempted itself from complying with federal radiation safety standards applicable to all other United States agencies and companies. As of this writing, the limits for astronauts are:

(1) 30-day exposure: 250 mSv

(2) Annual exposure: 500 mSv[50]

(3) Career exposure:

Career Exposure Limits for NASA Astronauts by Age and Gender*				
Age (years)	25	35	45	55
Male	1,500 mSv	2,500 mSv	3,250 mSv	4,000 mSv
Female	1,000 mSv	1,750 mSv	2,500 mSv	3,000 mSv

*Visit https://srag.jsc.nasa.gov/Publications/TM104782/techmemo.htm for more information on radiation limits.

Figure 6

However, on 24 June 2021, the National Academies of Science, Engineering, and Medicine released the Consensus Study Report *Space Radiation and Astronaut Health: Managing and Communicating Cancer Risks* which provided "the committee's recommendations and conclusions regarding the updated space radiation health standard, NASA's radiation risk communication strategies, and a process for developing an ethics-informed waiver protocol for long-duration spaceflight missions."[51] The Report recommended applying a single space radiation standard to all astronauts (regardless of gender), utilizing the most protective approach to do so. A 600 mSv career dose limit was recommended.

And calculating dose exposure for deep space missions is a bit involved. In the space environment, radiation exposure to astronauts in interplanetary space (1.18-1.84 mSv/day) as they travel to Mars is three times what they will experience once they land on the Red Planet. Standing on the surface, shielded from many of the cosmic rays by the thin atmosphere and the bulk of the planet beneath them, they will experience between 0.64 and 0.92 mSv per day. So, a round-trip to Mars will subject a crew to approximately 1200 mSv total exposure – which is 24 times the annual dose legally allowed for an Earthbound nuclear power worker.[52]

The current NASA radiation limits (unless they are revised to reflect the gender neutral dose limit recommended by the June 2021 NAS Report) are further complicated by astronaut gender. Women (XX chromosome) are significantly more limited in allowed exposure with greater subsequent cancer probabilities than their male (XY chromosome) astronaut counterparts (see Figure 6 above). This is because they face a higher risk of radiation-induced breast, ovarian, and uterine cancers that males are not subject to, plus increased risk to their lungs. Therefore, XX chromosome bearing astronauts are limited by flight rules to a radiation exposure level and subsequent mission opportunities of less than half (40%) of their male XY chromosome counterparts. How this is going to play out when it comes to a Mars expedition is anyone's guess as, by today's rules, unless the already risky standard is further relaxed (or just waived outright for cultural reasons), no female (XX chromosome) astronauts would be eligible to fly on such a mission. However, former NASA Administrator Jim Bridenstine stated 8 March 2019 in an interview with *Science Friday* that "the first person on Mars is likely to be a woman."[53]

Yet, of the nearly 50 astronauts in NASA's astronaut corps in late 2012, only three were eligible for NASA's historic one-year mission to the International Space Station (accomplished by Scott Kelly), "principally because of the stringent radiation limits imposed by the Space Agency," according to Robert Behnken, the Chief of the Astronaut Office at the time. "If the radiation limit was different, it might have doubled the number of eligible spaceflyers," Behnken added. "We were effectively limited to three candidates for the two positions, the prime and backup for that flight, based on the programmatic and health constraints that were in place," Behnken said. "We are running into these limits now with the exploration that we're doing, so it's not a stretch to assume that if you go further and farther and longer, you're going to run into them continuously with the exploration class missions."[54] Behnken, along with Doug Hurley, ultimately flew his own historic 62-day mission to the ISS in May 2020 as one of the first two astronauts to launch aboard a commercial orbital spacecraft – the SpaceX Crew Dragon Demo-2 (SpX-DM2) – in the history of spaceflight.

It is also important to recognize that the average annual radiation exposure on the International Space Station is approximately 238 mSv of less than full spectrum GCR. That is about a quarter of the two-way transit dose for a Mars expedition, without considering additional surface exposure on Mars. A round trip to Mars would require more than four times the exposure to cosmic radiation than astronauts currently experience on the ISS per year, exceeding their lifetime exposure by 40% – and that is provided they have never flown in space before the Mars expedition begins![55,56]

Why is a 600 mSv career limit established (currently recommended by the ICRP) for all NASA astronauts? Because exposure to radiation sources, regardless of the dose, permanently damages cells and signals certain cells in different astronauts to become carcinogenic later in life.[57] At some level, the probability that the chronic damage will morph into a fatal cancer is so elevated (3%) that it becomes unacceptable, and the Agency imposes the limit. That limit is called the Risk of Exposure Induced Death (REID).

This is an important time to redirect the discussion from raw data to understand its meaning in human terms and reflect on the reality of the exposure data just presented. These are hard numbers, not emotionally driven opinions. As mentioned earlier in this chapter, the full expectation is crew obligated occupational radiation exposure equivalent to 4.58 whole body CT scans per month – a total of 55 per year. That is exposure to rates of ionizing radiation 13 times more per year than what is allowed for professional nuclear power workers.

Imagine if some medical or nuclear power group on Earth wanted to "normalize" an occupational activity by exposing individuals to these levels of high energy radiation in order to accomplish their professional or corporate goals. There would be not only well-deserved outrage by all responsible professional organizations, but they would also almost certainly be shut down and lose their professional certifications *for even suggesting such a bizarre occupational exposure safety debacle.* And yet, for NASA, this exposure is being advertised as their professional norm, which at least subliminally suggests that all the other world nuclear organizations, who are uniformly headed in the opposite direction by systematically limiting their occupational exposure to lower and lower levels, are somehow being far too cautious.

Let us likewise examine the application of this identical statistic in the nuclear industry. If the limit on exposure inside United States nuclear facilities was established based on a statistical death rate (REID) of 3% of the nuclear workforce as the acceptable limit of death in nuclear power plants, then they would be advertising that it would be acceptable and completely normal for the industry to lose 14,200 workers to nuclear energy sources at their power plants. (Remember that REID is an acronym that means "Risk of Exposure Induced Death"). Of course, *any* suggestion of such an absurd, even bizarre notion in the nuclear power industry would cost the manager who even whispered it their job. And if any power company suggested this, they would immediately lose their NRC license, and an investigation would be launched. Yet, inside NASA, it is considered a license to kill 3% of their deep space crews and business as usual. The only difference between the nuclear power industry and NASA is that the number of dead astronauts down the road will be smaller.

It certainly appears that NASA has finally come to openly allow the mission objectives to override crew safety at any cost. They continue to deny that the time is long past to significantly overhaul their human space exploration architecture and, in so doing, place crew safety and wellbeing above mission objectives. A carefully calculated crew loss rate is not acceptable when it is possible to fully prevent such deaths. At this point, such a conclusion is no longer conjecture. It is deadly accurate, even if NASA is so blinded by their Apollo-days self-perceived invincibility that they actually do not recognize just how far down the rabbit hole they have descended. If NASA fails to address this critical issue, then it is clear that they do not maintain any ethical standards when it comes to flight crews outbound into deep space.

Shooting Them All Off Planet

"As long as you are healthy, we're going to load you up on a rocket and shoot you off the planet." This statement was made on 5 August 2022 by NASA's Chief Astronaut Gregory "Reid" Weisman before a crowded news conference. Weisman was speaking of the entire NASA Astronaut corps, both men and women alike, regardless of age or previous deep space GCR exposure. In confirming that NASA has accepted the National Academies of Sciences' recommendation of a unisex career limit of 600 mSv, Weisman called the former biological limits for men and women "draconian," noting that NASA has "...equalized all radiation limits. It doesn't matter whether you're a man, whether you're a woman, it is the exact same."

Unfortunately, while this comment may well represent prevailing social dogma, it is entirely biologically fraudulent and represents the Agency's failure to apply critical thinking skills by separating the socially engineered narrative from the realities of XX-vs-XY biology in deep space. This policy, as well as the National Academies of Science's newly recommended "gender neutral" deep space radiation exposure limits, places XX chromosome (female) astronauts in immediate danger. As was discussed a few pages earlier, when exposed to the unmitigated toxic effects of deep space, XX chromosome astronauts face a statistically determined and clearly enumerated greater risk of a fatal disease from radiation-induced breast, ovarian, and uterine cancers which their XY (male) counterparts are not subject to, in addition to an increased risk to their lungs.

Yet an absolutely insightful and brilliant statement was also made at the same moment by Chief Astronaut Weisman when he added, "If we **cannot make these spacecraft equitable**, and we can't fly any type of person on them, then we **need to look at our systems and reevaluate**." Known for his thoughtfulness and commitment to fellow astronauts, Weisman clearly states that for the good of all astronauts, it is past time to redesign the antiquated and dangerous Apollo model and implement First Principle Solar System Class Human Exploration

Engineering Systems equitable to both XX and XY chromosome astronauts; systems that not only assure that they survive, but that they will thrive throughout their missions – and for the remainder of their lives.

Having experienced an extended stay of 165 days aboard the ISS in low Earth orbit from May to November 2014, and with his dedication to the safety of the Astronaut corps as they advance to deep space missions, Chief Astronaut Weisman is no clueless bureaucrat. The top astronaut echoes the alarm of this book and highlights the imminent necessity for First Principle vehicles when he clearly sounds the admonition to the Agency – and the world – "*If we cannot make these spacecraft equitable, and we can't fly any type of person on them, then we need to look at our systems and reevaluate.*"

Final Warning

In the mid-1960s, when I was a young teen, a vast majority of adults and late teens smoked cigarettes, including most members of my immediate family – the majority of whom ultimately died from the effects of their addiction, including my own father and two of my grandfathers. But as I was entering adolescence, the American Cancer Society, and even Congress, were focusing on the problem with a serious intent of slowing the awful ongoing health disaster and were beginning to do the things necessary to attempt to stem the steady stream of dying and permanently injured.

One of the things discovered in all of this was that inhaled cigarette smoke bioaccumulated natural polonium-210 in epithelial cells of the respiratory tract. The deadly toxin polonium-210 is the same radioisotope that was used to assassinate Russian activist Alexander Litvinenko in London in 2006.[58] Polonium-210 emits alpha-radiation, which can seriously damage DNA on contact with sensitive internal membranes and is a powerful carcinogenic agent when inhaled directly into the lungs. But polonium is *just one* of the carcinogenic agents found in tobacco.[59]

And yet, I clearly remember the open ridicule of these cessation efforts and the results discovered in the scientific studies by the endless stream of seemingly hopeless nicotine addicts. For so many victims, it was just too difficult to replace the addiction with cold, unemotional facts and stop consuming – even if the data was backed up with millions of casualties and a sharply increasing rate of social awareness. I do not know of a single family in my circle of acquaintances who did not lose one or many members to this terrible scourge. But in the '60s and even beyond, the hazards posed by cigarette addiction were widely labeled a hoax.

As late as 1987, I complained to one of the directors of Kennedy Space Center's NASA safety organization that the offices of the Space Shuttle Orbiter Processing Facility (OPF) were so filled with the persistent haze of cigarette smoke that, as a chronic asthmatic, my health was truly suffering. A former smoker himself, he looked me in the eye and stated, "There is no proof that secondhand cigarette smoke is harmful to anyone." I remember privately asking myself how this individual was ever considered qualified to lead a safety office at all.

Of course, no one ever considered smoking in the shuttle Orbiter Processing Bay (OPB) within the OPF because, as well as being a fire hazard, the smoke particulates and aerosols were "harmful to the orbiter and its equipment." The NASA sentient employees, however, were left entirely on their own, casting about the bureaucratic wilderness for some shred of biological proof to protect themselves. Fortunately, the smoking policy was changed by NASA Headquarters 18 months later, and smoking was no longer allowed in any NASA offices or processing facilities.

There is another illustration that accompanies this wide expression of stubborn individual stupidity that demonstrates human resistance to any change, however simplistically beneficial. At the exact same time period that smoking cigarettes was formally labeled as dangerous to health, the U.S. government began to require seatbelts in new vehicles. Again, the overt, open ridicule of these efforts was loud and resonate. And again, as in the former case, I did not know of a single family, including my own, who had not lost loved ones that could have been saved had their automobile been equipped with seatbelts.

Society has profoundly changed since those days – at least in our ultimate recognition of the dire effects of smoking and the beneficial effects of safety engineering in automobiles. Cigarette smoking is recognized by most people now for the dangerous, addictive, personal, and societal health disaster that it really is. But in order to arrive at our 21st-century enlightenment and enabling of tangible measures to stop the ongoing health emergency, we had to endeavor not just to discover and document the scientific and biomedical facts, we also had to struggle against the cigarette lobby and the legions of the addicted in terminal denial. Meanwhile, the cemeteries continued to stack up the cancer ridden bodies, as they still do today.

Russian author Aleksandr Solzhenitsyn's *Cancer Ward* is one of his masterpieces on Soviet life. It is a brilliant, if not altogether depressing, semi-autobiographical tale in which he depicts the squalid conditions in the former Soviet Union's medical system cancer wards. Of particular note in the classic work is how many of the patients and cancer ward officials continued smoking in the ward itself.[60] This, and many other recorded instances just like it, highlight in particular the stubbornness of the human character, even in the face of near-certain death and assured permanent injury. Even when surrounded by others who are dying along with them from the same congenital defect of reasoning, the human mind is manifestly difficult to change once it is set. Indeed, such human intransigence is so legend that German physicist Max Planck said that "science itself advances one funeral at a time."

If you listen, most people are quick to express how intelligent the species is. But if you actually watch what they do and take notes long enough, in many very curious ways, humanity can definitely be considered a challenged species in this remote corner of the galaxy. And the discussion of cosmic space radiation by the world's most preeminent space faring organizations early in the 21st century is no different at all. The "adapt at any cost" lobby is not just alive and well in NASA – it is in charge.

Today, in the 21st century, there are legions and whole armies of individuals who are ready to fly out to Mars or the Moon – *right now, dammit* – and live functionally unshielded from any radiation on the way there or on the surface of these bodies. As is highlighted later in this book, they have convinced themselves that humans can somehow adapt to cosmic radiation. Or there is the belief that certain hand-wringing eggheads have simply overstated the issue and that after the first few hundred individuals have successfully established the interplanetary bases, their gut feelings will be fully justified. "Just leave us alone and let us fly out to meet the human destiny we have all been waiting for!"

These are the astronaut-wannabe patients who are smoking and joking together in the cancer ward of 21st-century human spaceflight. They have not received the memo that the old Apollo human spaceflight paradigm has absolutely run its course. It must soon be replaced by a new, far more failsafe, sounder, and more realistic model based on the actual data returned from deep space. Yet, even now, we are just beginning to recognize the limitations of 1960s technology to even fully record or report the actual data on the only deep space human crewed missions ever launched.

I close this discussion in this way so that I may finally conclude the chapter on cosmic radiation with a clear warning, much like the warning that has been a part of every package of cigarettes since 1965. It is referenced directly from a scientific journal article. Just like the cigarette pack label, it is an unambiguous, stark reminder of what risks we take with human lives if we do not depart low-Earth orbit actually prepared for the true hazards that we will face:

"The large ionization power of Galactic Cosmic Radiation (GCR) ions makes them a major health threat to astronauts and constitutes one of the most important barriers impeding plans for interplanetary travel by crewed spacecraft."[61]

Read the warning again.

Those words represent the handwriting on the 21st-century spacecraft walls. We need to toss the disgusting and anachronistic Apollo-era habits of smoking and exploring deep space, change our ways, rationally stop our rush out to infinity for a single sober moment, and just breathe through this. We must engage our minds instead of our hormonally driven Trekkie excitement and frankly read and believe the 21st-century deep space data along with all the lessons learned in the past 50 years of human spaceflight.

Remember the term "pernicious toxic milieu of cosmic radiation in deep space"? That phrase was not taken from an Internet meme. It was recorded word for word from a peer reviewed scientific paper.[62] Yet, regardless of what we face out beyond 36,000 miles, trust me, *we can still most assuredly* get to where we want to go from here. But it will not be in all the old school ways we have become so emotionally attached to that they are now etched deeply into the stone of our cultural zeitgeist.

And what is the human body's response to the waste products of supernovae and black holes? The physiological, biochemical, and cellular disease mechanisms of human exposure to cosmic radiation are discussed in detail in the chapter titled "Chronic Cosmic Radiation Exposure Etiology" in the next section.

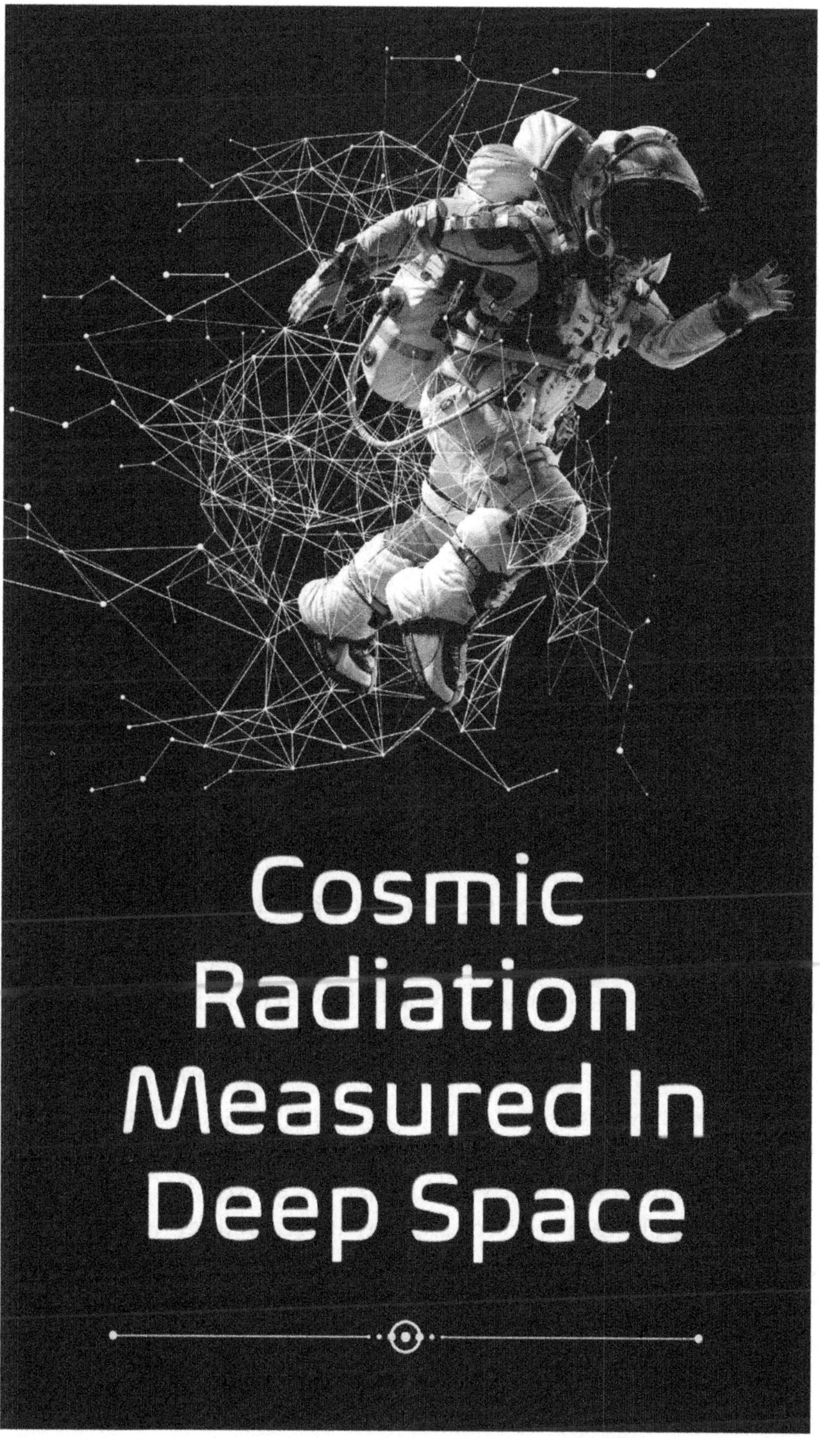

Cosmic Radiation Measured In Deep Space

"It is hard to imagine a more stupid or more dangerous way of making decisions than by putting those decisions in the hands of people who pay no price for being wrong."

— Thomas Sowell

"I don't care what the fanatics and the eggheads say, but how much cosmic radiation is actually *really* out there?"

TO ANSWER THAT QUESTION IN THIS CHAPTER, I AM GOING to mostly avoid the highly variable and typically unpredictable solar particle events (SPEs), which, on average, comprise only 5% of the total cosmic ray radiation dose, and simply focus on, relatively speaking, the "steady state" and far more dangerous chronic background galactic cosmic radiation (GCR). Therefore, refining the question, let us ask the very fair and straightforward query,

"How much GCR is *really* out there?"

When I climb into a spaceship and head out above the Van Allen shields, as I very much hope to do one day, what actual GCR exposure can I expect? (The follow-up question, "What will that amount of cosmic radiation actually do my body?" is discussed in the upcoming chapter "Chronic Cosmic Radiation Exposure Etiology.") As a human in the can riding away from low-Earth orbit at a minimum of Mach 33, I truly want to know the answer to that very precise question – the real deal, no bullshit, not politically correct answer, un-slanted by either a Mars settlement activist group or anti-deep space travel club. I just want the facts without all the hinky flags-and-footprints or "you're-taking-food-from-starving-children" sentiments. Just give it to me straight – please!

If you want the real facts, space explorer, you have definitely come to the right place!

In the end, you would appropriately imagine that this is actually a *science question*. But if you review the available popular and scientific literature, the casual investigator would think they were looking for the Ark of the Covenant. It is further complicated by a near religious furor if you disagree with anyone's prejudiced views on any side of the matter. The actual trend and outcomes of integrating science with political narratives these days are clearly evidenced by the Armageddon class circus generated by the COVID-19 pandemic storm that ended up looking a lot like a Monty Python episode.

If you decide to strike out on your own and delve into the available literature as well as the myriad internet resources (including NASA sources), I can assure you that obtaining the answer to that question is not an easy venture, especially if you do not have a modicum of training in the understanding of ionizing radiation as well as GCR and the relatively rare SPE storms. Further, there is a significant percentage of dishonest and inaccurate data – much of it quoting other dishonest and inaccurate sources in a typical Internet self-feeding, endlessly circling fake news cycle. In my research, I have discovered wildly inaccurate information published in peer reviewed journals as well as very skewed and intentionally cherry-picked data even from NASA publications. Or, I must more accurately state, *especially* from NASA publications. This should surprise no one. After all, NASA's job is to protect the current exploration model, particularly when it is starting to look very sketchy. (This is reviewed in more depth in the chapter titled "Lies, Damned Lies, and Statistics.")

In this chapter, I am going to give you all the available actual radiation data collected from deep space – as well as a few from Earth and low-Earth orbit. None of this is estimated or projections. It is all real-deal-data in all of its ugly variability. After the data is presented as it was collected, I will ask you to approach the table and watch how I deal the cards openly and freely in front of you so that you see how I came to my final conclusion, allowing you to analyze things before your very eyes and make up your own mind about my analysis and methodology. At the end of this chapter, I am going to take the data I have assembled here and use it throughout the remainder of the book. So, you will know *exactly* from whence it came – from this page forward!

When I was a PhD graduate student at Oklahoma State University, my major professor was teaching a class in Groundwater Hydrology. During the first three class periods, the professor filled innumerable chalkboards with complex equations hour after tedious hour. Near the end of the third day, he reached the bottom of one chalkboard and finally sketched out a relatively simple algebraic formula. "That," he said, "is where the last three classes have led us. This equation was derived from all those others. Since this is the singular most important equation in this field, I wanted to make sure you knew exactly where it came from!"

Truly, the exact same conclusion can be made here. These next few pages will lay out much of the only data we actually have from deep space – which is astonishingly little. And since this is a singularly important data point in this book and in the future of crewed exploration in deep space, I wanted to make sure you knew exactly where what we have *actually comes from!*

There are four primary sources of GCR data from deep space and the edge of deep space from which we can derive our answer with some level of confidence:

(1) the data collected during the Apollo Moon missions,

(2) the data collected from the International Space Station over decades,

(3) the current "gold standard" GCR deep space data collected from the United States Mars Rover Curiosity on its more than eight-month trek to the Red Planet, and,

(4) the current data from cislunar space gathered from the Lunar Reconnaissance Orbiter. All of these data points were collected in deep space, unfiltered by the Van Allen shield or Earth's atmosphere.

But before we begin, allow me to point out two very important factors concerning this data: The deep space Apollo data is highly variable. The deep space Curiosity data is not, relatively speaking.

The Apollo data was collected by 1960s vintage active and passive dosimeters. The passive TLDs (thermoluminescent dosimeters) were fixed to three locations on each astronaut's flight suits (ankle, thigh, and chest). Unlike Curiosity's fixed advanced detection instrument, the Apollo dosimeters moved with the astronauts inside and outside the spacecraft. Explained one NASA source, "Individual dosimeter readings have varied approximately 20 percent from the average because of variations in the shielding effectiveness of the Apollo spacecraft and the differences in duties, movements, and locations of the crewmen."[1]

Let me also be quite clear: the GCR data collected on the ISS, while of very good quality, is collected *below* the Van Allen radiation shield and is therefore incomplete, shielded, filtered, and does not represent the full spectrum of deep space radiation. While some elements can be indirectly inferred, the data is not directly valid or applicable to deep space GCR quality or quantity.

However, Curiosity data is the gold standard because of the GCR data collection instrument's fixed and unmoving position inside the spacecraft and its relative technological superiority. The Curiosity rover's radiation collection instrument is called the Radiation Assessment Detector (RAD). The only variability in Curiosity's data was relatively small changes in GCR flux due to solar wind and magnetosphere differences during its 254 days of interplanetary flight, accounting for less than 5% of the total variability. The Curiosity GCR data was collected by the RAD instrument, which was significantly more advanced than the technology that 44 years earlier collected the Apollo data, and which almost certainly accounts for the Apollo data's high variability.

The technological differences between Apollo radiologic dosimetry and that of the Curiosity rover are quite dramatic. Apollo dosimetry was taken more or less directly from off-the-shelf World War II and nuclear power technology available in the mid-1960s. It is much like comparing the Apollo Command Module and Lunar Module computers with the computers designed for the current NASA Orion spacecraft. Quite obviously, as historically important and impressive as the Apollo missions were, the IT systems were as primitive as they could possibly be and still do the job at all. There is just no adequate comparison between Apollo era IT and Orion era IT. The same is likewise precisely true for Apollo era terrestrial radiological dosimetry and Curiosity era interplanetary dosimetry.

Curiosity's Radiation Assessment Detector was specifically designed to quantify cosmic radiation – both SPEs and GCR – using the latest digital detection, resolution, and quantification technology available at the time. Yet, Apollo's back-fitted terrestrial dosimetry was decidedly none of these. While some may argue that it was just too simple to screw up, such a comment will reveal a decided lack of in-depth knowledge of thermoluminescent dosimetry and the application of appropriate quality factor corrections to the TLD light curves upon analysis of TLDs used in high mass, high energy proton (87%), alpha (12%), and very high energy (HZE) particles (1-2%). As recent as the first decade of the 21st century, this (refreshingly unemotional) technical discussion has been ongoing in medical radiology concerning neutron energy quantification in radiobiology using TLD technology. (Note that in the 21st century, there are many different kinds of radiometry besides TLD. But because this was the passive dosimeter used in Apollo, we will focus solely on that technology – then and now.)

Backing up just briefly, a thermoluminescent dosimeter (TLD) consists of a case holding a fixed chip that is coated with lithium fluoride (LiF) or calcium fluoride (CaF) as a solid crystal structure. When the TLD is exposed to ionizing terrestrial radiation or ionizing cosmic radiation at ambient temperatures, the energy interacts with the phosphor crystal and deposits all or part of the energy in the crystalline structure. Some of the atoms there absorb that energy and become ionized. In turn, this produces free electrons, plus areas lacking one or more electrons, called holes. Imperfections in the crystal lattice structure act as sites where free electrons can become trapped and locked into place.

To "read" the TLD, the crystal is heated, which excites and vibrates the crystal lattice, releasing the trapped electrons as photons. Released electrons return to their original ground state, issuing the captured energy from ionization as light, hence the name "thermoluminescent." Released light is evaluated electronically, and the number of photons counted is proportional to the quantity of radiation striking the TLD. The "glow curve" produced by this process is then related to radiation exposure.[2]

As in all technological assumptions, the real world is somewhat more complicated than at first glance. The TLD has a well understood error profile. Under the best conditions, the TLD is widely recognized to yield 93% accurate results when evaluating terrestrial ionizing radiation energies under ideal conditions between independent analog readers. From 1968 to 1972, 93% reliability was considered the industry standard.[3]

Energy deposition and interaction with the crystalline lattice and its subsequent release of energy via its glow curve is dependent on whether it was activated by neutron or gamma radiation (or a mix), and its energy discernment at higher energies and mass is *very limited*. Using new techniques developed in 2007, almost 40 years after Apollo, a process for discerning the highly variable effects from very low mass to exponentially higher mass particles deposited in the TLD crystals has been developed called "deconvolution" for *terrestrial* neutron applications. With this technique (the ability to quantify the differences peak by peak inside the glow curve, as well as characterizing the well-known neutron quasi-exclusive contribution to the high temperature region), it became possible to propose a method by which the respective doses can be estimated through a simplified analysis, not the traditional methods of peak resolution differences, which invariably led to errors in dose estimation.[4]

This 21st-century technique was, obviously, not available to the scientific cadre of the Apollo radiological analysis team. Therefore, since the TLDs crystalline lattice has a shelf life and must be read within six months, those dosimeters have long since passed their viability. And even if they had been retained, they are useless for submitting to any advanced analysis half a century later.[5]

The Apollo Era TLDs (1968-1972)

On every Apollo mission, each astronaut carried a personal radiation dosimeter (PRD) and three passive dosimeters. The PRD provided an active, visual read-out of accumulated radiation dose to each crewman. The PRD is about the size of a deck of cards, and pockets for carrying them were provided in the flight coveralls, as well as in the spacesuits.

The TLDs acting as the three passive dosimeters were placed in the garments that were worn throughout the mission. The passive dosimeters contained lithium fluoride thermoluminescent dosimetry powder, nuclear emulsions, neutron-dosimetry foils, and foils for detection (not quantification) of high-atomic-weight cosmic particles.[6]

The Curiosity Rover Dosimetry (2011)

Curiosity's RAD instrument is far more advanced and accurate, and is a high resolution, wide range digital instrument versus a chemical strip reader. It evaluates the incoming radiation stream using a stack of paper-thin silicon detectors and a small block of cesium iodide to measure high-energy charged particles streaming through deep space and through the Martian atmosphere now that it is operating on the surface of the Red Planet. As the particles pass through the detectors, they lose energy, producing electron or light pulses. An internal signal processor analyzes the pulses to identify each high-energy particle and determine its energy.

In addition to identifying neutrons, gamma rays, protons, and alpha particles (subatomic fragments consisting of 2 protons and 2 neutrons, identical to helium nuclei), RAD can identify heavy ions up to iron on the periodic table.[7] Thus, when evaluating the data as presented on the following pages, it is essential to keep these broad and significant differences in mind.

The Apollo Program

Between 1968 and 1972, nine 3-men crews totaling 24 humans (three crewmembers flew to the Moon twice) departed low-Earth orbit for the Moon and returned. Of those missions, three orbited the Moon and did not land (Apollo 8, 10, and 13), and six missions landed crew on the Moon and returned to Earth. Once again, it is nearly certain that when examining the Apollo mission data between flights that the first reflexive response is likely to be surprise at the striking variability between missions.

When reviewing the dosimetry data table shown in Figure 1, note that for each Apollo mission, these numbers reveal the average exposure of all three crew members based on their very limited dosimetry technology. On the missions that did not land on the Lunar surface (Apollo 8, 10, and 13), all crew members were exposed to the same relative stream of cosmic radiation. But on the missions that landed, the two crew members on the surface were exposed to about half the GCR that the Lunar orbiting astronaut was exposed to because of the shielding provided by the Moon beneath them.

While the orbiting astronaut was also shielded by the Moon to a limited extent, it was not nearly as effective as the shield provided for the astronauts on the surface. This also accounted for some of the Apollo dose variability between crew members.

Apollo Mission Deep Space Dosimetry Results

Mission	Mission Time	Ave. Crew Dose mSv/Mission	Ave. Crew Dose mSv/Day
Apollo 8	6d - 3h	1.6	0.19
Apollo 10	8d	4.8	0.60
Apollo 11	8d - 3h	1.8	0.22
Apollo 12	10d - 4h	5.8	0.57
Apollo 13	5d - 23h	2.4	0.17
Apollo 14	9d	11.4	1.27
Apollo 15	10d - 1 h	3.0	0.30
Apollo 16	11d - 1 hr	5.1	0.46
Apollo 17	12d - 13h	5.5	0.44

Mean 0.469
Standard Deviation 0.341
Variance 0.1308
95% Confidence Range 0.246 - 0.692 mSv/day

Figure 1[8,9]

Figure 1 shows the datasets collected from all Apollo missions flown into deep space based on their relatively primitive technology. This data table was assembled by the author from disparate NASA data compiled from pre-to-post Apollo mission reports. (References are provided at the end of this chapter). As the author, I independently conducted this statistical analysis in accordance with the Apollo data. The conversion was based on reported RAD exposure (neutron) converted to millisieverts/day at 100 millisieverts per neutron RAD. This analysis was conducted for all Apollo missions that departed Earth, traveled to the Moon, and returned to Earth.

As indicated in the Figure 1 table, according to the dosimetry instrumentation on the Apollo flight hardware, the highest exposure limit of 1.27 mSv/day was on the Apollo 14 mission. The lowest was recorded at 0.19 mSv/day on Apollo 8. The mean daily exposure on the Moon missions was 0.469 mSv/day. The standard deviation between missions was 0.341, and the statistical variance was 0.131. At the 95% confidence interval, the exposure range on the Moon was between 0.246 to 0.692 mSv/day. This represents a significantly wide variance of expected exposures mission to mission. However, the real surprise is not at all in its variability but in its exposure rate outside the Van Allen shield. That issue will be addressed later in this chapter and in the next chapter.

Again, NASA's official response to this variation from a document released in the early 1970s was, "Individual dosimeter readings have varied approximately 20 percent from the average because of variations in the shielding effectiveness of the Apollo spacecraft and the differences in duties, movements, and locations of the crewmen."[10]

Mars Curiosity Mission

Given the wide variability of the Apollo data, it is interesting now to compare the data that was returned from the flight of the Curiosity rover from Earth to Mars. As was stated previously in this chapter, the Mars bound Curiosity rover departed Earth on 26 November 2011 and landed on Mars 254 days later. While in flight, the craft measured the whole radiation spectrum using a device called the Radiation Assessment Detector (RAD) mounted on the rover itself, collected data all the way from Earth to Mars, and continues collecting data while on the surface of Mars.

It is important to note that the RAD detector was not fixed *outside* the spacecraft. The RAD was installed with its sensor pointing "up" from the deck of the rover nested several layers deep inside the spacecraft ensemble. Before it received radiation from the space environment, it was shielded by:

(1) the entire descent stage bolted to the rover just inches above it, consisting of the structural components of the stage, the propellant tanks, and all the associated electronic packages,

(2) then, above the descent stage was the aerodynamic Back Shell bolted to the heat shield below, and finally,

(3) the third layer of shielding included the Cruise Stage bolted atop it all, which provided the interplanetary flight management and heat radiators for the inflight phase.

Above the RAD instrument was a total of 5,293 pounds of shielding in the form of the combined interplanetary system itself. The geometric configuration surrounding RAD was complex and estimated to be 16 g/cm^2 – roughly equivalent to the Orion spacecraft design at 15 g/cm^2.[11] Yet, even while buttoned up so snugly inside the folded rover nested within its multiple layers of spacecraft, the RAD reported an average interplanetary space dose of:

1.8 mSv/day

95% of this was from galactic cosmic radiation (GCR). When the rover landed on Mars, the same instrument reported that the combined shielding effects of the Martian atmosphere *and* the planetary body below it resulted in a reduction factor of 2.7 on the surface of Mars for a decreased radiation exposure (again – mostly cosmic) of:

0.67 mSv/day

Contrasting Apollo and Curiosity Missions

The average daily GCR radiation doses reported from Apollo consisted of the eight days of flight in deep space to and from the Moon and time on the Moon. That average recorded radiation exposure was about 0.5 mSv/day, with some notable variability.

In the subsequent Curiosity interplanetary cruise between Earth and Mars, GCR was recorded at 1.8 mSv/day. Why the difference? On average, the Apollo missions reported about 3.5 times lower doses than the Curiosity mission – and yet both were sufficiently removed from the Van Allen shield to be considered deep space missions.

It is certain that the instrument that was recording the doses during the flight to Mars was shielded to some extent and thereby reported a reduced amount than the actual external cosmic radiation dose. To what extent is unknown since neither the Apollo missions nor the Curiosity mission had external instruments to measure the unshielded doses. Further, an analysis and comparison of the shielding characteristics between the Apollo hardware and the Curiosity in-flight hardware have never been conducted, so there is no basis for an analytical contrast between the two systems.

As the author, my considered opinion after all the information has been evaluated is that the Curiosity data was obtained over a period of nearly eight months using a state-of-the-art instrument designed specifically for the deep space environment. The Apollo data was obtained utilizing 1960s collection and evaluation technology from 27 manned astronaut opportunities (24 individuals total) in highly variable circumstances without employing any deconvolution analysis of the TLD data, thereby certainly skewing the high energy glow curve component, since it was designed to measure terrestrial ionizing radiation and not GCR.

So far, I have not heard of anyone choosing to apply only the Apollo GCR data to calculate Martian crew exposure probabilities because it is more "Mars mission friendly." But, although such a determination would be totally absurd, it would, none the less, be unsurprising typical.

Low-Earth Orbit (LEO) Missions – ISS

While the International Space Station (ISS) is flown below the Van Allen radiation shield, significant cosmic radiation exposure data has been collected over decades of continuous operations. The ISS GCR data from LEO is obviously the most extensive database ever accumulated on cosmic radiation beneath the Van Allen shield. While the dose is lower because of the effectiveness of the shield, it is significantly higher than at sea level on Earth due to the further shielding effects of the dense atmosphere of our home planet.

NASA has published a summary of these many years of ISS data in a very concise synopsis. The cosmic radiation (mixed GCR and SPE) experienced at ISS, orbiting approximately 240 miles above Earth in LEO, is (depending on Solar Activity) variable between 0.44 and 0.88 mSv/day, with an average of 0.66 mSv/day.[12] But note that the spectrum for the cosmic radiation exposure on ISS may be significantly different than that of deep space GCR (the distinctly unique health issues experienced by Apollo versus ISS astronauts are discussed in exhaustive detail later in the book).

Gemini – Skylab

Bookending the Apollo missions were Gemini (1965-1966) and Skylab (1973-1979). Here are their average cosmic radiation findings:

Gemini	0.42 mSv/day
Skylab	0.71 mSv/day

The Skylab data was skewed high by a solar particle event (SPE) during its final mission – Skylab IV – but it is still very consistent with ISS long term data.

Comparing the Data

Comparing all the mean exposure data in one snapshot:

Combined Mean LEO Data Beneath the Van Allen Shield

Gemini	0.420 mSv/day
Skylab	0.710 mSv/day
ISS	0.660 mSv/day

Combined Mean Deep Space Data

Apollo	0.469 mSv/day
Curiosity	1.800 mSv/day

Admittedly, for most people, data tables begin to blur together, and even statistical tools finally fail to make much cognitive difference in the attempt to resolve issues. But let us examine the accumulated data from another angle and see if it makes any sense.

As a young County Surveyor Assistant on the Oklahoma prairies of my Native American heritage, whenever it was even remotely possible, we would *always* begin every job by searching for and locating the United States Coast and Geodetic Survey Benchmark. All United States surveys are legally coordinated from these benchmarks. From there, local, state, and national surveys are oriented. In this case, I am going to select the gold standard GCR data reference, the Curiosity data, from which to draw my conclusions for all deep space cosmic radiation information. It is my benchmark for this discussion and for the remainder of this book.

Comparing Curiosity GCR Data to Apollo

Let us examine first the Apollo data from which there are a total of nine data points. According to NASA scientific literature, there is an internal variability of 20%. We will compare this to the Curiosity benchmark data.

Apollo daily GCR mean: **0.469 mSv/day** – a continuously moving data reference attached to the 27 constantly active and shifting Apollo crew positions for a total of 83.75 days.

Curiosity daily GCR: **1.8 mSv/day** – a steady state geometrically stable reference logging data for 254 days.

Difference: Apollo data calculated the GCR levels to be 3.8 times lower than the deep space benchmark.

A Long Look at the Data – A Critical Analysis

Obviously, recorded doses from all human space missions conducted beneath the Van Allen shield (extending to approximately 36,000 miles from Earth's surface) *must* yield cosmic radiation levels lower than the steady state cosmic radiation level in deep space outside the belts. If they do not, the collection and evaluation methodology has generated erroneous results, or there is a spacecraft configuration preventing the sensor from adequately recording cosmic radiation data accurately. This does not consider solar storms – solar particle event (SPE) – when it is possible for the Sun to unleash an event that may cause radiation levels to quickly climb past the GCR background. That was clearly the case with the Skylab IV mission in LEO when it recorded a total exposure of that crew to 2.04 mSv/day over 87 days. However, Skylab crew documents also report that, except for the solar storm peak, the recorded GCR more resembled the consistently recorded amounts reflected by all other LEO missions.

However, by simply lining up all the data on the same page, with conspicuous inconsistency, the record indicates that the Apollo deep space crew radiation exposure data is fundamentally the same as LEO data. And, when compared to the ISS, Apollo exposures were actually lower!

If you assume that the ISS data is the gold standard for LEO cosmic radiation monitoring for the same reasons that Curiosity data is the gold standard for deep space, then, when comparing Apollo deep space cosmic radiation to ISS, there is something fundamentally amiss. There is clearly an issue afoot that has not been discussed widely and, for all practical purposes, is fundamentally ignored by NASA's current documentation.

Humor me as I employ a logic tool known as *reductio ad absurdum* to the data sets. If the Apollo data is consistent and correct, then the Van Allen shield is not shielding anything below it at all, and, in fact, the shield is actually increasing radiation below it on a regular basis. That argument is fundamentally absurd, leaving only one viable logical conclusion: the Apollo radiation data *must be flawed* as it consistently recorded radiation levels 3.8 times lower than the benchmark data. Why?

As the author, it is my argument that the post-World War II 1960s era Apollo GCR capture and evaluation technology (TLD) was elementally substandard compared to Curiosity instrumentation and was not initially calibrated in its original design for cosmic particle energies but for terrestrial ionizing radiation. This fact must therefore account for the wide disparity. Further, the Apollo data was only consistent within itself to *within 20%*, lending even further doubts about its collection technology and evaluation methodology when compared to the relative gold standard of Curiosity. This has now been substantiated in the recent peer reviewed literature of cosmic radiation in the Lunar environment stating, "Measurements on the lunar surface performed during the Apollo mission cover only a small energy band for thermal neutrons and are not sufficient to estimate the exposure."[13]

Obviously, this discrepancy is well understood by NASA. But it would be helpful if the Agency would make a few annotations on the validity of the Apollo data that is still widely published by NASA without comment and has led to much public misunderstanding of the actual deep space radiation environment, particularly on and near the Moon.

21st Century Deep Space Lunar Cosmic Radiation Data

The seriously compromised Apollo cosmic radiation data set would represent a crippling loss of data for the Lunar surface and the near Lunar deep space environment were it not for the Lunar Reconnaissance Orbiter (LRO) Mission launched 18 June 2009. Included onboard the Lunar orbiting spacecraft was an externally mounted cosmic radiation device called the Cosmic Ray Telescope for the Effects of Radiation, or CRaTER. The CRaTER instrument was specifically designed and launched in preparation for future crewed missions to the Moon, and its data has become the gold standard radiation data for the Lunar environment. The orbiter's initial first-year exploration mission began in 2009, and subsequent scientific missions have been ongoing since the second year. Noah Petro, LRO's project scientist at NASA's Goddard Space Flight Center in Maryland, commented at the June 2019 mission 10th anniversary that all seven of its instruments were still collecting data, and "there's enough propellant left on LRO to continue the mission for at least seven more years, based on the current rate of fuel consumption."[14]

The CRaTER is an externally mounted, completely unshielded instrument (except for the instrument's detector covers) that has been directly sampling a continuous stream of cosmic space radiation since 2009. Indeed, it was the CRaTER data that identified the influx of GCR during the low point of the solar minimum of 2019 by detecting an incoming high energy radiation flux from outside the solar system at near record levels.[15]

Here is a summary of LRO's findings of the GCR levels in near Lunar orbital space and on the surface (estimated from orbit) from 2009 through 2019:

Lunar Surface GCR: 1.04 – **0.301 mSv/day** (per solar cycle)
Lunar Surface Average: **0.68 mSv/day**[16]

(Note – Curiosity recorded **0.67 mSv/day** on the surface of Mars)

Lunar Orbit Average GCR: **1.34/mSv/day**[17]

(Note – Curiosity recorded **1.8 mSv/day** in interplanetary space.)

Again, LRO GCR data is nearly three times higher than the reported Apollo data.

Racking and Stacking

The average background radiation for humans on Earth is 2.4 mSv/year – or 0.006575 mSv/day.[18] For our comparison, we will assign this a value of 1.0 Earth-Normal Radiation Exposure and relate this figure to the 21st-century GCR deep space measurements.

Deep Space Radiation Relative to One Earth-Normal per Day		
Surface of Earth:	1	x Earth-Normal Radiation Exp/day
Surface of the Moon:	103	x Earth-Normal Radiation Exp/day
Surface of Mars:	102	x Earth-Normal Radiation Exp/day
Lunar Orbit:	204	x Earth-Normal Radiation Exp/day
Interplanetary Space:	274	x Earth-Normal Radiation Exp/day

In the United States, the legal limit for exposure to occupational ionizing radiation for Nuclear Workers is 50 mSv/year. Again, for our comparison, we will assign this a value of 1.0 Legal Limit Radiation and compare this figure to the 21st-century GCR deep space measurements.

Deep Space Radiation Relative to Legal Limit USA per Year

Surface of Earth:	0.05 x Legal Rad Worker Limit USA/year
Surface of the Moon:	4.96 x Legal Rad Worker Limit USA/year
Surface of Mars:	4.89 x Legal Rad Worker Limit USA/year
Lunar Orbit:	9.78 x Legal Rad Worker Limit USA/year
Interplanetary Space:	13.1 x Legal Rad Worker Limit USA/year

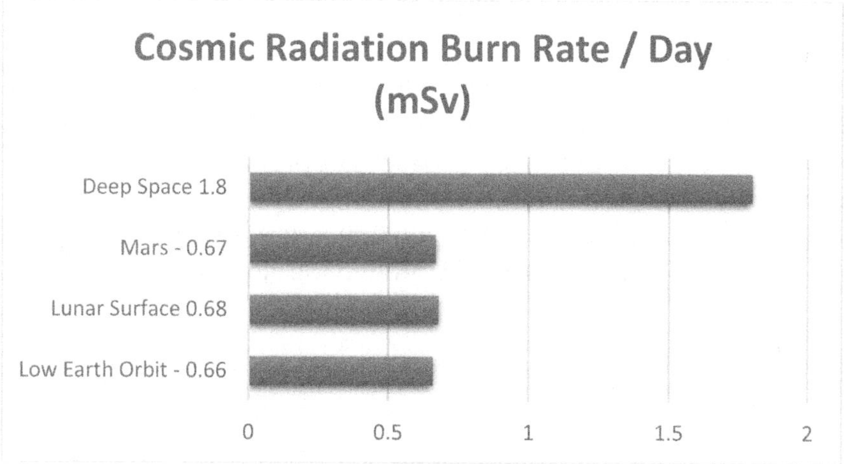

Daily GCR Burn Rate

As we have discussed, any galactic cosmic radiation received on the ISS in low-Earth orbit (250 miles) is pre-filtered and represents a significantly different spectrum than is represented 36,000 miles above the Van Allen shields. Thus, the average 0.66 mSv ISS daily burn rate must therefore be accepted as out-of-context data and with no small skepticism in the perspective of comparing it to GCR in deep space.

As is clear, GCR received in deep space is 2.7 times the rate that it is received in low-Earth orbit or on the surfaces of Luna or Mars. Thus, traveling in the open interplanetary void in functionally unprotected ships is 2.7 times more hazardous per unit of time than exploring on Mars or in LEO.

The Relevance of the Data

If this chapter does not trigger a "Houston, we've had a problem here" moment, then someone or some agency has been lurking near the government Kool-Aid kiosk for far too long. The European Space Agency estimates that "an astronaut on a mission to Mars could receive radiation doses up to 700 times higher than on our planet."[19] They rightly call this a "showstopper" – a certified trigger-word in today's space sub-culture. There is no historical precedence for this situation except, perhaps, for the 17th-century Jamestown expedition member who problematically discovered the native hatchet sticking out of the back of his head. This native artifact gifted in such a totally surprising manner was not exactly what he was expecting when he sailed west in absolute triumph and confidence concerning what they just *knew* lay over the misty horizon.

At least we have an abundance of warnings which, if the Internet space exploration forums are any indication, still may not be quite enough. But what is truly epically astonishing is how this real data and its inevitable biological sequel are actually spun into an opportunity for yet another round of very sketchy countermeasure challenges by NASA and other space exploration entities around the globe.

Lies, Damned Lies, and Statistics

"It is never too late to give up your prejudices."

— Henry David Thoreau

"For a successful technology, reality must take precedence over public relations, for nature cannot be fooled."

— Richard Feynman

Exploiting the Probability Game

THE TITLE FOR THIS CHAPTER IS DERIVED FROM THE enduring comment of British Prime Minister Benjamin Disraeli (1804-1881), "There are three kinds of lies: lies, damned lies, and statistics." However, this title, linked to Prime Minister Disraeli's quote, inches perilously close to finger pointing, and that is the very last thing I want to drift into. This book is not about accusation *at all,* but all about possibilities and intelligent exploration leadership that surmounts challenges based on a specific, unambiguous philosophical clarity mated with scientific and engineering solutions. Yet, no one can arrive at intelligent exploration decision making without first clearly identifying the false path one may be walking down and suggesting a new route up the ramparts to the endless frontier. But, in that humans tend to act like humans, even that dispassionate logic can, and almost always does, elicit an emotional backlash. This subject is no different at all.

In my 30-year career at NASA, I have never even one time met any scientist or engineer in the Agency who deliberately misled the public or their superiors in any way about *anything*. In total transparency, I did hear via well-placed sources of very limited activity of the sort through the proverbial grapevine, but I did not know the people who perpetrated it personally and, rumors being rumors, I had no idea of the actual validity of such gossip. Nonetheless, as far as my three-decade immersion in the Agency, I can confidentially assure the public that, specifically with respect to its individual employees, NASA is about as ethical as any organization in the world. I have never met a group of people so intelligently dedicated to the task of world class exploration and devoted to the advancement of humanity – even despite their quite ethically challenged Congressional masters. But NASA as a government entity has some deeply entrenched, self-serving, ethical issues, just like any other government bureaucracy, as this chapter lays out in detail.

As an example, the rather corrupt decisions that are being made related to the exposure of deep space flight crews to cosmic radiation are being conducted under the cover of a legal sleight-of-hand, which is the subject of this chapter. By taking this expedient route, the participants from multiple government agencies can somewhat legitimately claim a certain level of contrived innocence and probably successfully finger-point their way out in a public hearing after the fact when the future disasters are investigated. That, as multiple previous human spaceflight disaster investigations have clearly borne out, is NASA's ethical Achilles' heel and has never been adequately addressed, nor does it seem to ever change. As an agency, NASA successfully hides individual accountability and responsibility better than any other. And the Agency does it all in the open light of day – and even involves other groups – to obtain their requisite cover.

With that clear background into the prevailing ethos and culture at NASA, we can thus examine the reality of lies, damned lies, and statistics in space exploration without having to assign individual culpability – as the system is so carefully designed. In the end, it all comes down to exploiting the probability game to defend and even empower a mission objective. This is not at all about a secret plot to deceive. It is about a legitimate scientific process that has been legally distorted specifically to artificially prop up and protect an exploration philosophy that is in the process of wholesale breakdown. Indeed, such legal maneuvering would not even be necessary if the exploration model were not already in the self-destruct mode. After all, uncovering it and shining a bright light on that critically flawed process is the core motive for this book's existence – to expose the truth and save lives. If you are reading this and planning to get on board a ship outbound for deep space and another world, you really need to understand this chapter well in advance – as the life to be saved may be yours!

In this chapter, we will likewise examine one of the prevailing probable biological endpoints of cosmic radiation on the human body: cancer. Yet, concern for downstream cancer statistical probabilities is rightly eclipsed by increasing concern over pernicious radiological tissue degradation, discussed in greater detail in the next section in a chapter titled "Hitting the Physiological Wall." Keep in mind there has been effectively zero deep space research on any of this in either human or vertebrate animal model physiology. And, what few data points we actually have are not promising at all for sustaining the current human space exploration model.

Mandated by federal law in the assessment of cancer claims among radiation workers, the mathematical model that assesses legal and fiscal responsibility, and even culpability, 42 Code of Federal Regulations Part 81 (42 CFR 81), was created. From it, 33 cancer risk models were developed for single primary cancers, multiple primary cancers, and overall for all cases combined.[1]

While the law is concisely subtitled "Guidelines for Determining the Probability of Causation – Energy Employees Occupational Illness Compensation Program Act of 2000," it lent as much scientific clarity as humanly possible to the hitherto vague notion that "radiation exposure causes cancer." Through these government models, probabilities are rather precisely enumerated. Cases are routinely evaluated under the mathematical models, and financial compensation is clearly laid out on tables. It just could not get any more transparent than that! The government thus protected itself and the nuclear industry from bankrupting, billion-dollar judgments based on emotional testimony, persuasive lawyers, and "expert witnesses" with countless skewed models that no "jury of peers" would ever likely understand anyway.

And yet, even though 42 CFR 81 and 82 were carefully and vigilantly assembled by a virtual army of competent physicians, physiologists, and scientists of many stripes, any thoughtful individual is compelled to ask just how they were assembled. How were the formulae developed that populated the tables that assigned financial judgments to radiological exposure as related to the specter of cancer cause and effect?

Biologically, if any cell is irradiated with even a single "unit" of ionizing radiation, there is a probability, however slight, that it will biochemically provoke cancer. Statistically, there is no lower limit of exposure that is absolutely safe – *none.* Yet, countless millions of people have dental x-rays each year, and there is no definitive link at all between dental x-rays and brain cancer. A scientist or engineer would say, therefore, that the probability is very low that dental x-rays are the source of brain cancer. What they carefully *do not* say is that the probability is zero – since developing a fatal brain cancer from one dental x-ray of a single tooth is certainly a greater than totally zero risk. Statistically described, it is far safer to get the tooth x-rayed than it is to drive to the dentist's office. And physiologically, the combined statistical risk to your health is far lower receiving the x-ray than allowing any level of decay to exist in your mouth.

Cancer is a stochastic effect of radiation. That simply means "effects that occur by chance and which may occur without a threshold level of dose, whose probability is proportional to the dose and whose severity is independent of the dose."[2] In gentler terms, that statement reduces to this: the chance (mathematically captured as an enumerated probability) of an occurrence of cancer rises with radiation dose. Obviously, the energy associated with any exposure to ionizing radiation is not stored inside the human body – but the traces and damages it may leave behind certainly are. But that ends the radiation's responsibility. Once a cancer is generated – showing up today or four decades from now – it, unfortunately, takes on a life of its own. The severity of the cancer that has been generated is totally independent of the amount of radiation received. The rate at which the cancer advances, the patient's outlook and prognosis, associated pain, and all other aspects of the disease, including mortality, are not necessarily related to the initial flux of radiation the individual obtained. This is the stochastic effect of radiation.

Medical sources of ionizing radiation aside, we all live on a "slightly" irradiated planet. There are natural sources of radiation all around us that we call background radiation. Some geological locations have more of it than others. Some homes are built in places where a natural radiation source (radon) seeps up inside the homes and irradiates the residents 24/7. Radon progeny products (formerly referred to as "daughter products") are also a part of cigarette smoke. The National Cancer Institute estimates that 15,000 to 22,000 lung cancer deaths in the United States each year (up to 60 deaths per day) are related to natural radon and its progeny products exposure merely through smoking cigarettes and by exposure to secondhand smoke.[3]

As we have discussed previously, natural radiation is all around us. We even eat it as a part of our normal everyday diet. As was identified earlier, bananas contain a natural radiation source called Potassium-40 that can easily be observed on sensitive radiation monitors just after eating one of the yellow skinned fruit. Brazil nuts, lima beans, carrots, and potatoes are also measurably radioactive. Again, even the ubiquitous pile of some kitty litters is slightly radioactive because it includes bentonite clay, which contains small amounts of radioactive uranium (U), thorium (U), and potassium (K).[4]

Now comes the key question around which much of this book focuses its concern. How did the scientists link numerical probability to occurrences of cancer in 33 different cancer causation models? Here is the answer directly quoted from the law, 42 CFR 81:

"In 1985, in response to a congressional mandate in the Orphan Drug Act, a panel established by the National Institutes of Health developed a set of Radioepidemiological Tables. The primary source of data for the 1985 tables is research on cancer-related deaths occurring among Japanese atomic bomb survivors from World War II."[5]

There it is in a concise and clear statement. *All* cancer probabilities are assigned by observation of cancers that occur in unique populations by a process called "empirical modeling." It is not possible to assign *direct cause* because the causative action is submicroscopic and probabilistic. Therefore, the probability of radiation induced cancer can only be enumerated statistically. And the only possible method to link statistics to cancer is to "map" its occurrence as it actually happens in a given population. With that statistical map, you can intelligently infer causation based on a comparison to the general population. It is not perfect, of course, hence the liberal use of the word probability. No scientist ever assigned cancer probability based solely on the energy or action of ionizing radiation inside a human body without referencing actually observed cancers and comparing them systematically to many different factors in the process of developing the empirical model. Literally, there is no other way to get to the answer.

Obviously, it quickly became evident that the particular population of Japanese atomic bomb survivors was an inadequate single point data source for a multitude of reasons. According to the Guidelines for Determining the Probability of Causation, which is a part of 42 CFR 81:

"The risk models are further improved by being based on more current data as well. Many more cancers have been modeled in the revised report. The new risk models also take into account factors that modify the effect of radiation on cancer, related to the type of radiation dose, the amount of dose, and the timing of the dose. A major technological change accompanying this update, which represents a scientific improvement, is the production of a computer software program for calculating the probability of causation. This software program, named IREP (Interactive RadioEpidemiological Program), allows the user to apply the NCI risk models directly to data on an individual employee."[6]

This was mandated by the 1983 "Orphan Drug Act" (Public Law 97-414).[7]

I slogged through all this scientific methodology and public law in order to repeat a vital point made just above because it is critical to understanding your individual risk assessment in deep space:

"No scientist ever assigned cancer probability based *solely* on the energy or action of ionizing radiation inside a human body without referencing *actual* observed cancers and comparing them systematically to many different factors in the process of developing the empirical model."

Why is this single detail so powerful and important? Because NASA's estimates of deep space crew cancer risks are based on formulae and empirical models of ionizing radiation from *Earth based* empirical models. Why?

Because it's all they have!

NASA simply does not have data from full spectrum deep space galactic cosmic radiation on the human body. So, they use an exponentially weaker and qualitatively different species with which to guess the stochastic effect on humans. While the statistics hold up well on Earth using these terrestrial sources, they have a very limited and effectively unknown application to the vastly different cosmic sources and spectrum. How limited? No one knows – but they apply them anyway, with the full intent of sending crews out to discover the effects instead of vertebrate laboratory animals. This act alone defines the audacious collapse of ethics in NASA.

Currently, there are only 24 humans who have traveled for exceptionally brief periods outside the Van Allen shield. That population is too small upon which to build any reliable models. In order to make the ionizing radiation models fit cosmic radiation, the only tool the Agency has is to guestimate an increase in the projected incoming cosmic radiation energy by a single magnitude. In other words, they have no observational records from which they can measure their projections. So, in order to move forward and to satisfy the legalities of occupational safety, they have literally had no choice but to create a non-empirical high energy, high mass deep space model from a comparatively low-energy, low mass Earth based model. Which, by the way, is not even remotely feasible.

But it was all they had, *so they did it anyway* – even though it is very likely to be magnitudes off for many excellent scientific reasons.

The fundamentally serious problem with that approach should be more than obvious if you are considering boarding a spacecraft designed around an Apollo-era exploration methodology. You cannot accurately empirically model an effect for which there is zero data to derive the model. It should be painfully obvious to any disinterested observer that this back-fit is not only absurd but also functionally useless. It may satisfy the lawyers, but the scientists and – one would hope – the astronauts, not so much. Later in this book, I will share a relatively easy and inexpensive way out of this serious scientific dilemma.

The biological effects of galactic cosmic radiation cannot be compared to terrestrial ionizing radiation, as was discussed in the earlier chapter, "A Brief Primer on Cosmic Radiation." The two types of radiation are not the same, and their biological effects are radically different. To create a back-fit range of probabilities and then claim empirical modeling is, in the end, astonishingly misleading. The reasons for this are enumerated in detail in an upcoming chapter titled "Chronic Radiation Exposure Etiology."

Today, all ultimate decision making in the space agencies of the world on the limiting endpoint of cosmic radiation exposure is their versions of probability that a given exposure to cosmic radiation in deep space will eventually result in cancer to astronaut crews. What kind of cancer? There is a list of cancers assigned since ionizing radiation, and, by inference, cosmic radiation can induce many different varieties of cancer – from tumors to lymphoma to leukemia. While these cosmic radiation induced cancers have never actually been observed or formally diagnosed, it is clear enough that once radiation exposure is high enough, they will be. Based on the mechanisms of lower energy terrestrial radiation induced cancer, that is a very reasonable assumption – if you consider such use of human astronaut subjects ethical to collect data on a potentially fatal illness, which I do not, nor does *any* competent, non-conflicted bioethicist.

Obviously, cancer is certainly not the only risk of exposure to whole-body doses of cosmic radiation. So why is down-range cancer risk the key decision-making characteristic? Again, the answer is that what drives and ultimately assigns probability and risk to radiation induced cancers is empirical observation of outcome. But, as before, a legitimate, scientifically valid model cannot be developed until the population of observations is large enough. NASA proposes to use human astronaut subjects to collect that data in deep space because even the most egregiously conflicted government scientist regards the use of non-human primates (such as apes) as totally unethical for the same purpose. Such is the nature of the absurd breakdown of logic in massive bureaucratic systems.

For example, since there have only been 24 individuals exposed to this energy stream (the Apollo astronauts), we have very limited information with which to draw conclusions. In statistics, we would say that "the 'n' (sample size, number of data points) was very small." Further, their exposure was extremely limited. So limited, in fact, that with an 'n' of only 24, the data is not considered statistically significant.

The longest exposure outside the Van Allen shield was Apollo 17 at 12.5 days. In the short span of just 48 months, the world clearly witnessed Apollo crews make nine trips to the Moon and back surrounded by the full flux of cosmic radiation – and not a single astronaut was obviously affected in any visible way on their return to the home planet. All of them were jubilant, smiling, and happy when they arrived back on terra firma. It certainly seemed apparent that a few days of exposure to cosmic radiation had no immediate health effects whatsoever. Thus, in the only cohort of humans exposed to full flux cosmic radiation in history, the direct empirical observation of the outcome was that there were apparently no immediate effects – which was entirely accurate.

The only risk that remained in the Apollo era thesis was that long-term cancers or other radiation caused illnesses that may have been induced by the cosmic energy that streamed through the astronauts' bodies. That brief experiment very loosely defined the assignment of probability. Yet not a single scientist, physician, or engineer could possibly have been satisfied that Apollo was the final word on human effects of cosmic radiation. And yet, that empirical model has been established and persists even to this day – and now exists in the mind of the general public.

However, there were other fears. Many experts persistently continued to investigate the nagging fear, decades later, that there could be some significant long-term effects on the ultimate health of these courageous men – even after only a few days of exposure to deep space GCR. In the peer reviewed literature, two areas stand out: cosmic radiation induced cataracts[8] and cosmic radiation induced cardio-vascular disease – which, in the end, proved just as deadly as delayed cancer.[9]

In 2001, a research team led by Dr. Francis A. Cucinotta of the Radiation Health Office at NASA's Johnson Space Center in Houston studied 48 cataract cases in current and retired astronauts. The team discovered a significant increase in cataracts for those who had higher "lens doses" from space radiation. They also found those exposed to higher amounts of space radiation got cataracts at a younger age than astronauts who received lower dosages. The cataract study determined that there was an unambiguous connection to the Apollo astronauts who departed the shadow of the Van Allen shields for deep space. "We find a significant association between cataracts and high-inclination or lunar missions, with 35 of the 39 cases observed after space flight occurring in astronauts participating in these missions."[10] It was also Frank Cucinotta who subsequently stated in a 2010 article, "A newer finding is the large risk of heart disease death from radiation that appears in many exposed cohorts... the mortality risk for heart disease may approach that of solid cancers at least at older ages and research in the area will be important in the future."[11]

And in the study concerning cosmic radiation effects on the vascular endothelium of Apollo astronauts published in one of the world's most prestigious journals, *Nature*, it was determined that "the cardiovascular mortality rate among Apollo lunar astronauts (43%) was 4–5 times higher than in non-flight and LEO (low-Earth orbiting) astronauts."[12] And that was from exposure to only a maximum of 12.5 days out of Earth's protective sphere as compared to the astronauts who had orbited beneath it for hundreds of days! While the "n" (usable sample size) was low, the results have the leading scientific experts very concerned. Thus, in scientific terms, the "n" is not the problem. The lack of coherent, representative data on the effects of GCR above 36,000 miles is the problem going forward.

As this information is being released to the public nearly half a century after their exposure, not only are some medical professionals obviously becoming concerned, at least some of the remaining Apollo astronauts and their families are concerned as well. Further, there was no assessment of risk assigned to the Japanese nuclear bomb survivors for long term damage to their vascular endothelium because the effects of terrestrial ionizing radiation are not the same on the human body as exposure to the full spectrum of GCR.

The Apollo Astronaut's Widow vs. NASA

One evening after dinner, I received a telephone call from a retired NASA Astronaut Physician. At the time of the phone call, I was still an active NASA life sciences engineer and Chairman of the Kennedy Space Center's Animal Care and Use Committee. He asked me if I would do him a special favor as a professional colleague in the Agency, to which I replied, "Of course!" (I must protect names and times here because of the very sensitive and personal nature of his request.) He said that he had been speaking with a friend, the widow of a NASA Apollo astronaut who had previously walked on the Moon. She told him of what she considered a personal horror story and wanted to know if anything could be done about it so that it would not happen again to other Apollo astronaut families.

She shared with him that it was the long-standing wish of her husband that in the event of his death that his body be carefully examined in a specialized autopsy for the effects of cosmic radiation during his flight to and from, and time spent on, the Moon. Thus, on his passing, while he was still in the morgue, she attempted to contact someone from NASA to take care of the autopsy before she completed the arrangements for his funeral. She indicated that no one in the

Agency that she spoke with had any idea how to make that happen. She added that she also attempted to make the right contacts through her own Agency connections as well as through other active and retired astronauts and discovered that her request was simply not possible. In the end, she was forced to proceed with the funeral arrangements because NASA had no answer for her or any other Apollo astronaut family.

She remarked that even though that incident had occurred years before, she had been speaking with other Apollo era astronauts (all now retired) who unanimously stated that NASA must take care of this essential duty so that the invaluable medical data would not be forever lost. The Astronaut Physician asked me if I – inquiring as an active NASA employee involved in the medical contracts of NASA life sciences – could please inquire whether NASA would be willing to establish such an autopsy procedure for the surviving Apollo era astronauts specifically to discover whether they bore any permanent injuries from cosmic radiation that would appear in a detailed autopsy after their deaths.

I agreed to do what I could.

Of course, considerably less than a quarter second after I hung up, I was unsure of the wisdom of even taking on such a quest. In all the world bureaucracies, there is none more intimidating or steeped in royal protocol than NASA and the astronaut corps – particularly the Apollo era astronaut corps. While I was a career NASA engineer and space life scientist, I was not in any sense "positioned" to even suggest such a practice be adopted in the Agency, nor did I know anyone who was.

After a sleepless night, I realized that before I made any Agency calls at all, I would have to make some discreet inquiries outside the Agency. If I managed to set off any alarms by making the wrong call to the wrong office, I would be sitting in uncomfortable meetings for several days explaining myself. I had been with NASA long enough to know that you simply do not go up and poke the hornet's nest unless you are ready to deal with the onslaught and risk getting shut down before you even begin. The phrase "above your pay-grade" (the oft-used crutch phrase for upper-level management to avoid the responsibility of actually having to answer difficult questions) came to mind repeatedly.

My first email was to a colleague who, in my opinion, probably knew more about the effects of ionizing medical radiation on the brain than any other human, Dr. Dennis Steindler, former director of the McKnight Brain Institute at the University of Florida's Shands Medical Center.

After hearing of the dilemma, he agreed that such autopsies had the clear potential of discovering links to cosmic radiation damage that would linger and leave traces of pathology even decades later. He concurred with the assessment that it could provide invaluable data that otherwise would be lost. He also agreed to help in any way he could. At this point, I had two highly qualified scientists – the Astronaut Physician and Dr. Steindler – in agreement on a most critical medical procedure of great value to currently living future Mars explorers.

Yet, after numerous inquiries over many weeks, I ran into brick wall after brick wall. All of my Space Shuttle era astronaut contacts had no clue as to how to go about establishing such a procedure. They were also uncomfortable getting personally involved in such a sensitive quest. In the end, the consensus agreement was that this request belonged squarely in the court of the surviving retired Apollo astronauts and not outside their group. I reported my findings back to the Astronaut Physician, who responded that, according to his contacts, the Agency was essentially stonewalling them as well. The idea thus died between us with considerable disappointment in, and frustration with, NASA.

The latest research data now appearing in the most respected scientific journals on the planet tell a revealing back-story of their own. The families of the Apollo astronauts and Astronaut Physicians were trying to get the attention of NASA so that this valuable data that will directly affect the lives and wellbeing of a whole new generation of space travelers would not be lost. Indeed, I spoke confidentially with one researcher whose astronaut cosmic radiation exposure research ended up on the cover of one of the world's most respected scientific journals. He admitted that his work was triggered by the same line of questioning by the same individuals. But he also warned me that he could not remark on specifics and retain his Agency connections.

Did the astronaut family requests fall on deaf ears? Did NASA really stonewall them? Was NASA trying to hide something?

As I have stated clearly before, after a career in NASA, I am extremely confident that I know the answer to those three questions *very, very clearly*. The answer to all three queries is a resounding and unqualified NO! Indeed, such suggestions are utterly ridiculous fodder for internet conspiracies.

The request did not fall on deaf ears because NASA does not have any ears in that area to even hear the question! NASA did not stonewall the question. NASA was *literally incapable of answering it*. In order to conduct such a specialized autopsy, first (astonishingly), it would have had to occur to the NASA spaceflight physicians in 1960-70 that such a future autopsy would ever be necessary. Second, there would have had to be an approved procedure before the first autopsy even began. Third, it would all have to be continuously funded – over decades of time. Fourth, NASA would have to contract qualified pathologists to conduct the autopsies and keep them on contract for half a century. Fifth, what would the Agency do with the data? And sixth, Congress would inevitably have to get involved, further convoluting the issue and forcing a hue of politics onto this most sensitive process. That is the most unfortunate issue of all because it links directly to one of the Agency's greatest delicate concerns: the already super-sensitive politics of deep space exploration.

NASA certainly is not even capable of "hiding" anything. Such a thought is so ludicrous that it literally belongs (and usually is) on the cover of a supermarket tabloid. NASA employees vote NASA to be the Best Place to Work in the Government year after year after year for a reason. NASA is defined by its people. They are the best and brightest assembly of professionals in the world, and it is with some confidence that I have stated unequivocally that *everyone on planet Earth knows it*. In such an environment of the world's finest scientists, engineers, and technical professionals, hiding anything is not just unlikely – it is a ridiculous idea, period. And finally, just to put this issue to rest once and for all – NASA professionals like me eventually retire and write books! And, so far, there have been no men-in-black standing at my door… yet.

Still, NASA's people are, obviously, "led" by those in charge of them. And these leaders are just like any other subset of Washington bureaucrats, subject to politics if not altogether ruled by them. So, when it comes to the subject of deep space exploration, politics reign, just like in any other beltway adventure. It is a political reality that has been set into motion from the first moment Alan Shepherd launched into space atop his tiny Redstone. It has been slowly assimilated by the Agency, but also Hollywood and the public consciousness that has morphed into 21st-century NASA, space, and exploring.

And now that we have our sights on deep space once more, the power that drives the exploration of another planet is the power of expectations. It is the cumulative inertia of tradition that has established this moment in history to make the crucial decisions just ahead that will drive us forward and create the space architecture that will spell either success or disaster. What we do and how we do it in the next several years will either define a future of productive, healthy astronauts, or we will build a system that blindly but intentionally and permanently injures them. If we proceed on without changing our course, many of them will certainly die, while many more will return to Earth permanently injured and maimed.

The key idea here is *intentional*. The Agency knows the risks very well and is relaxing the safety rules specifically to protect their archaic mission model and *not* protect their crews, clearly knowing and openly enumerating the number of their astronauts who will eventually die specifically because of their flawed model. So how does that fact fit together with my best and brightest statement? Remember the comment made in a peer reviewed journal, "The large ionization power of galactic cosmic radiation (GCR) ions makes them a major health threat to astronauts and constitutes one of the most important barriers impeding plans for interplanetary travel by crewed spacecraft"? This warning has been issued in a very astute and professional manner – as a published, for the record comment in a public journal – and not just once. But the caution is separated by a vast bureaucratic gulf from the decision makers who agree that it is *acceptable* that a percentage of astronauts *will be killed indirectly by cosmic radiation*. These decision makers are absolutely insulated from the tragic results and any personal responsibility of their decisions by massive layers of bureaucracy. They even have a NASA acronym for their death-SWAG: the Risk of Exposure Induced Death, or REID.

In the inevitable investigations after the equally inevitable disasters (just as was clearly described in the Apollo 1, Challenger, and Columbia accident investigations), NASA the bureaucracy is and has always been a world class cover-your-ass organization with no names that can be traced back later to an individual or individuals in charge of decisions involving the lives of crews. That may sound like a bold statement, but research it for yourself. Which individual or individuals in the Agency were held responsible for the deaths of 14 Space Shuttle astronauts? In the decades since the Challenger and Columbia disasters, no one was held responsible – even though the investigations uncovered many individuals who made incompetent decisions that led directly to the deaths. And nothing in NASA has changed.

Elon Musk is quoted as stating with dead-on-accuracy, "So, yeah, there's a tremendous bias against taking risks. Everyone is trying to optimize their ass-covering." Thus, I will extend this challenge to any reader – show me one document that features a signature by an individual that authorizes and takes full responsibility for the current radiation exposure limit and results of any subsequent injury for U.S. astronauts. Good luck with that. I suggest looking in the crate underneath the Ark of the Covenant in the anonymous government warehouse.

At least a partial answer to the astronaut widow's question came far too late for her or the other Apollo families on 21 March 2017 with the passage of the To Research, Evaluate, Assess, and Treat (TREAT) Astronaut Act. The purpose of the TREAT Act is to authorize occupationally related lifelong medical monitoring, diagnosis, and treatment for former astronauts who have completed at least one spaceflight mission. However, it does not mention autopsies except for mishaps and accidents.[13] And, of course, this act *does not* cover private astronauts on private space missions.

Establishing Reasonable Cosmic Radiation Limits

All cosmic radiation dose estimations and recommendations are based on studies and statistics of ionizing radiation augmented by a "quality factor." In my career as a scientist and engineer, we call these "SWAG" factors – aka Scientific Wild Ass Guess. In establishing cosmic radiation limits and projecting human injury, there has never been a more appropriate use of the term SWAG. In the concise words of the United States Center for Disease Control, "We don't know what causes most health problems that could be linked to cosmic radiation. We don't know what levels of cosmic radiation are safe for every person."[14]

This statement by the United States CDC is vitally important:

"We don't know..."

Thus, it is blatantly and professionally dishonest to project cosmic radiation limits and exposure effects using terrestrial ionizing radiation models in the nearly complete absence of any scientific evidence on the actual short term and long term effects of exposure to cosmic radiation. And yet, if you review the information and professional papers of any international or private space exploration entity, you realize that they all use the same conclusions based on the same terrestrial ionizing radiation exposure modeling.

Again, there is no conspiracy here. The simple fact of the matter is this: without the SWAG, *we have nothing at all*. Today, as we look into the void, there is only the SWAG that we have to cling to. There just is no biological exposure data. Without the ridiculous SWAG on which we base all our projections of the effects of cosmic radiation on our deep space astronauts, we would only have a shrug of the professional shoulders with which to project our future mission planning. That, of course, is totally unacceptable. So, we gin up the tables using whatever we have available, with a near-fuzzy zero understanding of its probable accuracy.

As a NASA professional for three decades, it has always been absolutely mind blowing how staggeringly shortsighted such behemoth agencies can be. NASA literally moved mountains and commenced design and construction of the Space Launch System (SLS) and the crew system Orion. They embarked on this monumentally expensive project intended specifically for deep space with *zero human data* on the effects of long term exposure to deep space cosmic radiation. They invested tens of billions of dollars for the system but literally nothing at all for even a modest mission to investigate the most essential of all the questions about the actual feasibility of what they are building.

How did they get this far without a flag being raised by any actual official at 300 East Southwest Street in Washington, DC (NASA HQ)? Because they had the SWAG. The SWAG is permission. The SWAG implies knowledge. But, in the end, the SWAG is still just a SWAG. All the equations, all the tables, all the Japanese terrestrial ionizing radiation data are just unrelated data on which to build their house of cards and justify their SWAG.

Yet the actual legitimate accuracy of their SWAG to reality in deep space is as different as a terrestrial beta electron is to a galactic cosmic primary with a mass of 1,837 times greater traveling at near light speed slamming into a crewmember's visual cortex and taking out a neuron with a flash of light evident in the core of their brain. But a NASA bureaucrat will never have to "see the light" from the safety of their NASA headquarters offices.

Has it ever occurred to anyone in charge at 300 E SW Street that no one should actually launch human beings into deep space based on a SWAG?

Today, there still remains a perplexing, almost complacent, *laissez faire* attitude toward interplanetary mission planning that is currently underway to make crewed expeditions possible beyond Earth. It is astounding that no one has yet apparently seen the handwriting on the wall and come to the realization that the time for a radical change in exploration policy has finally come to pass.

But – they have not!

This is stunningly reminiscent of the Virginia Company's Jamestown planning. The appalling corporate arrogance bred by the specter of empire building led them to cavalierly accept the sacrifice of human lives to gain their foothold on the frontier, again, based on the near total absence of information. The truly tragic part of this repetitious historical cycle is that none of the deepest implications of what they did, nor what they are currently doing, almost certainly ever even occurs to the mission planners.

In the 17th century, adventurers were literally lined up to head off to the New World in search of glory and riches, with dreams implanted and encouraged by those who did not actually give a tinker's damn whether they came back or not – only that their dreams of wealth and empire were satisfied with the lives and bodies of the youthful dreamers. In the 21st century, the astronaut application acceptance rate is one in many hundreds.

Clearly, nothing has changed.

But we do not have to repeat the historic tragedy of Jamestown. Today, we can bypass that terrible example simply by gathering the data we need (before human launches) and shielding the crew from exposures to galactic cosmic radiation and microgravity that we already know will harm them. There is a way to answer the open questions and ethically design for adequate shielding – even against cosmic radiation – which is one of the major subjects of this book.

Derived Statistics

The birth of statistics is traced back to Londoner John Graunt in the mid-17th century. The subsequent science of statistics was popularized in 1770 by the German political scientist Gottfried Aschenwall, and Sir John Sinclair is credited with introducing statistics in English use in 1829. Aschenwall's science dealing with data about the condition of a state or community and Sinclair's numerical data collected and classified were specifically employed so that data could *never* be misrepresented. However, often statistics are purposely used today expressly to mislead, misdirect, and obfuscate. So how is it even possible that such a pure science could be hijacked and manipulated when it was barely even a century old, and such abuse became legend? The blame can be laid upon a seedy branch of statistics informally known as derived political statistics. These are statistics that are developed particularly to lead to a predetermined conclusion. Hence, British Prime Minister Benjamin Disraeli's statement is more than just a maxim, for indeed, "There are three kinds of lies: lies, damned lies, and statistics."

But when it comes to statistics and assigning probabilities pertaining to cosmic radiation exposure, there is a well understood and relatively straightforward methodology that is essential to comprehend. It is based on cosmic radiation data collected in deep space and exposure of actual vertebrate animal subject models in the genuine environment. (I mention such an endeavor established at Lagrange point 5 in "Why We Will Depart Earth Forever – Reason 2," touch on it in the upcoming chapter "Chronic Cosmic Radiation Exposure Etiology," and discuss it in full detail in the upcoming Module Four chapter "Making It Right.") The outcome of such a series of experiments would no longer be a derived statistic but an actual one. In just a year, we would intelligently transition from guessing and endangering to knowing far more than we do today. We could then make decisions based on actual gathered knowledge and develop a new set of tables based on authentic vertebrate animal exposure to the full spectrum and dose of GCR.

In a very small first step effort toward that end, a team comprised of members from NASA Ames and Johnson Space Centers has developed a Life Science Flight Project experimental payload called

BioSentinel with funding provided by Advanced Exploration Systems within the Human Exploration and Operations Mission Directorate. BioSentinel is one of 13 CubeSats scheduled to launch aboard Artemis I for a flight past the Moon and into an orbit around the Sun. It will be the first CubeSat to run a biology experiment in deep space during its 12 to 18-month mission. The researchers involved were right on for recognizing the immediate necessity for gathering valid data about the effects of GCR on astronauts in deep space. Regrettably, the currently predominant bureaucratic powers that be in the Agency and congressional funding channels severely limit the advancement of requisite deep space research. Instead of launching vertebrate animals to investigate a thorough physiological range of responses to microgravity and the synergistic effects of the added exposure to full spectrum cosmic radiation, NASA has decided to launch this investigation on the cheap – literally.

Although CubeSats are so tiny, low cost, and lacking in complexity that some have been devised and launched by high school students, BioSentinel is designed and will be monitored by legitimate life scientists. Unfortunately, instead of NASA investing the required resources to discover the effects of multiple exposures of GCR and microgravity on vertebrate animals and their complex physiology, BioSentinel is sending not vertebrate animals but *Saccharomyces cerevisiae* (yeast cells) to gather an exceptionally narrow band of accumulated DNA damage data on the single most important questions in human space exploration. Although the nanosatellite mission states that "human cells and yeast cells have many similar biological mechanisms, including for DNA damage and repair," and collected "data will be available for validation of existing models and for extrapolation to humans,"[15] the data returned from BioSentinel will be far too little, too late, and too removed from actual vertebrate subjects for determining the final questions that NASA will ultimately task its astronauts to discover at the direct risk of their lives and health.

If I were going to board a rocket for Mars tomorrow, it should be an ethical prerequisite for the scientists managing my mission to explain cosmic radiation and its effects to me in detail so that I know exactly what I am getting into based on actual data collected in deep space on something more than yeast cells! Instead of dazzling and reassuring me with extrapolated statistics, I want to delve a layer or two deeper than the DNA double strand breaks and repair mechanisms in unicellular beer fungus. I would demand to know just how these statistics were derived. And when it comes to human lives, this is certainly *not* an academic exercise. In the 21st-century astronaut corps, I have come to understand from private conversations that questions such as these are typically not openly asked, nor should they be asked of the "wrong people" (astronaut management), because it may be interpreted as timidity or a lack of trust in leadership, thus limiting the inquirer's future flight opportunities.

However, the good news in all of this is that there is another clear path forward that will guarantee success in our quest for interplanetary human exploration and settlement. But the successful path is certainly not going to be a political one, nor one based on past traditions. Instead, it will be founded on an intelligent and applied philosophy of science and space exploration that has yet to be employed.

In the end, the widow of the Apollo astronaut must have eventually come to realize that NASA did not provide for the request of her husband because NASA *could not*. The Agency was simply not ready for the question that should have been obvious to them first – of all people on Earth – 42 years earlier. Unfortunately, it was only made obvious by what, to the Agency, should have been monumentally embarrassing research articles released just two years after my involvement in the dilemma. But NASA did not have eyes to see nor ears to hear in the 1960-70s. And today, it is no different. NASA and private space exploration enterprises still know well where all this ultimately leads, as evidenced by the ever-increasing published research findings. Yet they all insist on pressing to MECO (Main Engine Cut Off) on an interplanetary trajectory using an antiquated exploration model with an inadequately protective ship filled with human subjects whose safety is predicated using fungus research data.

In the 21st century, as in all previous centuries, statistics can still prove to be damned lies when the best of starting out assumptions have been morphed into today's ridiculous deep space focused SWAG. What the future astronaut community needs most desperately is an actual seat at the table when it comes to deciding their ultimate fate. What *are* the actual effects of cosmic radiation on the human body and mind? How many precious lives are now poised and ready to fly out on a deep space SWAG? There are many government and private crews ready and eager to fly to the Moon and Mars who have been told just enough – but not nearly the whole story. Why? Because someone is hiding the information? No. Because no one has the information. And apparently, not one responsible agency or entity wants to spend the necessary resources to discover it by sending mice and rats before launching their astronaut crews to find out. It is somehow easier and "more ethical" to send out flight crews than animal cohorts. Besides, the humans have willingly signed away their legal rights to sue for the injuries that the Agency already knows are likely to occur.

It all represents a collapse of government ethics that is truly astonishing. Not since the ethical meltdown involving the Tuskegee Syphilis Experiment (detailed in an upcoming chapter, "Ethics and Blind Engagement") has the Unites States government acted with such openly audacious disregard for human subjects under its purview. At least at Tuskegee, the government hid its motivation. But in the 21st century, it brazenly advertises them in plain sight, just as the Academy of Medicine has aptly called NASA out for its open and very public failure of ethics.

In order to escape our self-administered prejudices, we need a way out. And that way out, as mentioned before, is through a deep space mission designed to answer the question using vertebrate animal models – *not yeast*. We need to back up and truly, honestly, understand exactly how we got where we are today, what philosophical drivers caused the current state of affairs, and understand the new, empowering philosophy that will literally propel us safely and intelligently into the infinite ocean of deep space.

Are we really in a monumental hurry to get out there before we fix the problems accrued over half a century of human spaceflight? If so, then we are in a deliberately calculated rush to use human beings as research subjects in a very dangerous gamble that may well take decades to resolve inside their bodies. To intentionally endanger the lives of humans and their long-term health based on lies, damned lies, and statistics is unprincipled by any professional standard.

A Pretense of Knowledge

"Nothing is so firmly believed as that which is least known."

— Michel Eyquem de Montaigne

"We are all born ignorant – but one must work hard to remain stupid."

— Benjamin Franklin

ON THE COLD, CLEAR EVENING OF 11 DECEMBER 1974, in Stockholm, Sweden, a thin, bespectacled but distinguished Anglo-Austrian economist by the name of Friedrich August von Hayek faced a hushed and expectant international crowd and began to deliver his Nobel Prize address. But while the innocuous professor looked anything but intimidating, his message to his peers and, by extension, the rest of the scientific world was anything but mild. In soft but resolute tones, Professor Hayek took his peers behind the proverbial woodshed and gave them the lecture of their lives. Indeed, nearly five decades later, the power of one of his major ancillary findings still reverberates as loudly as it did on that cold, dark evening in Stockholm. Friedrich August von Hayek was awarded the Nobel Prize for economics that night, but he also levied a major correction and a clear warning to his peers that applied to every scientific discipline just as much as it did to the discipline of economics.

While he spoke of his epic economic theory that would change the way nations would project their economic health, he also addressed something else more sinister – more elementally corruptive to science – which was deeply entrenched in his field and others. He called it a "Pretense of Knowledge." Like an insidious digital trojan and worm to a computer, this pretense of knowledge had crept into economics, was integrated into computerized models and complex analyses of economics worldwide, and had led to catastrophic failures in understating the actual health of the world's economies. Hayek squarely blamed a philosophical error he labeled as "Physics Envy," which directly led to a pretense of knowledge in global economic theory.

Monetary scientists attempted to develop complex, inherently chaotic economic models based on more predictably stable fiscal principles and laws, like those of their peers in physics. The physics models nearly always behaved as predicted. Conversely, economic models almost always failed outright or predicted very poorly. The difference, said Hayek, was the deep complexity and extreme variability of the economic models that did not at all apply to the more straightforward physics models.

But the resultant faith in the flawed models led to a psychological chimera that Hayek identified as the pretense of knowledge. He declared bluntly, "We have made a mess of things." He later stated that the illusion of a pretense of knowledge was "downright dangerous" and "it is an approach which has come to be described as the *scientistic* attitude – an attitude which is decidedly unscientific in the true sense of the word, since it involves a mechanical and uncritical application of habits of thought to fields different from those in which they have been formed."[1]

Hayek's statement also concisely describes the ongoing state of affairs in 21st-century human space exploration. We have seized on a pretense of knowledge about the long-term effects of exposure to cosmic radiation and simultaneous long-term exposure to microgravity – astonishingly – without *any* coherent or systematic actual scientific data on the impact to vertebrate physiology in the full beam of cosmic radiation in deep space. It is a clear and unambiguous illustration of a caviler *scientistic* attitude. This indiscriminating approach is decidedly unscientific in the true sense of the word since it involves a mechanical and uncritical application of a combination of wishful thinking and wholly unrelated guesstimates of hodge-podge approximations of probabilities based on the exposure to *terrestrial* ionizing radiation by atomic bomb survivors of World War II and leveraging the low-Earth orbit, filtered radiation experience, microgravity, and subsequent multimorbidities to the physiologically unsampled and unknowns of deep space. In essence, the effects to astronauts of exposure to deep space cosmic radiation and microgravity are based on fundamentally undelineated, unresearched opinions, which are, in turn, based on incongruent data used to design spacecraft systems and human exploration models. It is, in Hayek's words, *scientistic* and invalid as any respectable resemblance of actual science. Further, it is being leveraged to commit human lives to unethical research opportunities on behalf of the government seeking human health data outside otherwise rational, principled, standard biological data collection methods. And based on all of this accumulated scientistic babel, we are now ready and planning to commit the lives of astronaut crews "to see what happens." We find ourselves perilously near the inevitable endpoint of ethical collapse.

In the 21st century, as in the last, we have once again made a mess of things.

In a very real sense, we have developed, as Professor Hayek stated, an uncritical application of habits of thought to fields different from those in which they have been formed. Speaking as a professional nuclear engineer, NASA space life scientist, and bioengineer of macrosystems, no one should ever be so foolish as to actually build and project outcomes for a wholly unique cosmic radiation paradigm by applying the projection of physiological injury from a terrestrial ionizing radiation model. These two models are so distinct that they *must* stand independently! Such a comparison is not only absurd but also perilous for those lives to which it is applied. While some of my peers have utilized various SWAG factors to attempt to reconcile intercellular energy transfer (biological damage) in the absence of real data, it is an approach that is distinctly *scientistic* and, frankly, "downright dangerous." It is a clear example of a pretense of knowledge – and it is ultimately an indefensible model. When applied to humans sent into deep space, it is also absolutely, unjustifiably unethical.

How do we correct this misdirected course before someone, or many are injured or killed by our pretense of knowledge and hopelessly flawed and inadequate models? The solution is so painfully and embarrassingly simple that it defies common sense that humans can often be so stubbornly irresponsible and intellectually oblivious. Especially when the right answers are astonishingly straightforward to obtain without using human subjects in a hazardous, predominantly unknown environment.

It is corrected by… *actual data gathered in the actual field*! And, in so doing, by devising a stand-alone, independent model of the systemic biological effects of cosmic radiation and microgravity collected in deep space on vertebrate animal subjects – not beer yeast! *Nothing could be more straightforward!* If this simple fact does not grab the attention of NASA policy makers, then they are absolutely beyond hope. And yet, from NASA Headquarters and across the globe – *crickets.*

It is difficult to wrap my scientific and engineering mind around the current, unfolding scenario. The world is now totally enraptured with sending humans outside Earth's protective shield to Luna and Mars for long-term and permanent occupation, continuously exposed to microgravity and functionally unshielded from full spectrum galactic cosmic radiation. We are collectively building trillion-dollar systems to make this bizarre situation a reality. And yet, except for a few *scientistic* personal opinions on the major showstoppers (cosmic radiation and the conjoined multimorbidity of microgravity), not a single dollar has been spent on *any* planning, construction, or deployment of vertebrate animal models outside the Van Allen belts in deep space to actually understand the real data before we send humans into the radiation field. We seem to be absolutely willing to risk human lives on a hunch based wholly on a pretense of knowledge to gather this first data set on a human model that cannot be dissected and studied internally immediately on their return.

We already have some wholly inadequate data based on the Apollo astronauts, as I have previously mentioned. The flawed data set notwithstanding, after only *days* in deep space, they displayed a significant increase in pathological responses of the body's fast growing, sensitive tissues, such as a statistically significant prevalence of cataracts. Worse, according to the journal *Nature,* "Apollo lunar astronauts are four to five times more likely to die from cardiovascular disease (*which is the disruption and long term damage to the body's endothelial tissues* – definition added) than astronauts who never left Earth's orbit or who never flew at all."[2] The cause, according to the research? It was writ in large letters on the title itself: *Possible Deep Space Radiation Effects on the Vascular Endothelium.* While these effects could conceivably also have been caused by exposing their intellect too close to the functionally unshielded confines of NASA Headquarters, is it not worth a few dollars and the relatively small effort (in space mission terms) to discover the truth before we send whole crews out for more than just days?

It is mind blowing that the Jamestown effect is still resolutely (and with great inertial abandon) driving the human space exploration train four centuries later. The *scientistic* approach – described by Hayek as "scientific hubris" – is alive and well in the human exploration of space. And it is as dangerous today as it was when the *Constant, Godspeed,* and *Discovery* pulled into Jamestown and delivered the first victims of a 17th-century pretense of knowledge to a wholly alien shore that they speciously believed they thoroughly understood.

Thus, we must recognize that a pretense of knowledge is part and parcel of the 21st-century philosophic stoichiometry of the unchecked disaster now lying so close aboard. And how does this stoichiometric mix behave? Just like casually tossing 160 grams of pure sodium metal into a glass of clear water: explosively. Remember this explosion as we now pop on our hard hats and headlamps and lower ourselves into the 21st-century catacombs filled with the blind alleys of human space exploration.

The Failure is Not an Option Malpractice

"One reason for this (the documented miscalculation of risk to astronaut crews) may be an attempt to assure the government of NASA perfection and success in order to ensure the supply of funds. The other may be that they sincerely believed it to be true, demonstrating an almost incredible lack of communication between themselves and their working engineers."

– From Richard Feynman's Minority Report to the Challenger Disaster Investigation

"The most rigid structures, the most impervious to change, will collapse first."

– Eckhart Tolle

BLIND ALLEYS BEGET OTHER BLIND ALLEYS. A POORLY designed machine, even if it is a prototype, will eventually begin to vibrate and rattle itself to pieces on more than one axis. As we witness the very significant vibrations of an international human space exploration program breakdown, it is now relatively easy to look back and see early signs of its volatility that unambiguously anticipated a future disaster in the making. Yet, the first steps in exploration, just like the first steps in designing a new machine, always begin with the prototype. In this case, the prototype was the Apollo Lunar Exploration Missions model. And it worked beautifully. Its historic performance was nothing less than absolutely brilliant – its sheer luck notwithstanding! But it has incontrovertibly reached its useful design limit and requires immediate replacement before it comes apart in deep space and begins to injure and kill flight crews.

Learning everything we need from the first model is always a normal part of every iterative design process. Reaching the end of the first discovery phase and compiling a list of all that went wrong – or what almost went *terribly* wrong – is not at all a failure. It is simply part of the normal learning process of all successful cultures and authentically intelligent species. Eventually, the intelligent design process iteratively self-corrects as its engineering and science team plans and builds a more capable machine. Any competent design engineer desires to construct a model from lessons learned that make the next generation much better. And so, they redesign, retool, and reinvent. No intelligent and gifted designer is ever married to a trifling prototype, no matter how many years of effort they have poured into it. The creative act is a journey, never a destination. Ultimately the creator dies, perfectly happy to hand their eternal process over to the next generation...

Unless, of course, the creator is not an individual but a bureaucracy with the hulking mass of a brontosaurus limited by a walnut sized brain.

Iterative design evolution requires a process of iterative design evolution. *One would think.* And it certainly would never look like the previous model two generations removed but now on steroids.

Thus it was in September 2005 that NASA Administrator Michael Griffin heralded the new generation of deep space exploration machines as "Apollo on steroids." And with that comment, he unwittingly doomed the next generation of space exploration and their crews by introducing the deep space Titanic that ultimately morphed in the sausage factory into SLS-Orion. He and NASA's leadership dramatically failed for several reasons simultaneously. They failed to grasp and employ any coherent exploration philosophy whatsoever. And they failed to apply any lessons already learned in deep space toward a new exploration engineering model – instead regressing backward to the obsolete Apollo (albeit on steroids) paradigm. Notably, they were almost certainly also unaware that NASA ever employed an exploration philosophy at all. Many were PhD engineers, with no concept whatsoever in the essential nature or reason for the "Ph" in their degree titles.

But to be completely fair to Griffin, in late 2005, many of those essential lessons that would doom the Apollo model as a viable exploration paradigm still lay as much as a decade away from discovery in the long term findings to be realized aboard the International Space Station that was still years away from completion and itself already morphing into a kludged together albatross designed by serial government committees. It is highly unlikely that anyone in NASA leadership then or now has any clue about NASA's subliminal exploration philosophy and how it inevitably, powerfully drives all their mission objectives and hardware design.

If the example that follows was not so dangerous to the future of human exploration and crews awaiting launch into deep space, it would actually be hilarious on so many levels. But one of the more glaring illustrations of a deteriorating exploration model was actually foisted onto the space exploration movement by a movie, and very few people – even inside the space business – know anything about its history.

"Failure is not an option," as used in the blockbuster movie *Apollo 13,* has a fundamental, basic meaning, which is, "It is absolutely not acceptable to lose a flight crew." It is impossible to find an argument against that core sentiment. Hence, it has become a kind of unarguable tenant worthy of the stature of inerrant holy scripture. Because of the context of a Hollywood rewrite of human space exploration history, its deepest meaning has become, inside the NASA culture and the culture outside NASA, that "human life has infinite value." Therefore, no effort is too extreme, and no expense is too great to ensure mission success and that every flight crew comes home safely to their families. Thus, this effort of true inestimable value has been framed with these five Hollywood generated words. In a 30-year career with NASA, I have never met a single individual who would argue with that emotional statement. So then, why do I declare it to be malpractice by the seemingly imperious title of this chapter?

Many of the reasons for my view are clearly stated in the pages that follow. But those arguments are put forth as a secondary dispute against the actual downstream financial, human, and exploration costs of Biblical-stature aphorisms matched up and morphed to engineering systems that are then unlinked to the real world, as we have clearly discussed. But when someone espouses a sacrosanct philosophy of near religious significance, religious fervor is sure to follow. Hypocrisy and its accompanying shadow of deep cultural bias also inevitably trail close behind.

On two tragic occasions when spacecraft with human crews of seven astronauts each were lost, the entire human space exploration system shut down until a period of national mourning was over, the President spoke to the nation, investigations were convened,

Congressional hearings were held, NASA Administrators and Center Directors were reassigned or shuffled, and two years of engineering fixes were retrofitted. (Mysteriously – no one was fired… but that is another story.) This cycle is repeated after each disaster quite obviously since failure is not an option, and it is absolutely not acceptable to lose a flight crew because human life has infinite value. The tragic national cycle is thus framed, articulated, and enacted by the underlying philosophy based on the forged Hollywood phase, "Failure is not an option."

Now back to planet Earth and the real world…

If an Airbus A380 with 853 souls on board, including men, women, grandmothers, and infants, goes down at the end of a major airport runway on the very same day as the seven astronaut flight crew is lost, the result is radically different. Flights on adjacent runways often continue, and as soon as the bodies and wreckage are cleared from the affected runway, more aircraft depart from it as quickly as possible. The same is true on our national highway system, where in 2021, more than 40,000 Americans were killed. That is a rate equivalent to nearly 16 Space Shuttle crews *per day*. I have personally witnessed human blood and brain matter being washed off the asphalt with a fire hose so that the roadway could be reopened as soon as possible.

Yet is it not equally true in our national civil aircraft system and on our highways that failure is not an option, and it is absolutely not acceptable to lose a single passenger… *because* human life has infinite value? Is it not also equally true that no effort is too extreme, and no expense is too great to ensure that every passenger in the air and on the nation's highways come home safely to their families each day?

Hmmm… Is your "radical disconnect" alarm going off yet?

Obviously, because of what actually happens after each of the different accidents, there is a completely separate standard at work here. So, what is the difference? Following an accident on the highway or an aircraft crash involving hundreds of innocents, why are not the entire global airline and highway systems shut down until a period of national mourning is over, investigations convened, Congressional hearings are held, high level officials are fired or shuffled, and two years of engineering fixes are retrofitted before the next plane can fly again, or cars be reengineered, or the nation's highways opened back up for automobile travel?

Are astronaut lives, therefore, intrinsically worth more than other people? Are they elevated and special above the people who pay for their missions?

Again, I can solemnly promise you that not just in the American culture, but also in the hallways of every NASA Center, and in Congress itself, that answer is also easy: of course not! Show me one single NASA employee or astronaut who would make a claim to this level of self-worth, and I will show you a certified ass completely unworthy of their honored position in the employ of the United States government. So why the disparity of cause and of freely spoken intent, and why the dramatic differences in effect? This book is devoted to uncovering that answer. The fault lies squarely in the execution of a fatally flawed exploration philosophy – or the lack of a coherent one – that drives costs, assigns relative actual cash values to real human lives, encourages grotesque cultural hypocrisy, and one that eventually forces the systemic failure and breakdown itself, all as a result, ironically, of a Hollywood fantasy phrase's ultimate real-world effect.

Out of Balance

The philosophies driving crewed spaceflight are absolutely out of balance. NASA is busy peddling the self-aggrandizing Hollywood fantasy that failure is not an option with the unhidden implication that it is so big and so good that it cannot fail – and even refuses to fail. But, at the same moment, Elon Musk advertised in 2018 and again in April 2021 that the expectation of his space exploration company was that a significant fraction of his initial Mars-bound crews could very well perish on the first few missions.[1] The disparity of balance is vast, with unnecessarily tragic consequences the result of both flawed approaches.

NASA somehow claimed ownership of the expression failure is not an option after the Ron Howard movie *Apollo 13,* which was based on the autobiographical book *Lost Moon* written by Apollo 13 Commander Jim Lovell. The term "failure is not an option" was in fact *never* spoken by the real Mission Control Manager Gene Kranz at any time during the actual 1970 mission and was not even mentioned in Lovell's book. It was coined by scriptwriters Al Reinart and Bill Broyles *just for the movie* 25 years later.[2] The problem with this sliver of revisionist history and co-opting Hollywood expressions into an exploration philosophy is that these ideas are innately powerful and eventually drive actual engineering events and expectations in the real world.

In one sense, it is true that failure is not an option:

Failure is not an option, because...
failure is an inevitable certainty!

Failure is defined as a statistically derived outcome of cause and effect in any human engineered system.

Unfortunately, the movie phrase has become a NASA slogan and a statement of bravado and arrogance. Indeed, it was the phrase's founding purpose that was subsequently overlaid backward onto history. The fact is, failure in any human endeavor is inevitable – even with the real Apollo missions. Indeed, when the Apollo 8 Mission Director Chris Kraft went to meet with Astronaut Frank Borman's wife just prior to the first mission to the Moon, he confided to her that the chance of her husband even returning back to Earth at all was not good. The exact conversation went like this:

Susan Borman asked, "Can you give me the odds?"

To which Kraft replied, "I think there's a 50/50 chance that they'll come back."[3]

Hollywood slogans aside, the fact that Apollo was so astonishingly successful was due to authentic hand-of-God intervention, because the probability numbers associated with the actual engineering crapshoot called Project Apollo were actually never in their favor at all on any of the historic missions!

No organization is built and sustained, nor is its reputation enhanced by denying reality because of a grotesquely overinflated Hollywood derived self-ego or any other human fantasy. Use of the failure is not an option slogan is a kind of engineering malpractice since the ultimate cost of ego and its accompanying illusions are always extraordinarily high. And the greatest loser of all is the engineering system that is irreversibly locked into its primitive state by not being "permitted" to fail or allowed to evolve. And, yes, every engineering system risk can be bought down and not permitted to fail by design –

with the right amount of money and time. Why do you think SLS-Orion has a price tag that would literally bankrupt most nations on Earth?

In real human space engineering, risk must be literally bought down with dollars and time. Risk buydown in NASA is like in no other organization in the world. Hence, by adopting a zero-failure tolerance philosophy and then explicitly sloganizing it and openly advertising it to your customers (the American taxpayer), the already disproportionate cost of human access to space in the investment of both dollars and time is increased by orders of magnitude. The cost of risk buydown in crewed space exploration is nonlinear. At the far end of that spectrum, every risk probability point that was bought down increased mission costs and was also accompanied by an attendant loss of mission time. The result was many more times the projected costs of both time and money. Risk buydown is one of the primary reasons why NASA has been methodically pricing their participation out of crewed space flight and stretching their timelines further and further to the right.

Eventually, the army of safety and quality assurance personnel, and the avalanche of policies and procedures that accompany them, are so vast, and the system is so moribund that its cost to function and time between flights becomes prohibitive even for a government agency. And how is it that such disproportionate spending is acceptable? The entity that NASA works for (the United States government) merely expands into another round of deficit spending whenever its funds run out. Throwing away $4.1 billion rocket systems after only one use or buying down risk at the end of an exponential curve is always affordable when the money supply is functionally endless and financial decisions are based on pork barrel politics instead of fiscal responsibility.

Implosion of the Old Model

Thus, by taking the ego bait, NASA is in the process of blindly but deliberately pricing itself out of the crewed space exploration business. It is more than a little ironic that the Hollywood fantasy catchphrase that failure is not an option may be the principal agent of failure itself. In the end, the expression is, likewise, a philosophic malpractice – one that the engineering, science, and acutely socially sensitive (but palpably philosophy challenged) Space Agency is almost certainly not even conscious of.

The contrast between the Elon Musk style and the NASA style is radically night-and-day different. Musk is a world class millennial genius, to be sure. Not only does he possess a keen engineering sense, but he is also just as adroit at managing the pubic and public expectations – even of the inevitability of failure.

(1) NASA: The going in assumption in everyday NASA is that they are too big, too experienced, and too good at what they do to fail. Since failure is not an option, they employ their access to unlimited resources to buy down risk. And since they are the only game in town, they also operate day-to-day like they actually have unlimited time.

(2) SpaceX: In lean, efficient, and wildly innovative SpaceX, one key method of radical innovation is to deliberately learn from programmed failure. Therefore, failure is not just an important design process; it is a vital engineering procedure by which new systems and processes are developed. It is also true that at SpaceX, failure is not an "option." Instead, at SpaceX, failure is a process methodology.

The NASA style is clearly one of being "too big to fail," feasting off their astonishing and historic past accomplishments – which they legitimately earned. But their old-school heritage and extraordinary, colossal, expensive bureaucratic baggage have driven them to extreme engineering conservatism and a serious aversion to any risk at all. Such distaste for risk can be justifiably described as astonishing for an organization whose business model resides exclusively at the far end of the extreme risk zone.

NASA would *never* hold a press conference and openly declare as an Agency prime objective that they would be happy if their new flagship rocket just cleared the launch tower high enough so as not to take out the pad when it blew up, as it had a high probability of self-destructing seconds after launch. And yet, in July 2017, Musk said just that at the news conference before the maiden voyage of the first Falcon Heavy launch carrying his personal Tesla roadster to an orbit beyond Mars.[4]

This exceptionally unique SpaceX style unambiguously reflects the personality of its sole billionaire owner and single ultimate decision maker: edgy, trendy, courageous, absolutely unapologetic risk taker who publicly and frequently advertises the direct opposite of the failure-is-not-an-option mindset. SpaceX openly, even brazenly, publicizes its naked risk-taking and, in the process, brilliantly manages it by laying it out there just as it really is – wide open, vulnerable, very human, and very much like real life in every respect. The public does not recoil at this. It excitedly embraces it as real life and authentic exploration.

Yet, the contrast between NASA and SpaceX is not as simplistic as one might reflexively expect. NASA is a massive, layered, federal bureaucracy with tens of thousands of moving parts. It burns through its 2022 $24.8 billion budget at $67.9 million dollars every day.[5] NASA's decision making literally begins in the White House, filters down through the National Space Council to Congress, onto NASA Headquarters, and then down to one of seven NASA Centers spread out all over the United States. From there, it gets really complicated. Government oversight and coordination of a literal army of contractors and subcontractors are initiated, and the daily activity of the nation's space program is administered and managed.

For every decision made in NASA that actually filters down to some tangible piece of hardware, there are literally thousands of connections to that hardware from members of Congress to NASA bureaucrats to contract engineers. And all of that attention is focused on every nut and bolt, which in turn involves high-end salaries for a combined NASA civil servant and contractor workforce of about 80,000 employees (approximately ¼ NASA government employees and ¾ contractor).[6]

SpaceX is not nearly so byzantine – nor anywhere near as expensive. In a 26 February 2021 court filing response to a Department of Justice request, SpaceX stated it had over 9500 employees on its payroll.[7] At SpaceX, the exact same piece of equipment makes it to the launch pad costing many times less. Onto every piece of equipment at both NASA and SpaceX is the priced-out cost of risk, and the layers of risk management are not free. The more layers tacked on before launch, the more it costs. The smaller the organization, the fewer layers are added, and the cost is reduced. Ultimately, the number of layers of risk buydown is defined not just by risk management layers but also by the entity's philosophy of risk.

It is fascinating to watch the very dissimilar human space exploration paradigms come together with NASA contracting SpaceX to be one of the human launch providers to the International Space Station. NASA injects its failure is not an option philosophy and cost into the process, but the government funded contract pays for all of it.

Fortunately for everyone (especially the taxpayer), the net result is that the task is, in fact, accomplished with a tangible degree of increased effectiveness by implementing the lean SpaceX approach even though a part of it is still encumbered with the NASA philosophy.

The significant question that remains in all of this is one of balance – one best illustrated by that very odd-couple NASA-SpaceX relationship. NASA cannot maintain its failure is not an option fantasy unless they are planning their own obsolescence, unwittingly or otherwise. At the same time, it would be astonishing to think that SpaceX would actually integrate its difficult, dangerous, good-chance-you'll-die/guaranteed-excitement-for-those-who-survive thinking into its engineering process for human spaceflight.[8] Such malpractice would match NASA's on the opposite end of the risk spectrum.

But because NASA's self-adopted failure is not an option Hollywood philosophy has now become literally institutionalized, and because SpaceX's attitude is decidedly not risk adverse, it is my view that Elon Musk is merely deftly managing expectations for reasons of long-term corporate survival. But I also believe he is continuing his practice of cutting edge, even historic, engineering innovation and not actually establishing a philosophy of phased catastrophe when human lives are onboard his craft. Cavalierly killing off his flight crews is a guaranteed pathway to bankruptcy and loss of reputation.

Ultimately, is the NASA crewed space program essentially too big to fail? Time will tell as all these questions are in due course driven by each of the individual entity's exploration and risk paradigms. This also includes Congress, whose members are, collectively, almost certainly utterly clueless to the precise definition of its own exploration philosophy, as reflected by its Space Agency. If it is possible, the ones who control the purse strings are far more clueless than the ones executing the endlessly morphing, extravagantly expensive human missions to nowhere.

Nonetheless, there clearly is a way out of this morass of a dangerous, antiquated, low IQ, Hollywood celebrated exploration paradigm anchored firmly in tradition but lacking even a modicum of common sense. These fictions are propped up only by government and contractor billions and a Space Agency shackled by Congressional intervention that seems to attempt to teach science to the public through sloganistic cinematic wisdom bearing its logo.

But there is a way out.

And that way appears to be via the fresh and vigorous intervention of the sheer might of private enterprise that still flourishes in our capitalistic culture. Private enterprises will not and cannot waste their life's blood on Hollywood slogans, tradition, and anachronistic technologies when their very existence is at stake. Only private contractors that are compensated by the government in cost-plus, award fee driven enterprises and are paid to do it the government's way are permitted the luxury of pursuing such monumentally and extraordinarily financially wasteful endeavors. But, for them, the waste is passed along to the taxpayer by permission of the fundamentally unaccountable congressional dullards who literally write such waste into law.

Even as this book goes to print, radical, new advances are being made in technologies paid for through private ventures that will guarantee human access to deep space and establish a permanent human presence out of low-Earth orbit. These advances will be carved into human history as layers of powerful, innovative, and fresh creativity that will overlay the old and empower the new. It has already begun.

It was not long ago that Claudia and I watched the inaugural launch of the Falcon Heavy from Cape Canaveral as its two first stage boosters dropped out of the sky, returning from the edge of space accompanied by supersonic booms, and landed in formation, simultaneously, side by side, just a few miles away from us. I realized at that precise moment that human space exploration had permanently changed. The old models were dead or dying. All other exploration architectures, including NASA's over-budget and over-schedule Orion-Space Launch System (SLS), were rendered totally obsolete on 6 February 2018 at 3:45 p.m. when the first SpaceX Falcon Heavy with Starman in his Tesla Roadster aboard launched from historic Launch Complex 39A at the Kennedy Space Center. But as the next chapter details, "totally obsolete" does not at all also necessarily mean useless. As bleak as its future may appear, SLS-Orion still has a dramatically clear shot at its own splendid contribution to human expansion outward into the cosmos, as is described in detail in the next chapter.

The new human space exploration model is now in view. But only if those who dare to send cars and robots into interplanetary space fully understand that with the technological revolution, there must also be a philosophical revolution that accompanies it. Otherwise, we have merely created new technology that will ultimately guarantee the same certain failure at a lower cost. Only this failure will not be measured in flight hardware and launch pads – but in human lives.

The Intelligent Engineering of Risk Buydown

In the end, every person on the national highways, passengers in commercial and private aircraft, individuals in the care of physicians and hospitals, and people in spacecraft all have a statistical risk of inadvertently being victimized by the very organizations and systems that are charged with their care and safekeeping. Risk comes with waking up every morning. But there are two distinctly different kinds of risk: risks that *can* responsibly and ethically be bought down and those risks that definitely *must* be mitigated. The culture is undeniably obligated to intelligently engineer risk buydown and diligently enforce and mitigate risks that can be engineered for safety in every case.

But in the case of the "Failure is Not an Option" Agency called NASA, accepting significantly higher risks than industry while purposefully failing to correct the exploration model that is inevitably destined to cost lives is unethical and immoral. Thus, the Agency is now in the very peculiar position of *deliberately elevating risk* so that failure actually becomes inevitable – which they openly advertise is not even an option.

The duplicity is breathtaking in every sense.

And in the case of SpaceX's deliberate and surprisingly ingenious "yippee-ki-yay, mother f-ker," crash and burn approach to developing spaceflight hardware, the fun and games all end when human crews strap in these interplanetary machines. Like Apollo's Kris Kraft, Elon Musk is clearly attempting to precondition the public's psyche for the ultimate failure – the loss of a human crew in deep space. But no amount of greasing the skids will ever make any difference after the history-making, catastrophic fiasco of such a tragedy, especially if the loss of the crew was due to a failure to design for the pre-known hazards of galactic cosmic radiation appended to the multimorbidity effects of microgravity and other space exposures that could have been mitigated pre-flight. Whoever winds up departing Earth in deep space deathtraps (regardless of who designs and builds them), the probability excuse will not fly with the public, expert witnesses, or juries. Together, they will condemn and judge any level of this kind of deliberate criminal negligence beyond low-Earth orbit, just as a British Judge ruled against and dissolved an equally culpable Virginia Company in May 1624.

So, as we glanced back at Jamestown in the 17th century, let us now cast our eyes forward to the future and understand that just as Virginia was an alien place, so are Luna and Mars and the vast *spatium incognito* between worlds, if not even far more so. Over four centuries later, we cannot afford to make the same mistakes, and we likewise cannot afford to import an Old World philosophy into a new world. Death and failure always lie close aboard in any human exploration venture. And yet, it is critically essential that we learn not to intentionally invite death and failure onboard by jumping off the cliff of Hollywood inspired foolishness without a clear plan for building a new set of wings on the way down.

NASA's well respected, historically significant deep space progeny – Apollo – has finally, after half a century, birthed its own new generation of deep space offspring: SLS-Orion. Yet, as many next-gen family trees are wont to do (especially family trees with no branches), this new child of Apollo can best be described as NASA's dysfunctional red headed stepchild on steroids. The preeminent question is, can NASA afford the therapy to circumvent its predilection for self-destruction? And yet, what is it about this unlikely progeny that still holds so much historically significant promise?

SLS-Orion

NASA's Dysfunctional Stepchild on Steroids

"There is noting so useless as doing efficiently that which should not be done at all."

- Peter Drucker

Insanity is doing the same thing over and over and expecting a different result.

— Albert Einstein

The Exploration Model with a Life of its Own

THERE IS AN EXPRESSION THAT HAS ENDURED FOR AT least a century that states bluntly that many problems "take on a life of their own." This is equally true of personal lives as well as enormous federal bureaucracies, and especially epically massive and expensive federal projects. It is, in fact, an artifact of human nature engendered by various vague mechanisms of human psychology and survival strategies. Ultimately, this artifact is described as a cognitive bias – a powerful capacity of the human brain to exclude the flow of logical information that would eventually suggest more appropriate corrections in behavior, or a course that has become emotionally established and one to which bureaucracies and people have become inflexibly committed by rote legacy.

As we orbit slowly over NASA and the world's space agencies and look down on them from the advantage of the distance and total silence of space, we will see thousands of the very best and brightest people, all extremely busy and engaged in a set of tasks that are aligned perfectly with a well-defined, historic, highly successful mission history - but a not-so-well defined and largely unfunded future mission – the human exploration of Luna and Mars. The people engaged in this venture generally assume that their previous mission model is going to fit perfectly (nicely back-fitted with a nearly endless string of mostly vague, merely promised yet untested countermeasure fixes) and will ultimately permit them to explore outward into deep, interplanetary space without a significant, mid-course correction. They have become so convinced of the supremacy of their patchwork paradigm that they have not bothered to do basic science and collect adequate data on the environment toward which they are chomping at the proverbial bit to man up and launch outward! Yet, while there is abundant evidence streaming in real-time to the contrary, these individuals persist in continuing to burn through massive resources, establishing serial milestones linked irrevocably to obsolete hardware, and investing in human capital for which they have actually documented a pre-planned certain level of inevitable injury and death stemming from their collective deliberate incompetence. They even attempt to "normalize" such unethical policies by openly advertising and utilizing the statistically derived acronym REID (Risk of Exposure Induced Death.

This process of intentionally creating a system with all the defining characteristics of a multi-billion-dollar juggernaut that cannot possibly succeed in the long term (based on the data returned from deep space robotic missions and its sheer extravagant costs) is utterly illogical. Indeed, the level of its absolute absurdity is breathtaking, especially since an organization with a stellar, world class reputation is leading the charge out to certified crew disaster, giving themselves the REID hall-pass permissions. Yet, it is ongoing today without pause. How can the very best and brightest of humanity deliberately paint themselves into such a corner? Or, by yet another completely apropos cliché, so industriously saw off the very limb that supports them? It is because the historically successful Apollo model of space exploration – NASA's inordinately triumphant older son – has blindly taken on a life of its own, creating a personal, very real, and impressively powerful cognitive bias. It is one so potent and resilient that it will either require a historic tragedy of equal and offsetting proportions, or it will necessitate an extraordinary and absolutely uncommon level of self-examination to change course now. We are literally locked into an endless feedback loop defined by the success of the past exploration model and are sealed into its ultimate deployment by the billions of dollars invested in the architecture of the Space Launch System (SLS). Changing course now will be monumentally difficult and, if history is any judge, as each day passes characterized by obscene resource burn rates, altering the course is also extremely unlikely.

Apollo on Steroids – With Contraindications

We had a perfect opportunity to make the necessary course changes in our human space exploration model in 2005 when NASA Administrator Mike Griffin announced the follow-up program to the United States Space Shuttle. The Space Shuttle experiment itself barely supported *any* of the hypotheses it set out to establish at its inception. It proved to be preposterously expensive – not even close to the cost-effective access to space that was advertised early on. Indeed, the U.S. Space Shuttle missions became the most expensive individual events in human history. Launch for launch, they cost the taxpayer an average of nearly half a billion dollars per event.[1] But, not to be outdone in financial extravagance, SLS-Orion has already replied, "Hold my beer," estimating each event to cost $4.1 billion per launch – more than a billion dollars more than a Seawolf Class nuclear submarine – and nearly all the hardware is tossed into the ocean as waste after less than 10 minutes of use.

The Shuttle flights were anything but routine, as had been hoped for. Indeed, Space Shuttle missions were considered dangerous. The Shuttle Space Transportation System (STS) killed 14 astronauts in two crashes, which involved a loss of 20% of the Shuttle fleet and represents an astronaut loss rate of nearly two percent. Comparing that to daily air travel, a two percent loss rate would equal the lives of 160,000 people per day or 58.4 million people per year! While it is unarguably and abundantly true that all spaceflight is hazardous (as of this writing), the justification for the Space Shuttle program was that the STS was specifically designed to change that. It did not. Indeed, it proved its originating hypothesis resoundingly false at a relative staggering cost in human lives and resource capital.

At the end of the Space Shuttle program, the predominant engineering response to the lessons learned was to deliberately retreat from the massive, hideously expensive, airliner size, most complex technological system ever built to something far more simple and far safer. Administrator Griffin described it as "Constellation – Apollo on Steroids."[2] The Constellation program was to become a deliberately, very Apollo-like space launch system designed to take humans out of low-Earth orbit and establish a permanent crewed Lunar base then move outward to Mars.[3] Constellation was literally a hardware blend of the most reliable components of Space Shuttle and Apollo.

The Constellation system was the intentional reduction of a logical engineering response to the very difficult lessons learned from Shuttle. Smaller, leaner, less complex engineering architecture often does work better – at least to some degree. But Constellation was also a conscious, unambiguous revival of Apollo-era space exploration technology and methodology. More than just capitalizing on relative Apollo simplicity, Constellation was unabashedly exploiting Apollo's historic success and fame – both human celebrity and a well-earned reputation of what was apparently (though not in reality) a relatively safe, reliable system plan. But clearly, in 2005, not a single thought was evidently given to the two showstoppers outside low-Earth orbit: cosmic radiation and prolonged exposure to microgravity. And that was for a good reason. At that date, the International Space Station was still six years from completion and had not yet begun to gather the real-time data that uncovered the actual, documented limits of human adaptability.

But if the human capacity to squander billions of dollars on cognitive biases was not enough, just when the Constellation program was ramping up, presidential administrations changed, and Constellation was abruptly canceled, wasting another $13.1 billion and an additional decade on a politically motivated false start.[4] To be completely fair to the new President, Barack Obama, the ultimately ill-conceived Constellation program had in its first few years already run far over budget and was behind schedule. Thus, NASA conveniently handed the new President a ready-made excuse to cancel it, as he had already promised he would do in the opening days of his campaign.

For a period of nearly a year, NASA was without a follow-on crewed space mission, paying exorbitant and ever-increasing fees for rides to its own space station from the overjoyed Russian space Uber. (It is whimsically reputed that when the Russian Space Agency Roscosmos was asked how they slept at night for blatantly and unashamedly gouging the Americans for rides to their own space station, they responded, "On piles of bags full of American money with many beautiful women.") Meanwhile, the hapless NASA juggernaut waited on the Administration and Congress to come up with a Constellation follow-on plan. All the while, NASA had to maintain its massive bureaucracy staffed by the world's finest brain trust of tens of thousands of individuals waiting patiently for their leadership to somehow figure out what the hell was actually going on.

As I shared previously, during that period, I privately cornered a NASA Space Center Director (a former astronaut of multiple Shuttle missions) and asked respectfully if he could articulate whatever the new vision was sorting itself out to be. He looked me in the eyes as a fellow NASA employee that deserved a straight, non-Chamber of Commerce answer and said bluntly but honestly, "Dennis, we're going to Mars. NASA is going to Mars. That's what we're doing. Not an asteroid. Not the Moon. Mars."

In the intervening months and years, his words rang in my head over and over. His blunt confession ran completely counter to the plans that would be unveiled and funds committed over the next six years. But it was not at all a surprise to me. I was aware that the "deep-state-NASA" thrived well under the radar. It outlived, and out-maneuvered, mere administrations and congresses that came and went. What he had privately revealed to me was the actual goal of the deep-Agency over the long term. A decade later, every word he spoke was coming to pass and all the other programs (just as he confidently assured me), even before they were formally announced, were dead and history.

Out of the perpetual fog of Washington DC's indecision, the Orion Crew Exploration Vehicle was somehow resurrected from the ashes of Constellation. But to save political face in an attempt to permanently bury any connection whatsoever with the previous administration, Orion received a long surname: the Orion Multi-Purpose Crew Vehicle (MPCV). The barbarous dogma of political purge always demands a complete cleansing of the previous administration's ideas and creativity, including names of inanimate objects. While it was still exactly the same Orion crewed space capsule, it now had a new name: Orion-MPCV. How the crewed capsule-still-on-steroids first name was saved out of the madness is anyone's guess.

It was resolutely established by Kennedy Space Center locker room gossip that Senator Bill Nelson had President Obama's ear and that he singlehandedly salvaged the NASA crewed space vision, and specifically Orion itself. Two administrations later, with Nelson now the NASA Administrator, I never once had any doubts that the rumors were entirely accurate. While there was some internal NASA grumbling when Nelson used his Congressional clout to fly on Space Shuttle Mission STS-61-C, in my personal opinion, in the years to follow, he more than earned his ride to space many, many times over. It is one of the few times I have witnessed a congressional junket work out extremely well for the taxpayer and the nation in the end.

Except for an evolving engineering pathway, it was exactly the same Orion crewed space capsule – the same Constellation-Apollo on steroids – with an identical mission to depart low-Earth orbit for deep unprotected space. But it was now coupled to a newly designed biggest-rocket-ever delivery hardware called the Space Launch System (SLS).

Thus, in the divorce of Constellation and subsequent remarriage, SLS-Orion stepped into the immediate dysfunction of the inherited U.S. Government family genome, already awash with much predilection for familial insanity. SLS-Orion thus became the red headed stepchild – underappreciated, awkwardly accepted, and even wearing the elder brother's used hand-me-down Space Shuttle orbiter RS-25 rocket engines. In order to compel the stepchild to meet the family's magnified view of their reputation, they even found a suitably disreputable beltway physician to put him on a course of steroids to pump him up and prepare him for the rigors of deep space.

SLS-Orion's Unseemly Formative Years

After the embarrassment of aborting Constellation and dumping its very expensive but useless carcass into the watery deep off Cape Canaveral at the exit of Bush II and the entry of Obama, the renamed MPCV stepchild space capsule still called Orion somehow survived the whole affair, its fundamental purpose virtually unchanged. Yet, by the time of its official renaming, NASA still had no discernable mission in deep space, and the reigning presidential administration certainly had no intention of even mentioning the former administration's ambitions – much less pursuing them to any degree whatsoever. So, in the second term of the Obama administration, the administration found busy work for the awkward stepchild hanging out from the last marriage: "Go up in the attic and trap a bat, why don't you?" That should keep him busy, out of trouble, and mostly out of sight. Significantly few space activists, and even fewer American citizens, even remember this phase at all.

The bat-in-the-attic mission to which SLS-Orion was assigned was called the Asteroid Redirect Mission (ARM)/Asteroid Retrieval and Utilization Mission (ARU). As NASA ginned up scores of PowerPoint presentations for the bat in the attic mission, President Obama confirmed an unexpected reversal. In October 2016, just a single month before Donald Trump was elected President, Obama unexpectedly formally announced that the United States was headed back to Mars, and NASA added that one of the fundamental purposes of the ARM/ARU asteroid venture all along was to train and prepare NASA astronauts for Mars exploration.[5,6] It was just odd that no one happened to mention that connection until after the Mars mission was announced!

Surprising absolutely no one on the planet, in keeping with all past incoming presidential administrations, the Trump administration quickly pulled the plug on the asteroid mission in December 2017, relieving even the most ardent space fans and thus paving the way to give SLS-Orion something vastly more important to do. Months prior to that, on Tuesday, 20 March 2017, less than two months after entering office, President Donald Trump signed a new law authorizing almost $20 billion in initial funding for the Space Agency for the 2018 budget year. For the first time, the bill actually included money in place of rhetoric to develop human exploration of Luna and Mars as one of NASA's objectives. Named clearly as the prime exploration vehicle for all human deep space operations was SLS-Orion.

On 14 May 2019, in the early days of the Trump Administration, NASA's radically redesigned new mission was thus named Artemis, reflecting the Agency's self-enabling addiction to their glorious Apollo past (in Greek mythology, Artemis was the sister of Apollo). Thus, the dubbing of the new mission pathway retrieved the image once more of an Apollo on steroids exploration system that was headed for the inevitable unshielded and unprepared reckoning with the most powerful forces in the universe. One unfortunate and rarely appreciated caution in naming a human space exploration system after Greek Gods is their capricious temperament. In the stories of the goddess Artemis, she is described as a compassionless and vindictive force to all who crossed her path.

One of the Greek myths involves arrogant King Agamemnon, the chief commander of the Greek forces during the Trojan war. In Sophocles' *Electra,* as the assembled army, along with Odysseus, prepared to set sail to wage war on Troy, Agamemnon bragged that he was a better hunter than Artemis herself. In the process, the King insulted and angered Artemis. She lifted her hand and stopped the winds required to direct the King's forces to Troy, then added a plague for good measure. She would only permit the fleet to continue after a terrible price was demanded from the King – the sacrifice of his daughter, Iphigenia. Agamemnon agonized over the decision until he finally surrendered and slaughtered his daughter on the altar of Artemis.

Unfortunately for contemporary NASA, Artemis remains as compassionless and vindictive as ever and will mercilessly extract lives for carelessness and arrogance toward the immutable laws of the universe on which humans are set to sail.

But in the early part of a new Presidential Administration, the awkward, red headed stepchild on steroids was now officially out of the attic and standing in full view of the American public, on a new interplanetary mission renamed Artemis, having been handed a historic directive without any equivocation. Quickly, the Mars bound mission evolved into the development of a Lunar orbital outpost called Gateway, and a long-term human presence on the surface of Luna itself on the way out to Mars – a proposal President Obama had scrapped in January 2010, just a year after entering office for his first term.

The Potentially Unethical Use of SLS-Orion

But, just like any course of steroids, there are always contraindications such as steroid induced psychosis. And, if not administered appropriately, there will ultimately be dire consequences for a super-Apollo space exploration architecture such as the SLS-Orion, which is ultimately designed for long term missions around and beyond the Earth-Moon system. The new SLS-Orion mission directive forced broad systemic assumptions on an engineering system originally designed for mission durations of only a couple of weeks beyond Earth. But, in lockstep with the space exploration philosophy conceived during the first days beyond Earth's atmosphere, the Agency blindly projected the elemental supposition onto weeks, months, and even years of exposure of humans to the deep space environment that would thereby invariably demand nearly limitless human adaptability. They have literally up-designed a 1960s quick-look crewed deep space mechanical system hyper-charged with steroids yet with extraordinarily little thought given to the potentially long term devastating effects of the environment on the individuals inside the functionally unshielded, microgravity locked hardware. From day one in the U.S. space race exploration model, the crews were expected to fully adapt – and it was never a mere expectation. It was a literal demand levied upon the soft tissues, exquisitely sensitive neural network, and carefully balanced biochemical systems of humans. Crew adaptation was a proviso that was assumed in the design of every nut, bolt, and wire of each of our spacecraft – and is likewise reflected in the latest 21st-century spaceship version on steroids.

It may be startling to discover that even from plans developed during WWII, Wernher Von Braun's vision for the crewed space program revolved around an orbiting station – literally. Von Braun briefed his recommendation for a 1-g spinning wheel to the U.S. Army in 1946, complete with sketches, and the concept was familiar and popularized to the American public over the next 6 years. Although the discovery of the potency of deadly galactic cosmic radiation, and even the existence of the Van Allen Belts themselves, were still to be realized, it was a foregone conclusion for Von Braun that any space station would employ artificial gravity for practicality and crew health safety.[7] Even his plans for a future Mars voyage and eventual colony included spinning the attached ships in the plane of the longitudinal axis during the prolonged journey.[8] Von Braun and his NASA contemporaries brilliantly developed the specialized Apollo hardware in order to fulfill President Kennedy's directive to get a man on the Moon by the "end of the decade" but never intended for it to remain as the permanent exploration hardware system model, especially for deep space.

However, that warning is *only* true beyond the Earth-Moon system, for deep space voyages, for crews at Lunar Gateway or any other Lunar space station, or for unshielded sojourns on the Lunar surface greater than what we can now only roughly estimate with inadequate data as a period of about a month or less. The future of SLS-Orion in the Earth-Moon regions of deep space is not a lost cause or unethical by any means, provided absolutely that the crews are adequately shielded on relatively short missions – or fully shielded for indefinite stays on the Moon, as we shall discuss in a few pages.

The Cost of Keeping the Stepchild

Of course, none of these issues are the red headed stepchild's fault. Except for its marginally sane upbringing, SLS-Orion is truly a potentially magnificent human space exploration system for limited, very short missions of a week or two, capable of many future historic exploration accomplishments, even accounting for its significant design weaknesses, and regardless of its technically shaky pedigree and dissolute cost. There is a growing tendency in the media to turn on the poor hapless child and render to him the same fate as his aborted sibling, Constellation. But that would be a colossal error in good judgment and an idiotic move on the part of the U.S. Government, as we shall explore later in this chapter. Unfortunately, as expressed previously, such inbred insanity runs freely in the U.S. Government/aerospace contractor family, and idiotic moves have become the expected norm in Washington.

At the time of this writing, the cost to fly a pound of cargo to low-Earth orbit (LEO) on the new NASA Space Launch System (Block 1) is estimated to be $26,567. Let me break that down. In 2017, the cost for each SLS-Orion launch was advertised as $1 billion.[9] The cost per launch estimate increased every passing year with accrued investment costs and steadily accumulating infrastructure increases. By 2019, the price for each SLS mission was estimated at more than $2 billion, but that did not include a full cost accounting extended over decades. By November of 2021, that estimate rose to $4.1 billion per launch, a price tag that includes the rocket's development but excludes ground systems and Orion development costs.[10,11] Therefore, delivering a payload of the maximum advertised capacity of 154,324 pounds equates to a whopping $26,567 cost per pound to LEO! The cost of that same weight of cargo flown by the Space Shuttle system to LEO was a comparable $27,000 per pound.

But the equivalent pound of the exact same cargo flying on a flight-proven SpaceX Falcon Heavy is nearly 25 times less at $1,064 per pound cost to LEO ($150 million per launch delivering 141,000 pounds to LEO) even if the Falcon Heavy expends all its boosters rather than recovering them, and the necessary fuel to make it happen (with the customer paying for the difference). If the two side boosters and the Falcon Heavy core are all recovered, the cost for the Falcon Heavy mission is lowered to $90 million, but the max payload to orbit is also lowered to 66,139 pounds due to the additional fuel weight required. However, the result is a payload charge of $2,716 per pound, which is still almost 10 times less than the SLS-Orion cost. With a third option of just recovering the side boosters and expending the core, the max payload mass to LEO is 125,663 pounds, but there is currently no published cost projection from SpaceX for that profile. Considering the advertised capabilities of SpaceX Super Heavy Starship, the cost will again be slashed by whole magnitudes to between *$5 to $10 per pound* to LEO.[12] The difference is principally driven by the individual exploration philosophy. If Super Heavy Starship performs to expectations (and there is no reason to believe otherwise), it will cost any astronaut customer a small fraction of the total to fly on SpaceX than any current or proposed U.S. Government rocket.

To try and understand its outrageous costs, here is an example of how "simple" government waste evolves into borderline waste, fraud, and abuse of taxpayer money. On Friday, 24 September 2021, NASA announced it had awarded a contract to Aerojet Rocketdyne for 18 new RS-25 Space Shuttle main engines destined for use in the SLS rocket. (NASA is using the old Shuttle engines for the first round of SLS missions. But instead of reusing these engines as they did in Shuttle, these all will be tossed away after only one use. The taxpayer's cost for *each* engine? After start up and production costs are added, these throwaway engines will cost $146 million each. The original shuttle engines, because they were reusable, cost the government $40 million each. The new disposable versions for SLS Orion will cost 365% more. And the waste adds up. Throwing away four of these to conduct each mission will cost more than half a billion dollars *per flight* on a mission price tag that already tops $4-5 billion each![13]

Waste, fraud, and abuse is not at all an unfounded complaint when compared to the clear choice the government has to buy the same services from private enterprise. For example, a single SpaceX Raptor engine that is actually more powerful than one of the Aerojet Rocketdyne RS-25 engines costs less than $1 million per engine. That is 146 reusable Raptors for the price of one disposable RS-25. You don't see that kind of BOGO in your local store. Yet, apparently, the good-old-boy government contractor club is still alive and well, and the taxpayer picks up the bill every time. Common sense, fiscal flexibility, and ethical monetary policy is never a factor inside the beltway.

It is true that no matter how all these accumulated costs are spread out over the life of the SLS-Orion program, they will always be several magnitudes higher than the cutting-edge and technologically far more advanced SpaceX hardware, and perhaps even the upcoming hardware being developed by Blue Origin.

Despite all this catastrophic fiscal news for SLS-Orion, there still exists a compelling argument for continuing SLS-Orion after a pause and with tighter fiscal controls in place, as is detailed in the latter part of this chapter.

Let me summarize actual *existing* heavy lift capabilities in the United States as of this writing:

(1) NASA can contract for a ULA (United Launch Alliance) rocket to deliver a 62,500-pound payload to LEO using a Delta IV Heavy at about $5,600 per pound.

(2) Comparatively, SpaceX can currently deliver a 141,000-pound payload to LEO using an expended Falcon Heavy at $1,064/pound.

<u>Planned Near-Future Rockets</u>:

(1) NASA has already poured at least $11.8 billion into planning for the Block 1 SLS rocket that is currently billions of dollars overrun and five years behind schedule. When and if it does launch, it is reported to be capable of delivering a designed 154,324-pound payload to LEO at the cost of $26,567 per pound. NASA also proposes a larger version of SLS called Block 1B, which has not been funded for proposed missions until after the first crewed return to the Moon mission (now scheduled for 2025). That rocket advertises a payload capacity of 231,485 pounds to LEO for a cost of $17,711 per pound (at 2021 Block 1 rates), although the actual cost of the Block 1B per flight has not been released by NASA.

(2) NASA advertises a planned rocket called the SLS Block 2 with a design payload capacity of 286,601 pounds to LEO. However, Congress has not funded any work for Block 2 beyond the half-decade old drawings. And there are no NASA Artemis Lunar missions proposed beyond the end of 2029 using anything but the still-in-planning, unbuilt, and as yet unfunded Block 1B booster. For all practical purposes, an SLS Block 2 booster is (as of this writing) just vaporware, and Block 1B appears almost as iffy.

(3) Jeff Bezos' Blue Origin Company is advertising that its relatively large, reusable New Glenn rocket has the capability of launching 99,210 pounds to LEO. Its maiden flight is scheduled for 2023. The New Glenn rocket factory is built and has been in full operation at the Kennedy Space Center's Industrial Park since 2016. No launch costs have been released. Unfortunately for Blue Origin, anticipated launch dates are constantly drifting to the right. Indeed, several pre-contracted ULA flights using Blue Origin hardware have already been reassigned to Atlas V rockets because the promised Blue Origin engines are still not flight rated. *Amazon Unbound,* released in May 2021, lends an embarrassing light on the new rocket factory and its problems,

depicting corporate dissatisfaction virtually across the board at the unacceptably slow pace of rocket systems development, management, styles, failure to successfully win key contracts, and other Blue Origin issues. To underscore his all-in commitment to his space frontier dreams, Amazon founder and chief Jeff Bezos stepped down as Amazon Chief Executive and moved his full attention to the success of Blue Origin. When reusable New Glenn finally starts flying, it should inject a whole new round of competition into the space transportation and heavy lift marketplace.

(4) SpaceX Starship and its Super Heavy Booster have already been built and are in an audacious round of flight testing as of this writing. Elon Musk has "aspirationally" advertised its flight cost to be as low as $2 million per mission. With an LEO payload capacity more than any other rocket ever designed (even more than the Saturn V), Musk touts an ultimate 330,000 pounds to LEO in its fully reusable configuration compared to Saturn V's 310,000 pounds. Assuming Musk's incredible claim of $2 million per flight is off by a full magnitude and that the actual cost of each mission will be closer to $20 million, even then, the cost per pound to LEO would still only be between $50 to $100.

Going Ahead Costs – If SLS Is Not Cancelled

As of this writing, most of the up-front and development costs of SLS-Orion have already been spent, a considerable inventory of hardware is in the pipeline (principally because of the incessant program delays), and some of the required equipment has been recovered from shuttle. Thus, canceling SLS-Orion today after the extraordinary expense of building it would be altogether insane, tragic, and immoral unless the equipment already built can actually be utilized. Unsurprisingly, such cancellations are always a typical and frequent U.S. government option. While it is true that some mechanisms absolutely must be enabled *immediately* by a very dysfunctional Congress to keep its future costs under control, and while its mission profiles must be slightly modified for crew safety in the Earth-Moon system, SLS-Orion is by no means a death trap (unless it is flown with humans to Mars – in which case, under its current configuration, it is a certified flying coffin).

In fact, a hybrid approach to Orion was already recommended by former NASA Administrator Jim Bridenstine who suggested replacing the SLS rocket with Falcon Heavy. But dispatching an SLS-Orion outfitted with an unshielded inflatable crew module on a seven-month-plus voyage to Mars would be equivalent to launching a human barbeque pit for a slow roast mission to the Red Planet. And SLS-Orion will perhaps likewise not always remain a fiscal sinkhole – unless Congress continues to do nothing at all to cap its costs, in which case all bets are off.

As was reviewed in previous pages, the cost of each SLS mission is estimated at $4.1 billion. But, again, much of that cost was spent in development, as well as in building ground infrastructure and flight hardware that is now prepared for flight. As a prime example, the SLS-Orion Launch Tower Mobile Launcher–1 cost the taxpayer a billion dollars. That seemed criminally irresponsible in itself since, in the current mission planning, it will launch only three deep space missions – one of them uncrewed.

But, as if to underscore their propensity for shameless fiscal profligation, the government awarded a subsequent contract for the construction of a seven-foot taller version called Mobile Launcher–2. It is designed to accommodate the slightly taller Block 1B rocket for a handful of missions after the first post-Apollo Lunar landing. (Why the Agency did not build one pad that could accommodate both models is anyone's guess.) The original contract was awarded for $383 million in 2019. But in June of 2022, the NASA Inspector General weighed in after the cost-plus contract had ballooned to nearly $1.5 billion and was projected to be delivered two and a half years late in a mobile launch pad construction project that will run for a total of eight years in production.[14]

Meanwhile, on the adjacent Launch Pad 39A stands the SpaceX bigger and magnitudes more complex, cutting edge launch *and recovery* tower that will be completed from start to finish in no more than 6 months – ready to fly and recover the largest rocket ever built at a fraction less than the cost of the NASA hardware.

The SLS-Orion fiscal carnage is so extreme that, smelling blood in the water, the media has focused its spotlight on the problem, with *Forbes* openly calling for the system to be scrapped in a February 2021 editorial. The subtitle read, "The government's new heavy-lift rocket is over budget, behind schedule, and plagued with problems. It's also unnecessary."[15] Further, a September 2021 article in *Arstechnica* summarized the situation by stating:

"Pigs get fat. Hogs get slaughtered. So how come no one has taken the Space Launch System rocket behind the woodshed yet?"[16]

Yet, regardless of the almost mindless absurdity of such expenses, much of the cash has already been exhausted or legally obligated. Hence, it is a little too late to be getting SLS-Orion buyers' remorse sticker shock now.

NASA's SLS-ORION's Hail Mary

On 26 October 2021, NASA released a Contract Opportunity Request for Information (RFI) on the U.S. Government's sam.gov website titled "Exploration Production and Operations Long-Term Sustainability."[17] Reading between the lines in this 25-page RFI, the Agency was, in its typically weird contract-speak, fundamentally seeking to offload SLS-Orion to private industry with NASA oversight before it is summarily canceled by Congress. But what a tale NASA spun in these short 25 pages! The Agency openly revealed its vision for Orion – the stepchild adopted out to the private sector but with assured visitation and continued supervision by the Agency.

In the document, NASA refers to the SLS-Orion system as a "national capability" that should be a viable part of deep space human exploration "for 30 years or more." In its RFI, the Agency sees itself as the "anchor tenant" (exclusive customer) with plans of "procuring" one crewed flight per year into deep space. NASA also indicates they would be willing to market the massive system to the science community and other government entities.[18]

Bottom line – NASA wants to reduce the cost of SLS-Orion by half-transitioning it to private industry. If this RFI evolves and follows the similar course of the Space Shuttle program, NASA may well find many eager big contractor suitors, as did Shuttle, because, in order to attract any bidders, the government typically develops a no-risk, cost over, award fee contract that is attractive to any corporation ready to cash in on a sure, risk free, fiscally rich project with the potential of lasting decades. The taxpayer, unfortunately, winds up with all the risk in the end, and pays all the bills.

The question is, can a system that is so technically antiquated and designed on an exploration philosophy ready to catastrophically collapse actually be saved by such a Hail Mary play?

One Final Chance at Glory

Whether this strategy actually works out or not, SLS has certainly fully earned immediate cancellation for the single reason of egregious waste and uninhibited abuse of the taxpayer's money. At this point, cancellation is almost the only just way out, assuming that, if nothing changes, the money pit is only going to get exponentially deeper.

However, aside from the assumption that someone in government will actually be the adult in the room and truly manage the hemorrhage of resources, and in addition to the fact that much of the hardware is already built, there is one single overriding and critical reason why we ***must not*** cancel SLS-Orion and instead proceed at flank speed with its currently assigned Artemis missions.

The Treasures Inside Shackleton Crater

Someone in the United States Government was paying very close attention to the strategic realities in the near solar system – not just on Earth, but between the near worlds of the Earth-Luna system. While on 14 September 2020, previous NASA Administrator Jim Bridenstine equivocated on the ultimate landing site for the Artemis III Lunar Exploration Program mission, the stated goal has long been the Lunar South Pole to establish a permanent presence called Artemis Base Camp.[19,20] For three reasons, it is essential that the United States be the first occupants to arrive at Shackleton Crater. By accomplishing this, *the entire cost of the SLS-Orion system and missions are fully justified.* (For a more in-depth discussion, see the chapter titled "A Strategic Necessity" just ahead in Module 2 of this volume.)

Shackleton Crater on Luna is located so close to the Lunar South Pole that the edge of its northwest wall touches the Pole itself. Shackleton Crater is an ancient impact crater some 13 miles wide and 2.6 miles deep. Its interior is in eternal darkness, while its crater walls are nearly constantly exposed to solar illumination. Shackleton Crater thus has two vital resources unavailable anywhere else on that entire world: virtually uninterrupted power and, potentially, a significant, primordial water supply.

The choice of this destination was announced by former Vice President Mike Pence on 26 March 2020 when he declared, "We are in a space race."

This was not a randomly selected site of mere scientific interest, regardless of NASA's external public positioning. Pence's words were not hidden in political or pop-sci gibberish. He openly, and with frankness uncharacteristic of the Beltway, declared Artemis was a strategic claim on a natural resource on Luna that is more valuable than gold on Earth – water. And Shackleton Crater has the best chance of harboring this exceptionally rare and equally exceptionally valuable resource on the Lunar surface that we are aware of.

Pence declared in his announcement that our presence at the Lunar South Pole would be permanent. "Urgency must be our watchword," Pence stated at the U.S. Space and Rocket Center in Huntsville, Alabama. "The United States must remain first in space in this century as in the last, not just to propel our economy and secure our nation but, above all, because the rules and values of space, like every great frontier, will be written by those who have the courage to get there first and the commitment to stay."[21]

Once again, read between the lines. The Vice President and Chairman of the National Space Council, responsible for all United States space policy, Pence openly used the terms "race" and "urgency." When Pence said, "Urgency must be our watchword," he candidly declared that urgency in reaching the resource is essential and that the United States needs to get there first and claim it. And then he clearly referenced the truly horrifically messy issue of space law and claiming strategic resources in deep space when he added, "the rules and values of space… will be written by those who have the courage to get there first and the commitment to stay." Only the most hopelessly naive can miss this.

Permanence translates as a physical claim on a vital strategic resource that is exclusively requisitioned and protected by a human presence. There is no resource more valuable on Luna of which we are currently knowledgeable than the theoretically vast Shackleton Crater icefield on the Lunar surface. While Earth is nearly completely covered in water, between Luna and Mars, this resource is decidedly rare. And, whether maintaining a human presence short term or permanently, water is the single most essential resource to assure viability on any world distant from our home planet.

Thus, Shackleton Crater offers the fulfillment of two of the vital reasons to get to Luna first and settle in permanently: water and power. The third reason is strategic presence. Once the United States establishes a water mine inside the crater (if indications bear the presence of the resource is actually located there) and occupies the best solar power array on its rim, it will instantly achieve strategic dominance of the nearest world – Luna.

Obviously, as is discussed in the chapter "A Strategic Necessity," there are space treaties to consider, as well as overriding United States laws, before claiming jurisdiction. But no nation on Earth is going to ask permission first and wait for a good answer while competing countries press ahead to launch, land, and physically occupy.

It is true that a crater that is 13 miles in diameter can offer room for more than one national resource recovery venture and positioning of solar generators around a perpetually exposed crater rim some 41 miles in length. Yet, Shackleton Crater must be surveyed, and the best sites determined before being claimed and secured by the strength of a full-time presence.

When the Biden Administration took control in January 2021, the new President scaled back on the former administration's probably always unattainable and politically motivated return to the Moon landing schedule and date of 2024. But in a remarkable departure from past administration's space-cancel culture, Biden clearly articulated continuing support for the Artemis crewed Lunar missions, handing the Agency reigns over to former astronaut and U.S. Senator Bill Nelson.

In a surprising move in April 2021, NASA selected SpaceX's Starship as the vehicle of choice to fly its Artemis crews to and from the surface of Shackleton Crater. In the current mission design, astronauts will fly from the Kennedy Space Center on the SLS-Orion Capsule, dock with Starship in Lunar orbit, and from there fly down to the Lunar surface and back.

It just so happens that SpaceX and other private, incredibly innovative minds have changed the face of human space exploration forever in terms of cost, accessibility, and even crew safety. But right now, NASA's Artemis is the only game in town when it comes to a laser-like focus on Shackleton Crater at Luna's South Pole. NASA already has most of the hardware in hand, and the rest is in the development loop with realistic dates assigned for completion and future mission launches.

Regardless of SLS-Orion's sketchy pedigree, again, the much maligned and thoroughly abused little red headed stepchild is literally the only game in town when it comes to carving out the United States' strategic high ground on Luna. Regardless of its profligate waste, fraud, and abuse, in one single Lunar mission, the program will more than earn back every dollar that has been spent on it to date. And if the preflight architecture can come up with an adequate radiation shelter, it will be certifiably crew-safe. Not a bad finish for a space-kid with a questionable pedigree named after a bitchy goddess, adopted by three Presidential fathers that despised one other, and raised by a very dysfunctional Uncle named Sam who has a well-earned reputation for having far more money than brains. Let us just all pray that someone actually in charge will wake up and see his real potential before it is too late.

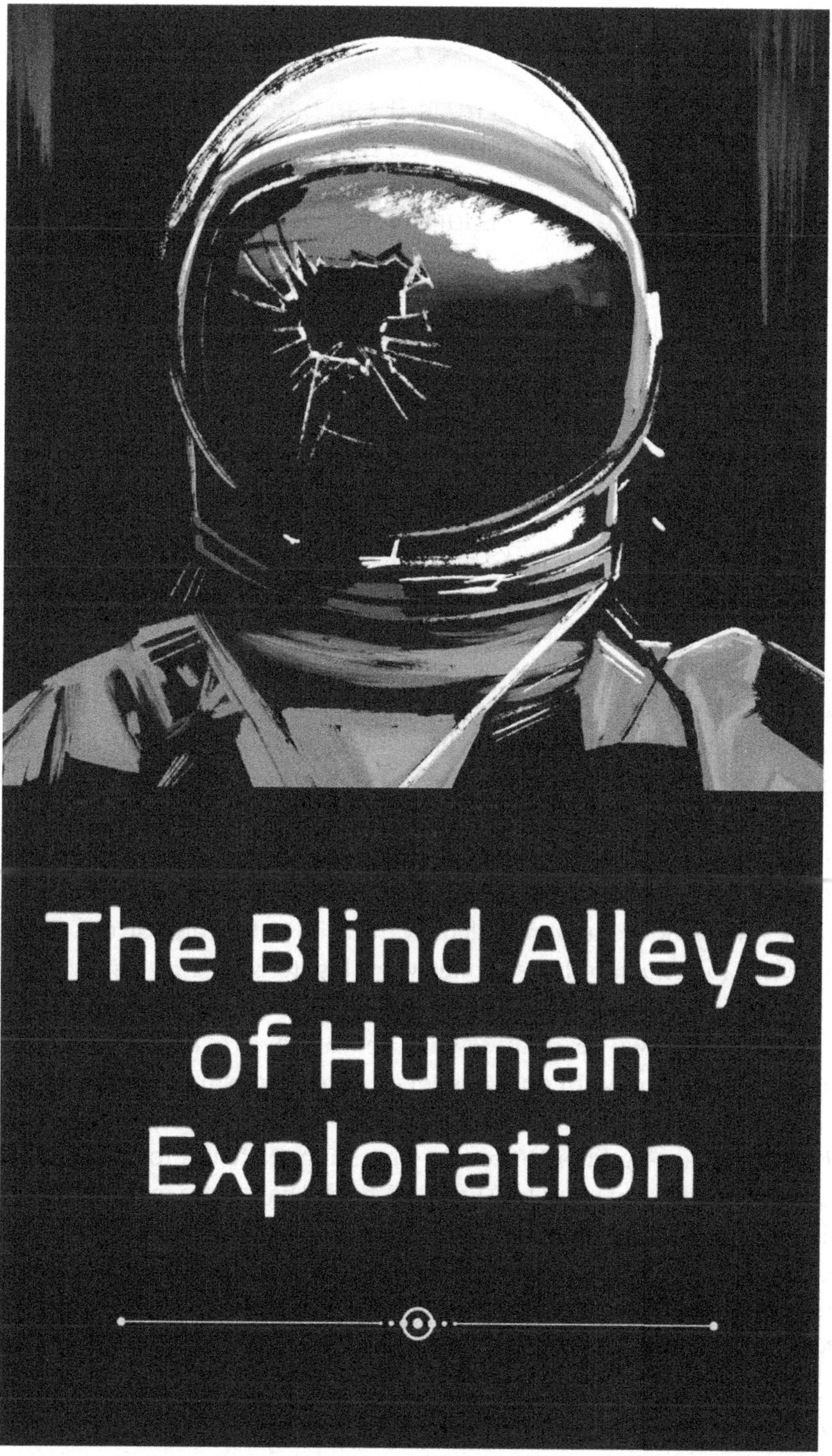

The Blind Alleys of Human Exploration

"As beautiful as simplicity is, it can become a tradition that stands in the way of exploration."

- Laura Nyro

"There is no harm in doubt and skepticism, for it is through these that new discoveries are made."

- Richard Feynman

THERE IS AN OFT TOLD STORY CONCERNING A FICTIONAL explorer named Bernard who wanted to find a mythical city of gold in the remote jungle on the continent of Africa. So, he gathered the resources that he needed by selling everything that he owned and going into a lifetime of debt, certain of the incomprehensible wealth that he was about to discover. And thus, he collected his equipment and a very large team of footmen to help him carry his belongings across the ocean and into this remote jungle. He was immediately beset with innumerable struggles. He and his companions were repeatedly attacked by the indigenous natives, and, on each attack, many of them were killed. Natural disasters also struck along the way, and much of his supplies were lost to raging rivers and fires while some of them tumbled down steep mountain escarpments.

Eventually, two years later, when he was down to his last surviving three footmen and with only one week's worth of supplies remaining, he serendipitously stumbled across another explorer from his country. His name was Algernon. After they exchanged stories for some time, Bernard finally confessed to Algernon that he had come into this jungle in search of the mythical city of gold, and, in the process, he had lost everything. Explorer Algernon sat quietly for a while, obviously struggling with his response. Finally, he looked at Bernard and said, "My God, man, that city was discovered a year ago in South America!"

Unfortunately for Bernard, he was not only exploring in the wrong jungle, he was also exploring on the wrong continent!

This story illustrates the perfect example of the penultimate blind alley of exploration. Bernard had all the equipment, all the resources, and all of the personal drive and sacrificial energy he needed to find his city of gold. But explorers that look on the wrong continent and in the wrong jungle will never arrive, regardless of their courage, good intentions, and virtually endless stream of resources.

Author Stephen Covey recounted this story with yet another spin:

"The leader is the one who climbs the tallest tree, surveys the entire situation, and yells, 'Wrong jungle!' Busy, efficient producers and managers often respond, 'Shut up! We're making progress!'"

Recognizing blind alleys in exploration requires a dual understanding of the merged human disciplines of psychology and sociology, for they work together in initiating the walk down a blind alley, and then in blindly and stubbornly perpetuating it. Obviously, every blind alley in exploration leads to a dead end. Once the dead end is reached, the explorer must back out and start over again, applying the lessons learned. This process requires two disciplines: flexible wisdom and humility. Without both working together, it is impossible to back out and start once more. Otherwise, the stubborn explorer typically stays and bangs his head against the far wall until he dies. Or she is forced to back out, regroup, and decide. That decision is: do I go back and try this again, or do I stop and figure out another way?

This thought process reflects who we are as humans: abjectly obstinate, proud, and typically unwilling to admit to any error – particularly one that we have invested our lives and careers in. In science, such obstinacy is legendary. Even in a discipline such as science, whose very organized methodology is to prove itself wrong, dogma rules the day and, more often than not, dominates even the process itself. Said the great Max Planck, "A new scientific truth does not triumph by convincing its opponents and making them see the light, but rather because its opponents eventually die, and a new generation grows up that is familiar with it... Science advances one burial at a time."[1] Because the discipline of exploration is made up of the same species, such pigheadedness is likewise legend. Literally, human intransigence is so fixed into our genetic makeup, it typically requires the grim reaper himself to become the final arbiter of truth.

Determining that a blind alley in exploration exists at all is manifestly difficult and, just as in science, typically has historically required that all the old explorers die off before the error is remediated.

This is because the average explorer's mind is not normally equipped for much humility, and their wisdom can be, on occasion, fixed and difficult to reprogram. Explorers require bravado and an overly inflated sense of self-importance, such that flexibility is rarely a part of the package. These traits, although they may sound negative, are indeed the precise characteristics that make explorers what they are and enable them to do what they do so well. It also makes them the very last individuals who would initiate change in their methods without a seismic event – such as a tragedy they are closely associated with or altogether responsible for.

Admitting a blind alley in exploration directly implies that false or incomplete assumptions were made, and the path to discovery is therefore going to be severely impeded to the point that it becomes unrealistic to actually reach the objective at the end of the blind alley. This is what Jacques Cousteau himself stated when he threw up his hands and walked away from undersea settlement because he too was lost in the darkness of his own blind alley. In fact, Cousteau's response was the very seismic event that destroyed not only the blind alley itself, but took his original dream down with it for members of his generation and the one that followed.[2]

With such a recent historical insight into these mechanics, it is imperative that we do not allow this to happen to crewed space exploration and settlement. If we continue to press on inside the current blind alley in which we find ourselves, it will soon invariably lead to disaster – whether in slow motion or all at once in the loss of an entire interplanetary crew or settlement. And, if the current pace continues, probably more than one crew or settlement simultaneously. Then the blind alley will make itself known with such unavoidable, startling impact that nearly every space exploration expert on the planet will solemnly declare in sworn testimony that they knew all along it was destined to fail. Such is the nature of perpetually timid "experts" emboldened to pontificate only by the glaring light of after-the-fact calamity.

Yet, we are all masters of blind alleys, for blind alleys are the way in which humans learn their way around the landscape of life and the universe itself. From the first day of our existence, we are creatures who move about, make decisions, and formulate our plans based on, of all things, blind alleys we have discovered and corrected for in life's journey. From our first clumsy attempts to walk that culminate in eating

Dad's shoes, we learn that some paths are profitable, and others are most certainly not. Our personal journeys are filled with obstructions and U-turns. Statistics for the United States demonstrate that the average worker changes career pathways from nine to 13 times in a working lifetime, just as students in universities typically change majors three times in the four years before graduating. These changes are for various reasons, but many of them are blind alleys – in career choices, company decisions, and other misdirected professional and academic pathways.

In science, the concept of blind alleys is actually engineered into the scientific method. Blind alleys are considered a legitimate part of the process itself. Indeed, were it not for blind alleys, no scientific progress could ever be made. A prime example is Thomas Edison, who conducted thousands of experiments to develop the electric light bulb, including testing over 6000 types of fiber for a functional, long lasting filament.[3]

And the number of blind alleys in exploration is legendary. Just to name a few, there are the blind alleys involved in the great Northwest Passage enterprise, both northern and southern polar explorations, undersea exploration, and most great mountaineering exploits.

Encountering any blind alley is always initially received emotionally as a failure – because, in fact, it is one. It is a failure to see the future in advance, which guarantees everyone receives their fair share of blind alleys in life – small and great alike. But if someone gives up and walks away from failure, as did Cousteau when he declared his dream impractical, then their vision is left for someone else to fulfill, walking down a new and refined path of exploration. Said Winston Churchill, "Success is stumbling from failure to failure with no loss of enthusiasm."[4] Or, in terms of this book, if I may be allowed to paraphrase such a great mind, "Success in human exploration is stumbling from blind alley to blind alley with no loss of enthusiasm." Trust me. Once we plant humanity's flag on the surface of Mars, there will be opportunities for success aplenty as we go stumbling from blind alley to blind alley with no loss of enthusiasm. Such is the future of every civilization we will establish on many worlds off our home planet.

Space exploration is even now involved in reconciling a greatly misdirected start of its own, which relates directly to the misdirected start in undersea exploration, for both were born out of the same culture, and both inherited the identical and altogether false expectation. It all began with the culture – the nameless culture that both inspired and motivated all the people who were responsible for the world's greatest period of human exploration in the 20th century. The culture had just defeated the enemies of freedom and unrestricted capitalism. Literally, the message was unambiguously communicated, by the strength of our free culture, "If we can dream it, we can do it."

Hinged to that absurdly simplistic false expectation was another just like it: humankind has nearly unlimited adaptability. And not only are we adaptable, we are strong, and we can adjust! Then, just for added emphasis, the steely eyed missile men rejoined with their personal addendum, "*...and we will damn well adapt to anything the universe can throw at us because failure is not an option*!" On these sloganistic expectations, the road to all the blind alleys in human exploration we now face was paved.

As we have discovered, both in space and in the oceans, humans are certainly not as adaptable as we once thought and defiantly advertised that we were. On the contrary, we have learned that the human organism has a sharply limited capacity to adjust to every environmental condition. Unfortunately, the expectation was freely communicated that the ever strong, ever capable superhuman hero could adapt to various high-pressure environments that would allow divers to interface with the oceans directly through an always open moonpool no matter what the depth. In space, we expected that microgravity would be a never-ending lark and that we could fly all over our spaceships whenever we wanted. Added to that was the belief that the trivial cosmic radiation damage would eventually heal itself and we could just get used to it.

All these expectations proved to be false – restricted by the human organism's adaptation limits. Such are the blind alleys in human exploration. Yet today, many people steadfastly cling to those old expectations. Missions are actually still designed around them, and we continue to plan with the idea in mind that someday, somehow, humans will adapt if we can just figure out that ideal countermeasure-counterpunch to offset the assault to the ridiculously audacious, antithetical thought that man's adaptation is not absolutely unlimited. We still resolutely set humans up in environments that are injuring them to prove a fatally flawed thesis that is the very foundation of the fatally flawed philosophy from which it was born.

The mental process at play is to simply demand that the plan work, come hell or high water, to preserve the initial expectation that man can endlessly adapt. The solution must, therefore, only be one countermeasure away: the right saturation gas mix or the magic pill that fixes muscle atrophy and permanent bone loss in space. So, if we can just get permission from enough international government and private agencies and committees, then even nature will not argue anymore. And there is always the thought that we can eventually genetically rewire our genes to become space humans, audaciously proposing a complete biochemical remix to mere DNA and those fickle chemical bonds which are routinely ripped apart by high energy cosmic primaries traveling at nearly the speed of light. And all of that reasoning is there just to sidestep the single inevitable conclusion: we have been exploring in the wrong way all along and trying to back-fit and patch-fix with ever-increasing lists of increasingly ridiculous countermeasures instead of just admitting error and walking out of the blind alley so that we can forge another, far more sane path ahead. The solution is achingly and ridiculously simple. Yet, we demand to do it the hard way and injure and kill our way to the mythical finish line on other worlds. Utilizing our current approach, it appears that the plan for humans is to build and fill advanced cemeteries on these other worlds instead of robust civilizations populated with intelligent and flexible humans.

It is endlessly fascinating to me how most people cannot step back, humbly admit they have made an error in building the exploration paradigm, and then flexibly retool the next generation of systems to meet the challenges and limits based entirely on what they have discovered. After all, is not discovery the ultimate purpose of exploration? Is not discovery the principal tool of change and progress? Indeed, is not science the discipline of continuous, seamless rediscovery and correction based on lessons learned and evidence compiled? Instead, they cling to old notions that the first generation of explorers discovered through trial and error to be multiple blind alleys.

Which is the more difficult process: inventing 5000 plus individual engineering and medical countermeasures to microgravity or engineering the single artificial gravity countermeasure by simply spinning the spacecraft? Therein lies the core discovery of the past half-century. And that discovery is one of just two new pathways required to assure our future success in deep space, as will be discussed in the following chapters.

Finally, why do the best and the brightest among all of the world's scientists and engineers actually perpetuate the Space Agency's blind alleys even after the obvious breakdown of their exploration paradigms is evident all around them? Simply because they are locked in a rigid system that does not accept any form of criticism or evaluation from the rank-and-file. There is no pathway up the line to hear feedback from their ranks except for very visible, but in reality, very ineffective, "safety reporting systems." In my personal experience from using them, these systems merely form an effectively useless loop from the reporting employee through some mid-level manager and back again to the reporting employee with no discernible authority and little or no change actually occurring in the middle. As is typical in huge bureaucracies, these are checkboxes that were derived after a disaster that are nothing more than window dressing patches that allow the old system to continue and give a visible but very false sense of system competence which is actually illusory.

Literally, an entire army of brilliant scientists and engineers can be held in rigid check by the weight of the massive, impenetrable bureaucracy whose favorite slogan down the line is, "It's above your paygrade." Reflecting on the Challenger disaster, Astronaut Scott Kelly described this very real and self-perpetuating systemic failure in NASA:

"It wasn't until years later that I understood that a management failure doomed Challenger as much as the O-ring failure. Engineers working on the solid rocket boosters had raised concerns multiple times about the performance of the O-rings in cold weather. In a teleconference the night before Challenger's launch, they had desperately tried to talk NASA managers into delaying the mission until the weather got warmer. Those engineers' recommendations were not only ignored, they were left out of reports sent to the higher-level managers who made the final decision about whether or not to launch. They knew nothing about the O-ring problems or the engineers' warnings, and neither did the astronauts who were risking their lives. The presidential commission that investigated the disaster recommended fixes to the solid rocket boosters, but more important, they recommended broad changes to the decision-making process at NASA, recommendations that changed the culture at NASA – at least for a while."[5]

It does not take a world class sleuth to recognize the implications of Astronaut Kelly's loud and clear final, minced "at least for a while" describing the sustained effectiveness of the presidential commission's recommendations on the decision-making process at NASA.

The world space exploration model is failing now because we can go no further down this blind alley into deep space without incurring extensive injury and significant risk of deep space mission loss. In the next section (Module 2), I will help you understand the true, detailed nature of that collapse.

Why We Will Depart Earth Forever

Reason 3

"... the greatest engine of innovation that has ever existed... is the product of a multitude of factors: extreme freedom of thought, an emphasis on independent thinking, a steady immigration of new minds, a risk-taking culture with no stigma attached to trying and failing, a noncorrupt bureaucracy, and financial markets and a venture capital system that are unrivaled at taking new ideas and turning them into global products."

— Thomas L. Friedman

When President Barack Obama canceled NASA's Constellation Program (the follow-up to the Space Shuttle Program) in February 2010, it left the United States just 17 months to figure out how to get our own astronauts back to low-Earth orbit and to its own space station. Surprising absolutely no one, it would actually require more than a decade to turn this dilemma around, stranding the United States on Earth, leaving us with no other choice but to hitch rides with a foreign power that had historically sketchy relations with the United States. Meanwhile, the Russians became the only space-faring Uber on the planet and immediately began jacking up the per seat fares to orbit – again, surprising absolutely no one. From 2011 until 2020, the price for a ride to ISS increased from $25 million per astronaut to $81 million, an average astronomical inflation rate (pun intended) of 32.4% per year. In that decade, the United States spent an estimated $4.3 billion dollars on the only ride in town to low-Earth orbit.[1]

This epically embarrassing ordeal reeks with inconceivable incompetence on the part of the United States and features not only a lack of even a modicum of a fundamental level of any coherent advanced planning but also a critical failure to coordinate the most elementary fiscal responsibilities and commitments from one administration to another. Were it not for an atypical lull in the conventional adversarial relationship between the U.S. and Russia, Boris and Natasha could easily have made access to our own historically significant and monumentally expensive national asset difficult, dangerous, and threatened. As it was, they exchanged their rights to seize an abandoned space asset (or just allow its orbit to degrade and watch it fall into the ocean) for absurd amounts of cash. Quite satisfied with the sizeable vacuum hose jammed into Fort Knox, the Russians sat back to enjoy a long and lucrative ride.

On the day after the only existing Constellation asset was launched into the ocean on 28 October 2009, everyone in NASA knew that Houston – and everyone else in the Agency – had a serious problem. The final space shuttle missions had been scheduled before even a hint of a replacement vehicle. For the first time since 5 May 1961, America had no crewed access to space and no follow-on program to replace the shuttle. At that moment, the only United States agency that had ever launched a human into space was completely out of options and out of American human access to space.

In stark contrast, the young NASA, seeded by the words and vision of President John F. Kennedy, transitioned from its first crewed suborbital launch through three individual crewed space test programs and the development of three separate space vehicles (Mercury, Gemini, and Apollo) to landing a human on the moon in just eight years and two months. The 21st-century NASA would be extraordinarily lucky to make it through the first set of meetings in an interval as brief as a single decade.

Yet from the loss-of-access emergency morphed an unexpected surprise in the form of an extraordinary instrument of legislative redemption – a law titled the SPACE Act of 2015 (also known as the Commercial Space Launch Competitiveness Act). SPACE is an acronym for Spurring Private Aerospace Competitiveness and Entrepreneurship. From that law, Elon Musk and his Space Exploration Technologies Corporation (SpaceX) were given a chance to innovate in ways never before witnessed in space exploration history. The truly astonishing power of creative, free, private enterprise was unleashed on the heretofore intractable problem of affordable access to space. Indeed, while it was driven forward by blinding genius, it was also powered by a single motivation: an opportunity for the development of a new economic empire fully energized by the people of a free capitalist economy.

Thus, the third reason why we will depart the Earth forever is something we as Americans are very familiar and comfortable with:

Economic Opportunity

The Commercial Space Launch Competitiveness Act opened legal channels that ensured equal access to NASA space contracting entities and enabled greater competitiveness in developing access to government contracts to revitalize and recover the U.S. crewed access to space. From it also came the motivation for SpaceX to exist and ultimately achieve access to and almost immediately dominate a competitive (albeit old school) international space business market.

But setting humanity's sights on deep space and other worlds is going to require many hundreds of billions of dollars and a single-minded approach to build a Mars colony. Thus, the way to forge ahead was made possible by President Obama's audacious sink-or-swim release of the formidable power of America's free enterprise system. Whether or not that was the President's up-front intent is irrelevant. He primed the pump, wittingly or not, and the waters of U.S. innovation gushed forth in spades.

And as the SPACE Act of 2015 was initially being forged into legal language, one of its ultimate congressional sponsors was drafting another act focused on protecting the interests of his Florida district. Representative Bill Posey (R-FL) of Central Florida was developing an Act that would create a legal framework to protect property rights for American companies that obtained asteroid resources in outer space titled the Space Resource Exploration and Utilization Act of 2015. Approaching Representative Kevin McCarthy (R-CA), the author of the Commercial Space Launch Competitiveness Act, for sponsorship, Rep. Posey's bill was finally included as a provision of the SPACE Act of 2015, and together they were signed into law.[2]

The work of rapidly expanding SpaceX as a low-cost space launch service became available to any entity that agreed to sign a contract and fly. And SpaceX has now initiated the Starlink planetary constellation of from 12,000 to 42,000 satellites in low-Earth orbit to provide global high-speed internet service to the entire planet. This income stream alone will dwarf SpaceX's launch service business by about 10 times. Musk announced at the Satellite2020 Livestream that the anticipated income from Starlink will be around $30 billion annually – significantly more than NASA's entire annual budget.[3] Morgan Stanley validated Musk's assessment (estimating the Starlink cash flow by the end of 2021 at 34 billion) and projected it would grow to *three times* NASA's annual budget by 2043.[3]

SpaceX plans call for the "aspirational goal" of simultaneously launching human colonists to Mars on the enormous Starship at a truly unbelievable rate to meet the company's design goal of *3 flights per day* on average during the window of opportunity, ultimately anticipating *one thousand* Mars-bound flights of crew and cargo per year.[4]

Musk stated in February 2019 that he estimated that each Mars-bound colonist passenger would pay between $100,000 to $500,000 for a ride to Mars (the return ticket is free) – yet another income stream for SpaceX.[5] How close that income is to breakeven for the company has not been advertised. Whether SpaceX can actually find the hundreds of thousands of Mars colonists with that level of income is also yet to be determined.

The economic drivers involved with the recovery and mining of asteroids have yet to be developed with anywhere near the fidelity as the SpaceX Mars colonization plans. And yet, the economic opportunity enabled by these activities dwarfs even SpaceX's ambitious economic plans for low-Earth orbit and deep space.

As will be discussed in the future chapter, "A Strategic Necessity," just one of the known asteroids contains enough precious materials to completely dwarf the world economies by 100,000 times the annual Gross World Product.[6] Another has more gold than has ever been recorded as mined on Earth.[7] And those are just two out of thousands of known asteroids.

Economic opportunity has always been a chief driver of human exploration in our history. But at no time has there been more opportunity than we see unfolding right now in deep space.

Of course, none of these activities are even possible without First Principle space exploration equipment and human-rated systems. The first nation or company that deploys them will command and dominate the economic engines of the space frontier. Likewise, the nation or the company that *does not* incorporate First Principle strategies will lead the first epically historic human disaster of the new frontier – and will likely lose the economic race altogether.

Module
Two
Ch.01
Ch.02
Ch.03
Ch.04
Ch.05
Ch.06
Ch.07
Ch.08
Ch.09
Ch.10
Ch.11
Ch.12
Understanding the
Collapsing Exploration
Paradigm

Radically Divergent

"We have minds that are equipped for certainty, linearity and short-term decisions, but instead, now we must make long-term decisions in a non-linear, probabilistic world."

— Paul Gibbons

"Some discoveries are so pervasively powerful that they all but destroy the cultures that birthed them."

— Anonymous

TO TRY AND COME TO GRIPS WITH THE COLLAPSING human space exploration paradigm as clearly as possible, it is essential to witness and compare the former model to the one that is quickly emerging to take its place. It is a new technology radically divergent from the old, and it is being created at a pace that matches and often exceeds the technology innovation surge that birthed the Apollo Moon program in the 1960s.

Much of it is led by SpaceX Founder and CEO Elon Musk, who, in my opinion, is truly an extraordinary millennial genius (which describes an individual who only comes around every 1000 years or so). These individuals so dramatically affect the human landscape that they leave it fundamentally and permanently altered. They do not merely improve the world, but they radically transform it. The key term here is radically. Their minds do not conform to everyday modes of doing business because they are not wired like most other humans. If they were, then these individuals would not be described as millennial geniuses. Their thinking is not alien in any sense – but if you compare the way they do business to the accepted norm, you will certainly think so.

As one example, Musk designed and built what was, on the day of its first launch, the largest rocket on planet Earth. He launched it off NASA Pad 39A at the John F. Kennedy Space Center (KSC) – the historic

Apollo pad from which all humans who previously visited the Moon had departed Earth. How Elon Musk planned, directed, and ultimately executed the launch offers a clear insight into his unparalleled innovative genius and astonishing, otherworldly chutzpah.

Elon Musk is a one-man, walking, mirrored image of NASA in the 1960s. NASA was not an individual millennial genius, obviously, but it was instead the *corporate embodiment of one.* Just after the Agency's founding in 1958, NASA and its people were anything but conventional. I was fortunate enough to personally know and count among my friends and family these real-deal, steely-eyed missile men who revealed plenty of stories about how NASA did business in those glory days when plans went from idea to paper to flying human crews in mere months, not decades. These were the days of NASA's radical creativity and spontaneous, inventive energy, which altogether fled as of the completion of LBJ's second term after fulfilling JFK's initial vision.

But the cumulative, debilitating damage caused by the self-centered, egotistical leadership flaw of the following nine presidents who insisted on wiping the national space program slate clean of all vestiges of a predecessor's plan while scraping the developments and ordering new hardware for a personal legacy in space, egregiously stifled the advancement of space exploration. Yet the Biden administration has so far broken the mold with initial announcements maintaining the Trump administration's return to the Moon plans virtually intact.

One late evening in 1988, just before midnight, as I sat in my office at Launch Complex 39, I looked up as an older, congenial man entered and easily sat down in the chair beside my desk with as much unconcealed familiarity as though he already knew me. I introduced myself, and he said, to my surprise and delight – his thick German accent clearly revealing his heritage – that he was Günter Wendt. Of course, I had known of the virtual legend Günter Wendt since I was a young boy. He must have seen and appreciated my smile as I acknowledged that I recognized that a legend had somehow found his way into my remote office and, at midnight, sat down to have a totally unexpected chat with me, a perfect stranger. To say we got along instantly is an understatement, as I welcomed the congenial icon with rapt fascination.

Although not a part of Werner von Braun's rocket scientist team, German born Günter, a former Nazi Luftwaffe flight engineer, came to the United States on his own terms and became the pad leader for every manned spaceflight from Mercury to Gemini and each manned Apollo mission. John Glenn affectionately called him "Herr Führer of der Launch Pad" because of his legendary German discipline that he personally brought to bear on the crew-loading process. When it came to the launch pad white rooms of Mercury, Gemini, and Apollo, no one – not even astronauts – argued with Günter because it was his pad, period. His face was the last human face any crew saw before flight as it was Günter himself who personally shut the hatch and latched it closed, leaving every flight crew during the Mercury, Gemini, and Apollo missions with his personal farewell blessing in his thick, trademark German accent.

In the wee hours of that morning, Günter and I spoke in my office for hours about the early days of human spaceflight. Then, together we walked over across the vast parking lot of the Vehicle Assembly Building (VAB) and rode up the glass elevator beside the assembled Space Shuttle to the top of the massive spacecraft. Günter and I then moved to the edge of the narrow, chained catwalk and looked down the length of the empty 184-foot, 165,000-pound spacecraft sitting still in its cavernous, dark bay.

In the dim light of the upper regions of the massive building, I looked into Günter's eyes and said, "I guess things have come a long way since the days of Mercury. You could almost stuff a Mercury capsule into one of the wheel wells of this shuttle!"

Günter gazed back kindly with a patient smile at the young, ever enthusiastic NASA engineer standing before him. He replied wistfully, "*Dennis, it tizz juz not da same.*" At the moment, I thought I understood what he was saying. But after a career with NASA at KSC, I now know *exactly* what he meant.

Several months later, I was visiting at the Gainesville, Florida, home of the legendary, disproportionately colorful, and equally excessively brilliant author of scores of books about space and the early days of space exploration, Martin Caidin. Martin was a take-no-prisoners gentleman who instantly counted me as one of his friends – and was not at all shy about bluntly informing me of this at our very first meeting and many subsequent visits. Martin was a very quick study who equated his opinions with fixed universal laws and was matter of fact about letting one know that. But he was also a man of humor and good will. If you were counted as a friend, you were also counted unequivocally as family.

I had not been in NASA even two years when I met Martin, and it was like having the heady opportunity of learning space exploration at the feet of a master – the original steely-eyed missile man, to be sure. And so it was that on the very first occasion of our meeting, Martin told me the fantastic story of his friendship with Werner von Braun and the day they stood far too close to a launch.

Martin said that he had been having an ongoing conversation with von Braun on his many visits to one of the early launch pads on the Cape, as von Braun was familiar with Martin's books on spaceflight. During one conversation, Martin admitted to having the desire to watch a rocket launch "up close and personal" from inside the evacuation zone. Martin said that von Braun was immediately interested in the idea. So, together, they plotted to "take a walk" to the scrub palms and high underbrush surrounding the pad as the countdown proceeded. Martin admitted that he had already scoped things out long before having the discussion with von Braun. The way Martin had it figured, if he got caught during a launch with none other than von Braun himself, they probably would not fire him, and certainly no one but the President was going to can the very demigod of American space exploration!

So, the pair literally hid in the bushes like a couple of schoolboys out on a lark. The countdown reached zero, and as the rocket roared to life, they stepped out of the brush to get a better view of it just hundreds of feet away. Martin recalled the awful, ear-splitting noise of the rocket's flaming engines as it lifted off the pad, and the blast of radiant heat from its tower of fire as it roared away into space. He described von Braun's face as a "frozen smile" as together they watched the launch from the forbidden zone. He also freely admitted that if anything had gone wrong, they probably would not have survived. But they were willing to take that chance to see something very few men had ever seen face-to-face before. Later that day, the skin of their faces turned bright red from the first degree burn they received in the full glare of the rocket's ascent fire at such close quarters.

It was von Braun himself whose team had launched the first manmade object into space on 20 June 1944 from the Peenemünde Army Research Center in Germany on a V2 missile just shy of a year before the surrender of Germany to the Allied forces. Wernher Magnus Maximilian Freiherr von Braun was the co-designer and architect of the V2 rocket that became the seed for the rocket family that culminated in the mighty Saturn V Moon rocket.

Over the years, I have recalled this story again and again and thought about the certified mixture of madness and bravado that must have driven those men to undertake such a risk. But now I fully understand it. It was, in fact, *absolutely necessary*. Both von Braun and Caidin were prodigies in their own right. Both were certified historic explorers of the new frontier. They felt as though it was their right to be there and see that – and the bureaucracy could… well, they could just deal with it! I believe that their reason is best summed up in the quote from Admiral Richard Byrd, penned in his journal while alone in the vast icy wildness under the fading Sun's deliberately slow prelude to the deep, unrelenting Antarctic night:

"I watched the sky a long time, concluding that such beauty was reserved for distant, dangerous places, and that nature has good reason for exacting her own special sacrifices from those determined to witness them."

I sat and listened in rapt, speechless astonishment as Martin told this story and so many others that first night of our friendship. I had to go to work the next morning, so I left Gainesville to drive back to the Cape sometime after 3 a.m. and arrived home just in time to shower and drive to my office at the Space Center. But the single thought that I remember so clearly was that as Martin recalled the early days of spaceflight in America, he concluded by saying, "*Dennis, it's just not the same.*"

Thus, Martin echoed Günter's sentiment word for word. In my youthful naivety, I initially thought these gentlemen were just lapsing into a good-old-days sentimentality. But now, three decades later, I understand it not as sentimentality at all, but as a literal declaration that something elemental – indeed, something seminal – had morphed into something else. And that something else was insidious: anti-exploration. A negative force had somehow overtaken the human exploration initiative.

Both of these extraordinary gentlemen knew exactly what that was. They had lived in both times. They had been a part of the metamorphosis from historic genius to business as usual in a massive government system. They had been eyewitnesses to a burgeoning organization with a small cadre of truly unusually gifted, world class scientists and engineers that had been so burdened and strapped by the overwhelming weight of bureaucracy and by endless policies, procedures, and federal codes that the lithe, gossamer persona of genius was systematically crushed by the overbearing juggernaut. That rare and ultimately gentle light of unchecked brilliance was slowly and systematically strangled into near darkness.

That is precisely what they meant, and I see it so clearly now. *It truly wasn't the same anymore.*

Years after these men had passed away, Elon Musk showed up on the Space Coast and rented the historic Apollo Pad 39A from NASA. He cavalierly announced that the first payload on his giant Falcon Heavy would be his own Tesla Roadster. The Roadster was a vehicle that he had personally built, and it was to be "driven" by a "robot" named Starman in a SpaceX spacesuit as David Bowie's "Space Oddity" played in a continuous loop. This whole payload – Roadster, Starman, Bowie, and all – would be flying out beyond Mars's orbit into the fringes of the asteroid belt and then looping back in toward Earth in a cyclic, billion-year orbit. And, after launching all that, Musk would land his boosters back on Earth in perfect, side-by-side formation, ready for another use.

And he did. Claudia and I watched it all with our own eyes, just a few miles from the action. I was quite literally personally moved to real tears. It was just something that NASA or one of its traditional contractors *would never do.* It was shockingly creative. It was impish and brash. I was incredibly inspired, and – *dammit* – it was just fun! I so very much wished that Günter and Martin could have lived to see it and watch it with us. Both men would have *understood* it immediately. In my mind's eye, I can envision them smiling together in the light of the boosters going up – and especially in the light of the boosters coming back down!

From my experience, NASA clearly had no sense of humor it was aware of, as spoken in clear, Men in Black terms. And with this ingenious endeavor, it was as though the staid, conservative halls of space had just been invaded by an alien no one expected. My astonishing sense of relief in all of it was altogether overwhelming!

And, as I would hear in the Kennedy Space Center gym day after day, many in the intensely competitive contractor cadre openly hated Elon Musk for his clear and unapologetic new-kid-on-the-block incursion on their holy ground. On 1 September 2016, when a SpaceX Falcon 9 blew up on a Cape Canaveral launch pad, the chorus of glee from a few KSC United Launch Alliance contractor employees echoed off the walls of the Neil A. Armstrong Operations and Checkout Building near my office. While standing out in the parking lot facing the black cloud drifting across the Banana River, a long-time acquaintance ULA contractor standing behind me said, "It's about time. SpaceX has it coming to them."

This was not mere parking lot sour grapes. At the time, SpaceX and ULA were in intense competition for military contracts, which included still unresolved lawsuits. The competition pitted Musk's relatively small upstart against the politically positioned, entrenched corporate giant. And when the rocket exploded during a routine preflight fueling test for mysterious reasons amid such unveiled animosity, suspicion immediately tilted toward industrial sabotage. Musk said later in an interview, "We literally thought someone had shot the rocket." It came to a head when SpaceX officials asked permission to inspect the roof of the nearest ULA building to the pad, about a mile away. Unsurprisingly, ULA refused, which forced the property owners (the U.S. Air Force) to step in and do their own investigation of the sabotage theory. In the end, the Air Force ruled out sabotage, and Musk admitted the fault was with SpaceX alone. He said, "It was a self-inflicted wound. It took us a long time, but we were able to re-create the failure."[1]

With the successful launch of the first Falcon Heavy's Tesla Roadster mission out beyond Mars orbit, Elon Musk's relentless, outside the box innovation had just sent a clear and unambiguous message. But that message was not to everyone. Just to those who had the ears that could hear it, decipher it, and assimilate it.

As Claudia and I watched that launch from the Cape, my emotions were real and triggered because I was an eyewitness to the glorious rebirth of one of humanity's most noble aspirations. As I stood on the shoreline that day, I blinked back tears for the memory of Günter, Martin, and Richard Schealer – my father-in-law and true inspiration, who was one of the early NASA pioneers at KSC – as well as the rest of the space men of their generation. These guys rode up to the top of the first unmanned spaceships on cherry-pickers and literally dumped buckets of hypergolic oxidizer in the rockets' intakes to prime the pumps. Yes, it is true, all of it. And they would be the first to admit that it was all certifiably crazy. But they were unwittingly walking down the pathway of human greatness – the greatest generation's wonderful encore to rescuing the entire world from certain doom. Physicist Niels Bohr described them all in this way: "We are all agreed that your approach is crazy. The question which divides us is whether it is crazy enough..."

We are blessed with genius so rarely that when it comes along, it frightens us. Other humans typically reflexively do everything they can to crush it, minimize it, and even ridicule it. The 21st-century experience is no different. But it is Elon Musk's constant self-confident smile that gives me hope. He knows the score – all of it. But he is going to singlehandedly colonize Mars and literally does not give a single tinker's damn about his envious and vocal critics. It always seems to me that in the Musk universe, those critics do not even exist. Yet, as I write this book dedicated to him and to his passengers, I pray he is intelligent enough to stop for just a moment and listen to solid scientific reasoning – for the sake of his ultimate success and for the sake of his crews' lives and wellbeing.

I give NASA almost no hope to seize this moment of change and to make this urgent design transformation. The Agency is literally too big and too fixed in its vast bureaucratic inertia to even slow down, much less to actually stop and correct its historic blind alley catastrophe in motion.

But Elon Musk is far too intelligent. He is significantly too drastically divergent in his radical creativity and in his engineering practice to fly his crews out in the vast cosmic ocean just to injure and kill them. The probability of him redesigning his model is very close to 100%.

Genuine genius is so truly extraordinary and rare that it senses these challenges and overcomes them because average human intellect is ultimately no match. In Elon Musk's case, he overcomes average human aggressive timidity with such alacrity and overwhelming, breathtaking creativity that its degree of surprising originality disarms and renders most people speechless with his rapid-fire and radically divergent mode of operation. Thus, because of such unexpected genius, we face an authentic and rapidly evolving sea of change in human space exploration that would be categorically impossible for any world government space agencies to either originate on their own, manage, or even begin to understand.

The vast moribund committees that design space systems are abjectly incapable of such radical, on-the-fly creativity, so, together they unwittingly distill brilliant new thoughts down to more conservative, far fewer adept likenesses of past innovation such as emerged from the NASA thinktanks of the 1960s. By this unwitting process, they thereby perpetuate mediocrity and carry forth with "proven technologies" – an established metonymical expression for politically safe – rather than attempt risky missions that might end up in failure (no longer an option in NASA) and unfortunate headlines from a media that has been deliberately taught by the Agency that NASA must actually be failure free.

Thus, they literally buy down their risk by intentionally stifling creativity as far too risky. Unfortunately, once the monster drawn up on paper is recreated in hardware at the cost of tens of billions of dollars, it is too late to respond if another approach emerges that is faster, better, cheaper – or even if one materializes before their eyes, as it clearly already has. And therein lies the reality we find ourselves trapped in today – the very quagmire of cultural and technological mediocrity, resisting and struggling against emergent genius and runaway innovation. On the fly creativity has long become antithetical and forbidden in the halls that were made hallowed by past genius and not by today's robed guardians of an expired legacy of formerly great audacity.

In 2014, once Elon Musk announced that he actually intended to recover his Falcon 9 boosters and thereby save billions of dollars and slash the cost of access to space, the headline "NASA, CNES Warn SpaceX of Challenges in Flying Reusable Falcon 9 Rocket" appeared in *Aviation Week*.[2] Seven years later, Musk had successfully returned nearly 100 boosters from space (a number that increases *monthly*), re-flown many of them, and landed two side-by-side simultaneously on multiple occasions. The cost reduction has resulted in an operational system that is 10 to 20 times less than equivalent government system launch-per-pound rates.[3]

Yet, just prior to Musk's next radical improvement in space access, on the occasion of the announcement of the Falcon Heavy launch, the former NASA Administrator somberly warned SpaceX once again that they were playing on fields reserved for governments only. He underlined specifically the astonishing degree of "abnormality" involved in even the *desire* to want to do so when he said, "normal people cannot, or should not want to, develop large launch vehicles."[4] Administrator Charley Bolden clearly underestimated the vast creative potential of "normal people." Then again, Elon Musk in no way can ever be considered merely "normal."

With notable irony, four years later, in the late summer of 2020, former NASA Administrator Charley Bolden became the first former high level NASA official to publicly announce the end of the NASA Space Launch System – which was rendered absolutely obsolete and unacceptably expensive, defeated by its reusable and far less costly competition built by those abnormal folks at SpaceX. Bolden admitted this in an interview, foretelling the inevitable demise of the ghastly overpriced and hopelessly behind schedule SLS-Orion program. In an *ARS Technica* article titled "Charlie Bolden says the quiet part out loud: SLS rocket will go away," the former NASA Administrator stated, "SLS will go away. It will go away during a Biden Administration… because at some point commercial entities are going to catch up… and they will fly for a much cheaper price than NASA can do SLS. That's just the way things work."[5]

Of course, these hard facts represent anything but NASA's messaging to the taxpayer public who is getting taken for an epically wild big contractor ride with SLS-Orion. NASA often serves up press releases propped up by its own perpetuated fairytale that failure is not an option. Because of this specific fable, they have permanently painted themselves into a corner and are totally incapable of admitting the true inherent dangers of what they are doing, not to mention the financial risk of imminent fiscal implosion. At the Agency, the coopted Hollywood decree rises up from the deep to bite them repeatedly in the rear. And when the occasional, inevitable failure happens, NASA is always seen as progressively incompetent because that is the only default expectation that remains after the failure is not an option target of perfection. It is the breakdown of their deliberate messaging and expectation of setting perfection as the everyday standard. It is the clear sign of a government bureaucracy that is often oblivious, brand blind, and incapable of separating their Hollywood persona from the real world in which they operate. Yet when the prince steps out on his balcony to wave to his peeps dressed in his tails and white tie and, inevitably, a passing buzzard unloads on his starched finery, what does he really expect the public's response will be?

But for SpaceX, it is an absolutely different story. They are totally free of self-aggrandizing, Hollywood generated slogans and lay it all out for the public in stark, ugly reality, actually communicating the raw truth of the risky ventures that represent all spaceflight endeavors. SpaceX deliberately communicates authentic, harsh reality as the full specter of failure on every flight. Said Elon Musk bluntly on the eve of his Falcon Heavy launch, "... the project has turned out to be more challenging than expected, and there's a lot that can go wrong here. There is a real good chance that that vehicle does not make it to orbit. I want to make sure to set expectations accordingly. I hope it makes it far enough away from the pad that it does not cause pad damage – I would consider even that a win, to be honest."[6]

Musk's uncommon genius extends quite beyond his corporate offices outward to a global public that he entirely understands. He communicates to them directly and clearly on their terms, in their language – tweeting his messages randomly across the social network fabric, presumably sometimes in his underwear on his way to bed, just like everyone else – not speaking down to them from lofty perches. His message, press releases, frequent interviews, and Tweets expose the surprising and quite unlikely: a humble billionaire's true face of human and unmistakably legitimate vulnerability – not infallibility. And it works. A truth starved public listens and understands genuineness. And Elon Musk speaks obvious truth – not preprogrammed legacy building empty nonsense so common to governments and federal agencies.

One overcast early morning in May of 2016, I stepped out of my office at the Kennedy Space Center and into the parking lot, facing southeast. A few minutes later, a brilliant orange flame erupted from a launch pad just a few miles away, and a squat capsule leapt off the pad and shot quickly into the cloud bank and out of sight. It was a SpaceX Dragon Capsule abort engine test. But as I watched the amazing sight and heard the brief scream of her engines, I also knew it was far more than that. It was another dramatic technological leap far ahead of NASA. It was engineered, created, and developed by SpaceX but paid for by NASA's Commercial Crew Program with the end goal that SpaceX could taxi astronauts securely to and from the International Space Station, yet safely abort on liftoff, if necessary.

But three years later, in April of 2019, during a very similar test, the same system exploded and instantly destroyed an un-crewed SpaceX Crew Dragon on the launch pad. It set back the national program to return access to space for American astronauts. Yet few rushed to judge SpaceX as incompetent, as they would have done if it had been a NASA capsule. The difference in the public response is clearly illustrated by the radical divergence between the two programs – messaging and plainly spoken expectations. The public is nowhere as stupid as the government expects them to be – as Elon Musk knows they are certainly not.

Of course, SpaceX recovered magnificently with their fully successful inflight test of the SpaceX Crew Dragon capsule just over eight months later, in January 2020. And the SpaceX team sealed the deal on 30 May 2020 by being the first private company to launch a crew of United States astronauts to the ISS onboard a SpaceX Crew Dragon riding a Falcon-9 rocket. That first crew to fly from U.S. soil into space in 8 years and 11 months returned safely to Earth without a single glitch on 2 August.

This example is the perfect illustration of Musk's technological and public relations brilliance compared to NASA's fail-safe, conservatively bound approach to technological progress. NASA's choice of an abort system on its ORION crewed system is a rehash of the nearly identical class of system used on the first manned space capsule (Mercury) in the early 60s and later on Apollo. In their defense, yes, it is simple. And yes, it does work. But it begs the question about the entire Space Launch System (SLS) that one NASA administrator described as "Apollo on Steroids": why are we moving backward instead of implementing new technological advances forward? NASA's clear message is that, apparently, Apollo was the ultimate space exploration paradigm, and there are just no improvements to be considered or developed. But *any* sexagenarian propped up on steroids – be it a human or engineering system – is on exceptionally shaky ground to begin with. Between the dollar signs, SLS-Orion has timidity written all over it. SLS-Orion is the hardware personification of "Failure is Not an Option" at an unspeakably profane cost.

Elon Musk is wont to disagree with NASA Administrators and their parochial rocketship system design philosophies on a regular basis. And he has clearly illustrated by his dual-purpose abort-landing system that he is entirely correct. His radically divergent approach is paying off not only in driving human space exploration costs down exponentially, but he is clearing the staid slates of proven-system imagination and foresight even while engendering and building radically divergent systems that are whole new technological generations ahead of any other space exploration architecture on Earth.

Meanwhile, SpaceX's competition, the Boeing Corporation, seemed to be in a neck-and-neck race with SpaceX to pull off the first successful launch of astronauts to the ISS using their parallel human access to low-Earth orbit system called the Starliner. On 20 December 2019, the uncrewed Starliner departed from Cape Canaveral Air Force Station's Complex 41 on a United Launch Alliance (ULA) Atlas V rocket, on what was presumably its final uncrewed test before actual crewed launches to space. Minutes after liftoff, Boeing and its Starliner dreams for human space access primacy began unraveling just as it touched the edge of space. The investigation determined its onboard computer clock was off by some 11 hours, and therefore the spacecraft failed to fire the thrusters that would send the craft to rendezvous with the space station. Other problems caused the spacecraft to fight to communicate with the ground. The spacecraft had wasted too much fuel and was unable to rendezvous with the ISS at all – its primary objective.[7] Elon Musk finally broke his silence on the cost and lack of reusability of the launch system utilized by the ULA (of which Boeing is a corporate entity along with Lockheed Martin) and declared on 13 August 2020, "Because their rockets are not reusable, it will become obvious over time that ULA is a complete waste of taxpayer money."[8]

As we look at Musk's Mars bound space system architecture, we are able to actually watch it morph live before our vision – yet another remarkable phenomenon totally unknown to NASA or any other world government space program. The SpaceX Mars exploration architecture announced in 2016 changed rather dramatically into a different one announced in 2017. And it changed yet once more into a drastic new vision in 2018, just before a truly radical departure from conventional space exploration technologies.

Meanwhile, the government contractor partnerships (many of them direct competitors to SpaceX) reply by dishing up their share of ridicule for the SpaceX team because Musk's very public methods are so outrageously bizarre to their boardrooms – government partnerships that thrive on conservative, proven engineering approaches with billion-dollar risk buy down programs. To the old school design teams, the SpaceX process simply appears chaotic and alien. And yet, it is driven by Musk's wild ADHD-esque fueled millennial genius that literally continues in real time to render every other design obsolete.

What follows here is a practically unimaginable illustration of his near supernatural process. At the start of 2018, Musk was openly advertising (and was actually in the construction phase of) the second iteration design of his carbon composite creation that he called the BFR (Big Falcon Rocket?). Atop it was mated the Interplanetary Transport System human capsule designed for transporting one hundred colonists to Mars on a single trip. But something changed his mind in mid-to-late 2018. By the first quarter of 2019, Musk had already killed the BFR design and totally scrapped its construction-in-progress. Then, he started and completed his preliminary design of a wholly new crewed interplanetary human space exploration vehicle and had a prototype flying at his south Texas facility that was renamed Starbase. The change from the old system in construction to a total redesign and new prototype flight article in the air was 10 months. *10 months!* The new rocket system was made from stainless steel instead of the far more expensive carbon reinforced fiber, and the Super Heavy booster with the crewed portion atop was renamed Starship.

Meanwhile, on the southern shores of Texas on the Gulf Coast beside a sleepy, tiny beach community called Boca Chica, Musk planted an impromptu army of workers to build his astonishing Starbase spaceport that will soon surpass any on the world. Just offshore, Musk intends to construct a fleet of launch pads for the largest rocket ever conceived – capable of launching nearly *four times* the payload of NASA's SLS system (like the SLS in its expendable mode). Pound for pound to low-Earth orbit, Starship cost is 8,000 times less. While every NASA SLS launch is priced at $4.1 billion, each SpaceX Starship launch is priced at $2 million.[9]

At SpaceX Starbase, Starship development is proceeding apace, fueled with astonishing innovation that even early NASA could never achieve. There, Musk is literally constructing a Starship assembly line, cranking out and testing prototypes in intervals measured by a few weeks. While NASA's Kennedy Space Center launched 24 individuals to the Moon in 4 years – unarguably the most amazing technological development in human history – Musk will launch four times that many humans to Luna or Mars in a single launch with a system he developed much faster.

As of this writing, SpaceX had crashed four Starships in flames in relatively rapid succession before finally sticking number five (Serial Number 15). The SN-15 trial resulted in a successful launch and landing from a relatively low altitude hop from the beach to the altitude of a commercial jetliner and back to the ground under the power of his unique methane burning Raptor engines. Musk aspirationally hopes to launch one of his Starships on his Superheavy Booster to orbit from Starbase by early 2022. The Superheavy-Starship vehicle is far and away the largest and most powerful vehicle ever built – designed to populate Mars with 1,000,000 colonists and their accompanying supplies in but a decade.

Surprising nearly everyone, SpaceX singlehandedly won the incredibly competitive NASA Lunar Lander bid to build the human landing system for the Agency's crewed lunar exploration missions called Artemis. In retrospect, while SpaceX had yet to perfect Starship, at least SpaceX was extremely far ahead of all the other competitors, who only had models and compiled wish lists to offer for NASA's contract reviews. Furthermore, SpaceX's Starship was far more capable than anything the competitors had to offer. So, why was it a surprise? Specifically because the SpaceX design represented a wholly new approach to lunar exploration while the competitors were Apollo technology look-alikes in function and in size. And while this award resulted in a flurry of pointless legal protests in the federal court system, by October 2021, the last suit was dismissed without changes to NASA's selection.

Nothing like this has ever been seen in any space development project in history – not even in the early NASA days. Musk passionately designs in his mind live as he speaks it in real time to large public audiences. Elon Musk's idea of the next engineering iteration is a new slide show at the next space conference. And that is what makes his genius work with such brilliance and power – absolute, unashamed, innocent, altogether brilliant fluidity. No one alive has ever witnessed anything like this. It is stunning in its capacity to create on the fly advanced, cutting edge, complex exploration technologies that work magnificently time after time.

Musk's soft spoken, humor studded presentations are fixed with the power of his ideas and the authority of his promises that are backed up with fully functioning hardware. Thus, his method is one of a radically divergent intelligence fueled by the authority of his in-hand, actually operational engineering creations.

But the most powerful feature of his method is his fluidity. And thus, as we observe his plan for exploring Mars carefully, it is more than obvious that he is not finished mapping his final architecture. And this is very, very good news!

Comparatively, NASA's laboriously slow-to-make-any-headway, absurdly expensive, over-budget, and extravagantly wasteful SLS-Orion has already left the station – and it is far too late. It seems that the only tool NASA has left in their bag of options is denial and implementing several hundred more countermeasures to try and offset the inexorable serial tragedies that inevitably lie just ahead.

But to be most fair, let us give NASA the benefit of the doubt and assume that Congress would be willing – and is even capable of – allowing the world's premier space agency the opportunity for a most dramatic wholesale redesign and mid-course correction. Perhaps we could imagine a savvy NASA Administrator – maybe even Bill Nelson – who appreciates the Agency's plight and the absolute need for this radical correction coupled with an understanding Congress. Together they could truly abandon and replace (after using the SLS equipment they have on hand to claim the resources of Shackleton Crater) the government's hideously expensive, as well as totally obsolete, launch system by teaming with private industry – SpaceX, Blue Origin, and others. Then, perhaps NASA could indeed get back in the game after a wholesale reset – starting at the drawing board. Logically, NASA could even swap a $4 billion per flight system for a ride on a friendly U.S. contractor's heavy lift system for 44 times less and achieve the same objectives without bankrupting the good will of their patrons in government, and simultaneously the American taxpayer who is funding this extravagant bureaucratic wet-dream.[10] NASA could still have their crewed programs, protect their astronauts, and avoid losing their stellar "failure is not an option" legacy in the sheer meltdown of a failure to keep up. Sadly, even this scenario looks increasingly doubtful.

But while the train has indeed long since left NASA's SLS-Orion station, there is still time to lift the bridge and let it slip unceremoniously off into the abyss without taking her crews down with her. NASA does have some experience with this approach....

By October 2009, it was well known throughout the government that Constellation's days were numbered, and that President Obama fully intended to make good on his 2007 promise spoken in the very first months of his race for the White House to kill the program.[11] In the hallways of the Kennedy Space Center, there was an exceedingly well-placed story that the Agency knew for certain that Constellation was doomed long months before the official announcement. But the Agency had already started stacking the first Constellation towering vehicle (Ares 1-X) and had scheduled it for launch. So instead of holding off on the $450 million launch[12] and waiting for the death knell, they concluded very soberly and unemotionally that sending the rocket off on a fiery goodbye over the horizon for disposal purposes would, in fact, be a far less costly alternative to disassembling the towering 327-foot rocket and then disposing of the useless solid booster segments.

At 11:30 p.m. on 28 October 2009, openly and unashamedly advertising Constellation as a still viable program, NASA lit the fires and let Constellation's first and last single inglorious mission fly off into the sky, and then unceremoniously dumped it into the abyss. The government saved a king's ransom in rocket disassembly and disposal costs. While it was true that the test itself was not at all a sham but represented the fully dressed-out test mission just as was originally designed, everybody and their favorite robot knew very well it was all a waste of time. And all that Constellation system data was thereupon faithfully collected for a system that would never fly again (and is rumored to also be securely stored in the same innocuous government warehouse in an unmarked crate somewhere near the Ark of the Covenant). On 1 February 2010, just 96 days later, President Obama formally announced the end of Constellation – as he said he would do nearly three years previously.

Today, Elon Musk is nowhere near such an impasse and hard decision. He still has plenty of time to integrate the First Principle of Human Exploration into his engineering paradigm and not lose *any* invested infrastructure at all. And if his tweets are any indication, he is already planning to incorporate the First Principle to some extent. In the end, he is far too intelligent to miss the obvious and give up the guaranteed success of singlehandedly moving humanity safely to another planet without an immediate Jamestown style die-off of his colony.

Finally, how about that cryptic message Elon Musk was sending by launching his red Tesla Roadster into deep space? Who was it that could hear it, and how many could understand its meaning when so many others were clueless? Musk is obviously no fool. He sent that subliminal message to the new generation of explorers who are in their teens and twenties today that are about to cast off their Earthly citizenships and become residents of new worlds. That subdued but clear message from the red Tesla Roadster was being sent to those of us who have ears to hear – the Martians who now live among us.

And the message? Well, it is one that has been received loud and clear! And if you are reading this book, future Mars colonist, that message was for you! I know I got the message – very, *very* clearly….

One Earth Normal

"The best way to encourage out of the box thinking is to draw the box correctly in the first place."

- Paul Gibbons

WE HAVE COME A VERY LONG WAY DOWN THE HUMAN space exploration blind alley, and, after half a century, we have finally reached its dead-end. The problem with blind alleys is backtracking and starting over. Yet, the starting over phase is considerably eased by the lessons learned getting there. With those lessons in hand, the retooling and reconfiguration phase is not just fast, it also allows the original goal to be attained much more quickly and far less costly than continuing down the false pathway. Continuing to pursue the dead-end is, obviously, an utter waste of money and human capital no matter how liberally the countermeasures are slathered on in a useless attempt to continue down the hopeless black hole to the entirely foreseeable disaster.

Said Agent Smith to Neo in *The Matrix*, "You hear that. Mr. Anderson?... That is the sound of inevitability…." Unfortunately, for those who have spent much of their professional lives in a race to see who can be first to reach the dead end, they have become both deaf and blind to the screeching sound of its inevitability.

In a true blind alley (like following an Apollo program style deep space exploration paradigm), the explorer can hammer on the dead-end wall all they wish. They can even try to systemically formalize it with new add-ons to the decrepit, obsolete ship of fools. But in the end, things can only get worse until that magic moment that the explorer looks about, takes stock, and uses that light bulb over their head to light their way back to the beginning of the blind alley so that they can truly find the correct path and get it right.

Such a do-over is embarrassing, to be sure. And it is difficult to reconcile with the explorer-hero persona. But, alas, it is also just another day in the life of real explorers. These actions are, indeed, defined as the process of authentic exploration.

In the case of human space exploration, the false path has led us down the unsustainable limitless adaptation trail. The further we inched our way down that blind alley, the more unworkable it became until ultimately the inevitable limit of our adaptation was reached, and now exploration under these conditions must cease or crews will pay with their lives – whether immediately or later in their lifetimes.

The inevitable – defined by the ever encroaching burden of countermeasures – began to engulf the program and to consume a massive amount of time and infrastructure so that the countermeasures themselves finally overwhelmed any progress toward the goal. Soon, as was evidenced on the International Space Station (ISS), the countermeasures, countermeasure research, and countermeasure revisions ultimately became the endpoint to the exploration venture, and the original vision of actually exploring space (instead of trying to fix the hopelessly decrepit model) was totally subsumed. Eventually, the mass and inertia of the workaround patches became the caution and warning light that signaled the end of the pathway had been reached, as discussed in detail later in this section in the chapter "Countermeasure Madness." So, as we back out of that false pathway, what is it that we need to correct first?

The very first thing we need to correct in a dramatic way is our understanding of basic human exploration in all venues. Like every other human activity, human exploration is driven by a well understood and clearly articulated human philosophy. Denial of that fact, or any attempt to replace that primal philosophical underpinning with tradition, engineering, politics, social engineering, or wishful thinking always, inevitably leads to a blind alley and collective death – as we are about to witness in deep space.

We must first correct our going-in philosophy, which defines all our engineering and mission designs thereafter, based on what we have learned. It is the philosophy that drives the train. That philosophy is the Sun in the center of the exploration system that everything else revolves around. Without that energizing, focal power and seminal understanding of self and its defining progress, the purposeless model will invariably branch off into chaos, as we are witnessing now. Every. Single. Time.

At the risk of overstatement, let us review that empowering philosophy writ large over the portal of our new human exploration paradigm once again.

First Principle of Human Exploration

"In every exploration system, we must require the systems we build to adapt to the human standard rather than expect the human to adapt to the machine or the environment – and in every design activity we will protect the human as a primary objective."[1]

Now that we deep space explorers have backed out of the dead-end pathway and have armed ourselves with permission to entertain this new concept, let us look around to further define it. As we do, we immediately discover that it has *already been engineered* by the United States Navy, who are highly successful explorers of inner-space. They have already specified their logical, thriving approach as ensuring a one Earth-normal environment in their submarines, thereby reducing the crew adaptation requirement to zero. Thus, the United States Navy submarine force has already fully adopted the First Principle of Human Exploration without ever actually articulating it. Neither a single cent of resources nor a single moment of time is wasted on adaptation countermeasures because they have planned a one Earth-normal environment directly into their ground-up engineering and thus have no need for any correction countermeasures – except for providing Arctic foul weather gear for the crews when they surface from under the ice at the North Pole!

Of course, those who are hopelessly committed to the old school exploration methods and the various romances associated with weightless spaceflight (and are also somehow inexplicably attracted to exposure to damaging levels of cosmic radiation) will immediately launch into a long list of excuses as to why U.S. Navy submarine standards are not even possible in space. In the end, all their "but-but's" are merely weak excuses to defend the status quo. Some applicable quotes that fit this case precisely are:

"Excuses are nothing more than another reason to fail." Mac McGovern[2]

"An excuse is worse and more terrible than a lie; for an excuse is a lie guarded." Alexander Pope[3]

"A life (*or, in this case, a system*) that is defended by excuses has become comfortable in defeat, it has settled for mediocrity and is destined for failure." Oscar Bogna Nomvete[4]

The bottom line here is direct and frankly quite simple: Had we not taken the false path at the beginning, and had we taken the philosophic path designed by the United States Navy to protect their crews, we would not be sitting at the wrong end of a blind alley in human spaceflight today. We would, in fact, today be sitting in permanent colonies on the surface of Mars and Luna – assuming, of course, that our ego infected Chief Executives would have reigned in their narcissistic dysfunction at the same time.

Interservice rivalries notwithstanding, the U.S. Navy provided the clear example of a First Principle exploration model that could easily be mimicked for space, and the fly-boy approach missed it entirely. This is typical of Navy superiority, as was clearly evidenced in the 1950s nuclear aircraft debacle. (Full disclosure and transparency disclaimer – I am a former United States Naval Officer and a student of Admiral Hyman Rickover.)

But NASA took the adapt while incorporating band-aid countermeasures route. And here we are, ramming our heads against a brick wall at the end of the wrong dark and seedy alley in human exploration. There is really no need to attack and kill the messenger (even though former Naval Officers are a tough lot).

Instead, those of us who have committed our lives to the human exploration of space need to collectively suck it up, retrace our path back out of the blind alley, and redesign our systems to be more U.S. Navy like – protecting our crews as the first priority of our engineering system designs and simply forgetting about adaptation to alien and dangerous environments that are proven to be inimical to life and living, such as are found in the deep oceans of Earth. After the initial emotional breakdown over the announcement of the finding, things are actually astonishingly easy to correct. It is not all that hard, especially when we discover how simple the solution and its standard really are.

David and Goliath

NASA, however, is not so quick to see the handwriting on the wall or know what to do with all those flashing caution and warning alarms. It has been said that there are no glass ceilings at NASA Headquarters.

That is true – for they are all made of titanium alloy. That comment has absolutely nothing to do with gender equality but with access into the administrative ivory tower at 300 E Street Southwest. Penetrating the tower is difficult enough. But when some scientist or engineer at any NASA Center dares to innovate under the Administration's radar, God help the poor civil servant when HQ finds out. Instead of sending a team down to evaluate their work, truly excited that someone may have a better way or an idea of genuine innovation, NASA Headquarters' reflexes often have the opposite effect.

As an example, I had been working on an innovative project in the late 1990s of a mobile laboratory-habitat to demonstrate advanced life support system concepts for a remote and extreme environment, with education and outreach components. My base of operation was in a shop complex on the Cape Canaveral Air Force Station on the other side of the fence from Kennedy Space Center. We were situated at the historic vacant Apollo Launch Pad at Complex 34, working mere meters from the remaining launch platform with its weather worn plaque bearing the chilling reminder of the ultimate cost of exploration:

LAUNCH COMPLEX 34
Friday, 27 January 1967
1831 Hours

Dedicated to the living memory of the crew of the Apollo 1

U.S.A.F. Lt. Colonel Virgil I. Grissom
U.S.A.F. Lt. Colonel Edward H. White, II
U.S.N. Lt. Commander Roger B. Chaffee
They gave their lives in service to their country in the ongoing exploration of humankind's final frontier. Remember them not for how they died but for those ideals for which they lived.

My work was thoroughly vetted through my boss, Dr. Bill Knott III, the Chief for the Advanced Space Life Support Systems Design Team, and had local approval through the Biomedical Office, which was my Directorate. Indeed, our NASA Washington D.C. Headquarters Chief had personally inspected our construction facilities. When the laboratory-habitat, the *Scott Carpenter Space Analog Station,* was completed, it was moved to Key Largo, Florida, to perform its mission located 30 feet under the ocean. The precursory shakedown missions were completed, and we successfully conducted our official SEEDS II (Space Exposed Experiment Developed for Students) mission concurrent with the Space Shuttle STS-86 mission, then took the Station on the road for a flurry of education and outreach visits to schools and public locations. This was followed by a formal Commissioning Ceremony in the Kennedy Space Center Rocket Garden, attended by Astronaut Scott Carpenter, the KSC Center Director, NASA Headquarters officials, and local politicians who addressed the crowd, which contained some of the classes of students who had participated in the program.

Our success led to scheduling a second round of missions the following Fall involving distinguished guests that would further our education and outreach impact. It culminated with the mission that was to run concurrently with the Space Shuttle STS-95 mission with Astronaut John Glenn. Senator Glenn recorded a short video supporting our concurrent work underwater.

We returned to Key Largo and once again carefully established the *Carpenter Station* in its underwater location. Eventually, the day came that I was scheduled to brief the Kennedy Space Center Director, Retired U.S. Air Force General and Space Shuttle Pilot Roy D. Bridges, on my now high-profile work. I invited Mr. Bridges to personally inspect the Station on the seafloor and to formally request the approval to proceed on the first phase of that year's operations, which included hosting numerous prominent guests who would connect with education and outreach locations across the continent. Well-known director-producer-writer James Cameron stood by to be our first guest. (Claudia and I had the distinct privilege of spending 24 hours with the personable and knowledgeable Cameron in the tiny Station as he earned his Aquanaut certification and generously shared with students over telecon links.)

As Mr. Bridges was also a certified SCUBA diver, he suited up in his SCUBA gear and swam down to the habitat for his personal inspection. I gave him the tour and explained our mission objectives, and he gave us his thumbs up to proceed. I surfaced alongside him and escorted the Center Director back to his car, sincerely thanked him for his time, and politely waved as he departed. But just as he drove off, I turned around in the parking lot to face one of the facility secretaries who handed me a telephone note marked "URGENT." It was a NASA

Headquarters telephone number. I gulped, closed my eyes, and whispered, "*Oh crap.*" This could not be good. I dialed the number on the paper. I was right – it was not good at all.

The extremely angry voice on the other end of the line ordered me without discussion to shut down my operations immediately and return to the Kennedy Space Center. When I inquired why, he replied that it was an "unsafe operation." Dumbfounded by the claim, I asked him what information he had that led him to that conclusion. He was not interested in discussing that and just directed me to terminate the mission forthwith and hung up.

I grimaced down at the yellow telephone note I still clutched in my hand. I had no idea what he was talking about as I knew the undersea Station intimately and better than anyone else – because I had designed it, had it built, and then successfully operated it for a season of undersea missions the year before. And my Center Director had just inspected it personally. Suddenly, I remembered the Center Director's business card that he had slipped into my hand just before he left and quickly fished it out of my pocket. Being a veteran of these kinds of Agency political machinations, he must have already sensed that something was coming.

I called Mr. Bridges on his cell phone and explained my dilemma. He had not even had time to drive out of Key Largo! His response completely surprised me. From my perspective in the NASA pecking order, I was standing between two angry elephants dancing around the tarmac, and I was just trying to stay out of their way. Getting crushed by one or both of them was high on the list of probabilities at that point. But Mr. Bridges said to me:

"I'll take care of NASA Headquarters. You carry on. It's my mission. It is a KSC mission, not a NASA Headquarters mission. I have inspected your operation, I have approved it, and I will take care of the problem. After your mission, make an appointment to see me in my office."

Thankful and relieved, I did as he instructed, and we had no other problems from NASA Headquarters. After the mission, Mr. Bridges ordered a top-down safety review of the *Scott Carpenter Space Analog Station* habitat for the record. It took NASA Safety six months to inspect it and review its systems. Not a single safety issue was discovered. The only finding was the recommendation to install a backup ground safety circuit inside the habitat to back up the one already on the pier – which our marine electrical engineer felt was counterproductive, but we did it anyway.

It was clear to me that Mr. Bridges probably already had wind of a problem even before he arrived to inspect the *Carpenter Station*. Then, as the layers began to peel away months later, and what had actually happened was ultimately revealed, things became truly bizarre. Please be aware that I am not seeking to defend myself or my actions – because my KSC team's mission was an outstanding, flawless success. The mission and the team need no defense.

I did not skip any step in keeping all of my superiors informed of every phase of my operations in nearly three years of planning and outfitting the undersea missions, which involved 15 weeks of underwater activity spanning two years. I was working with the local KSC group and with my Education and Outreach funding manager and her team, who were nearly all NASA Headquarters connected Civil Servants.

Obviously, the NASA Astronaut Office was not in any way connected because it would have been totally inappropriate to include them. But it was from the Astronaut Corps, of all places, that the problem began.

I personally met many astronauts in my 30 years at NASA, first as a Space Vehicle Safety Specialist, then continuing throughout my career as a life sciences systems engineer. I still have the privilege of calling some of those amazing individuals my friends today. Even before joining NASA, I had interviewed Mercury Astronaut Scott Carpenter for the U.S. Naval Institute's monthly journal, *Proceedings*. He became a true friend and personally contributed monumentally to my career. I must add that I have never met any astronaut that did not deserve their wings. I have found almost all of them to be well selected, enthusiastic, and perfect for the job. But there always seems to be one exception in every crowd...

If illustrating ego on a scale of 1 to 10, this astronaut would easily rate a 15 – quite different from his peers. (In order to protect this self-identified deity, I shall not name names.) Yet, because this book is already going to cause enough of a monumental stir in some Agency high places, I will refrain from the gross details of this story to at least try and preserve the equilibrium at NASA Headquarters, if that is even possible. The bottom line was, that as the layers were peeled back on this rather odd set of circumstances, it turned out that our small operation (that would not even appear as a microscopically thin line on any NASA budget) was innovative enough to threaten a much larger operation similar in concept to ours that, although it was far better funded, was still in the early design phase. Our mouse-sized innovation threatened the individual with all the money, prestige, and power in the Agency. And one phone call from him with allegations made up from thin air almost shut us down. Had not another astronaut (who also happened to be our Center Director and outranked the complainer) just inspected our laboratory-habitat minutes before, they would have been successful in shutting us down.

The fact is, authentic innovation is exceedingly threatening to the status quo in any human activity – even in the trillion-dollar game of human space exploration. More directly speaking, innovative individuals greatly threaten people entrenched in powerful positions. Innovation, as in this case, pits the Davids against the already well entrenched giants. Like the boy David's plan, a slingshot is a weapon whose applied power can be projected far beyond the length of a mere metal sword. Goliath ridiculed David until the totally unexpected and innocuous weapon held in the hands of a youth projected enough inertial mass to destroy his brain from a distance.

The Sunday School story of David is a tale of human innovation, and it features all the actors of 21st-century human spaceflight: the small guy with a genuinely innovative idea facing off with the truly formidable government agencies and their contractor empires. Over the years working inside the NASA behemoth, I have often discovered that it is the extremely innovative that most threaten the status quo. It is the small team working late at night and on weekends that most worries the established programs, for it is from these innovative, opportunistic skunk works – who, for good reason (as this story clearly illustrates) are working hard to stay well under the radar – that the creative impulses originate that often cause the most angst inside NASA.

What the Agency needs instead of a fire brigade to extinguish these tenacious brigand techno-rebels is a radar to detect, nourish, protect, and encourage them. It would be far better for NASA to innovate and change internally than to be displaced by innovation that has come from outside the ivory towers and causes their carefully guarded empire to experience overnight obsolescence – as is coming to pass before the eyes of the world today.

Technological Genesis of the False Path

The technical genesis of the initial false path (blind alley in human space exploration) makes perfect sense, even 50 years later. The race to be the first to land humans on the Moon was dressed in science, but it was in fact funded for (not by), and fundamentally powered for, strategic and military reasons. We had to be first because our mortal enemy (quite literally in 1960) threatened us with the military superiority of the high ground of space. Clearly, there was no cost too great and no effort too small to make it happen. It was just like a national war effort – because it truly was (or, in this case, a "cold war effort"). Our technology was not just stretched to its ultimate limit, we literally invented on the fly what technology we did not have. We learned how to do that from the greatest generation (who was still in control of our country), who in turn learned it in the race for the American Republic to survive the authentic Axis threat of World War II.

The Mercury-Gemini-Apollo technology was bare bones, hurry up and get there as fast as humanly possible. We literally did not have time to worry about nuances such as microgravity or cosmic radiation. Fortunately, we did not have to. Our initial crews were up for such short periods and returned with no immediately discernable negative effects that we were able to push on and win the technological race against our foes.

What no one ever envisioned in the fever to win the race across deep space to the Moon was the long-term negative effects of the victory itself. It was a singularly great moment in planetary history. Some still say it was perhaps the most significant human event since the dawn of creation. It was a project so incredibly successful and epically historic that it was framed and locked into perpetuity as the go-to model for all future human space exploration. Unfortunately, it was also fatally flawed for that future mission in ways no one could discern when the glory all ended in 1972. This created a pair of seemingly unlinked problems for all the human space exploration systems designed thereafter.

The first problem created was that Apollo did not at all represent a realistic model for long term space exploration in terms of ergonomic and psychological adaptation and, most particularly, exposure to the long-term damaging effects of microgravity. Secondly, the Apollo exploration model failed to account for the biochemical effects of chronic incident cosmic radiation and all the associated radiobiology issues. And, as if to magnify the threats, these issues synergize with one another to make the conjoined physiological effects even more crippling in deep space over long term missions.

Apollo's single-minded focus to be first was brilliantly met. Unfortunately, we continue to celebrate it as the final word in human space exploration systems – which is far more than it ever truly was, and which it is decidedly not in the 21st century. Apollo was never meant to be used as the definitive human space exploration design template for all future space exploration activities – even on steroids. It was simply conceived to be fast and first. It literally represents the bare bones limit of what we needed to make it to the Moon at all. In 1969 technology, the Apollo system was a flying miracle that arrived at Luna with its haplessly overloaded computational system alarms blaring all the way down to the lunar surface. Were it not for the pair of cool heads named Neil and Buzz who manned the controls, things could have easily ended far differently.

Remember from the earlier chapter "SLS-Orion: NASA's Dysfunctional Stepchild on Steroids," von Braun had plans for a spinning wheel space station with artificial gravity as the model and direction for NASA space activities long before Apollo was ever considered as a response to an emerging Soviet threat. And if we look ahead to the next generation of human space exploration systems after we finally disabuse ourselves of the current blind alley that we are now up against, we will be able to easily see Apollo as a thin-skinned aluminum tent created for a week's camping trip to the Moon. It worked great. Lots of photos were taken, souvenirs returned to the home planet, and the homecoming parades were huge. But it's long past time to get over it.

The next generation requires a full-up interplanetary home with protection from cosmic radiation and with all the amenities to make the yearlong stay on Luna truly enjoyable and, above all else, safe.

Clarke and Kubrick's Prophecy

Sir Arthur C. Clarke was a walking, innovative intellectual genius in his own right. I had the privilege of meeting him and speaking with him personally as a student at Oklahoma State University in 1971. His vision of the future of human space exploration was so expansive that he could see all the way beyond the impasse we are currently experiencing to the next generation of spacecraft just ahead of us. Clarke teamed up with motion picture producer Stanley Kubrick and released the epic film *2001 - A Space Odyssey*. I strongly encourage you to take another close look at the movie. Pay careful attention to the spacecraft *Discovery*. If the patients had not taken over the NASA Headquarters asylum, this would have been our next step after Apollo, following von Braun's initiative.

One reason this Hollywood project should be so incredibly embarrassing to space exploration mission designers is that a year and a half before we landed on the Moon, Clarke and Kubrick already skipped Apollo, fast-forwarded to the next generation of human spacecraft, and had it up and running on the big screen. Fifty years later, we discover that, obviously, no one leading any NASA design teams actually watched the film and integrated its approach to artificial gravity. Otherwise, we would not be in the fix we are today. (No mention in the film was made of cosmic radiation protection because in 1968 little was known of the cosmic primary environment in deep space and the Van Allen belts were discovered just ten years before the movie was released.)

The engineering point that Clarke and Kubrick were making was powerful and right on dead center. They had designed and built their cinematic spacecraft to mimic one Earth-normal for the *Discovery* crew's long interplanetary mission to Jupiter. It was considered by Clarke and Kubrick just as absurd then as it is today to even think of demanding that crews suffer the major human discomforts of living more than just a short time in the microgravity we have come to discover is not just uncomfortable, but toxic over extended exposures. And, in 1968, they had no data at all for the ultimate dramatic ill effects of prolonged microgravity on crew health! They gave *Discovery* artificial gravity simply because of the epically innate stupidity of not doing so! Ironically, the space system designed for the movie was far more advanced in a realistic way than any human system ever launched, even today. While that statement is also widely thought to be true of the Star Trek starship *Enterprise*, the difference is that we had the full technology to build and launch *Discovery*'s artificial gravity system in 1968 when the *Space Odyssey* film was released. In stark comparison, even today, well into the 21st century, the Star Trek spacecraft designs are hopelessly unrealistic on any level. (All fan rage accepted!)

But there is at least one human who apparently did pay careful attention to the design of *Discovery*: Elon Musk. On 18 June 2018, Musk tweeted a video clip of the interior of *Discovery* and said, "Running track in @SpaceX BF Spaceship will look something like this." He then added in his characteristic style, "And we're definitely going to have a pod bay door override button. Lesson learned."[5]

Instead of coming back home after Apollo and rationally discussing our real future in space and our long-term exploration plans and strategy, the Space Agency was tossed about like a political bone between incoming administrations and the unbroken line of successive presidential JFK wannabees. We ginned up space missions to match unused Apollo era hardware left after the unceremonious, premature cancellation of the program to pay for a tragic Southeast Asian war that epically blew up in our faces.

Quoting Vizzini from Princess Bride, "You've fallen for one of the two classic blunders! The first being never get involved in a land war in Asia, but only slightly lesser known: never go in against a Sicilian when death is on the line!" It appears the U.S. government fell for both blunders and mounted a challenge against deep space itself without recognizing that just like a land war in Asia, in unshielded deep space, death is always on the line.

Pay careful attention to the overriding exploration theme of these half century of follow-on missions. It was human adaptability to the space environment. Here is the thought process: The Apollo heroes fully adapted. They were the right stuff. Therefore, the incredibly successful Apollo model must be duplicated in all other follow-on human programs of all the nations of Earth. Nobody needed artificial gravity in the 1960s. Astronauts could adapt, and all we needed in the future to change or modify the realities of human physiology (and thereby force adaptation to a hazardous environment) would become known as countermeasures.

By the time the Space Shuttle began to fly, and the U.S. Space Station was on the drawing boards, the philosophical die had been irrevocably cast. Adaptation to the space environment was fully expected, even on 1000-day journeys to Mars. It is the way we do things in NASA. It has always been this way. If it was good enough for Apollo…

Ultimately, it is a useless exercise to what-if the past. The collective morass of erroneous decisions and bone-headed technical choices were made for all the wrong reasons, and almost all of them by clueless, self-serving Presidents and politicians making science decisions for the Space Agency in Congress. But today, we have the clear opportunity to reverse the mistakes of the past and launch our spacecraft designed around the First Principle, protecting our crews as a priority so they can productively and in full health do the explorer's job each day. After the mission, they can all return home with one Earth-normal physiology: good eyesight, first-rate cardiovascular health, nominal bone structure, normal musculature, and not loaded with ticking genetic time bombs for deadly cancers later in their lives.

Just think of it – we can explore like the United States Navy. Not spend a single second experimenting on human limits and the damage done to our crews because that is the job already taken care of by our engineering systems. In the future, we will spend our time in deep space actually exploring, not endlessly documenting our own philosophically failed processes and making that absurd quest our very reason for being.

Further, by the employment of First Principle mission architecture, female and male astronauts can be assigned with total equality and without risking damning an XX chromosome (female) crew member with downstream cancer just to fly out to Mars or stay longer on the Moon based on a decision of social engineering rather than medical facts. NASA crews can be guaranteed perfect health, just like sailors coming home from extended missions on United States SSBN submarines. How do we assure that for all future astronaut explorer crews? By engineering, as closely as possible, a one Earth-normal standard into their spaceflight systems.

The Standard: One Earth-Normal

What exactly does one Earth-normal mean? It is defined as the ideal environment for all human beings on Earth as derived from the average measurements of key conditions. It is comprised of the particulars of one atmosphere of pressure, a temperature of 72° Fahrenheit, 1.0 gravity of acceleration, and a background exposure to ionizing radiation of no more than 2.5 millisieverts per year (of which 11% of background radiation consists of cosmic radiation). Thus, it is important to realize that one Earth-normal includes some exposure to high energy cosmic radiation – about 0.0009 times that found when unshielded in deep space. It is also essential to understand that these parameters define the average conditions for people on Earth, not the exact environment experienced by *every* individual. Therefore, it is accurate to recognize that the ideal one Earth-normal conditions form a baseline for a slightly varying range in which all human life can safely and successfully function without permanent injuries.

Even with an exposure of an average of 0.28 - 0.33 mSv per year of cosmic radiation on Earth, the human body is not engaged in adaptation of any kind. Instead, it is functioning at one Earth-normal in virtual solid state. These are the conditions for which humans were specifically wired. If you alter any of these conditions, the body then begins to engage in an adaptation mode or an injury repair mode. As we have previously discussed, once the body achieves a point where it can no longer adapt, it has reached the Human Adaptation Limit – that point when the human body shifts from an adaption response to controlling and healing chronic injury.

Quite obviously, a vast majority of Earth's human population *does not* live in exactly one Earth-normal conditions. Instead, most of us live within what was previously defined as the Human Adaptation Zone: the zone at which the human is capable of functioning at near 100% physiological and mental capacities without any risk of permanent injury over indefinite exposures to multiple off-nominal one Earth-normal environmental conditions.

In this rather narrow band of off-nominal conditions, humans have developed various natural biochemical countermeasures to keep the body inside the Human Adaptation Zone via homeostasis. We wear warm clothing when it is cold. We wear light, short, and loose clothing when it is uncomfortably warm. We engineer our living spaces to keep us inside that narrow band of temperatures as close to one Earth-normal as possible. We humans are constantly busy keeping adaptation to an absolute minimum by our behavior and countermeasures. Thus, our very specialized human niche engineering process allows for a great majority of people to live very close to one Earth-normal every day by nominal achievement of many complex, integrated processes of physiological homeostasis. So, the collective purpose of all our engineering systems (such as air conditioners, heaters, fans, clothing, and all the rest) is to move us as close as possible back toward one Earth-normal conditions.

But there are whole population exceptions, of course. There are clearly situations where humans live in extreme temperatures and pressures far from one Earth-normal, and yet thrive. The Nepalese are one such example. Living in the very high Himalayas, they have adapted since birth to atmospheric pressures that are dangerous for those who live at sea level. It is said that a high percentage of the average human population, if plucked from sea level and flown immediately to Base Camp Three on Everest without any acclimation, would probably die in 24 hours. And even the hardiest world class mountaineer cannot compete with the Nepalese Sherpa, who routinely rig Mount Everest for climbers before the climbing season and then carry their packs nearly all the way up to the world's highest peaks. Their bodies have almost completely acclimated to the extreme altitude from the first day of their birth, and they can easily cope without injury below the Death Zone. But higher than 26,000 feet, at the designated Death Zone, even the Sherpa begin to lose the capacity to adapt and, just as all other humans, are in danger of succumbing to the critically low atmospheric pressure. Many totally acclimated Sherpa have died above 26,000 feet in the Himalayas.

The capacity of indigenous Eskimo adaptation to intense cold is another example of an extreme limit that humans have adapted to from birth and thrive. Greenland, the Yukon, Siberia, and Lapland, Finland, are further models of populations established in some of the coldest regions on Earth. Likewise, Athens, Greece; Jeddah, Saudi Arabia; and Sulaibiya, Kuwait, are some of the hottest recorded places on the planet, yet communities thrive even there – some since antiquity.[6]

As an aquanaut, I have lived for extended periods of time, totaling more than 30 days, under 1.67 atmospheres of pressure. That is right at the no-decompression limit. But at that depth, even over weeks of exposure for me and many others, there have been few reported injuries or long-term negative effects. It appears that at least 1.67 atmospheres of constant pressure may well be inside the Human Adaptation Zone. And yet, deep saturation divers living and working under hundreds of feet of pressure for days and weeks have experienced many of the exact same multimorbidity symptoms as do astronauts on prolonged space missions in low-Earth orbit. For example, they experience "airways dysfunction and changes in lymphocyte population numbers leading to a relative immunosuppression. Changes in neuropsychological function have also been detected, including altered memory and cognitive ability with corresponding brain stem reflex abnormalities and EEG changes[7]" (as I cover extensively in my book *Undersea Colonies*, where I also discuss the importance of First Principal habitats for exploration and permanent settlement in the undersea realms).

Beyond the nominal Human Adaptation Zone limit, even for those who have adapted at the extreme ends of the zone, deleterious symptoms are common to all humans living for prolonged periods in extreme environments. Inevitably, their bodies break down, struggling to maintain the healing process while trying in vain to cope with their chronic injuries.

We have almost no valid scientific information on living for prolonged periods in gravitational fields that are less than 1.0 gravity of acceleration but not at microgravity. Once again, this is thanks to NASA bureaucrats who caved in to political pressure and killed the *only* instrument on the International Space Station (ISS) that could have answered that specific vital question (this is discussed in detail in the upcoming chapter "The Partial Gravity Scandal.") It still ranks as one of the most important questions we have no answer to – indeed, one of the most vital questions for human settlement off Earth! It also ranks as one of the most astonishingly shortsighted technological failures of NASA – a politically contrived waste of a key potential of the International Space Station to discover the truly essential answers to the vital questions of human space exploration.

As a young space life scientist bioengineer, I remember a lecture given by my brilliant boss and mentor, Dr. Bill Knott III. He said that plants react to gravity in a biological response to the gravitational vector called gravitropism. Plants are engineered to grow "up" against the vector. They also respond to light (phototropism), and the roots respond by growing toward water gradients (hydrotropism). He noted that there had been studies in the 1960s that demonstrated that plants only needed a relatively small gravitational field to engage gravitropism and grow with some degree of normalcy.

Obviously, there is quite a span of differences between plants and human beings. Yet there is a fair amount of speculation that the gravity of Mars (38% of Earth), or even the Moon (16% of Earth), may be within the Human Adaptation Limit, unlike prolonged exposure to microgravity. But without research and data, we are only truly guessing. Today, no one really knows the answer to the question, which is tragically long overdue for the individuals who are even now readying their equipment to make the long term leap into deep space and other planets. What long term effects will reduced gravity have on crews and colonists? Not a single person has any clue because NASA was asleep at the switch, ignorant and blind to their most fundamental reason for being.

Finally, there is abundant evidence from exposure to microgravity that bone loss and disappearance of musculature begins within days, as we will discuss in a subsequent chapter. The speculation (without any research data) is that unless a rigorous, disciplined exercise and physical training program is required for the partial gravity of Luna and Mars crews, that some amount of musculature and bone loss will also inevitably occur on both of these worlds. That is because healthy skeletal structure is specifically designed to hold our bodies erect against 1.0 gravity – one Earth-normal. If you drop that gravity to near zero, as on the ISS, bone loss begins almost immediately, *even despite a training regimen*. The same is true for loss of musculature over the skeletal frame. Like the bones of our bodies, the musculature is specifically designed to hold our bodies erect against 1.0 Earth gravity. Again, we do not have any data at all concerning bone or musculature loss in partial gravity, as is found on Luna (0.17 g) and Mars (0.38 g).

As a logical extension of that speculation, it is believed by many physiologists that the bone loss and loss of muscle *may* (I would strongly emphasize this is sheer speculation) mirror the partial gravitational field. In other words, crews on Luna would experience about three times more loss than their astronaut counterparts on Mars. And if the Martian astronauts eventually do acclimate to Mars gravity, they may find a return to Earth very difficult as their physiology would have fully adapted to a new one-normal standard: the one Mars-normal standard. It is quite possible that just as plants can adapt fully to partial gravitation environments without injury, humans *may* also be able to adapt to partial gravity as well – but certainly not without a serious adaptation penalty on return to Earth. And yet, we do not know.

No one knows, or can even legitimately speculate on, the prevalence of puffy face syndrome, elevated intracranial osmotic pressure, or any other effect of microgravity occurring in a partial gravitational field. (These conditions are discussed in detail in the upcoming chapter "Chronic Microgravity Exposure Etiology.")

Again, humans have the capacity to adapt within the nominal adaptation range. Rarely, they can even do so with little or no countermeasures. But the primary concept to understand is that to prevent permanent injury or death, humans must live inside the Human Adaptation Zone as close to one Earth-normal as possible. Outside this zone, the process of injury begins, and will ultimately lead to unrecoverable injury or even death if it is not reversed.

Inevitably, intentionally depriving humans of protection to extreme, toxic environments that cause injury or a measurable increase of risk of death becomes a question of professional organizational ethics. Once we develop systems that are specifically designed to transition humans to environments, or integrated life support systems that exceed known limits outside one Earth-normal, it is essential – and in many cases, a legal requirement – to document the ethics of the program. But before that can take place, the organizations in question must actually have genuine integrity – and not have sold and altogether compromised that integrity by making intragovernmental deals or writing waivers to preserve their exploration paradigm at the direct cost of crew health and wellbeing.

The Ethics of Neverland

"Ethics is nothing else than reverence for life."

- Albert Schweitzer

On the Bleeding Edge of Insanity

IN THE SECOND WEEK OF JUNE 1974, AFTER COMPLETING A year of graduate school, I stood a mere pace apart from my exceptionally intelligent and quite world-wise father as I was about to board a bus to United States Navy boot camp. After he assured me of his love, Dad left me with a single snippet of advice: "Son, don't volunteer for *anything*." I pondered this morsel of wisdom during that bus ride to my destiny and a lot of hard-won maturity that lay dead ahead in my very near future.

That bus trip signaled the beginning of four decades of government service, during which I quickly came to understand his own hard-won insight. I had signed up to join the most powerful organization in the history of the world. It advertised itself as the bulwark of freedom, and that was often entirely true. But this government also has its very dark and sinister side as well. Its many basements are filled to overflowing with skeletons. Its countless secrets are deep and classified – not merely for national security but for self-protection. And not every one of them is decent, ethical, or pretty.

The deeper I have looked over the years, the more I have come to understand that my Uncle Sam has the congenital disease of latent insanity, and he often acts it out. And when he does – living out his fantasies in his own Neverland on the bleeding edge of his well-protected version of insanity – things get very ugly. That is when I came to understand the wisdom of my dad's parting words – that deliberate duplicity and ethical breakdown are part and parcel of Uncle Sam's personality. Therefore, to volunteer for one of his many schemes is to risk believing that the story he has told you is entirely true, and you have actually received the complete data set before risking your life.

In this breakdown of fundamental ethics, many trusting souls have been and soon will be lost. Heeding the call to follow Uncle Sam into the historically sketchy Neverland of government ethics, whether it be to participate in a medical study (Tuskegee), stand out on an open field before the mushroom cloud (Operation Desert Rock), or ride out on a functionally unprotected spacecraft into deep space (deep space expeditions), is to experience his latent insanity up close and personally, bringing home far more than merely a t-shirt in exchange for your trust.

Ethics and Blind Engagement

The terms "ethics" and "blind engagement" go together like gasoline and matches. And their catalyst is more often than not willfully sponsored by governments who have the unfettered power to use the unholy pair for their unrestrained purposes.

Blind engagement was a term coined by the human resource marketing expert Mark Willaman at the turn of the 21st century. Blind engagement describes the practice of artificial engagement through the casting of wide-net mechanisms of the internet with the followers of a brand – and that is where it gains its power and momentum. While the high-tech term has only been around for several decades, its practice is actually centuries old.

In the case of human space exploration, NASA relies heavily on blind engagement to maintain interest in the NASA brand. And NASA also requires a targeted blind engagement directed toward Congress in order to continue to fund its brand. In every case, blind engagement by NASA leverages its well-earned global reputation for technical excellence and maintains its role as the unchallenged leader of world technology development. Because of NASA's legitimate success, its use of blind engagement with the followers of its brand has been virtually effortless, carried forward for half a century by its unrivaled historic reputation.

But blind engagement is, after all, always blind. And that is where blind engagement and ethics have historically collided – and with tragic results.

Trust me – I'm From the Government

The phrase, "Trust me, I'm from the government," is a bit of dark humor very much earned by many governments throughout history. As is quite evident, the expression implies that there exists an unspoken blind engagement of the public toward the government's brand. And it is also evident in the actual use of the expression that this expectation always comes packaged with the real risk of catastrophic consequences. The United States government itself has fully earned its own dubious place in this hall of shame.

Below is a partial, dark laundry list of international ethics laws, regulations, and guidelines that have emerged since the mid-20th century from the civilized populations of the world in the aftermath of governmental betrayal of the trust of their people in brutally creative ways. And why did these governments do these awful things to their own people? Simply because they could.

Nuremberg Code (1947)

The Nuremberg Code was written as an ethical bulwark to document and prevent any western government atrocities in medical experimentation such as the Nazi government engaged in during World War II. This Code of medical experiment ethics arose in response to the Nuremberg Trials of Nazi doctors whose medical laboratory terrors were unspeakably cruel and, for the first time, documented the true specter of the power of government coercion on its innocent private citizens.

Of course, the United States government gets particularly annoyed when compared on any level with to the National Socialist German Workers' Party. Their annoyance would be well understood, but for a few of their own rather "annoying" exceptions, as we shall cover in this chapter.

Declaration of Helsinki (1964)

The Declaration of Helsinki began to sharpen the distinction between the raw horrors of the Nazi medical experiments and the more refined systemic ethical breakdowns of the late 20th century. Through seven revisions (1964-2013), it sharpened and codified a vast range of medical practices and ethics involving human subjects in research.

The Belmont Report (1979)

As was stated previously, the United States Government typically reacts with undisguised holy, self-righteous indignation when compared to the National Socialist German Workers' Party (Nazis). With some notable irony, the United States was itself engaging in its own abuse of its citizens nearly a full decade before the Nazis ramped up to create their horrific hell on Earth.

As nearly all governments are inclined to do, the Belmont Report (officially titled the Belmont Report: Ethical Principles and Guidelines for the Protection of Human Subjects of Research) was written to address the United States' particularly Naziesque practices launched against its own subset of what the government health service apparently categorized as "undesirable citizens" with zero rights. The Belmont Report was written by the National Commission for the Protection of Human Subjects of Biomedical and Behavioral Research, which was created as a result of the National Research Act of 1974.

The reason for the establishment of the National Commission for the Protection of Human Subjects of Biomedical and Behavioral Research stemmed from the barbaric, serial, grotesque failure and, again, self-policing of any semblance of ethical behavior on the part of the United States Government. The Commission was established in the aftermath of the infamous Tuskegee Syphilis Experiment – a clinical study conducted between 1932 and 1972 by the U.S. Public Health Service in which rural African American males who had tested positive for syphilis were studied for decades. Even though the government researchers knew and meticulously documented that these men had contracted the disease and clearly had the means to treat them, they did not inform them of that fact in order to study the horrific progression of their infection over many years at the cost of multiple lives of mostly African American citizens.

And the Tuskegee study was only one of a series of other such profoundly unethical government, self-policed studies conducted over the decades prior to that, such as the use of prisoners for medical studies with little or no consent and nearly always accompanied by coercion.[1] The Tuskegee subject men were initially told that the experiment was only going to last six months, but it was extended to 40 years. By the time the experiment was uncovered by the press, 28 patients had died directly from syphilis, 100 died from complications related to syphilis,

40 of the patients' wives were infected with syphilis, and 19 children were born with congenital syphilis.[2]

How did this happen? Because a self-policing government rarely acts with integrity in carrying out its own self-examination. As will be described in the next pages, the government simply exempts itself and its employees from the laws that they write to protect everyone else, thereby openly endangering those individuals' lives, health, and welfare to protect the government's schemes.

In a 21st-century redux of the Tuskegee experiment, the United States Government, through its Space Agency, now leverages its considerable power of blind engagement to protect its status quo, even if the cost is in human lives and wellbeing.

The Government's Neverland Army on the Desert

As tragic and truly unbelievable as the Tuskegee scandal was, it was typical of U.S. government overreach at the cost of life and permanent injury. But the tragedy of Tuskegee is only one small example. Between 1951 and 1957, the United States government launched a classified military operation named Operation Desert Rock, in which it is estimated that 750,000 troops participated. Their assignment was to go out and sit in a World War II era trench dug into the southwestern American desert and wait for an atomic bomb to go off just a few thousand yards from them. And then, while the mushroom cloud was still building over their heads, to actually march toward ground zero in order to let the fresh nuclear fallout fall directly on their bodies.

The narrator for the training film made as the exercise was actually unfolding relates the objective for the surreal actions in the single sentence, "The biggest value of this operation is to prove to ourselves that it can be done and to find any weaknesses."[3] Apparently, the government intended to prove that half a century of accumulated science of exposing humans to high levels of ionizing radiation somehow did not apply to their troops. And even while the same government was exhaustively documenting the devastating effects of exposure to this identical fallout in Hiroshima and Nagasaki, perhaps they reasoned that the highly radioactive residues of nuclear weapons were somehow more user friendly on U.S. soil. Or perhaps tough American soldiers could "adapt" to the onslaught of nuclear energy on their bodies.

It was, in a single word, insanity. And the price of this ethical collapse is still being extracted from the participants and even their children's lives decades later. Just how bad were the results of this abuse of U.S. troops and wholesale collapse of ethics? We will never know because much of the information is still classified. As I learned while serving as a commissioned officer in the U.S. military, the United States legitimately protects its security by classifying information – just as it likewise illegitimately protects its own unbridled unethical behaviors by the same classification system. Even worse for past military victims of Operation Desert Rock, it appears that the government did not expend the identical effort to carefully monitor and document our citizens and military members as they did the victims of Japan's nuclear exposures.[4]

But after the effects of the exposure began to sicken and kill veterans and their children, the government set up a fund for their care in a 1990 bill titled Radiation Exposure Compensation Act.[5] As in this example, the United States Government reflexively responds by attempting to correct their wrongs – but only after the damage has already been done to the people under their charge. In a few pages, you will read about the government proactively setting up a system to care for the victims they are aggressively planning to create in the future in deep space called the TREAT (To Research, Evaluate, Assess, and Treat) Astronauts Act. It is an interesting new twist in the cycle of government ethical response to a calamity they are *planning on in the future* but have not yet actually launched, and injuries and deaths that have not yet been documented but that they clearly know are coming.

Thin Ice – NASA and Human Subjects Research

As mentioned, the National Research Act was signed into law in 1974 and created the National Commission for the Protection of Human Subjects of Biomedical and Behavioral Research and spurred the subsequent release of the Belmont Report. This Act established the Institutional Review Board, or IRB, for studies involving human subjects. There is a similar review process and established Board for animal research subjects called the Institutional Animal Care and Use Committee (IACUC), for which I was the NASA Chairman at the Kennedy Space Center for 14 years covering U.S. Shuttle, Mir, and ISS payloads. Both the IACUC and the human IRB were managed out of the KSC Biomedical Office, where I was assigned at that time. In that position, I was very familiar with the workings of both Boards and, on three occasions, submitted my own research protocols and presented them to the KSC IRB.

Not only was I conversant with the process and both Boards' activities, but I also had day-to-day firsthand knowledge of the individual people involved. I cannot speak more highly of the personal integrity or the intimate commitment to compassion and excellence among those teams. While I was not as familiar with the same institutional teams at the Johnson Space Center who handled all astronaut protocols, I was nonetheless quite aware of their activities, knew many of their members during that time personally, and have witnessed that they shared the same process commitment to excellence and integrity as their KSC counterparts. However, these committees and their dedication to following the letter of the law and looking out for the interests of all research subjects are *not* the subject of this chapter.

All human research protocols conducted in space are reviewed by the NASA IRBs, and there are never any exceptions. The IRB is comprised of physicians, physiologists, members of the community, and professional ethicists. Every step of their research activities is reviewed in depth for scientific and medical necessity and requires certification by the professional submitting the protocol that the research is necessary, not duplicative, and minimizes the risk to the subjects involved. Further, the subjects involved are given full and complete disclosure in an exhaustive list, including any and all hazards the experiment might present to them while they are subjects. And they must sign a statement indicating they have been made fully aware of the experiment, its goals, and any risks to them before they are allowed to participate. There are no exceptions to this process, and the legal penalties for negligence on the part of the researchers or members of the IRB are severe. This is the view from inside the committees themselves.

But the view outside is an *entirely different* matter
and is radically dissimilar.

While the individuals who are a part of these committees are honest people with deep integrity, their world class excellence is unfortunately clouded by, and deliberately limited by, the limitless capacity of government to significantly interfere with and crush the ethical process by its interminable methods of justifying and defending its self-interests above all other considerations, even if those interests are not especially moral or ethical ones. This conflict of integrity perhaps lies at the very core of a most significant intuitional blind spot. And all of it is centered on NASA's exploration philosophy.

Giving Self-Permission to Change the Rules Covering a Flawed Exploration Model

Let it not be lost in all the superlatives written in this book about the Space Agency – NASA is still a part of the federal government. And it is the government that sets the rules for everyone covered by the laws they write and enforce. NASA, or anyone inside the Agency, is *never* permitted to break the law. *But* the Agency can grant itself exceptions and rewrite the law while acting in its legal capacity as an arm of the government that no one else is privy to. And, because its established exploration model is fatally flawed, in order for NASA to continue crewed exploration in deep space using the obsolete, antiquated Apollo human exploration model, they were "forced" to change or amend the law that protects all others in the United States so that they could "legally" cover themselves. Otherwise, they would not only never depart low-Earth orbit, they would be limited by law to how many days American astronauts are allowed to fly in space, including missions aboard the International Space Station (ISS).

When it comes down to publicly admitting to fostering a fatally flawed deep space crewed exploration model, starting over, and all the congressional and public horror that would inevitably entail, NASA bureaucrats apparently reasoned it was just easier to amend the laws that would never be amended for anyone else and just fly anyway – to hell with astronaut crew health. NASA had thousands of astronaut wannabes standing in line. So, as long as the current astronauts were permitted to fly, they would undoubtedly fly today and take their chances with the very dubious models based on terrestrial ionizing radiation – which was exactly what happened. The only problem with that kind of blind engagement to a failed model is that innocent (and very distracted) people are always injured in the process.

And here is the legal driver underlying this ethical failure: if the limits are changed, then the Agency IRBs can review the flight protocols with a clean conscience and let them fly, specifically because it is within the altered legal limits. Yet if the IRB reviews a protocol that exceeds a federal law or limit, they absolutely cannot let it pass. Hence the impasse. NASA has deliberately changed the ethical rules and thereby created a rubber stamp IRB process.

Today, NASA must fish or cut bait on their way out to deep space on long-term missions to Luna and Mars, and their regulatory watchdog has already seen it coming. The NASA Office of the Inspector General announced in October of 2015, "Based on current knowledge, astronauts on a mission to Mars would exceed NASA's career radiation dosage limits. Although the Agency plans to continue efforts to develop countermeasures to address the radiation risk, NASA is likely to *seek an exception* from the current standards for those that cannot be fully mitigated."[6]

The truly horrifying moment in this two-sentence revelation from the IG comes when it is ultimately revealed just from *whom* it is that NASA will seek the exception. But worse still, the Inspector General (the government's apparently toothless watchdog) and NASA *fully expect* to carry on with a fatally flawed model and risk injuring American astronauts *as their going-in plan*.

The entire text from 29 Code of Federal Regulations (CFR) 1960.18(a) – which is the current standard – reads:

"In addition to complying with emergency temporary standards issued under section 6 of the Act, an agency head shall adopt such emergency temporary and permanent supplementary standards as necessary and appropriate for application to working conditions of agency employees for which there exists no appropriate OSHA standards. In order to avoid any possible duplication of effort, the agency head should notify the Secretary of the subject matter of such standard when the development of the standard begins."

I am surprised that the legally designated "neutral voice" of the NASA Inspector General in the October 2015 announcement is so transparently biased toward propping up the faulty exploration model at the expense of the health and wellbeing of the crew – in clear contraindication of international standards of exposure to any source of ionizing radiation. Note the careful permissive wording: "NASA is *likely to seek an exception* from the current standards for those that cannot be fully mitigated." Really?! Since when is it *impossible* to fully mitigate cosmic radiation? The last time I reviewed the phenomenon from my very first class of radiation control technology 101, all radiation of any species can be very successfully mitigated with time, distance, and/or shielding. It is as though NASA has authored its own set of physical laws that are somehow different than those that the rest of the universe operates by. That apparently is what happens when lawyers and government bureaucrats wind up in the science business.

Since the source of cosmic radiation is the galaxy itself, we can go ahead and give up distance mitigation to the IG – but time and shielding, we will certainly not. NASA is up against its own self-designed, fatally flawed model and has somehow convinced the Inspector General of the "impossibility" of providing adequate shielding for their crews in circumvention of their rather dismal grasp of basic physics. While it will certainly demand a change to their process, their models, and the cost, it is a complete fabrication and denial of a rather straightforward scientific process to state in a public forum that it is *impossible to fully mitigate*. Obviously, the only reading of this is that the Agency is setting the stage to change the federal and international standards and limits *again* rather than their archaic human exploration model. The first change was to populate the ISS for long missions. The next change will be to enable astronauts to fly on extended Lunar or Mars missions – ratcheting up the permissible radiation limits once more to truly horrific levels.

The fox has been placed in charge of the NASA hen house, which is always the easy way out for the government when it inevitably winds up policing itself. If it were not so tragic and of such historical import, such hair-brained shenanigans could be comical. Yet, if they proceed with this plan, their historic imprint will certainly be anything but humorous.

Going forward in this book, I will refer to the OSHA (Occupational Safety and Health Administration) occupational radiation limits as well as limits on exposure to ionizing radiation placed on nuclear workers by the Nuclear Regulatory Commission (NRC). While there are some descriptive differences, the bottom line is that the annual limits are the same for both agencies. OSHA requires employers to protect workers from exposure to all ionizing radiation sources that are not regulated by the NRC. And the NRC regulates exposure to radioactive materials such as those found in nuclear power plants.

The NRC limit for radiation workers is regulated by Title 10, Part 20, of the Code of Federal Regulations (10 CFR Part 20), "Standards for Protection Against Radiation," which establishes the dose limits for radiation workers. Although the limits vary depending on the affected part of the body, the annual total effective dose equivalent for the whole body is 50 mSv.

The OSHA limits are defined by 29 CFR 1910.1096, and just like the NRC, although the limits vary depending on the affected part of the body, the annual total effective dose equivalent for the whole body is likewise also 50 mSv – described in the Federal Regulations as limits of 12.5 mSv/quarter.

Before NASA began racking up uninterrupted months on the ISS, NASA astronauts were classed together with all professional radiation workers under the Occupational Safety and Health Administration's (OSHA) classification system. They are the same legal limits that are placed on all workers at nuclear plants, and even upon commercial airline pilots. Limits for all other classes of nuclear workers proved far too low for NASA (whose implementation of adequate shielding was "impossible" and thus threatened their obsolete exploration model), so they petitioned a fellow government agency. And, of course, the government's right arm gave its left arm the wavier simply for the asking.

Government-agency-OSHA eventually even acquiesced to the point that it actually allowed government-agency-NASA to set its own radiation exposure guidelines! Under 29 CFR 1960.18, NASA was granted a waiver by OSHA to institute independent limits for ionizing radiation exposure for their astronaut crews.[7] Therefore, astoundingly, OSHA's occupational ionizing radiation protection limits no longer apply to NASA. NASA's own Office of the Chief Health and Medical Officer now establishes the radiation exposure limits. So, NASA will simply ask itself for permission for any exceptions they need when they fly humans outside the Van Allen belts, which they will have to do if they ever wish to depart the confines of the ISS and venture to Luna or beyond. While United States radiation workers are protected by regulations that require that their risk remains well below statistical significance, NASA crews are given a choice – either quit or accept the Agency's Risk of Exposure Induced Death (REID) that statistically determines a certain number of astronauts will die from cancer because of the government's self-exception to federal standards for worker safety in Astronaut crews.

It is essential to carefully note that according to the regulation (which in the United States carries all the weight of Administrative Law), NASA's right arm of the government had to declare a radiological exposure "emergency" to OSHA, its left arm, and then fully justify such an emergency as a permanent part of doing business. The wording of the regulation required NASA to support that the emergency claim exists forever "for application to working conditions of agency employees for which there exists no appropriate OSHA standards."

This is certainly a convenient plan for NASA. What entity would not like the ability to just call up the office down the hall and arrange a deal that would allow them to not only legally make up their own occupational radiation limits to cover their failed exploration model, but to do it on a sliding scale basis available for virtually unlimited escalation? And how about enlisting their own Inspector General to go along with the plan by "anticipating" yet another lifting of standards? Remember the IG's prescient statement looking forward to a crewed Mars mission? "NASA is *likely to seek an exception* from the current standards for those that cannot be fully mitigated...."

Consider now the plight of the poor nuclear power plants in the United States. They have both OSHA *and* the NRC watching their every move carefully and demanding not only strict adherence to the United States standard but also active ALARA programs (As Low As Reasonably Achievable) – which has resulted in annual reductions in exposure to levels far *below* the required standard! And if they do not get on this bandwagon and keep exposure low, unpleasant things occur in the industry. But not at NASA. They are headed dramatically in the opposite direction of allowing magnitudes of exposure over the legal limits established for the rest of the world and are using ALARA itself as another layer covering their radiological exposure excesses.

The alarming double standard between NASA and its nuclear power counterparts on the ground is more than conspicuous. Imagine that if *any* nuclear power plant got tired of the time and expense of implementing legitimate ALARA standards resulting in continuous *reductions*, so they just "declared an emergency" to OSHA, announced that adequate shielding is simply "impossible" for them to implement, and thereby changed their occupational exposure limit specifically so that they no longer had to mitigate employee exposure to radiation. After all, strict compliance to regulations and continuous improvement on the standard are expensive and require radical changes in the way they do business in order to protect their workers. The problem with even suggesting the implementation of that nifty song and dance would be the immediate response of its *other* monitoring agency – the Nuclear Regulatory Commission.

Administrative law required that NASA prove that they needed to eliminate the requirement to keep NASA employees and astronauts at the same radiological exposure level as all other U.S. workers specifically because "there exists no appropriate OSHA standards." They hung their argument, of course, on the term "appropriate," which is, in itself, extraordinary and does not especially exactly speak well of the regulatory body called OSHA that has absolutely no problem hammering everyone else on the ground for the slightest infraction. As a NASA Space Vehicle Safety Specialist professionally responsible for monitoring and enforcing OSHA standards for two years, I know whereof I speak. As a matter of fact, my NASA Safety mentor informed me early in my career that due to our responsibility to seriously and persistently monitor for any OSHA infractions, Safety Specialists were so disliked that you could not even find six other Safety Specialists to help carry your coffin. But it was the law, and the law brooked no flimsy excuses, especially the whine that anything was "impossible" to stay within the legal framework.

In granting the exception, did OSHA mean that the legal limitations on the ground were appropriate for nuclear power workers but not necessarily appropriate for government employees called astronauts? That could not possibly have been a part of the public discussion because that would brazenly reveal the astonishing weakness of the government's contrivance to exempt itself. The argument obviously centered around the lack of "appropriateness" of protecting astronauts in space with adequate shielding, with NASA arguing it would cost far too much, requiring radical changes in every aspect of human access to deep space, and *therefore was not possible*. In the end, the weight of the Agency's argument is logically thus deduced as "the safety and health of NASA astronauts is based on long standing and codified standards on the ground, which demand continuous reduction in the standard, which are not appropriate in space." Why not? Because NASA does not want (or does not realize that they need) to change their exploration model – and astronauts are volunteers (with people lined up in the wings to take the much coveted job) and are therefore quite expendable.

Obviously, NASA HQ will be infinitely outraged by such an inference, but it is impossible to escape the only remaining logical conclusion of the matter. Would any of those mission planners willingly allow their precious child to be exposed to the radiation equivalent of 55 whole body CT scans per year? I would certainly hope not. And if any NASA officials say that they would, I will invite them to have an interview with their state's child services authorities. And *why* would they be reluctant to approve such exposure? Because the damage to the bodies would be irrevocable – and they know it. Indeed, NASA knows this truth better than anyone and has taken the regulatory agencies out of the loop just so that they can fly with their sliding scale radiological limits under their own control. The only watchdog NASA had left to check their folly was their own Inspector General. And, by their October 2015 public statement, the NASA IG appears to have been compromised as well.

To actually witness this dramatic collapse of government ethics in layer after layer of regulatory oversight is quite extraordinary. It is the 21st-century repetition of Tuskegee. Except that the government is so certain of the power of their blind engagement (as evidenced by 22,000 applications of the nation's best and brightest for the latest astronaut class) that they need not worry about the little issue of deliberate injury to their crews who each have been promised the unique social distinction of becoming astronaut heroes just for trusting the United States government. And, to appease their conscience, they have enacted the TREAT Act to care for their future victims. It is rumored that one wealthy individual of the Renaissance, at the end of his life, built a hospital for the poor and that upon completion, the ongoing gossip was that it was far too small for all the victims he himself had created in his lifetime. NASA and Congress will soon be able to relate.

Keeping that in mind, recall the NASA IG's statement once more: "Although the Agency plans to continue efforts to develop countermeasures to address the radiation risk, NASA is likely to seek an exception from the current standards for those that cannot be fully mitigated."

And from whom do they seek that exception? *They seek it from themselves*. It is, at best, a clear case of egregious conflict of interest –

absolutely unethical and self-serving on the part of the United States Government.

The entire objective of the Institutional Review Board, in a method ensured by the individuals who comprise the Board itself, is to make certain that the legal framework developed specifically for the protection of research subjects is rigorously followed. But when the law is changed by the very agency that submits its protocols to the IRB for the purpose of advancing their activity, then their oversight role has been short circuited and corrupted. If the protocol complies with the law or government policy (however it was contrived), the IRB then becomes a rubber stamp for the Agency that made up its own laws and limits just to meet its mission requirements. In this case, the collapse of ethics that formulated the genesis of its oversight role has clearly imploded, and the entire purpose of the Space Agency's IRB system has become a part of the ethical sham.

Before the October 2015 statement, there was already a very clear, well justified, and *quite appropriate* OSHA standard in place. It just did not fit the NASA flight model that had already reached its operational limits and end of its functional life. So, NASA bureaucrats were set free to do whatever they wished. Thus far, NASA's skies have been clear to any remonstration. Evidently, the pair of government agencies get along quite well with one another, and, evidently, OSHA trusts NASA explicitly, even though OSHA would never allow such an agreement with the nuclear power industry. And, for exceptionally good historical and medical reasons, the NRC would certainly and absolutely never allow such cavalier institutional insanity to run free! Fortunately for everyone, the Nuclear Regulatory Commission is apparently not a participant in this appalling cabal.

This ethical meltdown fully embraced by several Executive Branch administrations defines the well documented, precise historical moment that the NASA human spaceflight model finally actually collapsed. It happened the instant NASA needed relief from the constraints of administrative law to push the antiquated Apollo human exploration model well beyond its ultimate limitations. But instead of pulling the plug and letting it die a natural death while replacing it with a better, more technologically advanced model, the critically ill space exploration model, already on life support, was propped up by the ultimate legal permissions to do whatever, whenever – ethics, propriety, and crew protection be damned. The stark tragedy of the government's legal shenanigans is only eclipsed by NASA's open and public use of the term ALARA to somehow justify this ludicrous abortion of individual radiological occupational protection.

The oversight and regulation of NASA were handed over to itself, and thus the end to their legal conundrum was finally safely in their own hands. The program was not forced to regroup, retool, and reconfigure until it met the legal standard for safety and health, as in every other nuclear and non-nuclear industry in the U.S. But, since the United States Government is not legally compelled to follow its own rules and guidelines, the NASA human spaceflight program thus continues today with only the oversight limits and laws that NASA set for itself with a wink from OSHA, its fellow government agency.

The problem with that approach is that once the end was reached (before the 29 CFR 1960.18 bailout), the NASA exploration model should have been allowed to collapse on its own. Then NASA would have set about retooling and rebuilding. Instead, NASA has been very busy planning and building more deeply flawed and dangerous hardware, with the requisite billions of dollars expended, for a series of missions into deep space, already knowing and anticipating (according to the NASA Inspector General) increasing the radiation limits one magnitude more with the stroke of a NASA pen. Their current plans include sending crews to Mars in conditions that so exceed ethical safety standards that a very statistically fuzzy subset of astronauts will ultimately wind up permanently injured, and many will die. Concurrent with the inevitable human tragedy will be the virtual ruination of NASA's ultimate "failure is not an option" credibility, along with the loss of all those billions of taxpayer dollars. Such is the typical shortsighted approach of a dinosaur with the footprint of a house but an organizational mind resembling that of a gerbil and the moral facility of a swamp crocodile.

The Trouble with a Second Opinion

With the evidence mounting from low-Earth orbit mission after mission that there was something seriously amiss in the expectation of adaptation to chronic exposure to the space environment, another outside organization conducted a review of NASA's crewed deep space mission plans and rendered an unwelcome assessment. In 2014, NASA received the review from the National Academy of Medicine (NAM), formerly the Institute of Medicine. NAM reviewed the Agency's plans for committing astronauts to long term deep space missions given the medical research accumulated in low-Earth orbit and the ethics of proceeding in the face of mounting uncertainty. It is my sincere hope that the Agency administration, or at least someone well placed in the hierarchy, began to realize the enormity of the looming crisis and asked for the review. But if NASA was looking for weasel words to justify their evolving ethical gray area, they were very much disappointed.

After examining NASA's long-term deep space mission intentions, the National Academy of Medicine concluded that any such endeavor under the stated parameters would involve an irresponsible amount of medical uncertainties. The report stated,

"These types of missions will likely expose crews to levels of known risk that are beyond those allowed by current health standards, as well as to a range of risks that are poorly characterized, uncertain, and perhaps unforeseeable."[8]

Its brief but powerful and unambiguous conclusion stated, "The committee finds relaxing (or liberalizing) current health standards to allow for specific long duration and exploration missions to be ethically unacceptable."[9]

That is the problem with a second opinion when one is in the business of self-regulation. You might actually get one you do not like. So, what does a huge government agency do with such bad news?

Nothing, of course.

I could not find a single NASA publication for the general public that even mentioned the report's conclusions even though there were many references to it in numerous other venues and publications. Interestingly, however, the Agency's Chief Health and Medical Officer cited the report in June 2016 before the Congressional Subcommittee on Space (Committee on Science, Space, and Technology) during their hearing on Human Spaceflight Ethics and Obligations: Options for Monitoring, Diagnosing, and Treating Former Astronauts. NASA's Chief Health and Medical Officer mentioned the NAM report in a four-line brief as a part of his testimony to the Subcommittee in a section titled "External Reviews."

Before I give the text of this testimony offered to Congress, please pay careful attention to what was *not* presented. Not a single word was given to the actual conclusion of NAM's report on the Agency's ethics. Again, the conclusion in the NAM report was, "The committee finds relaxing (or liberalizing) current health standards to allow for specific long duration and exploration missions to be *ethically unacceptable*."

But here is what the NASA Chief Health and Medical Officer summarized to Congress:

"(The NAM Report) states that NASA has an ethical and moral imperative to ensure access to health care, longitudinal follow up, and preventive long-term health screening of current and former astronauts."[10]

While the Chief Health and Medical Officer's testimony was given to encourage Congress to pass legislation that was ultimately enacted in March 2017 titled the TREAT Astronauts Act, it was typical of a government agency to leave out the key part of the NAM report that stated clearly that it would be unethical to actually put their astronauts in the position of needing much of that life-long monitoring and care in the first place. Even more astonishing, in a world class example of disingenuous testimony, the report's ethical conclusion was used in reverse: to pass a law to monitor the long-term health effects of deliberate injury to its crewmembers – the very outcome the report warned against! Apparently, no one in Congress that day challenged the testimony that selectively referenced the NAM Report findings in order to defend the Agency's request for legislation to monitor the extent of the damage and injury they intended to expose their crews to in the future.

The Nuclear Burnout Approach – The Lifetime Radiation Limit Quagmire

In June 2021, in response to an Agency request, the National Academies of Science recommended that NASA establish lifetime/career cosmic radiation limits of an equally applied 600 mSv for male and female astronaut crew's career lifetime.[11]

The Academies clearly pointed out that any other approach (the genetic and physiological exposure differences between sexes notwithstanding) is inherently sexist and, therefore, the male astronauts are to be constrained not by limitations of human genetics or physiology but by a prevailing cultural "fairness" doctrine. I have no problem at all with the recommendation specifically because no human being, regardless of sex, should be permitted to fly in these cosmic radiation ovens at essentially unshielded levels (significantly above federal standards for radiation workers) for *any* reason. Anything that limits the doses to any crewmember is to be applauded – social construct or not.

In the Academies plan, once the limit is achieved, the astronaut is no longer eligible to fly... unless, of course, the self-regulated and self-policed Agency conveniently writes itself a wavier (but we shall discuss that in a few pages). However, NASA, encouraged by the National Academies of Science, is merely returning to an old exposure problem in the nuclear industry called "burnout."

It is useful to frame this discussion in its historical context. Until 1994, the nuclear industry's maximum lifetime exposure rule was expressed in the formula (N – 18) x 5 in rem, where N is the current worker's age in years. For example, a worker who was 40 years old was limited by this formula to a lifetime exposure of 1,100 mSv:

$$(40 - 18) \times 5 = 110 \text{ rem } (1{,}100 \text{ mSv})$$

This indirectly resulted in what we referred to in the nuclear industry as "burnouts." A burnout was a floating contract individual who selectively worked hot jobs at their maximum annual exposure rates, running as fast as they could to their lifetime limits at various nuclear plants around the country. This was done for very lucrative salaries. Once they reached their burnout level, they quit and found another profession. The problem with the burnout approach was that the workers were receiving acute doses up to the government regulated max limits over relatively short periods of time. But (even in the early years of commercial nuclear power) lifetime career limits were designed for doses received over a career that had the potential of lasting decades – not just a few years. Therefore, the burnouts were literally abusing their bodies in the mercenary approach, swapping easy money for the probabilistically determined elevated health risks that may not show up for decades or perhaps never at all. Unfortunately, they were not only playing Russian roulette with *their* future health but with the health of their future progeny.

The Nuclear Regulatory Commission changed all this in 1994 with 10CFR 20 and, ultimately, the implementation of ALARA empowered by federal statute to put regulatory teeth into finding ways to continuously mitigate and *reduce* the dose – no longer permitting policies in the commercial nuclear industry that encouraged what is now the NASA approach of running the dose as high as the regulatory limit allows and then adding a waiver for even more dose (which is by any conceivable measure unethical). But the U.S. government, with its agencies, has the power to obviously do whatever it wishes, and then periodically runs Baghdad Bob out into the public square to cover themselves by unashamedly claiming adherence to ALARA's benefits of continuous reductions of exposure to terrestrial ionizing radiation.

While the allusion to the infamous Iraqi government truth spinner may sound excessively harsh, Baghdad Bobs are not created overnight. They are always creatures of a creeping deliberateness born of a methodical process of the government's evolution of a declared "state truth." In this case, the Agency is so inured to protecting their flawed model and so accustomed to the widely assumed veracity of the great lie, that adequate shielding of crews in deep space is declared "impossible." They are duty bound to literally obscure the certainty of physical law as though just by speaking an alternative reality into existence somehow gives it the veracity of authoritative, sacrosanct truth. In the inevitable round of ensuing investigations, they always seem surprised that they did not actually have the authority to dictate physical laws as well as government policy that ended up costing lives.

Since 1994, the new lifetime and annual dose limits in commercial nuclear power plants have been designed around observed dose reductions specifically ensured by ALARA mitigations. The astronaut's terrestrial nuclear professional counterparts on Earth in the United States are constrained by a 50 mSv annual limit so long as the average annual dose over five years does not exceed 20 mSv per year. Under these constraints, based on Scott Kelly's exposure during his 340-day mission, no NASA astronaut would be able to spend more than 47 to 116 days on the ISS, depending on their cumulative radiation profile.

Remember that Kelly received a whopping dose of 146.34 mSv of cosmic radiation (not lower energy terrestrial source ionizing radiation) in 340 days (just 26 days less than a year) on ISS – an average of 0.43 mSv per day. This dose was over three times the legal annual limit for commercial nuclear professionals on Earth. And after applying the dose limits for the consecutive years rule, in less than one year, Kelly received more than *seven times* the limit allowed annually over a five-year span on Earth! All this is perhaps invalidated to some unknown degree by the as of yet unaccomplished actual comparisons between the higher energy cosmic radiation and terrestrial ionizing radiation.

Unfortunately, there is almost no data to make these comparisons and with which to knowledgeably and ethically send human crews out into the cosmic radiation onslaught in deep space – a conclusion also voiced by the U.S. Centers for Disease Control and Prevention, as will be discussed in a few pages.

But remember that on Earth, in the United States, these doses are the maximum legally allowable exposures for workers in the nuclear power plant (NPP) industry who (because of the legitimate application of ALARA's programmatic dose mitigations) have *pushed their dose down* to around 4% of their maximum allowed dose. The average NPP worker receives 2 mSv of exposure per year, or 0.005 mSv per day.[12] That is some 86 times *less* than the average daily rate of exposure of 0.43 mSv that Kelly received in the functionally unshielded ISS.

In 1993, the National Council on Radiation Protection and Measurements (NCRP) published its Report 116, "Limitation of Exposure to Ionizing Radiation," in which it recommended that a cumulative lifetime effective dose for an individual from occupational exposure should not exceed 10 mSv multiplied by the individual's current age in years.[13] In other words, a 45-year-old astronaut, NPP, or radiation worker (since the research that accompanies the recommendation applies to all humans) should not exceed a lifetime effective dose of 450 mSv rather than the previously accepted 1,100 mSv – a reduction of almost 2½ times. Using this formula, a 27-year-old astronaut should not exceed a lifetime effective dose of 270 mSv (a lower dose due to their ability to develop more radiation induced maladies or cancers over their average longer lifespan), which is over 4 times less than the 1,100 mSv limit.

The significant spectral differences of cosmic and terrestrial ionizing radiation notwithstanding, in deep space on board an Apollo class interplanetary ship, any astronaut with no previous spaceflights on their record would exceed that dose in 250 days – almost exactly the average time required to reach Mars per the Hohmann catalogue of transfer orbits. So to do any surface work on Mars or return to Earth, they would be out of luck!

Apparently, NASA is approaching these lifetime limits as shifting objectives via its sketchy and indisputably unethical waiver process – not limits to be reduced per any legitimate use of ALARA. This way, they are not forced to call the Apollo deathtrap model what it really is and have to redesign it. This very public ethical sham was not invisible to the National Academies, who flagged this approach in their June 2021 report by warning the Agency in clear and unambiguous terms (emphasis added):

"Recommendation 5 (regarding NASA's waiver process): NASA should develop a protocol for waiver of the proposed space radiation standard that is judicious, transparent, and informed by ethics. To avoid the perception that *an exception to the standard is built into the space radiation standard itself,* NASA should follow the ethics decision framework in developing a waiver protocol and it should provide supporting analysis and explanation justifying any waiver to the standard."[14]

In their report, the National Academies also suggested a lifetime cosmic radiation exposure limit:

"Conclusion I: The committee concludes that astronauts who travel on long-duration spaceflight missions are likely to be exposed to radiation levels that exceed the proposed new space radiation standard of an effective (career) dose of 600 mSv. Unless technological advancements and engineering controls provide improved radiation shielding or other protections to astronauts, for a mission to Mars to proceed, NASA would need to seek waivers to the radiation health standard both for the mission and for each astronaut."[15]

This is fascinating for many reasons. According to the Centers for Disease Control and Prevention, we do not have enough information on the effects of full spectrum cosmic radiation to make *any* conclusions or *any* recommendations – especially not those based on the significantly different, lower energy species of terrestrial ionizing radiation sources. Said the CDC, "We do not know what causes most health problems that could be linked to cosmic radiation. We don't know what levels of cosmic radiation are safe for every person."[16]

Yet based on zero evidence concerning the actual effects of full spectrum cosmic radiation on vertebrate animals in deep space, the National Academies of Science tossed out a unisex 600 mSv lifetime limit, knowing that astronauts flying functionally unshielded Apollo model space systems that NASA is about to crew up and send into the void will *require waivers* to their stated limits *even before the launch!* Their only caveat to Recommendation 5 is something along the lines of "NASA, when you bust this lifetime dose limit exposures of your crew (as you will do to fly your obsolete exploration model out on long duration missions), at least be ethical about it...." The reality is that the entire process of deliberately injuring humans and kicking off a probabilistic death-SWAG inside their bodies is just as unethical as the Tuskegee experiments. Only this time, it is being done by a powerful government Agency out in the open with written permissions and as many waivers as they can find gullible astronauts to sign.

The recommended unisex limit that places cultural fairness over the reality of binary biology quite obviously places women under the age of 35 immediately and directly in harm's way. As was discussed previously in "A Primer on Cosmic Radiation," XX chromosome humans (females) have specific organs that XY chromosome humans (males) do not have and that are especially sensitive to the potentially carcinogenic effects of terrestrial ionizing radiation. While NASA and the Academies' discussion never even mentioned the complete lack of data on full spectrum cosmic radiation on vertebrate physiology (for which there is almost zero accumulated data), they freely opened the door to exposing younger XX chromosome astronauts to magnitudes of even more powerful cosmic radiation to satisfy cultural demands of "equality" rather than inescapable XX-XY chromosomal realities. Instead of a relatively simple exploration system redesign, they opted to up the risk even more, targeting XX chromosome astronauts (females) under 35 in their suggested "politically sensitive" plan.

Said former NASA Space Radiation Project Manager and Chief Scientist Dr. Francis (Frank) Cucinotta about this issue, "I think this change is problematic." Cucinotta likened the idea to playing Russian roulette with our XX chromosome astronauts, and that the suggested standard career dose actually places them at increased risk from radiation exposure. He expressed that exposure consequences for females are not at all biologically the same as those for their male counterparts, as the proposal implicitly implies. But the former NASA Space Radiation Project Manager and Chief Scientist did not stop there. He remarked to Space.com in July 2021 that focusing solely on cancer risk also totally ignores other problems linked to radiation, such as heart disease and cognitive issues. He replied that the Agency should be looking at more data before they make these kinds of comprehensive changes. Cucinotta said, "I don't understand why they're doing it now... There's plenty of time to do research, this decade, for longer lunar missions or a Mars mission."[17]

We need to pause for a single moment and listen to Dr. Cucinotta's carefully worded and layered warnings that precisely amplify the chapters and detailed discussions in this book. The early data is clearly suggesting that cosmic radiation is not just a cancer risk to human physiology, but likewise a threat to the sensitive, fast growing tissues of the body in the limbic brain, the human heart, and the cardiovascular systems. While in our primitive early grasp of the cosmic radiation problem, we have limited its effects to what we know from our experience with terrestrial ionizing radiation and focused on its carcinogenic properties, GCR is a much different and far more dangerous beast. It is considerably more detrimental, and insidiously layers the human organism with latent damages that may not show up for decades after the exposure. This is discussed in depth in the chapter "Hitting the Physiological Wall."

Just as this book proposes, we must gather actual data on the effects of full spectrum galactic cosmic radiation on vertebrate physiology. Projections based upon our far weaker and fundamentally different terrestrial sources do not appropriately apply. And exercising the license to apply those SWAGs to living precious humans is unconscionably unethical.

But one physician member of the Academies had another view, revealing NASA's own input and influence on the committee's ultimate deliberations even before they were engaged. "We agreed with NASA's conclusion, that the best thing to do was set it according to the standard for the most vulnerable population, which is the 35-year-old woman," they said, and added, "that NASA's exposure limits are significantly lower than those of other space agencies."[18]

So, was the NAS acting as an independent voice based on a NASA request to examine and impartially recommend? It rather seems that the Academies were determined to carry our NASA's recommendations to them instead of acting objectively as they clearly went in with instructions from the Agency – at least on biological radiation policies.

Further, notice how the NAS participant joins the prevailing NASA voice on normalizing harmful levels of ionizing radiation. "Harmful" is very specifically defined by the Agency's own statistics that these levels (even before the inevitable waivers at NASA's lowest acceptable levels) will kill at least three in every 100 astronauts based on lower energy and far lower kinetic damage potentials of terrestrial radiation sources. But the participating physician noted that NASA's levels were "significantly lower than those of other space agencies." This is not at all accurate based on NASA's levels prior to this 2021 recommendation. Until the

Agency actually accepts these recommendations, NASA's peak level for a 55-year-old man is 4,000 mSv lifetime exposure (see chart below [19-20]) while the European Space Agency (ESA) and the Canadian Space Agency (CSA) use a single career dose limit of 1,000 mSv for all genders and ages.

Career Exposure Limits for NASA Astronauts by Age and Gender*				
Age (years)	25	35	45	55
Male	1,500 mSv	2,500 mSv	3,250 mSv	4,000 mSv
Female	1,000 mSv	1,750 mSv	2,500 mSv	3,000 mSv

I have never seen or heard a NASA representative compare their magnitudes-higher levels of exposure to U.S. NPP or any other radiation profession. They almost always choose extreme examples like Chernobyl exposures or other national limits. Would it not be more instructive to Americans to hear the other side of the story, such as the average of 2 mSv/year exposure for the average U.S. nuclear power plant professional with a 50 mSv/year legal limit compared to the ISS astronaut who receives 86 times more? It would also be instructive to describe the extensive legal framework that protects NPP workers from exposure to much lower energy terrestrial radiation based on the medical and physiological fact that, as described by the United States Nuclear Regulatory Commission:

"The radiation protection community conservatively assumes that any amount of radiation may pose some risk for causing cancer and hereditary effect, and that the risk is higher for higher radiation exposures. A linear no-threshold (LNT) dose-response relationship is used to describe the relationship between radiation dose and the occurrence of cancer. This dose-response model suggests that *any increase in dose, no matter how small,* results in an incremental increase in risk. The U.S. Nuclear Regulatory Commission (NRC) accepts the LNT hypothesis as a conservative model for estimating radiation risk."[21]

At least in the 1960s Apollo era commercial nuclear power industry, the burnout limits were fixed and immovable. Then the Nuclear Regulatory Commission moved on toward more responsible and far more ethical protections by implementing ALARA and giving it legal teeth to mitigate (which led to reductions) – not granting increases nor waivers for excessive exposures to fit industry objectives. The NRC required continuous redesign practices and process improvements until ALARA had its intended effect of protecting NPP professionals. The burnout process was not just eliminated; it was prohibited because the science of the lifetime dose limits is not negotiable in a legitimate and ethical framework of human protection from ionizing radiation of all forms and sources.

But in NASA, astronauts have now taken on the role of 21st-century nuclear burnouts. The difference is that in the Agency, there is no non-negotiable limit. The waiver process assures that there will not only be whole crews of radiation burnouts, but added individuals flying with even higher radiation burnout permission slips! With the presently proposed deep space flight hardware, no astronauts can go to Mars on the current limits suggested by the National Academies of Science in June of 2021. Therefore, the waiver process will be required, as the Academies plus NASA's own Inspector General openly admit. It begs yet another ethical question – why even set a lifetime limit if you know it cannot possibly be adhered to if you continue to use an obsolete engineering model that will *always* require waivers to fly to Mars? Would it not be far more ethical to recommend a truly fixed limit for the welfare of the crew based on solid science (not a convenient SWAG with zero data) and then fix your recommendation on the ethical imperative to redesign your space system *rather than* expose the nuclear burnout crew flying on additional waivers to an ever upward spiraling Risk of Exposure Induced Death sentence?

Has NASA really convinced their fellow humans that it is actually impossible for the Agency that brought mankind Apollo to fix and redesign their systems? The answer is that apparently, yes, they have. And the equally astonishing conclusion is that most of the world actually believes this to be the truth.

But it is not, as the remainder of this book will describe.

At the End of Ethical Collapse

The living, soft tissues comprising human organisms do not give a whit about politics, government agencies, fair warnings, waivers, or radiation limits. They accumulate damage, and they multiply injury. And, if pushed beyond their limits, they are irrevocably harmed. Extending the limits simply to sustain a human contrived model or the stellar reputation of a federal agency that simply does not abide failure of any kind does not change any of this equation at all. Human tissue damages easily in the face of cosmic energies expelled from supernovae or the fringe regions of black holes. These real humans with beloved families will come back home, and some will get sick while others will die. And the United States Government will assuredly follow its historical role of handwashing, finger pointing, and shrugged shoulders. Or maybe they will pass yet another law, and then another, by which they will perhaps ultimately exempt themselves or create a substantial superfund for their victims. Indeed, the TREAT Astronauts Act clearly anticipates the need to treat those whom they have deliberately injured. In the end, after it is all over, no one who loved those affected will really give a damn whether they flew into deep space or not. They will just mourn those who are suffering or forever gone.

We already know fully and completely how to shield a human being from cosmic radiation in space. But how is the human organism shielded from the dreams of empire swirling about in the brain and heart of a massive, powerful, but essentially mindless bureaucracy? Ask the Tuskegee innocents forever injured by the U.S. Government's self-policing regulations in their vile quest for the elusive Holy Grail of research, comprised of some tidbit of superfluous knowledge apparently worth more than their subject's lives and wellbeing. These people were the poorest, weakest, and least noticed in our society. They were also the least likely to push back – and that is precisely *why* they were selected.

Governments excel at blind engagement of their goals – often at the cost of highly variable ethical standards and precious lives – in order to maintain "the mission." Of course, these same agencies would fully prosecute and destroy any non-government entity who conspired to do the same thing they practice in the open. At the end of the day, all that any citizen and their family have going for them when they walk into that government office is a clear understanding of the fundamental definition of human ethics. Whether it is into the Tuskegee Health Department or the NASA Astronaut Office, it is always ultimately the same. In the words of Albert Schweitzer,

"Ethics, too, are nothing but reverence for life. That is what gives me the fundamental principle of morality, namely, that good consists in maintaining, promoting, and enhancing life, and that destroying, injuring, and limiting life are evil."[22]

Astonishingly, the very law that was passed to honor the Tuskegee victims and to ensure such a horror would never happen again has already been short circuited by NASA and OSHA. The Institutional

Review Board itself is now entrusted with wielding the government's rubber stamp with the full, clear consent of the NASA Inspector General.

Regrettably, now, it seems, many Americans, astronauts, and astronaut families still actually believe that the United States government and its agencies will do whatever is necessary to safeguard and protect their lives, health, and wellbeing – and those of their loved ones – when they entrust themselves to their care.

Sadly, the current NASA exploration model, and the Agency seeking to relax the industry radiation exposure standards by truly extraordinary magnitudes, are, in so doing, destroying, injuring, and limiting lives. As Dr. Schweitzer has said, such deliberate actions (particularly by a powerful government that has the capacity to change their model specifically to protect lives) at this level of premediated engineering malpractice are quite simply unethical by any definition.

It is also, as he stated, evil – and evil to the deepest core.

Expendable Astronauts

"Man has gone out to explore other worlds and other civilizations without having explored his own labyrinth of dark passages and secret chambers, and without finding what lies behind doorways that he himself has sealed."

- Stanisław Lem

IN 1994, I SAT IN THE OFFICE OF MY BOSS, DR. WILLIAM "Bill" M. Knott III, together with our Controlled Ecological Life Support System (CELSS) team and our NASA Headquarters' Chief, Dr. Maurice "Mel" Averner, Program Manager of NASA's Advanced Life Support Program. Mel seemed to feel quite at home with us, his own underground science enclave, hidden out between the cold war relics of leftover missile hangers across the river from Kennedy Space Center on the Cape side. Two of his favorite sayings were: "Working in Washington is like being dragged down by ants" and "It must be pretty late in the day when ants cast such long shadows." This latter was an observation about the legendary NASA egos.[1]

That morning, we were celebrating record back-to-back harvests from our biomass production chamber. We had even integrated test subjects into the chamber to measure their carbon dioxide uptake by the plants, integrated fish into our plan, and even built a kitchen to make edible products from the food we had harvested for future space colonists. It was a heady time, doing groundbreaking research on a historic requirement for humankind's ultimate permanent break from Earth.

Mel patiently listened to our progress summary, for it was Mel who funded us. Without Mel, we would have long since turned the lights off and gone home. He was not just our Headquarters' champion – it was Mel alone who kept us anchored to NASA's Mission Plan and in business. It was not much of a task, ultimately. Our budget was almost insignificant in the scheme of Headquarters. But it was incredibly visionary, and Mel knew it as well as we did. It was Mel Averner who was NASA's first champion of terraforming Mars, and it was Mel Averner who was the first Program Manager for NASA's Global Biospherics Program. This program funded some of NASA's initial research on global climate. Mel had a loose attachment in the Headquarters crowd but the firmest of grips on where NASA was really headed in the distant future.

One of Mel's signature attributes was that he rarely ever wore a tie in NASA Headquarters and never on the road. He traveled comfortably and relaxed. Yet his stare was intense. I interpreted his initial gaze as an evaluation, and I am certain I was not wrong. But the preliminary penetrating stare loosened into an agreeable smile, although even that did not hide his intensity. Mel was a walking encyclopedia of scientific knowledge, and he was willing to share it with anyone who wished to listen. Yet he was in no sense boorish and was so engaging that, when he spoke, everything he said was not only easy to fathom but was also always powerfully insightful.

After listening to us sell our program (to the only one on the planet who actually did not even need the pitch), we finally sat down and waited for his analysis of our progress.

Mel did not stand up. He sat in the far corner of the office, silent for a moment of reflection, and then said to our bewildered surprise, "Don't be too proud of what you have accomplished. I know it's fantastic, even historic, work you all have done here, and I will continue to do all I can do to keep you in business. But this can all go away tomorrow. Some new President or Administrator or Center Director can come in here and turn the lights off without notice. And here's how it happens in NASA. Somebody comes in here and shuts you down after over a decade of hard, essential groundbreaking work vital to the future of manned exploration of Mars. And guess what? It's all gone overnight. The company brain has no long-term memory. Ten, twenty years from now, someone will come into the Agency and realize this kind of research is vital to press on ahead to Mars or the outer solar system. So, they will reinvent this laboratory and start completely over and do your work again, just like you were never here. That's how it works in the government."[2]

Mel went on to urge us to immediately implement a data storage system to preserve our work so that it would not be lost for the future and not have to all be done over again. And we did.

Unfortunately for humankind, Mel's prophecy came to pass exactly as he said. A decade later, just as we were relocated into a brand-new Space Life Sciences Facility, the CELSS work was canceled, the building given back to the state of Florida, and all our work was unceremoniously scrapped. Understanding the astonishing complexities of an advanced life support system that is an absolute requirement for the sustenance of human life beyond Earth would have to wait for *"...someone to come into the Agency and realize this kind of research is vital to press on ahead to Mars or the outer solar system. So, they will reinvent this laboratory and start completely over and do your work again, just like you were never here."*

We were forced to close because the "next generation" rocket booster and crewed capsule – based on an obsolete and dangerous exploration model – was to suck up every cent of available resources. Until, ultimately, in the next Presidential regime change, even that too would be canceled – its billion-dollar hardware scrapped and dumped into the ocean. After the dust had settled and the thrones had changed hands multiple times, our work was gone with no corporate memory that we had ever existed at all... Just like Mel had said.

Mel Averner was a brilliant scientist and an even more brilliant observer of the Agency and the federal government. I thought of his prescient warning as I was packing my boxes to leave the brand-new Space Life Sciences Laboratory and move on to another assignment. I watched NASA convulse with each new Administration and the incredible waste of the past Administration's programs with the scrapping of billions of dollars in flight hardware after each new inauguration. I witnessed Mel's DC ants firsthand, hard at work clogging up an old tunnel and making a new one that would also be subsequently clogged with each new passing queen of the mound.

Yet, the NASA people are individually much like Mel Averner – brilliant, enthusiastic, and all-in-devoted, with the single vision to explore space safely and with enthusiasm even despite the ever-shifting mission. But the dreams of empire will remain perpetually that – just dreams – if those who come to lead the empire with their own fantasies of personal legacy systematically destroy all vestiges of the vision of the people who came before them.

Mel Averner was explicitly right about corporate NASA as an entity totally different and apart from the individuals inside the Agency. Corporate NASA is often dumbfoundingly witless and absurdly illogical, driven by the ever-shifting demands of revolving door administrations and a shepherding congress with the corporate foresight of a Barbie doll. Corporate NASA often either has no vision at all, or it demonstrates negative vision in lockstep with its administrative and congressional "leadership." And, just as the US Public Health Service demonstrated with their syphilis study in the forty years from 1932 through 1972, NASA, the government Agency, can operate completely in the sphere of an astonishing lack of corporate ethics, as this chapter will describe in detail.

Meltdown at 300 E Street Southwest

In the previous chapter, I made this statement regarding the invocation of Public Law 29 CFR 1960.18:

"This ethical meltdown fully embraced by several Executive Branch administrations defines the precise historical moment that the NASA human spaceflight model finally actually collapsed. It happened the instant NASA needed relief from the constraints of administrative law to push the antiquated Apollo human exploration model well beyond its ultimate limitations."

And so it was that one collapse followed another in rapid succession. The unheralded collapse of the engineering model was followed in a few years by the slow-roll collapse of Agency ethics. This meltdown of ethics did not happen overnight, nor was it entirely the cause of NASA's in-house secrecy. NASA cannot do anything in secret, nor does the Agency sit on UFO files, hide faked Moon landing evidence, or have secret refrigerators stuffed full of alien corpses.

There are no conspiracies I have ever even heard whispered about or have witnessed even obliquely, as I previously expressed. NASA does everything out in the open – even their ethical meltdowns, in which they always seek help, advice, and inter-agency cover. In this case, NASA knew as well as anyone that they were about to step out on very thin ice, so they went beyond the Agency and sought outside justification for their intent to openly and publicly injure their astronaut crews. And, as this book goes to press, they are still very busily engaged in the same process of CYA ethical meltdown.

Nowhere to Go

Literally, by the turn of the century (hopefully well before), NASA knew it was headed for an unmovable legal barrier with cosmic radiation (even in low-Earth orbit) by using their antiquated and very flawed exploration model. On the International Space Station (on board which the crew receives an average of 0.66 mSv of cosmic radiation every day), each astronaut will break the OSHA annual radiation limit for a professional radiation worker in just shy of 76 days. To date, some 220 astronauts from 17 nations have visited the ISS. And, of the crews of more than 60 missions, only one U.S. Astronaut has spent less than 76 days on orbit. All the rest have significantly exceeded the US annual 50 mSv radiation legal limit for professional radiation workers. The International Commission on Radiological Protection (ICRP) and the International Atomic Energy Agency (IAEA), whose recommended annual norm is only 20 mSv (the equivalent received in 30 days on ISS), have stated that any individual nuclear power plant (NPP) worker actually racking up as much as 50 mSv/year exposure is described as an emergency condition, allowed only as a "short-term dose for emergency workers dealing with nuclear emergencies."[3]

What NASA deems business-as-usual on the ISS, the world nuclear power industry has classified as an extraordinary catastrophe. For example, for the nuclear disaster at the Japanese Fukushima Daiichi power plant, only six of the hundreds of workers responding to the disaster received "extraordinary doses," which were roughly equivalent to the crew exposure of ISS Expedition 46 (Scott Kelly and Mikhail Korniyenko). What the world NPP industry considers calamitous and is the Hollywood gist for hero movies such as *The Fukashima 50,* NASA accepts as nominal everyday exposure. The estimated doses of Kelly and Korniyenko exceeded the U.S. annual limit by 4.48 times and the ICRP/IAEA limit by a factor of 11.2 times.

Indeed, the rest of the world is heading in the opposite direction in radiation exposure safety by a legitimate use of ALARA (As Low As Reasonably Achievable) principles. According to the Nuclear Regulatory Commission, while the U.S. nuclear industry exposure limit is 50 mSv/year, the average nuclear worker actually receives only 1.5 to 2.5 mSv/year. And that dose is declining because of the appropriate deployment of ALARA across the nation.[4,5] On the ISS, the average astronaut will receive more radiation in *two days* than the average U.S.

NPP professional receives from industrial ionizing radiation sources in an *entire year*. That NASA brazenly waves the ALARA banner is disgracefully duplicitous.

The documented reason that ALARA standards are so rigorously enforced by the NRC is for worker protection, since, as they advertise, "The radiation protection community conservatively assumes that *any amount* of radiation may pose some risk for causing cancer and hereditary effect, and that the risk is higher for higher radiation exposures. A linear no-threshold (LNT) dose-response relationship is used to describe the relationship between radiation dose and the occurrence of cancer. This dose-response model suggests that *any increase* in dose, *no matter how small,* results in an incremental increase in risk. The U.S. Nuclear Regulatory Commission (NRC) accepts the LNT hypothesis as a conservative model for estimating radiation risk."[6]

It is on this assumption that the OSHA guidelines and limits protect U.S. radiation workers, and the ALARA standard was developed to *reduce* exposure downward *as much as possible* from the legal maximum limit, never increase it. The world legal standard is moving in quite the opposite direction of NASA standards for two well documented reasons:

(1) A linear no-threshold (LNT) dose-response relationship (or dose and cancer) and,

(2) Any increase in dose, no matter how small, results in an incremental increase in risk.

These are the precise reasons the National Academy of Medicine deemed NASA's deep space mission plans unethical. The Agency's extraordinarily reckless plans are moving exponentially in the opposite direction of the rest of the world's ever decreasing exposure of their workers to ionizing radiation. NASA now routinely exposes its ISS astronaut crews to 122 times more radiation *per day* than the average nuclear power plant worker – 0.66 mSv/NASA vs. 0.0054/NRC.

Thus, the ISS radiation environment is a true incomprehensible horror show from *any* nuclear professional perspective. NASA has always known this. If the ISS was under the jurisdiction of the NRC, its missions under the current exploration paradigm would have been limited long ago. Except for NASA posturing to the contrary, the ISS is demonstrably unsafe (and simply illegal for everyone else per ground radiation limit standards) for missions of over two and one-half months (76 days) duration.

By the second decade of the 21st century, NASA finally acted to get around the barrier for which there were no competent countermeasures or combination thereof and kept exploring space by setting its own employee radiation limits. This action averted a shutdown and desperately needed redesign. Unfortunately, it also only kicked the can down the road. Regrettably, the Agency's astronauts are living inside that can.

But NASA did not show up to the office one day with a letter from OSHA handing them the proverbial keys to the safety cupboard. OSHA had to be convinced in writing to document their legal abdication to their government brethren. OSHA has butt-covering responsibilities, just as does NASA. Ultimately, a typically hapless Congress supplied the blessing.

In the larger view, NASA had few choices. The Agency either had to change the law, make a new law, shut down long-term space exploration altogether, or finally admit their exploration model was broken and subsequently fix it. In other words, redesign or take the damn-the-cosmic-torpedoes-full-speed-ahead approach. Naturally, NASA usually takes the bureaucratic path of least resistance, so they moved to make their own laws concerning exposure to cosmic radiation. The only problem with that plan is that NASA cannot make its own laws, and the government agency to whom the law has been entrusted cannot abdicate to another agency. So, NASA needed a loophole that resembled the law-making approach. All loopholes in Washington DC are begotten with a series of labyrinthine meetings that would put any *Maze Runner* adventure to shame.

Yet, before the first meeting, NASA had to leave the office with their scientifically sound plan carefully documented and ready to present. Fortunately, NASA's radiation professionals are literally peerless in the world. Further, NASA either directly funds or has access to every significant cosmic radiation study in the world. Hence, in the process of preparing their brief to the outside agencies for review, NASA combed these documents – carefully – to make their ultimate point, which was eventually directed (via these outside agencies) to OSHA. NASA could not alter, fake, or even begin to hide the horrifying numbers. But they had another strategy.

They would normalize them.

Normalizing a Horror Show in the Beltway

Somehow, NASA would have to walk away from the table with permission to make their own radiation limits and fully justify that exposure of NASA employees to radiation in space was somehow "different" than all other global ethical exposure considerations for every other employee on Earth. In so doing, NASA would make the case that an annual dose hundreds of times greater than the average exposure of a U.S. professional NPP worker was fully justifiable and an accepted standard.

Yet, totally missing in their data set and eclipsing all other concerns, one of the most essential realities that overshadowed all other considerations was somehow, unbelievably, lost in the shuffle. All the numbers they had assembled – that anyone on Earth held – were based on research only from terrestrial ionizing radiation and limited, synthesized, low-fidelity, low-energy, single particle models of cosmic radiation. Therefore, embedded in many of these reports (even from NASA's own experts) was the explicit and repeated warning from a multitude of studies that the estimates of the radiobiological effects on humans were speculative with no possible determination of an accurate error of estimation. This was so because no research on the full spectrum and energy of galactic cosmic radiation (GCR) had *ever been conducted* by anyone in history *in deep space*, and the simulated GCR created on Earth was very low fidelity, at best.[7-17] That particular caveat must have escaped virtually unnoticed (or unheeded), much like the burdensome financial realities hidden in the fine print of a tantalizing new car ad.

So, once their presentations were prepared, the appointments were scheduled across the beltway to make the case of normalizing injury to astronaut crews.

The Long Road to Ethical Implosion

Never forget the sheer magnitude of the Agency bureaucracy nor the serpentine catacombs of other federal and non-governmental organization (NGO) bureaucracies with whom the Space Agency interfaces daily. NASA's task was to negotiate all of these in proper sequence, including all those whom the Agency gamely refers to as "stakeholders."

As was reported clearly in both the *Challenger* and *Columbia* investigations, this is the process where big government agencies routinely and deliberately hide responsibility. It is not a coverup by any means. But when there are multiple government agencies involved in a momentous safety question such as this one, no one individual or individuals in the process are signatories on any final agreement, accepting personal and legal responsibility for the outcome. Therefore, when the inevitable injury occurs, the usual investigation will conclude that responsibility has been cloaked by the process of committee so that no individual is held responsible. This is the fundamental nature of government management. The process exempts itself from oversight and responsibility. So, when the predictable disaster finally comes, do not bother going door to door at NASA HQ looking for a civil servant belly button to push, because one does not exist – by design.

One of the first in the series of discussions began with a 1989 meeting with the NGO National Council on Radiation Protection and Measurements (NCRP). More accurately, the NCRP is a congressionally chartered non-profit organization without government oversight. Yet, please be aware that as an NGO, the NCRP is not funded by Congress so that it has at least the illusion of independence. But the NCRP could not exist at all if it were not for majority government support through its "sponsors." As shown in this illustration is a cover slide from an NCRP meeting held in April 2021. Be careful to note where its major funding comes from (and from which a large percentage of its volunteer committee members are drawn). Also note that a vast majority of the NCRP support is from indirectly funneled taxpayer dollars, and that at the very top of this sponsor list is NASA. The purpose of this slide is to clearly demonstrate that the "independence" of beltway NGOs is always suspect, at best. And while the NCRP is apparently not guilty of any level of corruption in this affair, its capacity to remain ethically detached and without conflict of interest or sponsorship bias may not even be possible. This issue of impartiality is so critical and sensitive in the federal government that

NASA and all employees of the Executive Branch are obliged to attend ethics refresher courses annually. One of the requirements of the Standards of Ethical Conduct for Employees of the Executive Branch addressed at 5 C.F.R. § 2635.502 is the sensitive issue of avoiding "...even the *appearance* of a loss of impartiality" or conflict of interest, including private affairs.

The NCRP committee that accepted this review apparently immediately understood its role in helping normalize the extraordinary radiation risk NASA desired instead of its role as acting in an advisory capacity. Instead of professionally evaluating the question, they went quite over and above, actually supplying the way out for NASA. This unusual response to providing a straightforward evaluation suggests there are truly ethical consequences associated with asking for an opinion from an NGO that receives a high level of support from the requesting entity. Chances are that you will obtain a favorable opinion appended with the method you will need to go forward to the next level, which the NCRP dutifully provided in whole cloth.

The NCRP suggested a statistical comparison of the risks on orbit to accidental deaths on Earth in the "safe," "less-safe," and "hazardous" occupational risks of industries in the United States for which there were statistics available – however unrelated.[18] These were not NCRP stats, of course, but were harvested from the National Safety Council and integrated into the NCRP/NASA equation.

To begin the comparison, the NCRP referenced U.S. ground-based radiation workers who feature lifetime radiation risks no greater than risks experienced on average in the "safe" industries, which was estimated as 0.5%. However, the NCRP ignored that Risk of Exposure Induced Death (REID) probability altogether and, by a bizarre and somewhat illogical judgment, superimposed the REID from the next occupational category, the "less-safe" industry. Here is how that very critical decision was justified in the peer reviewed literature: "However, the dose limits for ground-based radiation workers were similar to the accidental deaths probability in the "less-safe" industries were [sic] lifetime risks of up to 3% were estimated at that time."[19]

That statement is fundamentally untrue, as well as nearly impossible to decipher from its convoluted narrative. As was stated before, U.S. radiation workers' lifetime radiation REID probability was no greater than risks experienced on average in the "safe" industries, which was estimated as 0.5%. But the NCRP, for rather ambiguous and very poorly stated reasons, selected the next category of industrial safety, "less-safe" industries, to leverage NASA's desire to alter their responsibilities by a very significant magnitude – the only way to preserve long term missions in the ISS, much less on an extended voyage to Mars. In so doing, the assigned REID increased six times – from 0.5% to 3.0%. The difference in that single judgement was that NASA would not ultimately be required to fall under OSHA standards for radiation workers because NASA was no longer counted as a "safe" industry for purposes of on orbit radiation safety.

The logical gymnastics here are astonishing and absolutely unethical, but typical in the government. NASA went to NCRP to find cover for long space missions on ISS. The only way they could find that cover was to slide out from under the government limits set for radiation workers by which NASA at the time was bound. The NCRP is a body that monitors and comments on radiation limits, not probability of death in rocket explosions or explosive decompressions on orbit. And yet, their recommendation to assign NASA a "less-safe" industry designation based on other totally unrelated industrial activities, and then affix that designation to the NCRP's radiation limit recommendations, is as duplicitous as it sounds. As you will see in just a few paragraphs, that decision later rose up to bite everyone involved in the proverbial rear – to the distinct embarrassment of absolutely no one (yet another government attribute).

The single critical conclusion by the NCRP to slide NASA up one notch to the "less-safe" industry category in 1989 ultimately gave NASA permission to normalize and enumerate GCR exposure injury to their crews as a significantly elevated "acceptable" risk of death to the Agency. NASA walked away from the very first series of meetings with the Holy Grail of permissions, sealed with the imprimatur of the NCRP, with which the Agency would successfully leverage their argument until they achieved their intended goal of legally normalizing cosmic radiation injury to their astronaut crews.

If the NCRP had more circumspectly assigned NASA's on orbit missions their actual designation of a "safe" industry (as it did nine years later), then NASA would have been required to remain under the same limits as all other occupations exposed to occupational doses of ionizing radiation in the United States and no OSHA exemption would have been necessary. But in giving NASA permission to be classed as a "less-safe" industry, each NASA astronaut was thereafter assigned a

Risk of Exposure Induced Death of 3% rather than <1% – again, a sixfold increase of contracting a fatal cancer in their lifetimes. The tragic part of that decision was that it was absolutely unnecessary and had no bearing on NASA's ability to fly safely to the ISS for missions under 76 days. NASA was given permission to be "less-safe" because they wanted it – not because any NASA mission required it, much less required an emergency permanent supplementary standard (which they eventually adopted).

The emergency was, of course, not having to responsibly redesign their exploration system to protect their crews for any mission over 76 days – such as a Mars expedition; NASA's long dreamed of next stop. Yet NASA denies even this, preferring to hide behind the cloak of impossibility, thereby deliberately injuring their crews rather than admit the limit has been reached, and respond with the ultimately unavoidable redesign.

Now that NASA held the permanent hall pass for freedom to do whatever they wanted, the Agency placed the 3% REID in effect a year later for all astronaut crews. The next stop was OSHA. And, with the NCRP data in hand, the agency responsible for safeguarding every Unites States worker enthusiastically caved. Based on the declaration of a permanent emergency in low-Earth orbit (as is required by federal code), OSHA handed NASA what it demanded – control over their own radiation exposure limits. As a result, unknown to virtually anyone, every NASA ISS mission is classified legally as a "radiation emergency" per 29 CFR1960.18(a). While the text specifically reads "...an agency head shall adopt such emergency temporary and permanent supplementary standards as necessary…," NASA, of course, justifies their exception as a "permanent supplementary standard." And yet, the intent of the law was to cover extraordinary circumstances, including emergencies. Thus, the fair response to NASA is to ask whether their permanent supplementary standard ethically belongs in a standard that also covers emergencies. The answer should be obvious.

Seven years later, after much had been committed by NASA and OSHA based on the 1989 NCRP recommendation, the twisted logic of the very sketchy house of cards inevitably collapsed in on itself. The single decision to classify NASA as a "less-safe" industry, thereby assigning the Agency a REID of 3%, changed dramatically. In a 1997 reevaluation, the National Safety Council (NSC) had modified its own stats rather radically. The nation's "less safe" industry category was now listed as a REID of < 1%, just like the nation's nuclear industry.[20] NASA, however, had already established their going-in plan all the way to Mars based on a REID of 3%. Yet even the NASA Inspector General warned in 2015 that the 3% death rate was insufficient for deep space expeditions based on their hopelessly archaic interplanetary human space exploration plan. "Based on current knowledge, astronauts on a mission to Mars would exceed NASA's career radiation dosage limits. Although the Agency plans to continue efforts to develop countermeasures to address the radiation risk, NASA is likely to seek an exception from the current standards for those that cannot be fully mitigated."[21]

The glaring, unanswered question is, why did not OSHA immediately recalibrate NASA's rating to the <1% national standard? Yet now, a full quarter century after the NSC's change to its industry designations and without any apparent oversight at all, NASA continues to cling to its self-appointed REID of 3%, still assuming the "less-safe" industry mantle of 1989, which has long since been invalidated. But its continued cavalier attitude about the safety and long-term health of its astronaut crews has certainly not gone unnoticed.

A 2001 report by the Institute of Medicine titled *Safe Passage* reaffirming an earlier report states, "Until the radiation hazards to astronauts can be controlled or otherwise mitigated by physical shielding, a 1998 National Research Council report states, 'long duration space travel should be postponed'."[22] Seven years later, sensing NASA's unbridled permissions to set its own limits at any level it chose, the National Research Council announced, "The committee strongly recommends that the permissible exposure limits specified in the current NASA radiation protection standards not be violated to meet engineering resources available at a particular level of funding."[23] Obviously, even the NRC has obliquely recognized that attempting to explore space using an obsolete model will necessitate more and more injury to astronaut crews as the engineering limits are inevitably exceeded, with the exposure limits to match. But with the 2014 National Academy of Medicine's declaration that "relaxing (or liberalizing) current health standards to allow for specific long duration and exploration missions to be ethically unacceptable," NASA's expendable astronaut plans were coming under review, being scrutinized, and receiving condemnation from other agencies as irresponsible and unethical.[24]

Yet, the nation's Space Agency continues down the road to normalizing crew injury with every public release of plans to fly into deep space with their antiquated equipment designed to injure their expendable astronaut crews. In fact and in reality, *any* nuclear power plant in the United States that exposed its workers to the same levels of the far less dangerous terrestrial ionizing radiation for two months would be immediately shut down by the same government that has given NASA carte blanche permission to expose their workers to cosmic streams of exponentially higher energies while calling it business as usual.

Moon to Mars Mission – Culpable Negligence in Action

Culpable negligence is a legal term that is defined as "more than a failure to use ordinary care for others" and is instead "a course of conduct showing reckless disregard for human life" which can include "a conscious indifference to consequences." [25] But because NASA, as an arm of the government is legally self-shielded from consequences by the government's byzantine self-policing and ability to write itself out of regulations that all others must follow for the public good, the Moon to Mars deep space mission is the ideal example illustrating NASA's push to finally drive the Apollo mission model to its ultimate catastrophic end in a breathtaking act of culpable negligence toward the welfare of its crews.

NASA announced the Moon to Mars Deep Space Exploration Objectives mission plan 22 May 2022 that proposed an extraordinary departure from previous NASA Design Reference Missions (DRM). [26] Instead of traveling round trip to Mars along Hohmann transfer orbits, NASA suggested a pair of highly elliptical orbits between planets that utilized a gravitational assist by Venus to and from Mars. Skipping the Hohmann orbit requirement that the crew reside on Mars between launch windows for a year or more, this specific mission plan allows NASA astronauts to spend only 30 days on the surface of the Red Planet before reboarding their spacecraft and returning home.

But as a career NASA space life sciences engineer, I can read between the lines.

Both publicly and privately, NASA has washed its hands of all responsibility for space toxicity outcomes (microgravity, cosmic radiation, and associated multimorbidities) by openly assigning their death-SWAGs (REID) and requiring their crews to sign the waivers with full disclosure of the very lowballed risks (see the chapter "Against the Physiological Wall"). Likewise, NASA has handed over to Congress the responsibility to care for the wave of inescapable lifetime medical sequels under the TREAT Astronauts Act.

But that is only one part of NASA's sleight-of-hand behind the scenes announcement of the May 2022 Moon to Mars plan. Since NASA has cleverly written itself out of culpability for the inevitable injuries, it is free to gloss over yet another stark limitation of its readiness to explore Mars and other deep space destinations in the Solar System – advanced life support. The driving reason behind Moon to Mars is that NASA is nowhere near ready to send crews out to Mars for a 916-day interplanetary mission. So the next best thing to assure that the U.S. flag is planted on Mars first is the Moon to Mars 650-day consolation prize. Unfortunately, in so doing, NASA has literally forced their deep space crews to leap out of the proverbial pan into the cosmic fires.

Since all crewed 21st-century spacecraft are currently still modeled after the Apollo paradigm, they are functionally unshielded and unprotected against space toxicity. Therefore, it is relatively straightforward to calculate crew exposures based on previous interplanetary cruise data. So, with that framework, let us compare the exposures of previous NASA Design Reference Missions with the 2022 Moon to Mars proposal.

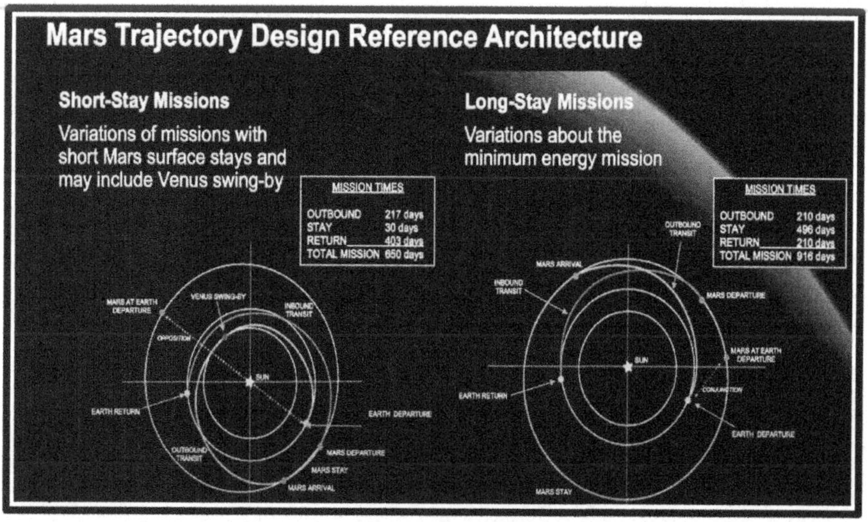

Using Hohmann interplanetary transfer trajectories, one previous NASA DRM had astronauts in transit between planets for a total of 420 days, and time on Mars at 496 days. But Moon to Mars has astronauts in transit for a total of 620 days with only 30 days on Mars' surface.

There are four issues to consider when comparing these DRMs:
(1) exposure to full spectrum deep space cosmic radiation,
(2) exposure to Mars surface radiation,
(3) exposure to microgravity, and
(4) fundamental scientific productivity.

Flying in deep space in functionally unprotected ships exposes astronauts to around 1.8 mSv/day of galactic cosmic radiation (GCR). But each day on the surface of Mars inside a fully shielded habitat (which can be relatively easily accomplished with abundantly available Mars regolith) allows crews on the surface of Mars to live and work *below* the legally acceptable annual limits for nuclear power workers. With that in mind, let us compare missions:

NASA 2022 DRM Moon to Mars Mission

Total Mission Days	650
Mission Days on Mars Surface	30 Days (for half crew)
Functionally Unprotected Crew Exposure to Deep Space Toxicity μG + GCR	635 days/avg. per mission
Average Crewmember GCR exposure	1,230 mSv/mission
Mars Surface Science Productivity Ratio	0.046

NASA Original "Long-Stay" DRM Mission

Total Mission Days	916
Mission Days on Mars Surface	496 days
Functionally Unprotected Crew Exposure to Deep Space Toxicity μG + GCR	420 days/avg. per mission
Average Crewmember GCR exposure	856 mSv/mission
Mars Surface Science Productivity Ratio	0.541

With just a glance, it does not take a NASA rocket scientist to raise the Whiskey Tango Foxtrot flag. The nearly scientifically useless and exponentially riskier NASA 2022 Moon to Mars Mission lasting far fewer days on planet still *increases* the GCR exposure by an additional half the career limit suggested by the National Academies and exposes crews to microgravity toxicity for an additional 215 days more while *giving up 466 days* of science and research on Mars. Such a mission is irrational on every level. Unless…

Unless one considers that NASA, as the undisputed leader in world class science, is also utterly controlled by politicians and lawyers, many of whom never entered an advanced science classroom a day in their lives. And, once more, NASA is keeping a secret. Few people outside NASA's hallowed halls understand that today literally no one on Earth – private or government – has any idea (yet) how to build a bulletproof advanced life support system for interplanetary travel or Mars settlement. *No one.* So, between the outside leadership of the Agency narrowing mission decisions to political ones and NASA internally knowing full well that even a 30-day campout on Mars is technically iffy today, they still stand between the devil and the deep blue sea. Hence, the Moon to Mars "let's-be-first" lunacy was thus contrived.

Because of the Agency's careless and legally responsibility-free assignment of their crews as expendable, something as bizarre and wholly irresponsible as the 2022 Moon to Mars mission plan materialized out of the ether and was actually made public. One must ask aloud, is there even a single individual at NASA Headquarters with a clue as to the trainwreck-dumpster fire they are planning and proudly waving at the world's cameras?

But if *someone* at the Agency does not eventually wake up and see the immediate need to capitulate to the laws of nature and human physiology and then redesign the archaic Apollo era deep space death traps, then private sector lawyers will ultimately record this mission for history in the inevitable investigations as:

> "…more than a failure to use ordinary care for others" but instead "a course of conduct showing reckless disregard for human life" which included "a conscious indifference to consequences."

Sadly, culpable negligence does not have to be the pathway to humanity's future as a multiplanetary species. No astronaut should depart Earth as some agency or company's expendable fodder.

Breaking Down
Multimorbidity

"The biological effects of a toxicant are amplified when an individual is weakened by additional, unfavorable factors."

– Pilippi Theophrasti Paracelsi
(aka, Theophrastus von Hohenheim)

Volumen Medicinae Paramirum
- Circa 1520

IN MID-DECEMBER 1972, ON THE MIDWESTERN PLAINS AT Oklahoma State University, the northwestern skies were deep blue and ugly. I had just completed my last final exam of the first semester of my senior year and was frantically packing my 1963 Ford Galaxie 500. It was a beast – an eight-cylinder gas guzzling big-boy who could move down the highway with a purpose. And on that day, I definitely needed purpose. The first flakes of snow were falling just ahead of a certified blizzard that loomed behind the dark clouds hanging over the northwest campus sky that was undeniably headed in my direction. The last thing I wanted was to be snowed in with no place to go on a nearly empty campus over Christmas break. At the invincible age of 21 years old, I had no way of knowing that I was about to learn all about the ins and outs of multimorbidity in a blizzard.

I headed out of town due east in a big hurry, my wipers slinging layers of icy sludge off the windshield. I had a two-hour drive ahead of me. But, with any luck, I could completely outrun the snow in an hour of committed, non-stop driving. As I cleared Stillwater, things seemed to be lightening up, and I sighed contentedly in relief, turning up my radio for an expected comfortable drive home. Of course, after a mere two decades of existence, no one had yet informed me that the gods of fate listen very intently for any sign of contentment so that they can throw everything they have at the hapless satisfied human. And, of course, riding just ahead of an Oklahoma prairie kick-ass blizzard, they were acutely tuned in to that contented sigh…

Once well past the city limits of the only significant urban presence between Tulsa and Oklahoma City, on the rolling grasslands that were steadily transforming from a dull winter brown to pure white Christmas cheer, the old Ford unexpectantly decided to quit. Just quit. One moment it was delivering all 200 horsepower from its classic 288-cubic-inch V-8, cranking out a steady 282 foot-pounds of torque, and then... nothing. One second, I was cruising with the steady purr generated by the results of the great sucking sound out of my gas tank. The next, I could feel the old Ford's increasingly limited inertia sucked up in total stunned silence by the negative g-s pulling me to an inevitable stop in the very last place I wanted to be.

Fortunately, I spotted an unpaved county road, made a right, and coasted to a stop, safely parking not far from the blacktop. I turned off the wipers to save precious battery power just in case I could restart the beast. I just sat there, stunned that my well-planned escape had apparently failed – and failed spectacularly in the rapidly approaching maw of a white-out, blowing, howling, horizontal snow-hurricane. As my windshield rapidly turned totally white, I decided that the only available option was to get out of the car and try to somehow convince myself that I knew what I was doing – which I certainly did not! Car repairs were not my forte. I was a geek long before the word geek was invented. I was certainly not an auto mechanic – not by any means.

In the snow, facing the black beast, I did the only logical thing I could do... I opened the hood. I prayed that there would be some obvious thingy hanging loose next to its thingy socket. No luck with that. Several weeks before, I had watched my friend in the dormitory parking lot blow into his gas hose because he said that sometimes the fuel lines get clogged. Since that represented the totality of my auto repair training, I concluded, "Why not?"

So, in the heavily falling snow, I pulled the gas line loose from the carburetor, just as I had watched him expertly do. The plan was to blow into the hose, just like he did. But the hose was not cooperatively long enough. It was hanging down next to the carburetor fairly close to the hot engine. "Whatever," I muttered in total frustration, feeling the snow melting on my neck. So, I climbed up on the front fender and leaned all the way over inside the compartment containing the stinking and still very hot engine to position my virgin lips around the dirty rubber gas line. As I got closer and closer, I could feel the heat from the engine rising near my face. I stretched in as far as I could. Then, right at the moment of dirty intimacy with this gas line, I had my first memorable, lifelong lesson in multimorbidity – which lasted about two unforgettable, and seemingly endless, seconds.

Just after my pristine lips made contact with the nasty black rubber fuel line, I accidentally touched my forehead to the engine's hot manifold. Instantly and reflexively, I jerked my body upward away from the hot engine – and soundly banged my head on the raised hood. This, of course, dislodged its prop and the hood fell heavily on my upper body, with the latch jamming a nice contusion in my back, once again pressing me in toward the hot, stinking – and still very dead – engine. It was sweet revenge for the black beast who had just responded most appropriately to the college boy getting fresh with her intimate parts without her permission.

Making the broad assumption that I am in polite company, I, unfortunately, cannot share my very creative expressions of surprise at what I had learned in those two seconds. Of course, I had no way of even imagining that it would provide the perfect illustration of multimorbidity in a book decades later about the hazards of deep space travel. And, in case the suspense of my fate on that blustery winter day has distracted you beyond further concentration, I will allay your fears. In 1972, since there was no such thing as a cell phone, I walked down that rural road to the nearest home and, using their dial phone, called one of my dormie companions back on campus to come pick me up. The next day, the car was towed back to campus, I got a ride with a friend, and I made it home just a day late and after the blizzard had safely passed.

Much like my incident in the blizzard, we have all either personally experienced multimorbidity or have witnessed it in one form or another. In this case, the original injury of a burn to the forehead caused a whole-body response (the reflex) that dislodged the hood prop, which then caused the hood to fall, resulting in the added injury of the latch bruising my back. In the process of the reflex and impact with the falling hood, I also mildly strained my neck. The original injury cascaded physiologically into a full suite of others.

While this is a chapter describing multimorbidity, there is another common term for the same set of conditions called "comorbidity," although, technically, that often refers to only two synergistic disease processes. However, it has also come to be defined by some as "all morbidities coexisting at the same time," which makes comorbidity and multimorbidity interchangeable terms. But since the term multimorbidity unambiguously covers all ongoing synergistic disease processes regardless of number, as the author, I have chosen that specific terminology for this work.

The 16th-century Swiss physician, alchemist, lay theologian, and philosopher of the German Renaissance, Pilippi Theophrasti Paracelsi (aka, Theophrastus von Hohenhei) first described multimorbidity in his classic work Volumen Medicinae Paramirum. Paracelsi (most often referred to by his single designated name Paracelsus) was the first to document multimorbidity. Often also referred to as the Father of Toxicology, he described it thus: "The biological effects of a toxicant are amplified when an individual is weakened by additional, unfavorable factors."

As a career physician - alchemist - lay theologian - philosopher and, by modern extension, the first toxicologist (what a no-prisoners resume!), from the eye of a physician, Paracelsus was the first to describe what happens when the human organism is exposed to, and thereupon must respond to, multiple toxins of any form simultaneously in order to maintain homeostasis. Following are some examples of fully physiological courses of multimorbidity to help establish a better understanding of the process.

It started with a viral head cold. After days of misery, it transformed into bacterial bronchitis. Eventually, the infection became something far worse: pneumococcal pneumonia.

In another case, a simple broken bone resulted in a bone infection called osteomyelitis, which then eventually cascaded into a whole-body blood infection called septicemia.

In yet a third example, simple chronic inactivity and poor dietary habits resulted in obesity, which developed into Type II diabetes, which eventually led to amputations and blindness.

Finally, one of the deadliest instances of multimorbidity is called a cytokine storm. For example, an individual is invaded by the viral agent responsible for the COVID-19 pandemic. The body's defenses kick in to meet the viral challenge. But, in some cases, the agent is so virulent, and its spread is so rapid that the body's defenses multiply so quickly that it produces a flood of immune cells and their activating compounds, which are called cytokines. This surge of cytokines results in a rush of activated immune cells into the lungs. The lungs become inflamed by this overproduction of a natural protein, and the subsequent fluid buildup will lead to respiratory distress, followed in many cases by a secondary bacterial pneumonia and, often, death.

Each of these examples is an illustration of multimorbidity. On Earth, when the human organism is simultaneously exposed to multiple disease or injury vectors, a synergistic decline in overall health may result in a suite of often unrelated vectors that are described as multimorbidity.

The word synergy is key to understanding multimorbidity. It is often accompanied by an internal process of effect amplification, just as Paracelsus described, as the body actively drives its processes seeking physiological homeostasis. As I will cover here, multimorbidity may also be described by linked, broad systemic responses to the failure to adapt to an off-nominal environment, such as microgravity or incident cosmic radiation. In many cases, it results in cellular, subcellular, and genetic responses – all of which together often magnify the organism's reaction as a suite of responses to correct the off-nominal condition or injury.

Further, in the space environment, these exposures often lead to human responses that fully earn them the label of "toxins of space." For example, prolonged, unprotected exposure to microgravity will eventually lead to "musculoskeletal toxicity," as is described in the next chapter titled, "Chronic Microgravity Exposure Etiology." Further, functionally unshielded galactic cosmic radiation has been described in the scientific literature as a "Pernicious Toxic Milieu of Cosmic Radiation," as was mentioned previously and will be thoroughly examined in the chapter just ahead titled, "Chronic Cosmic Radiation Exposure Etiology." Prolonged exposure to these combined attacks of toxins and/or toxic environments will result in a multimorbidity response by the human organism in *every single case*.

In the previous chapter, we reviewed the Human Adaptation Limit, which defines the extent to which the human organism can adapt to exposure to adverse external, environmental conditions – such as unprotected exposure to the suite of conditions present in the interplanetary space environment. There is also the similar analog on Earth of the human body's response to multiple, simultaneous illnesses. Again, this effect, as described above in several examples, is called multimorbidity – "multi" meaning more than one, and "morbidity" meaning a state or symptom of injury and disease. Once again, in the case of functionally unprotected spaceflight, any condition that leads to the breakdown of the human physiological capacity to adjust and achieve whole body homeostasis outside the Human Adaptation Limit becomes an injury/disease process or pathology.

Thus, exposure to the effects of living in functionally unprotected deep space may result in an adverse, multi-level biochemical adaptation response as the human body seeks physiological balance for individuals outside the Human Adaptation Zone. This may ultimately lead to the Adaptation Breakdown Syndrome in more than one area of an organism's physiological failure to adapt. Again, these processes are collectively known as multimorbidity.

Ultimately, one essential finding aboard the International Space Station was the demonstration that the combined, interactive toxic hazards encountered by the human physiology in space are not isolated as separate physiological manifestations. In other words, the human body does not exhibit isolated injury profiles for long term exposure to microgravity, then another distinct set of isolated injury profiles for exposure to cosmic radiation, and yet another distinct set of isolated injury profiles for chronic stress (to name a few of the threats.) We have observed the human physiology merging the collective responses of non-adaptable injuries together, amplified as a conjoined physiological process called multimorbidity. Of all the discoveries on the ISS, the observation of the exceptionally complex interplay and synergy of human systemic malfunction in the space environment was one of the most singularly important, and yet also the most unacknowledged, discoveries of 50 years of human spaceflight.

Given the current exploration model, the human body is challenged one second after the boosters shut down and the spacecraft enters space in low-Earth orbit (LEO). But the human body is also somewhat adaptable (within specific limits), depending on the individual and the exposure intervals. The human body instantly rallies to the off-nominal environment of space. But it is not just one system that rallies and attempts to achieve homeostasis. There are many (not just individual) systems that work together as a whole organism response, as the body does not seek to adapt one system at a time in isolation. It adapts as a conjoined system response along many linked biochemical pathways.

That off-nominal environment can be highly variable. Key to a full understanding of these interlinked phenomena is this:

The environment that an astronaut encounters in space is completely and very precisely determined by the engineering system that protects them. The astronaut is *never* at the mercy of space but at the mercy of the engineers that designed the systems that surround them!

As a career advanced space life support system bioengineer, may I please repeat this essential point:

The environment that an astronaut encounters in space is completely and very precisely determined by the engineering system that protects them. The astronaut is never at the mercy of space but at the mercy of the engineers that designed the systems that surround them!

Thus, it is not the deep space environment that is primarily responsible for the injuries and deaths. It is the human spacecraft systems and the philosophic model from which they evolved that damages, injures, and kills astronauts. Therefore, if future deep space crews are permanently injured or killed by exposure to the space environment, it will be the direct fault and probable criminal liability of the organization that deliberately set them up and knowingly paved the way for the functionally unprotected toxic exposures.

At the funerals, they will praise their bravery in exploring a deadly frontier. But they should instead hold the victims accountable for their own stupidity and naivety in departing Earth on functionally unprotected and obsolete deathtraps. But that will be for the juries to decide.

Today, the only engineering system available for access to deep space is the antiquated, obsolete, hurry-up-and-beat-the-Soviets Apollo human space exploration model projected onto the new 21st century "on steroids," and with a radically different mission. But soon, if the warnings and advice of this book are heeded, much more advanced space systems will provide artificial gravity and adequate cosmic ray shielding so that the need for adaptation is minimized and potential injury is largely avoided. Therefore, the adaptation profile is wholly dependent on the engineering system itself. (That idea is greatly expanded in Book Two - Module Five of this work.)

As was specifically stated in the preface material of this book, in half a century, we have discovered that riding into space for a weekend, or even a couple of weeks, of fun, frolic, and vomiting in LEO certainly *does not* pose a human adaptation problem. Upon return to Earth, the human body reverts to 100% normal in a relatively short period.

Although there is a small level of an increased probability of cosmic radiation-induced cancer exposure, it is currently considered statistically minimal for short jaunts in LEO. The human body has demonstrated this fact over and over after nearly 600 astronauts' exposure to low-Earth orbit for relatively short periods.[1] But by extending the off-planet period to months, things invariably become a different story altogether.

In deep space, we have zero data on the multimorbidity of total exposure to full spectrum cosmic radiation and microgravity over a period greater than twelve and a half days. There have been only 24 astronauts exposed to this environment, and then only for an average of less than 10 days per crewmember.

Obfuscating A Broken Spaceflight Model with Acronyms

Multimorbidity from all space hazards to human health is described in Space Agency parlance as RIDGE, which is an acronym for Space **R**adiation, **I**solation, and Confinement, **D**istance from Earth, **G**ravity fields, and Hostile/Closed **E**nvironments. Most government agencies are habitual abusers of acronyms. For example, the U.S. Navy has cleverly located all its many acronyms in a book titled *Dictionary of Naval Abbreviations,* which is referred to by its own acronym DICNAVAB. But as bad as the Navy acronyms were, in my 42-year combined government career, I have never encountered one this contrived, convoluted, and obfuscatory. It would have been more genuine to clearly spell our RIDGE's more precise meaning: "Unprotected Space Is Hazardous to Human Health" or USH^3. Or perhaps "We Need a New Exploration Model" or WNNEM. In the end, RIDGE, USH^3, and WNNEM all mean exactly the same thing. As was described in detail in previous chapters, NASA's REID death swag, or Risk of Exposure Induced Death from human exposure to cosmic radiation in deep space, is yet another example of shielding rotted and archaic space exploration models behind acronyms that all clearly define the end of the deadly and anachronistic practices currently engaged by human spaceflight entities. Apparently, NASA has not learned that their old, obsolete systems cannot be propped up or otherwise hidden by complex acronyms that no rational human could ever decipher without attending a four-day government symposium.

The human body has been accurately defined as a bag of fluids held upright against the force of gravity by a rack of bones, fibrous tissues, and an enervated layer of skin, whose suite of gravitationally supported processes are coordinated by a complex interlinked dynamic of electromechanical stimulation and biochemical events altogether limited by a very small range of environmental conditions.

Note that this deliberately simplified description depicts an organism that is very specifically constructed to operate and thrive in a gravitational gradient, featuring all the necessary physiological design characteristics to make this possible – bones, attached musculature, a circulatory system that features gravity linked check-valves, and numerous similar gravity linked design characteristics. Many other human physiological roles are also reliant on that gravitational field to function properly. Obviously, we have clearly discovered that gravity is not an urgent physiological demand over short periods of days or weeks in microgravity. But, beyond that, depriving a human being of gravity exceeding the Human Adaptation Zone *will* spiral into systemic toxicity and multimorbidity, and induce permanent injury.[2]

That is why the preceding statement that *the human body is challenged one second after the boosters shut down and the spacecraft enters space in low-Earth orbit* is so wholly accurate. The space frontier, even within the micro-thin slice of low-Earth orbit, challenges the human organism because the human organism's processes, physiology, and interlinked mechanical and biochemical systems were never designed to nominally operate in that environment. Wishful thinking (nor the projected philosophy and all the social milieu linked to manifest destiny rubbish) does nothing to change this scientific and physiological fact. The human body is specifically intended to operate nominally in a gravitational field, perpetually and by design seeking homeostasis in a nominal one Earth-normal environment as was discussed three chapters previously and will be discussed in greater detail in the next chapter, "Chronic Microgravity Exposure Etiology."

We will actually begin to challenge our physiological systems and set up the inevitable suite of multimorbidities to the toxic environment of deep space even before the boosters ignite in Earth orbit and propel us to Mars. Once thus committed, there is no ability to return to our home planet for nearly 1000 days in most exploration models (an average of 973 days per the Hohmann Earth-Mars/Mars-Earth transfer catalogue).[3] As all of our studies in low-Earth orbit have demonstrated, the human body begins to attempt to adapt to microgravity. But homeostatic adaptation to microgravity is not the only thing going on. The body is also simultaneously dealing with the assault of many off-nominal conditions (from stress and isolation to cosmic radiation), which ultimately sets the stage for the deleterious effects of multimorbidity in every human organism.

As was discussed in "The Limits of Human Adaptation," it is essential that no one lumps adaptation and nominal healing from injury into the same category – because they are certainly not equivalent. But, in the case of the body's inability to adapt to an overwhelming environmental challenge, when the adaption limit has been exceeded, the stress morphs into injury as the body's systems break down – progressing from the adaptation response to injury and multimorbidity.

Exceeding Human Adaptation Limits
Multimorbidity at the End of the Human Adaptation Zone

Again, a close analog on Earth to exposure to the deep space environment is Everest Base Camp, which is right at the limit of the Human Adaptation Zone. Climbers go there to acclimate for days, or even weeks, before their push to the summit of Everest. For the professional guides, Everest Base Camp is also a field of testing and observation of neophyte mountaineers. Many climbers never ascend above base camp because they simply cannot acclimate. Ask any climber who had invested nearly $100,000 or more in the trek why they did not just press on to the summit anyway. They will be happy to give you a very personal lesson in multimorbidity. They will point to the hundreds of unrecoverable bodies littering the death zone on the way to Everest glory and let you know they preferred to go home rather than to tempt the limits of their physiological adaptation on the desolate side of a Himalayan peak the natives reverently call Chomolungma. Pissing off or underestimating the "Mother Goddess of the World" is just bad juju – 100 grand or not.

Everest Base Camp is right at the limit of the Human Adaptation Zone for many physiological functions. But the limit of adaptation is highly variable. The physiological response to exposure at Base Camp can certainly differ significantly depending on each individual. In this specific terrestrial analog, it is fascinating to watch the change in the Human Adaptation Zone statistic as the altitude begins to increase toward the summit. While perhaps half the mountaineers can fully acclimate at Everest Base Camp located at 17,700 feet elevation, as the climbers continue upward, the human physiological adaptation response starts to change quite rapidly as more and more individuals begin to experience the Adaptation Breakdown Syndrome and injury.

There are four more camps as the mountaineers make their way up to the summit. Three days after departing Base Camp, climbers reach Camp 3, also known as the Lhotse Wall, at 26,300 feet. Everest Camp 3 is the absolute end of the Human Adaptation Zone. Above 26,000 feet, all climbers, including the indigenous Nepalese guides, will begin to experience fully developed Adaptation Breakdown Syndrome, which includes bodily injury and broad symptoms of multimorbidity. While some of the native hikers can climb higher and stay longer without any supplemental oxygen, even they will not survive long here above this limit of human adaptation. Because of this, the region is called the Death Zone for exceptionally good reasons.

After reaching the Death Zone (the absolute limit of the Human Adaptation Zone above 26,000 feet), it becomes a literal race for survival. Even with supplemental oxygen, the climbers begin to lose lucidity, reflex, and stamina at an alarming pace. Above 26,000 feet, the danger becomes so extreme that even if a climber succumbs and dies here, no team of mountaineers – or even Nepalese locals – can safely recover their body. In the Death Zone on Everest, even the long-accepted mountaineer creed of bringing back their dead from the mountain's slopes is abandoned.

It is an environment so close to the limit of human adaptability that just making another 3000 feet to the summit is an agonizing trek, and one that is accompanied by great physiological danger and an exceptionally high fatality rate. As the human organism fights to adapt to these toxic high elevations with very low partial pressures of oxygen, several physiological breakdowns begin to affect the body simultaneously as the process of physiological multimorbidity progresses. The body is unable to maintain full oxygen supply to the blood, which results in chronic anoxic physiological conditions and their associated responses. Supplemental oxygen is an absolute necessity for nearly all climbers to offset this chronic anoxia at or above the Death Zone. At the same time, the body's circulatory system is sufficiently challenged so that fluids begin to accumulate in the lungs of some climbers in a condition known as High-Altitude Pulmonary Edema. This illness can occur to some mountaineers (and even tourists) as low as 8,200 feet, and worsens above that altitude.[4]

At the same time, yet another condition may simultaneously develop in addition to the chronic anoxia and high-altitude pulmonary edema. It is called High-Altitude Cerebral Edema, which results in the brain swelling with fluid because of the physiological effects of high altitude.[5] The consequences are disorientation, lethargy, and nausea, among other symptoms. If left untreated (by descending rapidly to a lower altitude), the stricken individual will not survive.

All of these conditions may assault the climber simultaneously in an ailment called Acute Mountain Sickness. This is actually a single term defining a process of multimorbidity – the product of multiple physiological anomalies that result in the very rapid synergistic decline of the human organism outside the limit of the Human Adaptation Zone. For some, even Acute Mountain Sickness can be fatal if not treated appropriately. After living at sea level in Florida for over 30 years, I personally have suffered this miserable and debilitating ailment at an altitude of 10,000 feet. The solution was simple – a leisurely drive down from Breckenridge to Denver!

Obviously, hundreds of humans have accomplished the successful ascent of Everest with full recovery. But over 200 have died trying, and many others have returned with permanent injuries. It is a tribute to human adaptability that a few of the Nepalese Sherpa who have lived at these elevations since birth occasionally complete the trek without any supplemental oxygen. Yet even they still suffer symptoms of multimorbidity above the Death Zone and cannot remain there long. The Everest analog offers the perfect model for space exploration and what happens to humans who are forced to try to adapt to an environment at the end of the Human Adaptation Zone. It is also an intriguing model illustrating the effect of exposures beyond the Human Adaptation Limit (here called the Death Zone): the onset of multimorbidity and its synergetic results.

The physiological response to exposure beyond the Human Adaptation Limit will certainly differ significantly depending on each individual, just as the Everest "laboratory" clearly models. But outside the Human Adaptation Limit, permanent injury begins inside *every* human body. In the case of a protracted period of these adverse challenges to the body, even changes on a cellular level begin to manifest themselves. Whether it is an acute onset (as in the case of a subset of aggressive disease pathologies on Earth on the slopes of Everest) or a much longer onset (as in the case of unprotected human spaceflight – which is itself an example of multiple ongoing pathologies), the end result is the same: multimorbidity ultimately leading to permanent injury or death.

Cellular Pathology in Multimorbidity

Regardless of the underlying factors leading to multimorbidity (again, defined as multiple assaults on the human organism from environmental stressors to psychological stress, disease, or injury), it is impossible to limit the pathological process to a single part of the organism. Just as chronic stress may lead to ulcers, heart disease, and clinical depression, it is impossible to separate the whole-body effects of multiple assaults on human physiology. The same is true in spaceflight. It is impossible to limit the impacts of microgravity on astronaut vision, cerebral swelling, or toxic bone loss. All of the effects result in a whole-body response in the human organism's attempt to adapt, even down to changes in the organelles within individual cells.

For example, subsequent to Astronaut Scott Kelly's 340-day mission to the International Space Station (2015-2016), it was determined (after comparing his cells to his twin brother who remained on Earth as his "control") that prolonged spaceflight in low-Earth orbit below the Van Allen belts resulted in changes in Scott's DNA *expression* (not changes in the DNA *structure*, of course.). In fact, the astronaut's white blood cell chromosomes grew longer while he was in space. The alterations occurred in a cellular structure called the telomere, a fragment of genetic material that lies at the tip of each chromosome string. Shortened telomeres are often thought to be accountable for age-related cellular dysfunctions. But, instead, Kelly's telomeres actually elongated in space. Yet once he returned to Earth, about 93% shrank again back to their original pre-flight lengths.[6,7] This specific example of alteration in gene morphology and expression clearly illustrates how the body responded to off-normal conditions at the limit of adaptation – even down to the sub-cellular level – and then, in this case, reverted to normal upon returning to Earth.

But that finding was not surprising, as years of previous research had demonstrated many of the same intercellular effects of changes in prolonged, unprotected spaceflight.

Under the stress of multimorbidity, the endocrine system becomes immediately involved, signaling various pathways in the body through hormonal interactions called a neuroendocrine response. Those hormones are very powerful signals to the body that elicit a reaction down to the intracellular mitochondrial level. The mitochondria play a pivotal role in reaction to whole-body stress responses (physical and psychological). Once the hormonal signal is received intracellularly, the mitochondria produce what are called Reactive Oxygen Species (ROS). At low levels during low to moderate stress, these ROS are considered to be important for metabolic adaptation.[8] But during prolonged periods of physiological stress from one or many stressors, these ROS signaling biochemicals activate apoptosis/autophagy (programmed cell death) pathways capable of inducing cell injury or death.[9]

Eventually, as the stressors continue to mount their challenges on the whole organism, the biochemical stress also increases within the cell structure. The ROS ultimately cause damage to the mitochondrial DNA (mDNA).[10] As this onslaught proceeds and worsens, every cell's energy producing capacity is reduced, and, again, susceptibility to cell death is increased, and the whole human organism begins to fail to thrive as the stresses continue for days, weeks, and months. Thus, the gene expressions themselves are changed by the stress of prolonged exposure to unprotected spaceflight, as was clearly documented in Scott Kelly's mission.[11] In the end, this condition leads to what is called epigenetic deregulation, or wholesale changes in genetic expression (again, which do not involve changes in the actual DNA sequence itself). These alterations involve modifications in histones that comprise the chromatin and DNA methylation, as well as an ever-expanding array of other widely pathogenic epigenetic processes.[12]

In another key peer reviewed study in the *International Journal of Molecular Science,* it was established that the dual effects of simultaneous microgravity and radiation exposure had an additive effect on the process of bone loss and resorption in microgravity, otherwise known as space induced osteoporosis. They discovered that the twofold effects of radiation exposure and microgravity synergistically worked together to accelerate bone loss in space.[13] In another research article published in the *Nature Partner Journal of*

Microgravity, it was determined that there was clear evidence of interplay between cosmic radiation and microgravity in DNA damage, and DNA damage response.[14]

In the *Journal Life* article titled "Space Radiation: The Number One Risk to Astronaut Health beyond Low-Earth Orbit," it was recorded that cosmic radiation and the synergistic effects of spaceflight (microgravity, stress, and the combined suite of physiological challenges) can place a crew at significant risk for blood changes and immune system dysfunction. The *Journal* also pointed to accentuated degenerative tissue responses, cognitive effects, and higher brain function impacts.[15]

Thus, multimorbidity ultimately expresses itself in the whole body as intracellular, genetic expressions that are invariably linked to physiological symptoms and observable pathologies. As these whole-body stresses accumulate, all the way down to the genetic expressions of the organism, other disease processes are inevitable as the interconnected system breaks down.[16]

Unraveling Cosmic Radiation Effects on Multimorbidity

Yet quite unlike the process of "adaptation" to microgravity, under no conditions can exposure to the heavy primary particles of cosmic radiation above levels documented in low-Earth orbit over relatively short periods *ever* be considered a process of "adaptation" – under *no* conditions. As documented in the previous chapter, "Expendable Astronauts," and stated by the United States Nuclear Regulatory Commission, "A linear no-threshold (LNT) dose-response relationship is used to describe the relationship between radiation dose and the occurrence of cancer. This dose-response model suggests that *any* increase in dose, no matter how small, results in an incremental increase in risk."[17]

The human body has demonstrated a capacity to "absorb" the injuries elicited by these doses of cosmic radiation for brief intervals. However, it cannot be depicted by any scientific description as a "process of adaptation," but rather accumulated cell death and the storing up of genetic changes, some of which are permanent. Beyond this, particularly in deep space above 36,000 miles, the cosmic flux is sufficient to initiate a process of immediate and latent gross morphological pathology, i.e., detectable injury, as statistically derived by recent documented Apollo findings. The physiological sequel thus is a process of "healing," certainly *not* adaptation! In confusing the two processes, NASA has inadvertently established a purposeful injury profile required by the exploration model as the accepted norm. It is not. Indeed, such inevitable injury sustained by deliberate exposure of their astronauts to their obsolete and functionally unshielded engineering design is unethical by any measure, as unambiguously described by the National Academy of Medicine.

If I abrade my skin, it will almost always heal if I am healthy and if the scrape is treated properly. Afterward, my nominal human function will not change. But, in healing, it is common that the body will create a lasting scar as a reminder of the injury. And if the injury is internal – perhaps linked to the cardiovascular system – the acute pernicious endothelial tissue degradation will cause a lasting injury generating unfortunate cardiac sequels decades down the road. Likewise, if I am chronically exposed to cosmic radiation at the rate observed outside the Van Allen belts, my body will heal the injuries to a limited extent but may store an accumulated risk of cancers as the unseen but very real, permanent scarring from the injuries. There is a significant question posed here because, without adequate research data, we are literally flying blind above 36,000 miles. (The upcoming chapter "It's the Biology, Stupid" in Book Two, Module 5 is devoted to a very detailed solution to this issue.)

Acute injury due to exposure to cosmic primaries is focused on cellular structures and DNA. Cosmic primaries are very large relative to a DNA strand (as compared to a discrete, massless particle of non-primary of ionizing radiation), and their Linear Energy Transfer (LET) is so high that there is direct photo-micrographic evidence that they can literally obliterate a cell nucleus or other vital cell organelle when passing through soft human tissue. The damage they possibly pose to a strand of DNA can be potentially significant. The immediate direct injury is at the intracellular biochemical level. The process of healing thus becomes the elimination of the necrotic (dead or dying) cellular structures and/or the active repair of individual strands of damaged DNA.

The high energy particles interacting within the human body at the cellular level can directly or indirectly damage a suite of biomolecules (complex or simple proteins and lipids as well as DNA) and, again, also have a very high probability of directly damaging cellular organelles because of the exceptionally high LET energies. These chronic assaults on molecular biology within the human organism result in a radiation-induced increase in oxidative stress – in a sense, a subcellular expression of multimorbidity. It is a cumulative effect that exacerbates degenerative tissue changes associated with accelerated aging, such as cataracts and cardiovascular disease – a finding directly related to the increased incidence of both injuries in the Apollo Lunar explorer cohort.[18,19,20]

The effect of terrestrial ionizing radiation compared to cosmic radiation on soft human tissue is profound because of the vast differences between inertial energies and direct damage potential. Unfortunately, because there have been no in situ studies in deep space to understand the effects of cosmic radiation on vertebrate cell biology, we have nothing but limited anecdotal evidence with which to project the effects of chronic exposure and – by definition – injury to astronaut crews by whole-body cosmic radiation exposure over extended periods.

But, by using animal models, research has conclusively demonstrated that the effects of spaceflight stressors (microgravity, environmental constraints, and emotional stress) combine in the organism synergistically with radiation and create a systemic multimorbidity that expresses itself as increased susceptibility to infection, delayed wound healing, and decreased survival potential.[21-23] Observed one researcher, "There is a need to understand the mechanisms behind the synergistic lethality observed with radiation injury."[24]

Breaking Down is Not an Option

Multimorbidity is the manifestation of interconnected physiological network processes within an individual which include genomic, metabolomic, proteomic, neuroendocrine, immune, and mitochondrial bioenergetic elements – all interconnected and all involved in the whole organism response and expression to the attempted adaptation (or failure to adapt) to the space environment. Thus far in human space flight, we have operated a kind of space Base Camp at 250 miles altitude, under the protection of the Van Allen cosmic radiation shield. And yet, even there, we have exposed our astronauts to a range of physiological challenges fully expecting, and perhaps even physiologically demanding, adaptation – much like Everest Base Camp.

The prevailing philosophy is that we can now freely and without expectation of fatality or permanent injury, push on up the slope into the deep space Death Zone – defined by even longer exposures to microgravity and a considerably harsher cosmic radiation environment above the Van Allen shields – and just record the data without physiological sequel. Only astronaut Scott Kelly has publicly documented rather alarming evident symptoms of multimorbidity to prolonged exposure to spaceflight that has been released openly in his own rather blunt words. But in Kelly's case, it was clear enough (and by his own admission) that he had reached the very end of the Human Adaptation Limit and was even experiencing the painful result of the Adaptation Breakdown Syndrome.

His suggested conclusion? There obviously need to be more countermeasures.[25] Instead of effecting a change in the core exploration philosophy itself, the established space agencies' more band-aids approach is powerful, persistent, and ultimately will prove to be just as deadly as pressing on up to the summit of Everest while blithely walking by the string of human bodies that litter the upper slopes.

At this point, we need to stop for a moment and appreciate the illustration provided by the Everest analog. If the ISS is Base Camp 3, then the deep space Death Zone lies above the Van Allen belts at 36,000 miles. We just do not know anything about the details above the shields – not the full cosmic spectra effects on vertebrate physiology, the limits prior to systemic injury, or the long-term medical sequelae after another unknown period of exposure in unprotected spaceflight in deep space. Thus, ethically and professionally, we are left with only two choices. And neither of them is "press on and see what happens," which, incredibly, is, in fact, the current plan!

The first choice is to go ahead and build the advanced space systems that eliminate harmful space radiation exposure and microgravity completely and thus professionally, technically, and ethically end the discussion entirely.

It sounds so simple... because it is.

The second choice is to send up animal cohorts to cislunar space or solar-centric orbits to discover precisely the effects and the limits before we are forced to build the advanced systems anyway. As a member of the human race, the necessity of even having this sub-basement-IQ conversation is embarrassing – but here we are. What should be more than evident, more than elementary, more than crystal-clear logic has somehow escaped those in charge of making the fundamental, logical decisions at the highest levels. And yet, in the premier Space Agency in the world, the discussion is not just absent, it is being deliberately avoided to defend the SLS-Orion system and the inadequate human exploration paradigm.[26] Imagine the moral and ethical scope of that decision. Hideously expensive equipment is being safeguarded while altering the laws themselves that were specifically designed to protect their astronaut crews just to ensure there is always an available cadre of living human shields to protect the failed paradigm.

In the end, Scott Kelly's personal narrative should be a powerful wake-up call to NASA and every agency, plus public and private enterprises, which expect to commit human lives to the exploration and colonization of deep space, Moon, and Mars. Kelly has clearly illustrated that we have already reached the Human Adaptation Limit in space. Based on what we have learned, to go any higher or stay any longer, we will need to change our exploration philosophy and rebuild the model. It certainly *does not* signal the end of human space exploration!

Only NASA will protect that lie specifically because it protects their model, which they have now spent incomprehensible resources to build. It indicates the need to understand the lessons learned and rebuild the model based on the sanity of collected ISS data matched with an engineering system that respects the human's design limits. It is just not all that difficult – and certainly not worth all the Kafkaesque machinations we have gone through to reach this historically absurd, severe, ethically compromised position that we find ourselves in today.

We have been lulled into a fairy tale of our own making. This self-deluding story is a manifestly powerful saga because it is all about us and our total awesomeness on steroids. Unfortunately, the universe could care less about how awesome we may think we are. The countless supernovae and PeVatrons that have long disappeared and the liberal scattering of black holes that spread their discreet cosmic particles over our galaxy at near light speed have no knowledge or concern about us mere specks of fluid held tenuously together by simple soft flesh. But that is not how the human mind thinks about itself. Regardless of the power of a supernova, we have declared with raised fist, "Bring it!" In our delusion, we have somehow convinced ourselves that breakdown and failure, well, that's just not an option we will accept, and there is always another box of countermeasure band-aids somewhere about on our ships...

Until, of course, the great cosmic shark of multimorbidity rises up from the black eternal deep and bites us on our collective morbidly reckless rears. If we truly prize the failure-is-not-an-option sloganized version of human spaceflight, we have certainly selected the absolute least viable pathway to making that a reality as we enthusiastically plan to send our unprotected human astronauts to the Death Zone above 36,000 miles and keep them there for long periods of an ethically challenged plan of "observation." And just what is our going-in plan? We want to send them out specifically just to see what happens to them so we can cleverly document our own governmentally protected latent immortality. We have staked their lives and wellbeing by going ahead and scheduling $4 billion missions on the answer we think we may-or-may-not know – with no data whatsoever to back them up. And if we permit them to fly to Mars, the bodies will lie on the Martian deserts just as they lie on Everest's slopes because there will be no way to bring them back home.

So how do we recover from this ultimate catastrophe even before it begins? That answer is found in the key paragraph from this chapter:

The environment that an astronaut encounters in space is completely and very precisely determined by the engineering system that protects them. The astronaut is never at the mercy of space, but at the mercy of the engineers that designed the systems to protect them....

Chronic Microgravity Exposure Etiology

"The only thing worse than being blind is having sight but no vision."

– Helen Keller

FROM THE BEGINNING, A GOOD WRITER ALWAYS KNOWS who their target reader is. And who do I imagine you are? Besides many of my very pissed-off colleagues by this point in this book, there are astronauts, future astronauts, and rocket company executives. There are high school students, career PhDs, university professors, English teachers (OMG), and every educational level in-between. But my prime target is those of you who will also look out your spacecraft windows and see the plains of Mars with your own eyes. I am aware that this work will be included in the electronic libraries of the Red Planet, just as today, my other books also reside in the libraries of a few Earth-based Mars analog stations.

I may have been a career life sciences engineer/scientist, but etiology is not a word I use frequently. However, after much deliberation, I also realized that no other word fits the need any better – and my readers can definitely handle it! Of course, I could have titled this chapter "Chronic Cosmic Radiation Exposure: The Cause, Set of Causes, or Manner of Causation of a Disease or Condition" – but then, that would have been a ridiculous acquiescence in maintaining and reinforcing public illiteracy. See how simple the decision was? Etiology does fit better, after all!

After extended contact with NASA and crewed space missions since my early teens, I have come to realize that it probably would have come as a major shock to nearly everyone in the 1960s and 70s that microgravity was anything more serious than a nuisance and mild discomfort after the first several days of space sickness wore off. But then we began to maintain crews in low-Earth orbit in the Russian and American space stations (Mir and ISS) for extended periods. The discomforts continued to morph into increasingly and more severe manifestations the longer the unprotected astronaut was exposed to the microgravity environment.

Eventually, after a period of months of exposure, the laundry list of documented microgravity induced aliments was becoming alarming. NASA immediately followed up with aggressive on-orbit medical investigations. Ultimately, the research suggested the undercurrent of growing concern about permanent injury was justified.

Not to waste the lessons we have learned in the previous chapter on multimorbidity, it also became clear that all of the physiological stressors of spaceflight were inseparable in the human organism's cumulative response. It became evident that, because of the process of multimorbidity, microgravity and cosmic radiation did not exist as independently manifested stressors at all. Rather, they synergistically magnified one another.

In this chapter, we will review the details of those physiological effects induced by prolonged exposure to microgravity that, over time, always wind up at the far limit of the Human Adaptation Zone at different times, depending on the individual exposed. But each astronaut always winds up at the limit and begins to incur injury. As we review them, always keep the aspect of multimorbidity in mind. While this chapter is focused on the physiological breakdowns induced by unprotected chronic exposure to microgravity alone, many of these ailments are exacerbated by other maladaptation vectors (such as cosmic radiation), which is covered in further detail in the following chapters.

A method for tracing the effects of microgravity on astronaut subjects is by tracking specific variables. One is defined as the highly variable Microgravity Recovery Time, or MRT. And, within the total span of MRT, there are two sub-periods embedded: Functional Microgravity Recovery Time (FMRT) and Total Microgravity Recovery Time (TMRT).

Think of it like this: if you stand out in the center of a large empty room and turn in a circle very rapidly for 90 seconds, then abruptly stop, you will quite obviously be very dizzy. It is likely that you will even lose your balance and not be able to stand upright on your own, so you will probably sit down on the floor. You may also feel ill. But if you sit patiently on the floor and wait it out, eventually, the world will stop spinning in your head, and you will be able to stand unaided again without risking falling over. In a handful more minutes, the queasy feeling will subside, and you will have returned to the full normal status you were experiencing before the spinning.

The moment you stopped spinning, you started the clock on your full recovery time (like Microgravity Recover Time – MRT). When you were able to stand up unaided and walk around without falling, you had achieved functional recovery from the spinning (analogous to FMRT). When nausea and dizziness finally totally departed, you were fully recovered to your status before you started the spinning (equivalent to TMRT). This is a very basic example of what happens to humans who are exposed to microgravity for any length of time.

In the beginning, the body attempts to completely acclimate. There is almost always vestibular dysfunction in everyone (disorientation, dizziness, nausea), but typically that resolves in a few days, much like seasickness is settled on an ocean liner or a Navy warship as the body's vestibular function adapts. That time of adaptation is individually highly variable. Some astronauts adapt completely in hours, while others never really fully adapt at all.[1] There are even a few astronauts who never feel any discomfort at all. At least, that's what they say…

In the case of long-term exposures to microgravity (as most notably described in graphic detail by Astronaut Scott Kelly in the preface to his book *Endurance*), after extended exposures, the Functional Microgravity Recovery Time (FMRT) – or that time where normal functions are greatly restricted – may persist for days or even weeks. And the Total Microgravity Recovery Time can require many months or years when returning to Earth or landing on the surface of another world. If there is significant bone or musculature loss, or long-term alteration to the optic nerve or shape of the eye, the former state of health and physiology *may never return fully at all*. If that is the case, then the astronaut crossed the line at the end of the full adaptation limit and began to experience permanent injury for which no healing process will return their systems to pre-flight normal, and homeostasis cannot be achieved. Again, these permanent injuries have even been observed in a few astronauts who have participated in extended missions (>6 months) in low-Earth orbit.

This data alone will ultimately define the functional usefulness of astronauts who fly on 180 to 270+ day missions to Mars in microgravity using the current antiquated, injury inducing, dangerous deep space Apollo era space architecture. Under this spaceflight mission model, not only will they arrive nearly maxed out on their permitted radiation doses (unless they are somehow granted waivers to the existing waivers - which their body's biochemical and genetic systems will not "recognize" anyway), they will land on Mars perhaps willing to go to work, but unfortunately also just starting their Functional Microgravity Recovery Time clocks. Common sense would assume that they will immediately seek deep shelter under the Martian regolith from additional exposure to cosmic radiation to short-circuit multimorbidity. But if structures have not been robotically created and shielded before their arrival, the first crews may well have to use heavy equipment to move multiplied tons of regolith to create their necessary accommodations, further exacerbating their exposure injuries. In the end, if these crewmembers possessed any level of "common sense" and a functional level of intelligence, they would have waited for a correction to the human spaceflight architecture before departing Earth forever!

Obviously, the specter of multimorbidity effects kicking in during this nightmare scenario is a real threat. With the collective effects of FMRT appended to cosmic radiation exposure far beyond the limits, it is almost certain that the astronaut crews on their first few days to a week on Mars will not be physiologically capable of engaging in any useful work at all except helping one another slowly walk to the cosmic radiation shelters.

To their distinct short-term benefit, in the Martian partial gravitational field (38% of Earth's), their FMRT should be accelerated to some degree. Of course, no one knows what this will be since there has never been a gravitational laboratory in space to understand this effect.

Expanding the secondary question – perhaps of even more importance – are these:

Will prolonged "acclimation" to 0.38 of Earth's gravity lock a new physiological standard into their bodies as a "new normal" that continues to injure with prolonged exposure? And how will this resolve itself when the person returns to 1.0 g on Earth? No one has any idea. And, except for the preposterous guesswork reassurances by the Mars faithful crowd, it remains a significant unknown health hazard. A new artificial gravity space station capable of altering gravitational gradients will have to be launched outside the Van Allen belts to understand this with enough fidelity to ethically commit humans to the frontier.

Unfortunately, none of that commonsense research was funded in the 1970s after Apollo, so the current crop of wannabe Martians expects to fly off to the Red Planet and find out the hard way on the planet itself. But once there and the biochemical realities begin to express themselves deep within their internal physiology, asking a Mars expedition member who flies out on an unprotected ship how they are doing on any given day will be about as socially acceptable as asking a member of the Donner party whether they think they made the right decision to head to California.

With all the phases of a round trip from Earth to Mars and then back home again encompassing around 1000 days, how will multiple adaptations and exposures to microgravity and partial gravitational fields affect the crew's capacity to fully adapt to one Earth-normal when returning home after the mission? Continuing that thought, after 1000 days in reduced gravity and the absence of gravity, will total recovery even be possible? The world's space agencies are betting that the answer is irrevocably positive.

But that assumption is based on the current fatally flawed exploration philosophy and runs afoul of their own spotty (or altogether missing) accumulated research data. So, they are cavalierly making unsubstantiated bets with the lives and long-term health of their astronaut explorers with almost zero data on these extended periods between multiple gravitational fields. The tragic truth is that they do not really have any answers. They do not have to explore space in this way, but they have appallingly chosen to do so anyway, wagering the lives and long-term health of their astronaut crews. (See the chapter titled "The Partial Gravity Scandal" for an expanded discussion.)

Not one competent scientist would ever bet anything on any outcome of a future Mars or long-term Lunar mission whatsoever based on these wholly data-deficient scenarios. We just do not have the information from which to make a qualified, responsible decision – and, with those decisions, protect the crews. Any scientist knows that some basic information other than wishful thinking and opinions is required to develop a competent hypothesis for any outcome, particularly with crewed systems. Yet, in this case, the world space agencies have transformed a mere hypothesis firmly anchored to a faulty paradigm into a fully developed working model – sans a complete suite of research data – and are preparing to commit lives to their scarcely educated guesses. And they have gambled nearly a trillion dollars in space hardware that was wholly obsolete before they built it to set up the perfect disaster in deep space of historic magnitude.

Scott Kelly's Experience

If you wish to know the intimate details of what it is actually like to live in space for nearly a year (340 days), there is absolutely no better source than NASA Astronaut Scott Kelly's book, *Endurance*. It is an excellent, well written, and easily readable resource for all future space explorers. It is also eye opening.

But of greater importance than what it was like to live there, Scott Kelly's description of what it was like to return to Earth was more than just a little informative. It tells an uninhibited story *no one has ever described before* with detailed, authentic, and candid clarity. There has never been a better description of the results of long-term exposure to microgravity than the one Kelly provides in his own words.

From the opening pages of *Endurance*, Kelly summarizes what he experienced upon return to Earth. On a former mission, Kelly had been to space for 159 days. But he stated that this time it was "much, much worse." He related that as he walked to his bed on his first night at home, that all his muscles were protesting the crush of gravity. And, as he lay in bed, his pain worsened, and he became delirious and feverish. He stated that when he slid to the side of his bed and stood upright that the pain in his legs was "awful."

Kelly also described that he could actually feel the tissues in his legs swelling. When he looked at them, they were "swollen, alien stumps, not like legs at all." He added that they were so swollen that his partner, Amiko, could not even find his anklebones when she squeezed them. He also said that his skin was burning, and everywhere his body contacted the bed, it was covered in a red, burning rash. Kelly confessed that normally he would have gone to the emergency room for help, but he knew that such a trip would only result in confused and puzzled physicians.[2] I highly recommend reading the full account at the opening of *Endurance*.

Kelly's book speaks volumes about the International Space Station itself and its inestimable importance to understanding what we face in deep space. Regardless of its shortcomings and failures as have already been discussed in this book, the information in the front several pages of Scott Kelly's book *is worth the entire cost* of the International Space

Station! The ISS was built to study microgravity. Tragically, for the future of human expansion to other worlds in our Solar System, all other considerations were fundamentally set aside. But in a few pages of his personal memoir, Scott Kelly delivers the hardcore summary of the endpoint of human adaptations to microgravity. Scott Kelly describes in graphic detail the limit to human exposure to microgravity – the core purpose for building the ISS platform in the first place.

What follows in this chapter is a detailed list of the adverse physiological effects of extended says in microgravity – a catalogue of effects and of multimorbidities.

For the purposes of this book, I describe an *extended stay* as exposure to microgravity *on the order of months.*

Microgravity Induced Musculoskeletal Toxicity - Bone Loss

What have we learned from half a century of exposing humans to microgravity? A *Nature* metadata analysis of the problem concluded in May 2021, "microgravity-induced bone loss is a significant and unresolved health risk for space travelers."[3] Another publication summarized it rather bluntly: "Astronauts returning from long missions on the International Space Station have suffered massive muscle and bone loss."[4] Any competent space life scientist I know would immediately reply with equal candor, "Define 'long' and define 'massive.'"

Each astronaut's physiology is unique, and their response to the exposure is somewhat different. And there is a variance between how male (XY chromosome) and female (XX chromosome) astronauts respond as well. But, okay, I accept the challenge. I define "long" as greater than three months and "massive" as spaceflight induced osteopenia and permanent bone loss that would be described by any competent orthopedist on Earth as "advanced pathology." This is a concise description of what we have discovered happens to both male and female crewmembers onboard the ISS for "long" unprotected missions.

As was described previously, throughout this book I have used the expressions *"functionally unshielded"* and *"functionally unprotected."* A functionally unshielded human spacecraft exposes its crews to cosmic radiation levels greater than the current United States limit for commercial nuclear power workers, which is 50 mSv/year (or 0.14 mSv/day). Likewise, when the term "functionally unprotected" is used in these volumes, it refers *only* to human rated spacecraft on long term deep space missions (not tourist class missions). A functionally unprotected human spacecraft exposes its crews to the long duration toxic mix of cosmic radiation over the federal limit for radiation workers and the synergistically toxic effects of microgravity, as is discussed exhaustively in this book.

In a presentation to the National Council on Radiation Protection and Measurements (NCRP) 57th Annual Meeting titled "Radiation & Flight: A Down-To-Earth Look At Risks" in April 2021, a series of studies were presented concerning the "musculoskeletal toxicity" of microgravity induced bone disease during prolonged spaceflight. Musculoskeletal toxicity is the term given to describe the adverse effects to the structure and/or function of the muscles, bones, and joints caused by exposure to any adverse toxic substance or environment. The descriptor is quite accurately placed for victims exposed to prolonged microgravity induced bone, muscle, and joint pathologies.

Spaceflight osteopenia refers to the characteristic bone loss that occurs during extended spaceflight missions. But not only do astronauts lose their bone density (which may be recoverable after several years), but they also lose bone structure itself. Astronauts lose an average of more than 1% bone mass per month spent in space deprived of artificial gravity.[5] There is also a concern that during long duration flights, excessive bone loss and the associated increase in serum calcium ion levels could interfere with the execution of mission tasks because, in many cases, it leads to debilitating kidney stones.

The loss of *bone mineral density* on a six-month mission may be reversible for some crewmembers after 1000 days post flight. But changes in the *bone structure* are totally irreversible and seem to mimic changes manifested in the elderly. Irreversible skeletal damage results from missions approaching six months or more. [6]

NASA itself publicized a somber warning startlingly similar to the results of the peer reviewed research mentioned. Said the Space Agency, "There is concern that long duration flights... will result in irreversible skeletal damage."[7] NASA has attempted to use rigid on-orbit exercise regimens and devices of various designs to offset this bone loss, as well as drug therapies established for post-menopausal women who have developed osteoporosis on Earth, but none have proved successful.[8]

This level of osteopenia, if observed on Earth, would inevitably be diagnosed as accelerated osteoporosis. Severe bone mass loss cases in menopausal women are typically only a loss of around 2% per year as compared to an astronaut's six to nine-fold bone mass loss in microgravity of 12% to 18% in the same period.[9] And finally, as we will discuss in detail in the upcoming chapter "Chronic Cosmic Radiation Exposure Etiology," there is evidence that the multimorbidity effects of double exposure to cosmic radiation and microgravity work synergistically together to accelerate bone density and structural loss with even greater severity.[10]

The "Effects of the Spaceflight Environment on Bone Health" presentation at NCRP's April 2021 meeting graphically demonstrated this documented bone loss in astronauts onboard the International Space Station.[11] Figure 1 depicts the astonishing loss of bone mass in 6 months due to the immediate toxic effects of microgravity on otherwise healthy bone structure compared to 50+-year-old humans on the surface of Earth. At this loss rate (restating the above), even a relatively young, healthy male astronaut will immediately exceed the severe bone mass loss of a menopausal woman (which is typically only around a 2% loss per year of whole body bone mass as compared to an astronaut's bone mass loss in microgravity of 12% to 18% in the same period).

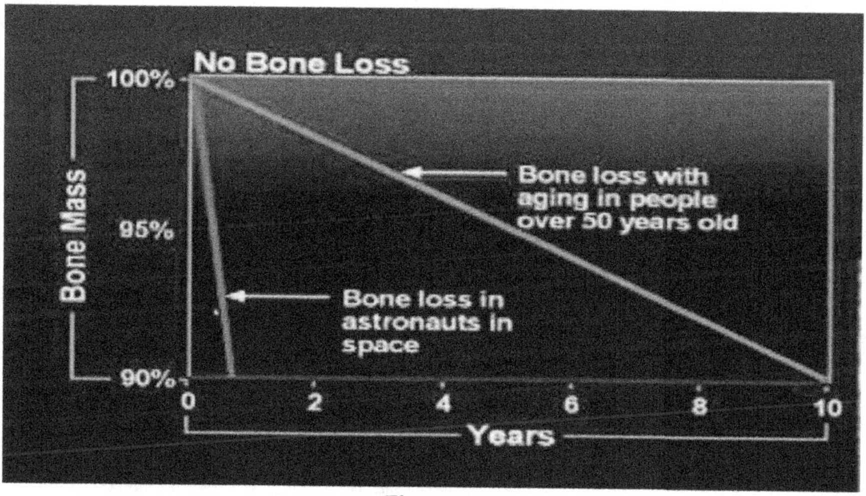

Figure 1

Figure 1 unarguably illustrates musculoskeletal toxicity in action – and the toxin to the system is microgravity. As an interesting note, this data was gathered in a research collaboration including the work of former NASA Payload Specialist Dr. Robert "Bob" W. Philips, DVM, PhD.

After 20 years of teaching and conducting research in Physiology and Nutrition at Colorado State University, Dr. Phillips was selected by NASA for the first mission dedicated to biomedical research in space, space shuttle mission SLS-1 (Spacelab Life Sciences 1). Prior to launch, he was removed from flight status due to a cardiac arrhythmia but supported the mission from the ground as the principal radio communicator between the on orbit payload crew and the ground based scientists with experiments being conducted in Spacelab.

From there, Dr. Phillips joined the NASA HQs team in Washington, DC, as the Space Station Chief Scientist for 3 years before moving to the Life Science Division in 1994, supporting Education and Outreach. It was in that position that Bob spent a 24-hour aquanaut mission with Claudia and me in the NASA *Scott Carpenter Space Analog Station* on the seafloor off Key Largo in 1998. It was a memorable experience that spawned a treasured friendship with this amazing man who did so much to focus public attention on life in space.)

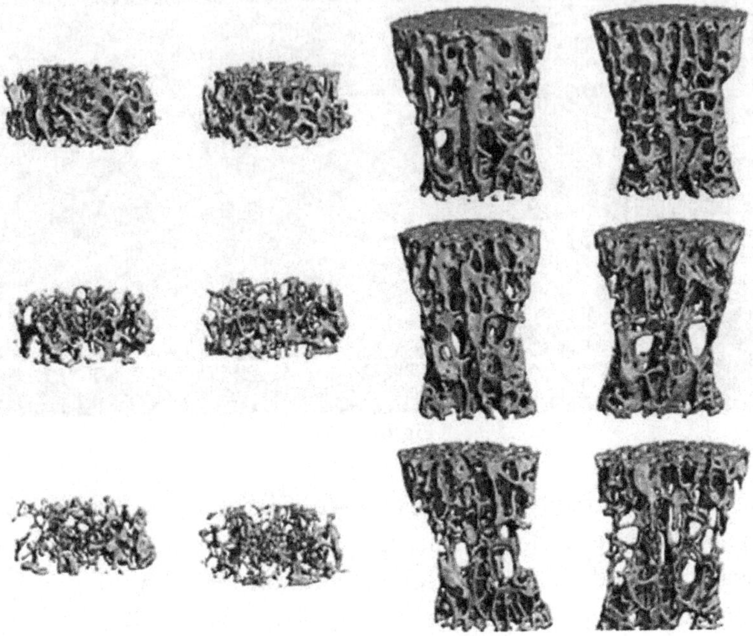

Figure 2

Figure 2 depicts the progression of bone loss on vertebrate skeletal systems over extended periods (months) in microgravity.[12] These micrographs clearly indicate a pathogenic progress that on Earth would be diagnosed as advanced osteoporosis.

Overall, we have gathered enough data on chronic bone loss in microgravity to document a 1–2% *per month* loss of bone mineral density during a typical extended mission (3-6 months). Interestingly, a greater percentage of the loss occurs in the weight-bearing sites.[13,14] In a 30-month mission to Mars, crewmembers not protected against exposure to microgravity could lose from 20-30% total bone mass.[15]

Again, as a reference point, the rate of bone loss on every NASA spaceflight on the ISS is 6 times the loss typically observed in Earth based elderly menopausal women for the equivalent time period.[16]

Countermeasure Backfire

Instead of redesigning their deep space architecture and building ships with artificial gravity specifically to protect the crews from debilitating injury such as the immediate induction of osteoporosis, NASA adds layers of countermeasures that often do not work in space as they do on Earth. One of the countermeasures is to attempt to load bones by the employment of vigorous exercise utilizing a machine called the Advanced Resistance Exercise Device, or ARED (seen left), which has shown itself to be of very limited usefulness to stem bone loss during long periods of toxic exposure to microgravity.

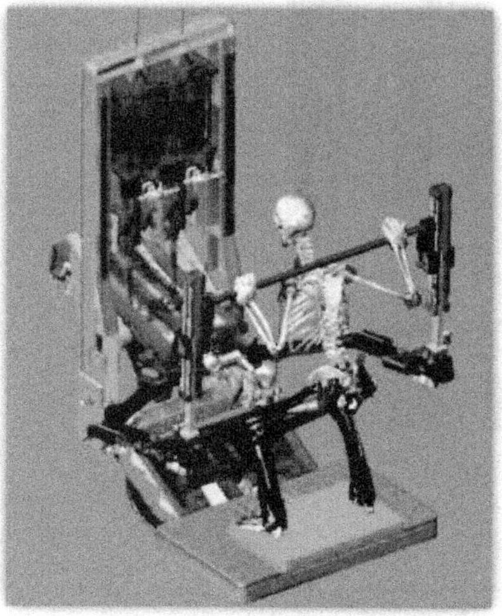

In addition, in the case of alarming bone demineralization in space, NASA has also conducted considerable research in using drugs intended to offset osteoporosis in order to mediate the damage done on board their wholly unprotected systems. One of these drugs is known as Bisphosphonate, a class of pharmaceuticals that are designed to reduce the loss of bone density in osteoporosis patients. The drugs have a good reputation in mediating the bone density loss on Earth and were therefore shipped to the ISS to use on NASA's astronaut crews to mediate their runaway osteoporosis on orbit.

The good news was that the combined countermeasure suite of daily and disciplined use of the ARED combined with the drug regimen appeared to lessen the effects of the overall loss of overall systemic bone density. But when detailed research was conducted on vertebrate animal models, it was discovered that outside of normal gravity, in the microgravity-radiation environment, the Bisphosphonate caused the bone density to be added in non-uniform layers (see Figure 3), resulting in abnormal layering of bone material inside the bone itself. While the bone's overall mass was greater, its density strength was *actually reduced* due to the bizarre, uneven internal deposition.[17] Figure 3 shows this asymmetrical distribution of bone mineral filled with open pores (called stress rises).

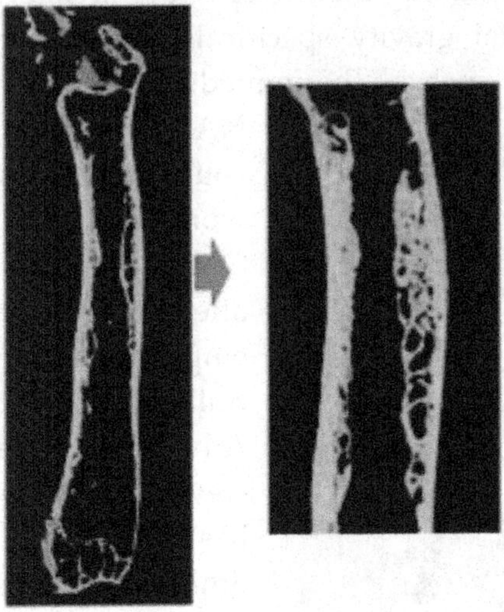

Figure 3

Such dissemination is likely to serve to actually weaken the bone overall (despite the fact that there appears to be more bone) because it is less dense along its stratified layering. Asked one perplexed researcher, "Why would you get progressive decline and yet more bone, even after preventing the net bone loss in astronauts with the combined countermeasures (drugs combined with activity on the ARED) during the course of their mission?"[18] Apparently, from this study, the drug and activity countermeasures appear to be creating bone mass but distributing it in such a way that it actually weakens the bone structure overall in this rather strange deposition pattern. Further discouraging news was reported during May 2021 in an article in the journal *Nature* which stated that the countermeasure drug most often used in space, Bisphosphonate, actually "...may interfere with the slow and often incomplete post-flight recovery" of the victims of microgravity's chronic toxicity on bone health while in space and, in many cases, after return to Earth.[19]

The Countermeasure Fog of Harold Frost's Theory

Similar to a stationary mechanical bicycle, the Cycle Ergometer with Vibration Isolation Stabilization (CEVIS) is another type of exercise equipment used to maintain and evaluate astronaut fitness in reduced gravity. NASA provided crews on the ISS with the ARED and the CEVIS for an excellent reason based on what they thought was sound theoretical advice. The idea behind the ARED and CEVIS is that if you "load up" the skeletal regions of the body with resistance type exercises, that the bone loss will be significantly reduced. It was hoped that use of the ARED and CEVIS would cue the body to keep the skeletal system loaded with calcium and strong against the gravitational gradient.

The basis for the concept is Frost's Mechanostat Theory. It involves the use of a system that monitors bone metabolism (longitudinal growth, bone modeling, and remodeling activities) as the bone is "loaded" by a mechanical system (in this case, the ARED and/or CEVIS) – hence the derivation of its name mechanostat. In his theory, Harold Frost concluded that the bone reacts to its environment by controlling its mechanical response by utilizing a natural skeletally activated mechanostat whose underlying purpose is to keep bone tissue strain at an optimal level through the hemostatic altering of bone structure.[20] Frost's mechanostat theory, introduced in his *The Utah Paradigm of Skeletal Physiology* in 1960, was the first to recognize the bone's organic-mechanical response to its environment and remains a cornerstone to much of osteology.

But instead of working as intended in space by simply coupling Frost's theory with the ARED and CEVIS machines, the actual results were skewed by the fog of reality. The fundamental task of "preventing bone loss" by loading the musculoskeletal system in microgravity was not as straightforward as anyone had initially supposed. For example, as was stated previously, the overall skeletal loss of mass in low-Earth orbit has been observed from 12% to 18% per year. But the human skeleton loses calcium differently in different regions – skull and cervical spine, upper limbs and thorax, pelvis and lumbar spine, and lower limbs. Microgravity induced loss has a tendency to be greater as one descends "down" the skeletal frame to the weight bearing bones. But, oddly, the human skull itself significantly *gains bone mass* in microgravity – documented as up to 2.2% per month – adding to the intracranial pressure problem as was previously discussed in this book![21]

What do these quite unexpected findings suggest? Of course, the first thought is that humans living full time in the microgravity of space may end up with quite large and heavy heads. Assuming that the brain does not grow inward at the same rate that it grows outward and squeezes into a radiation baked mush, humans could ultimately look exactly like the bug-eyed creatures from the movie *Mars Attacks* (Ack! Ack! Ack! – and never forget that all of Earth's creatures have an indomitable spirit that is made even stronger by the music of Tom Jones!)

But on a more serious note, it is clear that the broad application of Harold Frost's theory in microgravity is not entirely applicable when planning an effective countermeasure strategy. Kudos to NASA for a great try. But as in all scientific research, there is always the negative result to the hypothesis to contend with – the dead end of the inadequate approach (something NASA has not shown itself very capable of dealing with). It is much easier to just throw more taxpayer dollars at the dead end and hope the issue will go away after the next countermeasure fix. First, there is the countermeasure strategy followed by the fix of the countermeasure – the very typical band-aid on the band-aid approach. After all, Congress apparently will believe and fund just about anything pitched from a NASA bureaucrat.

However, the ISS data throws some remarkably interesting wrenches into Harold Frost's theory so universally applied in microgravity countermeasures. First, the human skull is a completely "unloaded" skeletal member. According to Frost's theorem, it should theoretically lose bone with at least as much rapidity as other artificially loaded bones in the body. Yet, the reverse is actually observed instead. Thus, without any involvement at all with a mechanostat countermeasure device on orbit, the skull significantly gains bone mass while the lower skeleton that is vigorously engaged with the ARED and CEVIS mechanistic devices still alarmingly loses bone to the point of toxicity. And while many rigorous hours spent daily on the ARED and CEVIS do demonstrate some limited level of bone preservation in the microgravity of space, it is spotty at best, with a wide range of results per individual astronaut.

All of this effort and expense just for the application of Frost's very iffy theoretical application in microgravity merely to try and prevent permanent crippling effects to astronauts upon return to Earth significantly reduces crew time spent on other research on orbit. Frost's mechanostat theory even predicts that after a period of time in the microgravity environment, the body would "adapt" and "normalize," and the resorption of calcium from the bones into the body (loss of calcium from the bones) would stop. But the vast evidence of many peer reviewed studies have shown clearly that there is no evidence at all that the adaptation process actually takes place in human bone loss in the microgravity of space.[22]

The concept of bone loss in microgravity is further obfuscated by a bewildering range of other interacting factors, including the whole-body system response to stress, altered metabolism, calcium homeostasis, and, of course, the most serious problem of all, mostly unfiltered doses of galactic cosmic radiation. And all of these issues somehow just must be only one single countermeasure away from a sweeping, effective, proven solution (the one that eliminates all the other hundred more poorly applied band-aids atop the other band-aids) – artificial gravity. But after more than 50 years, such a "miracle" has not appeared out of the void.

Persistent Toxicity After Arriving on Earth and Mars

The almost universal misunderstanding of long-term exposures to the toxic effects and broad spectrum of the often-pathological multimorbidities associated with long term functionally unshielded and functionally unprotected exposure to microgravity and galactic cosmic radiation is that after landing back on Earth or on another world, the effects of the toxicity are somehow turned off like a switch and everything returns to normal. This misunderstanding is truly unfortunate. To make matters even more confusing and prone to easy misrepresentation is the fact that we only have data on this return-to-the-planet scenario for astronauts who transition from long periods in microgravity to a one-Earth-normal environment.

Because of the failure of NASA to provide a biological space-based centrifuge for vertebrate animal model experiments, we have zero data on what to expect when crews leave deep space after a long interplanetary voyage and land for an extended stay in a partial gravity of 0.38 g (Mars). As we are about to discuss, the persistence of bone loss after return to Earth from prolonged spaceflight is concerning, but what happens after crews land on the surface of a world that is only one-third of Earth's gravity after a prolonged cruise-state exposure to microgravity without the protection of artificial gravity? Will the persistent decline in bone loss mirror the loss on return to Earth, or will it be magnified, say, three times? Without any research data, no one knows. What we do know, and have documented thoroughly, is stated unambiguously in a *Nature* journal article: "...several studies reported that 2–5 years are required to recover microgravity-induced bone loss, and that in some individuals the complete recovery was not achieved."[23]

United States astronauts who return from extended spaceflights are monitored very closely. It was discovered that for crew retuning from space who were loading their bones normally (exercising, walking, and conducting their life normally in the 1.0 g environment of Earth), there was a continued loss of bone and bone strength one-year postflight. This held true even for the astronauts that had preserved their bone during the course of their mission with the drug and exercise countermeasures while in space.[24]

It is essential to be aware that these changes in bone density, distribution, and overall morphology are almost certainly synergistic with the toxic effects of cosmic radiation dose. Summarizing once again, both microgravity and cosmic radiation work to disrupt normal human physiology and express layers of toxicity in human physiology.

Microgravity Induced Musculoskeletal Toxicity
-
Loss of Musculature

There is also a concurrent loss of musculature in microgravity alongside bone loss. While astronauts exposed to microgravity experience a 1% loss of bone mass per month, the loss of muscle mass is much greater. According to NASA studies, they can experience up to a 20% loss of muscle mass on spaceflights lasting five to 11 days.[25] Obviously, the rate of loss levels out substantially as exposure to microgravity is extended. In the first cellular analysis of the effects of long duration space flight on human muscle study ("Prolonged Space Flight-Induced Alterations in the Structure and Function of Human Skeletal Muscle Fibres" published in 2010 in the *Journal of Physiology),* researchers discovered that during a six-month mission on the International Space Station (roughly equivalent to a one-way trip to Mars), astronauts' muscles lose more than 40% of their capacity for physical work, despite in-flight exercise. Regardless of their physical condition upon departure, the astronauts all returned to Earth after 180 days with a musculature tone that resembled an average 80-year-old.[26]

This data demonstrates that if we continue to follow the current antiquated exploration paradigm of unprotected spaceflight, we are actively planning to send astronauts to Mars who depart Earth in excellent physical condition, knowing from our on-orbit research that by the time they arrive and land, they will only have the physical capacity of 80-year-olds! It is not at all impossible to imagine high-tech titanium walkers stationed at the landing pads. (I suppose that would be "One small shuffle for humankind...") And yet, everyone in the Agency and in private enterprise is enthusiastically ready to fly to Mars today, even under these astonishing injury profiles. The lack of situational awareness and associated blindness to which the Apollo era exploration paradigm has infected the thinking of even the most professional NASA flight planner and manager are astounding.

Quite surprisingly, it was also discovered that those individuals who were the best conditioned before being exposed to microgravity were also the most likely to come back with a withered or atrophied musculature. More succinctly, on an average comparison, their muscles lost 35% of their capacity to produce force and 20% of their velocity. Synergistically, these combined effects resulted in a 45% loss in power required for strong, quick movements.[27]

Circulatory Maladaptation

Exposure to microgravity also annuls a degree of the human body's circulatory design that literally *requires gravity* to work properly. We are equipped with check valves in our veins and arteries that prevent blood from pooling in our lower extremities as we stand and move about. Further, our much thinner capillaries are designed to pass blood and fluids by the powerful strength of osmotic pressure intended specifically to offset gravity's constant force. If you remove gravity from the physiological equation, the body's systems almost immediately begin to malfunction. Just like any machine, if you eliminate the conditions for which it was designed, it cannot function in its nominal operational modes.

In microgravity, these effects are not debilitating at first – just annoying. But after weeks of exposure, the effects become more physiologically significant and eventually injurious. The pages that follow describe the impacts of a maladapted circulatory system that is present in every case of human spaceflight when astronauts are exposed to microgravity for extended periods. They begin with a serious complication that leads to observed blood clots on orbit and eventually strokes in microgravity, as described later in this chapter. Literally, in the event of human exposure to microgravity for even a few days, the circulatory check valves work against nominal circulation, short circuiting proper blood flow, and leading to the conditions described in the pages ahead.

Reverse Blood Flow

If a farmer digs a deep water well, they will install a check valve in the bottom of the well's intake pipe to keep the water from flowing backward due to the pull of gravity into the well after the pump is turned off. If the check valve is missing, the pump would have to be re-primed after each use, and the system would be rendered effectively unworkable. The human body is designed to function in a gravitational gradient. Thus, there are thousands of check valves in the venous circulatory system to prevent blood from being pulled down by gravity between and during each cardiac pump cycle. Yet, although there are thousands of check valves in our veins, there are only two arterial check valves: one in the aorta and the other in the pulmonary trunk. The heart's main valves also act as check valves to control the flow of blood between the chambers. Likewise, the lymphatic system features unidirectional valves (called intraluminal valves) to assure proper flow. Without these mostly gravity dependent check valves, the system cannot operate properly and ultimately breaks down.

In any complex biological or even non-biological interconnected system designed exclusively to function in a gravitational field, removing one of its chief design parameters will inevitably lead to malfunction. For example, in a system optimally designed to operate in an environment that requires check valves, if you suddenly remove the parameter that makes the check valves operative, the system cannot function optimally or, more typically, it will not function at all.

In 2019-2020, medical researchers announced that they had discovered that when a human is in space, in some instances in certain systems in the body (specifically because of the human body's maladaptation to microgravity), blood actually flows backward on each heartbeat. In this case, the check valves cannot function nominally without the constant backpressure of gravity on the venous plumbing. In space, this was observed in a vessel called the left internal jugular vein. On Earth, this same malady has been observed in cases of serious pathology such as a blockage lower down the system, like a tumor growing in the chest.[28] Of course, as one might expect, if blood is flowing backward anywhere in the human circulatory system, eventually, things are going to start going wrong.

The Microgravity Stroke and Embolism Machine

In the same research published in November 2019, appearing in an article in the *Journal of the American Medical Association* (JAMA), the team on ISS also recorded other alarming effects of long space missions in microgravity. The article was titled, "Assessment of Jugular Venous Blood Flow Stasis and Thrombosis During Spaceflight." The medical term *thrombosis* refers to the formation of blood clots, and *blood flow stasis* refers to the failure of blood to flow properly. In this study, 11 healthy astronauts (nine men and two women with an average age of 46) onboard the International Space Station for six-month missions were studied using Doppler ultrasonographic machines for circulatory system adaptation. Six of the astronauts experienced stagnant or reverse blood flow, one had a blood clot, and another was found to have a potential partial blood clot. In the case of the astronaut with a blood clot and the other with a potential clot, they were immediately started on blood thinning medications which eventually resolved the problem while in space.[29,30]

The threat of such conditions existing in the body is stroke and pulmonary embolism, both of which are downstream effects of a blood clot. Retrograde (backward) blood flow and stagnation (stasis) anywhere in the body can lead to serious problems such as stroke or pulmonary embolism. For example, the term *blood flow stasis* is commonly used to describe what happens in the left atrium of the heart during chronic or episodic atrial fibrillation (Afib). In this condition, the Afib (abnormal heartbeat) causes blood to pool in the left atrium, increasing the possibility that this pooled, static blood will clot in the heart chamber and be released into the circulatory system resulting in a downstream clot in the brain or other part of the body. Most patients diagnosed with chronic Afib are immediately prescribed long term blood thinners because the blood flow stasis threat is significant enough to warrant these powerful medications.

In the JAMA study, the cause of both the retrograde flow and stasis was the human body's maladaptation to microgravity over extended space missions. Unfortunately, NASA's official response was to protect their flawed exploration model rather than expressing a primary concern for the long-term health and wellbeing of their crews. Instead of using the important finding to confirm the need for an exploration paradigm change and immediate redesign, Michael Stenger, the study author and director of NASA's Johnson Space Center (JSC) Cardiovascular and Vision Laboratory, stated,

"As potentially scary as this may sound, this novel and interesting finding isn't terribly concerning. The reality is that this has probably been happening since we started flying in space, we just never looked before. This gives us the opportunity to now conduct further research to determine what is causing this before speculating too much on potential consequences."[31]

If I were a physician and *any* of my patients exhibited retrograde blood flow, blood flow stasis, or clotting enough to order blood thinners, I would be very concerned. And, after treating the symptoms, I would immediately order their exclusion from the circumstances that caused such serious and potentially life-threatening conditions. I would certainly not use the cavalier diagnostic expression "novel and interesting finding (that) isn't terribly concerning." This comment is typical in the Agency as NASA deliberately lowballs *any* risk for which they cannot change the legal limits that would endanger the perpetuation of the Apollo exploration model.

If it is true that this malady may have been going on all along in long-term missions in microgravity, it is yet another urgent reason to create a new pathway to human space exploration! What the 2019 JAMA article findings described is that if the gravitational component is removed from the circulatory system, the human body has been observed and documented to become a thrombosis creating machine (as was detected in multiple crewmembers) and, therefore, potentially also a stroke and embolism machine.

If that is not the case – if the blood clots discovered on the ISS were "not terribly concerning" but merely "novel and interesting" – then why were the affected astronauts given powerful prescription blood thinners while still on-orbit for the remainder of their mission? The answer is quite obvious. If not treated (as on Earth), the probable sequels to the observed thrombosis are pulmonary embolism or stroke. That was the reason for the administration of the drug. Again, it is deeply troubling that to the Agency none of this is actually "terribly concerning."

But the *New England Journal of Medicine* stated the legitimate concern more candidly in early 2020, clearly differentiating between sending crews to the low-Earth orbiting ISS treehouse at merely 300 miles versus a distant planet:

"Although cervicocranial venous overcapacity is a known effect of spaceflight, the resultant changes in flow organization, local whole-blood viscosity, and prothrombotic risk need exploration. Active astronaut surveillance and experimental models are critical to the development of prevention and management strategies for venous thromboembolism in weightlessness, especially with future plans for prolonged space travel to the Moon and Mars."[32]

The Oxford University Press *European Heart Journal* labeled its peer-reviewed piece in the spring of 2020 with the even more straightforward question, "The thrombotic risk of spaceflight: has a serious problem been overlooked for more than half of a century?" It then added, "Evidence for thrombotic risk in spaceflight is unsatisfactory. This issue deserves further study... to enable safe governmental and touristic human spaceflight."[33]

Puffy Face Syndrome

Puffy Face Syndrome was the first clear evidence of circulatory system maladaptation and is typically the first visible symptom during spaceflight. Without gravity, the human machine literally cannot function as it was designed, and fluids begin to build up in the upper body, becoming particularly evident in the face. The powerful human osmotic pumps are working just as designed to keep blood circulating in the head, neck, and brain area, but gravity is not present to offset the osmotic pump effects.

Literally, the body is osmotically pressurizing the human head, and its offsetting pressure-balancing physiology has been rendered functionally useless by the absence of gravity. Likewise, the intraluminal valves of the upper body lymphatic system are also malfunctioning. Therefore, excess fluids immediately begin to collect in these areas without gravity to move them out in the typical pathways.

The effects are noticed as an uncomfortably stuffy head feeling, much like a head cold, complete with watering eyes. But in the absence of gravity, the watering eyes do not clear themselves by the fluid falling away as tears. Instead, the fluids coat the eyes with a tenacious liquid lens that makes it difficult to see.

In the end, Puffy Face Syndrome manifesting itself as prolonged intracranial hypertension is merely an annoyance; but it is unfortunately just the early symptoms of an extremely dangerous subset of underlying systemic breakdowns in progress that worsen over time. An anatomic cause of these microgravity related intracranial hypertension and visual disturbances has been termed Space Obstructive Syndrome, or SOS. This anatomic failure of adaptation to the microgravity environment leads to other conditions in a cascading effect throughout the human cranial regions.

Vision Loss

One serious SOS effect of potentially irreversible injury from long exposures to microgravity is called the Spaceflight Associated Neuroocular Syndrome (SANS), formerly known as the Visual Impairment and Intercranial Pressure Syndrome (VIIP). It affects some 80% of astronauts dramatically, while the other 20% apparently do not experience the effect at all. SANS results directly from the same physiological maladaptation that causes Puffy Face Syndrome – the buildup of intracranial pressure from a dysfunctional intracranial circulatory system. But instead of expressing itself as a mere annoyance, the constant abnormal cranial pressure begins to squeeze the optic nerve leading into the eye and simultaneously exerts pressure on the eyeball, deforming its shape.[34]

Astronaut John Phillips departed Earth with perfect 20/20 vision but returned after the completion of his 180-day mission on the ISS with 20/100 vision. The reason was the effects of SANS. Ground exams discovered that not only had his vision changed, but the shape of his eyes had also been affected. The backs of his eyes were flattened from the pressure, extending his retinas forward. He also had choroidal folds, which are like stretch marks in his interior eye, and his optic nerves were inflamed.[35]

According to Dr. Noam Alperin of the University of Miami, who authored a paper on intracranial hypertension and space effects, "People initially didn't know what to make of it, and by 2010 there was growing concern as it became apparent that some of the astronauts had severe structural changes that were not fully reversible upon return to

Earth. If the ocular structural deformations are not identified early, astronauts could suffer irreversible damage." He added, "as the eye globe becomes more flattened, the astronauts become hyperopic, or far-sighted."[36]

The summary of the direct eye effects from microgravity includes papilledema (swelling of the optic nerve), globe flattening, choroidal folds, "cotton wool" spots, thickening of the optic nerve, and decreased visual acuity. NASA admits that the vision of some astronauts upon return to Earth may never revert to baseline.[37]

The most concerning aspect of this microgravity-induced malady is the risk of crew blindness during long duration missions in deep space with no hope of rescue until end-of-mission. The choroidal folds induced in some astronauts by the SANS have been implicated as a direct cause of blindness in the general population.[38]

Alteration of Brain Structure

Puffy Face Syndrome, SOS, and intracranial hypertension are merely the external structural manifestations of a wholesale internal modification of structure, physiology, and intracranial circulation in the microgravity environment. The brain's physical form itself is also significantly altered over time. Once more, with the nominal function of the whole body interrupted and eventually altered by exposure to microgravity, the negative effects on the upper extremities are particularly negative.

Because of the significant role of osmotic pressure systems in the head and brain to maintain fluid pressure against gravity in one Earth-normal, in a microgravity environment, these systems continue to function as designed by pumping fluids against a gravitation gradient that no longer exists. This situation pressures the head, altering its physiology. Thus, the entire cranial morphological system begins to significantly change at once in the absence of gravity.

While the sinuses, eyes, and external features of the head swell as they accumulate fluids in the microgravity environment, the core of the head is also collecting fluids and swelling. But because the skull is rigid and cannot expand to release the burden, significant changes in the structure of the brain itself commence as the soft tissue is exposed to the chronic pressure, and substantial morphological transformations commence.

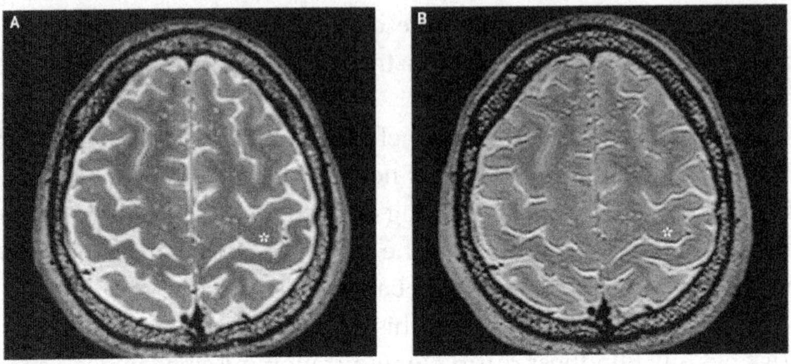

Figure 5

Brain scans of astronauts returning from extended periods in space reveal that the convoluted brain structure at the top of the brain begins to change. The folds of the brain (called the *gyri* and *sulci*) begin to narrow, creating a distinct alteration of its appearance. (Figure 5) There was also noted additional evidence of brain-shifting and a narrowing of the space between the top of the organ and the inner lining of the skull that is normally filled with cerebrospinal fluid (CSF).

A study was conducted examining the MRI scans of 34 astronauts, 18 who had been on short missions plus another 16 who had been in space for missions of six months or longer. 94% of the astronauts on extended missions exhibited a distinct narrowing of the space between the top of the brain and the external lining of the skull. But only 18.8% of the shorter-term (less than six months) missions in microgravity had these changes. The research article concluded, "Narrowing of CSF spaces at the vertex occurred frequently and predominantly in astronauts after long-duration flights."[39]

My wife, Claudia S. Chamberland, speculates that this intercranial pressure may be in part responsible for the observed 2.2% per month increase in bone density in the cranium in the microgravity environment. Because the skull, like all other human bone structures, is a living system continuously seeking homeostasis, it is very conceivable that one of its homeostatic mechanisms is to counter an increased internal pressure loading by compensating with increased bone density. Such a concept even complies with Frost's Mechanostat Theory. Perhaps one day, research will be conducted to shed light on the probability of this hypothesis.

What does this mean for astronaut health on an extended mission such as a Mars voyage? These structural changes may ultimately impede the flow of cerebrospinal fluid and exert pressure on surrounding tissues. Explained physiologist Donna Roberts from the Medical University of South Carolina, "We hypothesize that upward brain shift and expansion of tissue along the top of the brain may result in compression of adjacent venous structures along the top of the head. While we cannot prove it yet, we suspect that this may ultimately result in a decrease in the outflow of (cerebrospinal fluid) and blood from the head."[40]

Immune System Dysfunction

The human immune system is comprised of many layers of astonishingly complex subsystems that must work perfectly and in concert, or human life is at immediate risk. These intricate orchestrated biochemical subsystems overlay and diffuse through every micrometer of the human body from just beneath the skin to the core of the brain. Therefore, it is absolutely no surprise that the immune system is one of the first sentential systems to break down and show signs of dysregulation when the body is chronically stressed by any external or internal unusual environmental condition different than the one Earth-normal standard – like chronic exposure to microgravity. Indeed, that is exactly what NASA has concluded.

A NASA study conducted onboard the ISS discovered that microgravity (combined with the added multimorbidity-inducing effects of exposure to cosmic radiation streams in low-Earth orbit and chronic stress) "confused" the human immune system. It was discovered that the immune systems of 28 astronauts were at least temporarily changed during six-month extended missions in low-Earth orbit. While clearly detecting immunological variations, the study suggested that on longer-term missions in deep space, these alterations could ultimately have more critical impacts on the astronauts than just a confused immune system.[41]

NASA biological studies and immunology expert Brian Crucian, who directed the study, said, "Things like radiation, microbes, stress, microgravity, altered sleep cycles, and isolation could all have an effect on crew member immune systems. If this situation persisted for longer deep space missions, it could possibly increase risk of infection, hypersensitivity, or autoimmune issues for exploration astronauts."[42]

These studies suggest a continuous maladaptation of the human body's immune system that does not appear to adjust in microgravity at all on prolonged space missions. According to Johns Hopkins University Scientist Dr. Mark Shelhamer, "there is no place during spaceflight where we see stabilization of the immune system."[43]

Finally, it has also been discovered in human spaceflight systems that microgravity apparently also increases the virulence (capacity to become infective to humans) of some microorganisms. An article in *Applied Microbiology and Biotechnology* that focused on microgravity-induced alterations of virulence in various microbes stated, "Interestingly, microgravity appears to induce a plethora of effects (on bacteria present in the space environment) ranging from enhanced pathogenicity to enhanced biofilm production."[44]

Recreational Exposures to Microgravity

After a continuous stream of exceptionally bad news about extended exposure to microgravity, please note that these findings *do not* apply to recreational exposure. Quite unlike any exposure to cosmic radiation, the body is apparently not injured by short term exposures to microgravity and fully recovers when reentering a one Earth-normal environment. The limit of that exposure that results in full recovery without physiological sequel is uniquely defined in this book as the *recreational limit* to microgravity.

It appears that the limit of recreational exposure to microgravity without risking permanent, irreversible injury is on the order of a few months. Evidently, the limiting factor and most immediate risk is loss of musculature and bone density. (Exposure to cosmic radiation is *not* factored into this discussion.)

Whole body muscle loss, cardiac atrophy, bone loss, intercranial pressure, and vestibular dysfunction all begin on the first day in microgravity. But then, realistically, many of the same effects (to some lesser degree) also take place when any human takes a two-week lazy vacation and only sleeps, reads, lounges in front of the fireplace, and watches movies. The body is not designed for a sedentary lifestyle and will break down in many of the same ways it does in microgravity. But even in the most sedentary lifestyle, the individual must continuously move against the gravitation gradient – just to go to bed, the bathroom, or the refrigerator. (And, regardless of the effects, such vacation breaks are quite healthy in the long term for a variety of reasons.) Therefore, physiological breakdown occurs much more rapidly in microgravity.

And just as we all eventually totally recover from our two-week couch potato breaks, astronauts on space missions lasting no more than a few months also fully recover when they return home. I value my health and wellbeing as much as anyone. But even knowing what I do about the long-term effects of spaceflight, I would volunteer TODAY for a three-month mission to the Moon living under a fully shielded habitat structure! That is what I define as the recreational limit to microgravity exposure. And I would certainly expect a full recovery and an almost immediate return to normal life when I arrived back at my home. Of all the bad news presented thus far, none of it would significantly affect any short-term space mission lasting three months or less... *provided adequate cosmic radiation shielding is available!*

The Partial Gravity Scandal

"If you see fraud and do not say fraud, you are a fraud."

- Nassim Nicholas Taleb

"Fraud includes the pretense of knowledge when knowledge there is none."

- Benjamin Cardozo

ALL SCANDALS ARE NOT CREATED EQUAL. QUITE UNLIKE a purely scientific issue, a scandal is not built upon hypothesis and testing, but it is instead typically a social artifact of failure, ethical collapse, or immorality. Every scandal lives and dies in the public zeitgeist by exposure – through media attention or inattention. Every scandal is about as nebulous and difficult to wrap your arms around as hugging the jellyfish's "cousin," a Portuguese Man of War – gelatinous, devoid of structure, slimy and hazardous. But this story of the partial gravity scandal must be told in order for you to understand one of the reasons for the imminent collapse of the current human space exploration model. Its sordid tale has left humanity in quite a difficult and potentially dangerous position as we aim to send humans to settle other worlds. It is a story of politically inspired engineering and even medical malpractice on the part of the United States Government and its Space Agency. And while it is true that Congress and NASA may plead ignorance, even the plea itself would represent breathtaking incompetence of a historic magnitude. It is most likely that they may also plead *nolo contendere* (no contest) because it would be quite impossible to legitimately plead innocent to such egregious scientific and engineering negligence.

Some scandals are merely political theater. (I will sidestep many violent arguments here by deliberately not giving any examples.) Other more relevant scandals – such as the Tuskegee Syphilis Study and its outrages that initiated the Nuremberg Code, the Declaration of

Helsinki, and the Belmont Report (as was discussed in the previous chapter, "The Ethics of Neverland") – resulted in the discovery and full exposure of egregious government malpractice that cost the lives of thousands of innocents. These scandals were certainly not political theater in any sense and revealed the astonishing capacity of governments across the globe to screw up a two-car parade, resulting in historic levels of death and carnage. The partial gravity scandal has already happened, but its subsequent carnage lies still just ahead in the future – as these pages will describe. And since it has little value as political theater, it has received zero press outside a few passing mentions in the human space exploration forums.

I liken the partial gravity scandal to a parachute company that sold $1 billion worth of parachutes yet never bothered to test even one. Or a pharmaceutical company that manufactured vaccinations for an entire nation but failed to test them before shipping them to the entire population. Or a person claiming to be a surgeon who attempted brain surgery but had never been to medical school. Each of these examples represents the degree of scandal that lies just ahead of humanity as we begin to build our machines to send humans out to live on other worlds completely unprepared for what we may find there.

As the previous chapter clearly demonstrates, after half a century of research on human physiology in space, prolonged exposure to microgravity is known to be toxic and is an unmitigated disaster when considered for long term space travel and exposing crews to lengthy voyages between worlds. Therefore, except for the most clueless and ill-educated astronaut wannabees, it has become abundantly clear that no future Mars voyage should be conducted with the crews in microgravity.

Unless, of course, those in charge are planning for the loss of the mission and disaster. But what about *partial gravity* – such as on Luna and Mars? Surely, in the past half-century, we have conducted extensive laboratory animal and human research in partial gravity gradients so that we are fully knowledgeable and prepared to settle Luna and Mars! Guess again. You will probably find the ugly truth far worse than you can imagine. This chapter will spell out the details of how we arrived at this astonishingly perilous moment as we prepare to depart Earth forever.

Gravity is described as "force" and is enumerated in force terminology: 9.8 meters per second/per second (m/s^2), which is equivalent to 1.0 g (or one gravity). Force is the push or pull on an object with mass that causes it to change velocity (to accelerate).

In Lunar gravity, my Earth weight of 175 pounds would be perceived as a weight of 29 pounds, because the force of gravity on our Moon is only 1.62 m/s^2. And on Mars where the force of gravity is about 3.7 m/s^2, my perceived weight would be 67 pounds. But as a NASA aquanaut living on the seafloor, when I came out of my undersea habitat, I selected my "weight" to be deliberately 10 pounds negative so I could sit on the "front porch" in the mornings and observe the undersea "sunrise" without floating away. There, I "weighed" only 10 pounds – about three times less than I would have "weighed" on Luna. I could easily have added another 20 pounds of lead to my belt to equate to a Lunar weight, but there was no need. I required just enough weight to comfortably remain in position on the habitat step. Therefore, at the habitat entrance below the sea, I was experiencing a partial gravitational load on my body of 0.0625 g. Again, it was never my intent to simulate the partial gravity of Luna or Mars – but I easily could have with the right amount of additional weight.

I describe my simulated gravity on the seafloor as partial because it is not equivalent to 1.0 g. On Luna or Mars, the same will be true. Since one Earth-normal is our standard reference, these other worlds will offer the human physiology only partial gravity by that definition. One of the most essential questions we need to understand *before* committing humans for long duration assignments in partial gravitational fields (such as on Luna or Mars) is still unanswered. As was previously stated, we already know that extended exposure to microgravity (the near-complete absence of any gravity) is not just toxic, it is crippling, and will result in permanent injuries such as bone demineralization resembling osteoporosis and bone loss which, in many cases, cannot be completely recovered from after returning to Earth's 1.0 gravitational field.[1,2]

As covered in the previous chapter, in microgravity, bone loss occurs at a rate of 1 to 1.5% a month, which would equate to a total loss of up to one-quarter to one-third of a person's bone mass on an average Mars expedition, not counting time on the surface.[3] Muscle atrophy is also a part of the total toxicity of extended microgravity exposure. The cardiac muscle loses its shape, becomes more rounded, and the heart's walls thin from lack of a gravitational field challenge.[4] How much of that cardiac musculature returns to pre-flight norms is not yet known. These are just a few of the physiological malfunctions that occur in prolonged exposure to toxic levels of microgravity.

But what happens when the astronaut's body is exposed to a partial gravitational field? Lunar gravity is about 1/6 that of our home planet. Mars' gravity is about 1/3 that of Earth's. But without any data, we can only speculate how the human body will respond to these partial gravitational fields. Forming a hypothesis in the vacuum of real data, here are three fundamentally different possibilities – aka wild guesses. I will focus on Mars for these examples:

(1) Theoretically, the human body may react with a fractional, linear response. In other words, in 1/3 g, bone loss and demineralization, as well as cardiac atrophy, may be 1/3 less than in extended exposure to microgravity. Of course, there are many other pathological abnormalities associated with long term exposure to microgravity, and according to this theory, all of those physiological responses would also be decreased by 1/3 compared to extended microgravity. There is some evidence to support this theory in that the amount of exercise you participate in on Earth or in space is directly proportional to bone density, muscular density, aerobic output, and absence of atrophy. But those outcomes are also affected by many other variables, such as age, hormone balances, and nutrition. So, to make the leap that this theory is actually correct and proceed with a mission based on these assumptions is dangerous and unacceptably risky.

(2) In yet another theoretical possibility, you could guess that the body only needs a "little" gravity to give it all the biological cues it requires to function normally. It is unlikely that this is true for the reasons stated in theory one above. However, consider that plants on Earth respond to three environmental vectors: hydrotropism (plant roots grow toward water), phototropism (plant leaves point toward light), and gravitropism (plants grow up against the gravitational vector). In experiments with plants in Earth-based laboratories, we discovered that they only need a fractional amount of gravity to respond appropriately and grow toward the gravitational vector.[5] It does not take much to cause the plant to respond exactly as it would in a one gravity field. Of course, a human being and Groot are two completely distinct biological organisms with rather extensive differences in complexity.

(3) Finally, in another theory, the body's response in a partial gravitational field could vary system by system. In only 1/3 gravity,

some effects exhibited in microgravity could perhaps cease altogether (such as perhaps facial and brain swelling), while other effects would only diminish linearly by 1/3. Again, these are *all* unsupported wild guesses with zero scientific evidence!

And there are also hints of serious problems for reproduction and viable offspring in a partial gravitational field. In the 2021 article "Reproduction and the Early Development of Vertebrates in Space"[6] from the peer-reviewed journal *Life,* the paper reviewed multiple studies, including the two Space Shuttle experiments concerning pregnant rats – the National Institutes of Health, Rodent 1 and 2 (NIH.R1 and NIH.R2) missions on STS-66 (1994) and STS-70 (1995). The comprehensive study concluded that the data is "incomplete and restricted," that "development in microgravity can lead to many neurological alterations... ranging from major and massive brain abnormalities to subtle delays in the development of the vestibular system," and that "there are (still) many... questions to be answered about vertebrate development under space flight conditions." It would be tragic to invest trillions of dollars to make humans a multi-planetary species and then not be able to sustainably reproduce the colony in the event that it ultimately became the only surviving human presence.

So, to you, reader, "Let me have your undivided attention," says the professor to the student. Partial gravity may be *the hidden reef* in the dream of interplanetary empire. Cosmic radiation is very serious, but there is a relatively easy, nearly complete solution to that problem, as described previously. Microgravity likewise has a relatively simple fix – spin the spacecraft.

But there is no such easy fix for partial gravity.

The answers depend entirely on the results of the extensive research that we needed decades ago but squandered on the blind tunnel of microgravity we have been wandering about and lost in for over half a century. If partial gravity on any off-Earth world is linear with respect to our discovery of the physiological effects in microgravity, then the idea of permanent or even long-term settlement is in serious trouble. Why?

Consider the average example of a typical Mars mission on a Hohmann transfer orbit expedition using the antiquated Apollo model of exploration. The average time spent in two-way transit is 523 days in microgravity. The average surface time on Mars in 1/3 g is 451 days (a 297.66 day in microgravity equivalent). Therefore, if the partial gravity of Mars is linearly expressed on the human body, upon return to Earth at the end of the 974 days, the astronaut would have received a net 821 days of equivalent exposure to the toxic effects of prolonged physiological influences of microgravity. Therefore, *at a minimum,* 1/3 of the astronaut's skeletal structural density would be missing on return to Earth – and that does not even factor in cardiac muscle loss!

Obviously, no responsible government agency or private enterprise would even consider subjecting crews to such permanently debilitating missions without any data to understand the long-term effect of such exposures. And we must add the consequences of functionally unshielded exposure to GCR for that period along with the effects of multimorbidity in such a known and documented toxic environment. Let us assume that these government agencies and private enterprises are indeed ethical entities who will certainly do the right thing by requiring proper shielding to OSHA radiation standards and by also providing artificial gravity for both legs of the expedition voyages. We still have a potential limiting problem. No one has ever addressed long term exposure to partial gravity once the crews arrive on the other worlds.

Imagine that I am determined to live on Mars, riding out there and then living in a First Principle Solar System Class system (which is a true statement!). So, my wife, my dog, and I ride out on a Mars Cycler ship or a Modified Starship Pont to Point Ship (see Book Two – "Advanced Human Space Exploration System") with OSHA compliant radiation shielding at one gravity. I land on the surface of Mars and am escorted to my astonishingly beautiful private residence that has been created with perfect ergonomic considerations for all human requirements.

As I arrive at my perfect home on the Red Planet, I am certain that I have finally attained that sweet spot I have always dreamed of. I made it over 285 million miles in a First Principle Solar System Class interplanetary ship in health and in safety, and now I begin my new life on Mars. I have radiation shielding, and my exposure is carefully monitored so that I do not exceed the legal limits. But I am exposed 24/7 to a partial gravity of 1/3 g. I go about my business, day after day, year after year, only challenging my bones and musculature at 1/3 g. What are the resulting long-term effects on an organism designed for 1.0 g?

No one knows.

No one knows what it means for one year or five years or permanently. Will it lead to an enduringly debilitating condition as my body declines below its designed limits? Will I no longer be able to function normally, even in 1/3 g? Again, no one knows. Would there be a turning point (as many sci-fi novels have speculated) where I can no longer return to Earth and survive under 1.0 g conditions with my permanently compromised physiology? Even the most enthusiastic future Mars resident (such as me) can only guess at the answer in the complete vacuum of any data.

What will we do if we discover that the human body can only thrive over the long term in partial gravities that exceed, say, 80% of 1 g? All of the scenarios are only guessing because of the spectacular failure of the world's government space agencies to even look into one of the most important questions in all of human spaceflight. If the 80% statistic was the verifiable case, then a line of research investigations would have to be launched into the effects of periodic exposure to 1.0 g in a partial gravitational field. In this event, on Mars or Luna, perhaps colonists would have to sleep for 8 hours in bed chambers spun at 1.0 g. Or would the entire habitats or even the colonies themselves have to be spun?

It has been widely suggested that determining the limiting gravity gradient for humans, if one even exists, is essential for validating whether successful long term space habitation is even possible for protracted periods.

If partial gravity is a limiting barrier to permanent habitation of Luna or Mars, these ideas would have to be incorporated, or colonists would either become debilitated by human limits to exposure to partial gravitational fields or have to return to Earth after a predetermined period. In that case, permanent human occupation of other worlds would not be possible without some radical engineering solutions that today are not even on anyone's radar.

The Smoking Gun in the Japanese Parking Lot

Any reasonable person would ask the very logical question: after 60 years of human spaceflight, "Why do we have zero information on human exposure to partial gravity?" The nascent scandal began in the byzantine halls of NASA Headquarters in the early planning days and transformation of Space Station Freedom to the International Space Station.

During the early years of ISS planning, the plant scientists and other space biology disciplines opted to build and place a 2.5-meter centrifuge on the Station in a Gravitational Biology node to answer critical questions about gravitational gradients on orbit – questions that were *absolutely essential* to future human space exploration beyond low-Earth orbit. The centrifuge would have answered decisively important questions in vertebrate human physiology via appropriate animal model studies involving decades of research onboard the ISS under partial g loading or intermittent g loading. This one device would have been crucial to humanity in the second and third decades of the 21st century.

But there were intense arguments and vociferous objections within the microgravity research community, which leveled warnings that just firing up the centrifuge would induce microgravity perturbations and ultimately would *ruin all their experiments*. In so doing, they somberly warned, it would thus destroy the cash-cow for the ISS itself – microgravity research.

Their persuasive and politically loaded arguments as the "big boys" on the research block reverberated up to NASA Headquarters in ways that would make a racketeer's dream come true. Hanging in for several years, the Gravitational Biology node eventually morphed into the Centrifuge Accommodations Module (CAM) and was then farmed out to the Japanese Space Agency (JAXA) for construction. A NASASpaceflight.com forum concisely describes what happened next: "The CAM flight disappeared from the manifest in spring 2001, when George Abbey's 'surprise' $4.8 billion ISS overrun exploded in the face of the incoming Bush administration, and the administration responded by cutting ISS back to 'U.S. Core Complete.' It (CAM) was eventually canceled altogether in 2005."[8]

The fate of the centrifuge and all the vital questions it could have answered for artificial gravity in space and partial gravity gradients on Luna and Mars were sealed by overruns and politics.[9] The truly astonishing thing about this tale of woe is that perhaps *the single most vital piece of scientific equipment* on the International Space Station (in terms of the eventual fate of long term human habitation of space) was canceled specifically in order to preserve a microgravity-research-only platform. On the day of the CAM's cancellation, the ISS became a space station of profoundly limited usefulness. That platform itself would ultimately injure its human tenders who failed to understand the key reasons for its initial inception and ultimate existence.

That single, $100 million CAM device that would have answered so many of the vital questions for the future of interplanetary space settlement and exploration is, as of this writing, "on exhibit" (abandoned) in a parking lot at the Chofu Space Center in Japan.

Centrifuge Accommodations Module (CAM), Chofu Space Center, Japan

Now we are poised to send humans to live on Luna and Mars for long periods of time, having *zero data* on the effects of reduced gravity on vertebrate organisms. All of that data was lost to trivial intramural skirmishes and a stunning failure of long-term Agency vision – a scandal of historic magnitude.

The Most Important Unanswered Question - The Etiology of Gravitational Gradients

Let us assume that today there is a world class Lunar research facility perfectly shielded from galactic cosmic radiation buried under two meters of Lunar regolith. Receiving my orders to fly to the facility called Port Eisenguard,[10] I ride out to KSC Launch Complex 39 A, board a Starship HLS, and launch to low-Earth orbit. When the window opens, I fire my engine for a trans-lunar injection burn. I fly quickly out from under the Van Allen belts into deep space and into the full spectrum of cosmic radiation. Fortunately for me (not a total starry-eyed fool), I am an intelligent, knowledgeable, steely eyed missile man riding in a fully shielded Starship and gathering only a small fraction of the cosmic radiation levels that the Apollo astronauts had to endure. As such, I am very much unlike the not-so-bright knuckle dragging "you can't hurt me" idiots riding about in unprotected space systems.

Four days later, I touch down on the surface of the Moon and enter the remarkable Port Eisenguard base located beneath the surface.[11] It is at a depth under the Moon's regolith that shields me from more than 99% of the cosmic radiation, so I am even more protected than an airline passenger. And under Luna's surface, I return to a fractional gravitational gradient after almost five days in microgravity.

But wait... What does that mean for me physiologically? In the gravitational field of the Moon, I am living each day at only 17% of the gravity of Earth. As I step out of my spacesuit and stand in my skivvies, I weigh in at exactly 29 pounds. So how does my body respond to a weak gravitational gradient as compared to microgravity – or – as compared to one gravity as on Earth? To answer that question, I immediately consult the world's space agencies' biological database on research conducted on astronauts and animal models carried out in partial gravitational gradients for the past six decades of the most expensive research ever conducted in human history. I open the book and – to my complete astonishment – the pages are totally blank! We know that there may be a limiting gravity gradient, but we do not have a clue what it is. *How can this be?*

That is indeed, literally, the question of the century! Why is the most essential data concerning astronaut exposures to partial gravitational gradients completely missing? It is more than just astonishing and disheartening to realize that after more than a half a century of space data collection we have actually squandered every single opportunity to understand one of the few vital questions concerning human exploration of space (rivaled only by the equivalent complete lack of data on vertebrate exposure to cosmic radiation in deep space). And now that we literally stand at the threshold of the exploration of interplanetary space and establishment of off-planet settlements, the pages we absolutely need and literally cannot ethically fly without – *are completely blank!*[12]

In 2017, five researchers performed a comprehensive study in the peer reviewed literature in pursuit of *any* research conducted in the area of partial gravitational gradients, and here is what they discovered: "The initial search generated 1,323 publication hits. Out of these 1,323 publications, 43 studies were included in the present analysis, and relevant data were extracted. *None* of the 43 included studies investigated long-term effects (of biomechanical and cardiopulmonary responses to partial gravity)."[13]

Bion 3 – Kosmos 782

As one lame exception to this, a single Soviet mission was ostensibly devoted to the partial gravity question in a solo mission lasting an altogether useless 19.5 days. This distinctly politically flavored flight flew onboard the Soviet life sciences research craft Bion 3 (also designated Kosmos 782 and Biocosmos 782). Bion 3 does feature its socially noteworthy geopolitical place in history. Launched in 1975, it was the first space mission on which the Soviets allowed a United States payload to fly. The 13,230-pound spacecraft consisted of a standard Soviet design Kosmos frame that was spun in flight, creating artificial gravity for the payloads. It was launched from Plesetsk Cosmodrome on 25 November 1975 and returned on 15 December 1975 after remaining in an elliptical low-Earth orbit of 140 by 252 miles for 19.5 days.[14]

18 payloads from 7 countries flew on Bion 3 and were selected by the Soviet space life scientists in a process similar to the NASA RFP, or Request for Proposals. Thus, Principal Investigators responded to a flight opportunity that could either utilize the gravitational gradient… or not. In fact, nearly all the experiments were flown as a generic spaceflight opportunity and did not use the gravitational gradient provided by Bion 3 as an integral part of their experiments.[15] The only

payload onboard the 1975 spacecraft that utilized the partial gravitational gradient was an experiment whose results were published in a paper titled "The effects of microgravity, artificial gravity and the null-gravity environment of a ground-based clinostat were studied on carrot tumor growth."[16]

Yes, future space colonists – your eyes are not deceiving you. The only documented scientific investigation of partial gravitation gradients ever conducted in space was a 19.5-day investigation of tumor growth… *on a vegetable*! Over half a century of human space exploration has been squandered and missed the apparently politically unsavory questions that will determine the future of human exploration and settlement beyond low-Earth orbit. Regardless of its highly publicized, internationally flavored dedicated purpose a full decade before the dawn of glasnost, the 19.5-day mission merely focused laser sharp attention on carrot tumors and thus was not even close to making Bion 3 a useful platform to investigate the long-term effects of artificial gravity or partial gravity gradients on vertebrate species. That is, unless our astronaut subject is Baby Groot, in which case apologies are extended.

Even in 1975, while the United States was resolutely married to counter measuring the living daylights out of microgravity and making it and human adaptation work come hell or high water, the Soviets were apparently open to at least a quick look at another path. And while the U.S. was firmly committed to the glittering blinders of their glorious Apollo model already nearing obsolescence, the Soviets were at least considering the fundamentally groundbreaking idea of giving the biology of artificial gravity a quick glance. Unfortunately, Bion 3 was just that – an absurdly abbreviated 19.5-day glance from which not a single useful scientific outcome was obtained which benefited humans in 21st-century spaceflight.

But alas, there remains an empty folder of experimental evidence on the effects of partial gravity on human crews. None. Zero. Zip. The data does not exist. The data collection device built and paid for by the American taxpayer for this task lies abandoned like a smoking gun involved in a world class crime in a Japanese parking lot. In 2005, multi-agency pin-headed politics abandoned the only piece of equipment ever designed to begin to examine this most essential question. Obviously, there is little need to belabor that scientific tragedy of historic magnitude at any greater length. Suffice it to remain as one of the most significant scandalous scientific blunders in the history of human space exploration.

But why is knowing the limiting gravity gradient, or whether there is one, so absolutely vital? It is so because, as we begin to populate the Moon and Mars for extended periods in our current fatally flawed human space exploration models, we have absolutely no idea how partial gravitational gradients will affect human physiology. There is some exceptionally limited evidence with plants and earthworms in mimicked partial fields that they respond to partial gravity in positive physiological ways.[17] But the evidence is so scarce, completely devoid of any vertebrate physiology experiments, and its application so wholly unrelated to our current exploration models, that we may as well have none at all. Thus, we depart planet Earth entrusting our lives and health to an environment that we know in many ways as little about as did the Virginia Company of the New World when they carelessly tossed up so many colonists to the whims of undocumented chance and the horrible deaths they each met in a truly alien world.

Because of the human penchant over every century for extreme shortsightedness, visionary blindness, and acting out the role of beholden sycophants bending the knee to political pressures, after almost sixty years and more than a trillion dollars spent, we still have zero clues how humans will adapt to partial gravity at the mid-point of their expeditions to Mars. Whether or not they would even be able to return to Earth except as functional invalids is just a terribly uneducated guess – but apparently a fully "acceptable risk" to those who do not have to take it. Yet these agencies have the power to commit human lives to the vast greyness of this ultimate endpoint of scientific hubris and vanity. Any hint of ethics has long since departed the Agency under these astonishing circumstances. NASA turned the lights out in the ethics office when the Agency made the deal with OSHA to ignore the rest of the world's ionizing radiation limits to suit their desperately flawed human exploration model.

Thus, if a completely unbiased, totally objective observer from another intelligent civilization would come to our solar system, stand aside, and view the big picture of what we are about to embark on and our overall preparedness, it is wholly conceivable that they would declare our species as intellectually unfit to expand our presence in the universe at all – and for very good reasons! We have absolutely and abjectly failed to obtain the fundamental data points on partial gravity and cosmic radiation effects on vertebrate animals in deep space – the two most essential questions that give us the requisite and most fundamentally basic data before we even think about departing from low-Earth orbit. On these two unanswered questions, many lives depend. And yet, we still fully intend to proceed anyway, which indicates a frightening level of cognitive deficiency at best, and from its most disturbing component, an evident lack of any real concern for the lives and wellbeing of our people. NASA has set itself in the same seat as the USDA and intends to conduct its own 21st-century version of the Syphilis experiment – this time using microgravity and cosmic radiation on astronaut crews. Why? Because the government writes its own checks and its own rules and there is literally no one to stop them from their insanity.

What if We Cannot Adapt to Partial Gravity?

There is an unknown probability that when we finally get around to investigating this question using vertebrate animal models instead of humans that we will learn that we are not able to adapt to long term exposure to partial gravity and remain healthy. Maybe we will discover that it is doable on Mars but not on Luna. No one can say for certain today without exposing their complete naivety in science. But let us say for argument's sake that there is no way humans can permanently live in relatively low gravitational gradients and stay healthy. What then?

There is yet another way to depart Earth forever, as suggested by my wife, former United States Army Major Claudia Chamberland, a military logistics and security officer. She proposes that the solution may be not building colonies on the surface of Luna or Mars at all. Her idea is that these First Principle colonies be constructed in Mars and Lunar orbit. There they can be adequately shielded and spun for 1.0 g of gravity. Crews can commute to the surface daily or be flown down for variable stays of days, weeks, or even months in shielded habitats where they can comfortably and safely explore, work, or play on the surfaces of these other worlds before periodic return to the orbiting prime colonies and their necessary one Earth-normal environments. After a time of recovery at 1.0 g, individuals may then safely return to the surface for more research or recreation. And when they decide to go back to Earth for a visit, their bodies, having been safely protected in a one Earth-normal environment, will be healthy and ready for an immediate transition back to the home planet. What would that cycle of gravitational rehabilitation and recovery look like? No one has a clue.

Major Chamberland's idea is probably the most feasible and workable plan for permanently colonizing other worlds just in case human physiology is not adaptable to partial gravity. Yet, without adequate research data in partial gravity, *any speculation at all* is just another egghead guess. We will not know until this quintessential vertebrate biological research is finally on someone's radar and is launched into deep space... sixty years late. Oh well… As my aquanaut pals frequently remind me, *late is better than dead.*

These ideas are radically different than the current plans nearly everyone is working toward and developing today. To many, the concepts presented in this book will be a bridge too far and will probably elicit an emotional response. But what all the schemes currently floating about as real hardware and internet vaporware individually failed to consider was the First Principle philosophy of the limits of human adaptation. In the future of human exploration, all planning will *begin* with the First Principle and will logically unfold from the seed of that primal, rigid, incontrovertible design constraint.

At the turn of the 21st century, humanity is much like a child riding in the family car headed toward Disney World. (I know this story very well – I am one of them.) In the child's completely inexperienced mind, there are only visions painted expertly by the Disney commercials: a place of eternal happiness, talking mice, dogs and chipmunks bigger than Dad, stores full of candy, ice cream, and toys, and parades every hour where every single human is smiling and having a truly magical day. There is even a talking rat in a Disney French restaurant that will wish you a happy birthday by name – an animatronic animal so realistic it is impossible to tell him from the real thing. The Imagineers of Disney are truly students of Merlin himself. Who, of any age, would not yearn to go there?

Only after landing on the property does reality set in for the child. The place is so crowded it is difficult to see the Magic Castle over and around the thousands of adults towering above them everywhere. The temperature could easily broil a chicken on the sidewalk. The line to get into one of the floating boats at the "It's a Small World" ride is 60-90 minutes long, with similar waits at every attraction. Dad has an attitude, is incessantly complaining about the cost for all this "fun," and the pesky older sister keeps whining about being so hot she could die. Mom's trying to be positive, but her hair is plastered over her forehead with sweat, and her fake smile is gradually fading into a grimace. The child had no idea that this was going to be the actual Disney World of his dreams. And yet, as I (your author) will attest, less than 24 hours later, the dream of Disney World is back in force, and the family goes back again every day for a week. Why? Because the magic that the dream Disney so flawlessly represents *is real,* and no amount of reality can overcome it. As Disney season pass holders over many years, we know. But we are not just Disney pass holders – the Chamberlands are Disney *dream holders.*

Dreams of Mars World are kind of like that. Today, no human who has ever lived has any experience of real life after landing on Mars. But those people who will live there are alive today. And some of you are reading these words either before your ride out to the Red Planet, or your eyes are scanning them on your way there or after you have arrived. By the time you read this, you may already know the answers we do not have today. Although it is a certainty that you cannot broil a chicken on the sidewalks of Mars, perhaps you can freeze dry it. And if your pesky sister is there annoying you, consider that you are even blessed just by her presence. Mars World, just like Disney World, can never be all that we have dreamed it to be. But once we discover what we need to know before we can live there safely (without becoming the lab rats expended in finding it out), then we can all sing together,

"What a beautiful world this will be
What a glorious time to be free..."

This is true because the magic that the dreams of Mars and Luna so flawlessly represent is real, and no amount of reality will ever overcome it. Unless, of course, that reality involves injury or death. But if we depart Earth prepared and with the First Principle philosophy, then just as our bodies are protected and nourished in micro gravity or partial gravity on any worlds on which we have set our sights, so then our dream is adequately nourished and protected by our understanding and intelligence working on all cylinders and fueled by intellect, not emotions.

Truly, 400 years later, the owners of the Virginia Company would feel fully vindicated by our own breathtaking incompetence and lack of concern for our use of people. When the English ships *Susan Constant, Discovery,* and *Godspeed* departed Blackwall, London, England in 1606, they cast off into the unknown with about the same level of preparedness that we are armed with today. The parallels are truly startling. Their crews were deliberately lured by the promise of exploration and fame to become commodities to be wasted for the greater corporate need in a completely alien world. And, in the end, no one was held accountable… at that moment at least. But there is always, of course, the slow, deliberate court of history, which has a very long memory.

However…

Very fortunately for everyone involved, there is a way out of this embarrassing conundrum without even revealing our astonishing incompetence to future generations. Zero lives need to be sacrificed to the reckless, capricious gods of exploration machismo, and no one's health needs be ruined. That way out is the purpose for which this book was written! We still have time…

But do we have the will to stop the train and fix it right this time?

The ALARA
Range of
Cosmic
Radiation

In court, there is the legal doctrine of *falsus in uno, falsus in omnibus*, which means untruthful in one part, untruthful in all.

UNLESS YOU ARE A PROFESSIONAL NUCLEAR WORKER, PART of the food irradiation industry, work in medical imaging or R&D, or are employed by the Department of Energy (DOE), it is probably unlikely that you have encountered the acronym ALARA. It is a legal definition found in Title 10 of the United States Code of Federal Regulations, which addresses energy related topics. Two chapters of Federal Code Title 10 impose ALARA regulations: Chapter 1 applies to regulating ALARA in facilities licensed by the NRC, and Chapter 3 Part 835 targets Occupational Radiation Protection for DOE facilities. 10 CFR Chapter 2 Part 50 addresses domestic licensing for medical therapy, R&D facilities, and commercial and industrial facilities. Part 52 pertains to nuclear power plants, and Part 70 relates to special nuclear material.

ALARA is defined in Title 10, Section 20.1003, of the Code of Federal Regulations (10 CFR 20.1003)as "an acronym for 'as low as (is) reasonably achievable,' which means making every reasonable effort to maintain exposures to ionizing radiation as far below the dose limits as practical."[1] Therefore, the health and safety of individuals are assured by maintaining their doses *within the limits* prescribed in 10 CFR Part 20 and by implementing ALARA (as defined by the Code) to maintain exposures as *far below the dose limits* as practical. This official technical term with a legal definition, rigid requirements in the nuclear industry, and application to the medical imaging and R&D world is a sobering reminder of what we have already reviewed several times in this book – that *all* exposure to ionizing and cosmic radiation comprises a potential injury to the human body, and that there is *no acceptable acclimation threshold above background dose.*

ALARA formally recognizes this reality specifically by *not acting* to establish any lower-level acclimation guidelines or acclimation windows to the exposure of ionizing radiation by any federal workers – which includes NASA astronauts. Instead, in professional nuclear parlance, *every exposure*, no matter how small, always results in a probability of downstream cancer and sensitive tissue pathology because of inevitable DNA injury. The ultimate purpose of ALARA is, therefore, to minimize this injury as much as is reasonably possible and to continuously and systematically *drive down* exposures to ionizing radiation as significantly *below* the established limits as is achievable.

The underlying philosophical construct of ALARA recalls an obvious principle: when the human being is exposed to any amount of ionizing radiation, something negative happens to them – every – single – time. That is, the penetrating ionizing radiation represents a potential injury to the human body and may result in numerous downstream consequences, including death from many possible causes (all of which originate in the subcellular regions of the human body). Every exposure to radiation is like getting shot by microscopic bullets. They are so small that most of them will pass through the body with no sequel. But, if you increase the exposure (if you are repeatedly shot, or shot multiple times in the same area), your probability curve starts to bend in the wrong direction. ALARA highlights that according to the laws of probability, the serious or fatal injury may occur on the very first exposure, perhaps after 20 years of exposure, or anywhere in between. Therefore (restating the public law), "As defined in Title 10, Section 20.1003, of the Code of Federal Regulations (10 CFR 20.1003), ALARA is an acronym for 'as low as (is) reasonably achievable,' which means making every reasonable effort to maintain exposures to ionizing radiation as far below the dose limits as practical."

Note that the philosophy of ALARA is crystal clear. In the United States, expecting employees covered by the law to "adapt" to any level of ionizing radiation, however small, is not acceptable – and it also contravenes public law. Such an idea is so outrageously unethical that ALARA was specifically created to build a legal firewall to protect American workers against it and thereby actually radically reverse any such notions in the workplace. ALARA states unequivocally that any exposure, however small, is dangerous and will be stridently reduced as a matter of law and principle.

So not only is ALARA required by federal law, it is also linked to compliance with the FDA's Center for Devices and Radiological Health (CDRH) and its Initiative to Reduce Unnecessary Radiation Exposure from medical imaging, with OSHA, and (as mentioned) with the NRC. The U.S. nuclear industry has implemented ALARA and, through its designed mechanisms, has *reduced* the exposure of the average nuclear power plant (NPP) worker to an average of 2.0 millisieverts (mSv) per year, down from their limit of 50 mSv per year – a targeted *reduction* of 2500%. During recent years, NPPs have recorded exposure *reductions* from 62-23% specifically due to ALARA program management techniques.[2]

On the ALARA medical front, in the March 2020 issue of the Radiological Society of America's journal *Radiology*, a summary of NCRP Report 184 was presented, reflecting the work of a 2017 subcommittee convened to evaluate changes in medical radiation exposure to patients. The work states that a "nearly sixfold increase in average per capita radiation effective dose from medical sources excluding radiation therapy" seen between 1982 and 2006 has been reduced such that "the collective medical radiation dose to the population has decreased by nearly 20%," and the "average per capita dose has decreased by 27%."[3]

In comparison, since 1989, NASA *increased* their dose limit from 50 mSv per year to an allowed annual maximum of 500 mSv for all astronauts to a career maximum of 4000 mSv for a 55-year-old astronaut male – an *increase* of 1000% annually.[4]

	Career Exposure Limits NCRP Report No. 98 (1989) Sv / mSv		10 Year Career Exposure Limits NCRP Report No. 132 (2000) Sv / mSv	
Age (yr)	**Male**	**Female**	**Male**	**Female**
25	1.5 / 1500	1.0 / 1000	0.7 / 700	0.4 / 400
35	2.5 / 2500	1.75 / 1750	1.0 / 1000	0.6 / 600
45	3.25 / 3250	2.5 / 2500	1.5 / 1500	0.9 / 900
55	4.0 / 4000	3.0 / 3000	3.0 / 3000	1.7 / 1700

Table-2: Career Radiation Dose Limits for Late Effects Defined a Decade Apart

Figure 1

Figure 1 presents Table 2 from NCRP Report 132, which depicts the career exposure limits in force in mid-2021 for NASA astronauts by age and gender. The table displays the comparison of career limits for astronauts from the 1989 Report 98 and the 2000 Report 132 (adjusted to include conversion to mSv). The values are lower for younger astronauts because it is understood that although they may live longer than older astronauts, exposure to higher doses earlier in their careers could present greater health risks during their later life. Monthly and annual exposure limits primarily exist to prevent the short term physiological effects of exposure. However, career limits exist to contain radiation risk within a maximum 3% increased lifetime excess risk of cancer mortality.[5] As discussed before, female (XX chromosome) astronauts' exposure limits are lower due to increased susceptibility to breast and ovarian cancers, plus a higher incidence of radiation induced lung cancer.[6]

But because the various international space agencies involved in human spaceflight use a variety of methods to assess risk and dose to their astronauts, and also a variety of dose and risk protection quantities and restrictions, in December 2017, the International Systems Maturation Team – Radiation (ISMT-RAD) experts agreed to compare risk estimation methods of the different agencies involved in the ISS (Canadian Space Agency [CSA], European Space Agency [ESA], Federal Space Agency of the Russian Federation [FSA], Japan Aerospace Exploration Agency [JAXA], and NASA) to reach a consensus on dose limits for multilateral human space exploration, to include missions beyond low-Earth orbit (BLEO).

In February 2018, ISMT-RAD proposed collaboration with the International Commission on Radiation Protection (ICRP). The goals of this collaboration were to examine effects that may impact mission success and crew health, and to develop recommendations on exposure limits for exploration-class human spaceflight missions along with developing a common health risk assessment framework. The Main Committee of ICRP established Task Group 115 Risk and Dose Assessment for Radiological Protection of Astronauts (TG115) in May 2019, to include input from international experts in radiation risk and dosimetry. The plan was to build on the work of the 2013 ICRP Publication 123, *Assessment of Radiation Exposure of Astronauts in Space,* not only for application to astronauts but for data that may also be of relevance for space tourism.[7]

Meanwhile, in December 2020, at the request of NASA, an ad hoc committee of the National Academies of Sciences, Engineering, and Medicine on the Assessment of Strategies for Managing Cancer Risks Associated with Radiation Exposure During Crewed Space Missions convened to "review and assess NASA's processes for long-term risk assessment and management for currently anticipated crewed missions with respect to cancer (excluding current and post mission effects of radiation) due to exposure to space radiation."[8]

A public session webinar on 14 April 2021 allowed Committee members to hear from the ICRP Task Group 115 and discuss international space radiation dose limits. The subsequent closed session fifth meeting of the NAS Committee on Assessment of Strategies for Managing Cancer Risks Associated with Radiation Exposure During Crewed Space Missions convened 19 April 2021 to finalize findings, recommendations, and report text.[9]

In June 2021, the Consensus Report *Space Radiation and Astronaut Health: Managing and Communicating Cancer Risks* was released, which provided "the committee's recommendations and conclusions regarding the updated space radiation health standard, NASA's radiation risk communication strategies, and a process for developing an ethics-informed waiver protocol for long-duration spaceflight missions."[10]

The Consensus Report highlights offered intriguing insight into NASA's intent for an updated space radiation health standard, citing, "NASA is proposing to move from a standard conveyed as a risk limit to a standard still based on risk but conveyed as a dose-based limit for radiation exposure during spaceflight over the course of an astronaut's career. The proposed limit of approximately 600 millisieverts (mSv) was determined by NASA by applying NASA's cancer risk model to the most susceptible individual (i.e., a 35-year-old female) to calculate the mean REID (risk of exposure-induced death), which was then converted to an effective-dose value. NASA proposed that this limit be applied universally for all ages and sexes."[11]

Other Industries and GCR

There is clear evidence that studies of exposures to ionizing radiation cannot be backfit to cosmic radiation, even with the so called "quality factors," because of the total *lack of any evidence* to validate this process. This has been acknowledged before in this book, but it bears repeating. The United States Centers for Disease Control and Prevention (CDC) states bluntly, "We don't know what causes most health problems that could be linked to cosmic radiation. We don't know what levels of cosmic radiation are safe for every person." They continue, "We don't know what causes most health problems that could be linked to this radiation, including some forms of cancer and reproductive health issues like miscarriage and birth defects."[12]

How much exposure do commercial aircrews and pilots receive on average, since (besides astronauts) they are exposed to the most GCR? According to the National Council on Radiation Protection and Measurements (NCRP), they have the largest non-space professional average annual effective dose to cosmic radiation (0.2-5.0 mSv/year) of all U.S. radiation-exposed workers.[13] Yet on NASA's International Space Station, the exposure of NASA astronauts is a whopping 1,240-1,248 times higher at 241 mSv/year.[14]

The CDC has determined by their statement that there is no known "safe limit" scientifically established for cosmic radiation as there are currently no documented effects on the human body except for suspicions of potential cancer and reproductive injury. This is not at all surprising since the CDC clearly identifies the database on the effects of exposure to Earth-bound ionizing radiation as coming from a study of World War II nuclear bomb survivors – whose participants' radiation experience was not even close to the models of cosmic radiation exposure! Hence, the CDC truthfully admits there is no cosmic radiation database from which to draw conclusions and make reliable projections *of any kind*.[15]

What then is the conclusion to that rather dramatic lack of information as we are poised to deliberately expose astronaut crews to hundreds of times the airline flight crew exposure that comprises our sole terrestrial database? It appears that the response of the space agencies of the world is to just go ahead and fly humans in this environment *and see what happens* – clear evidence of an ethical meltdown in the self-policing modus operandi of the United States government – again. But what effects will chronic exposure to unprecedented levels of cosmic radiation on the human body have when conjoined with limited adaption or maladaptation processes such as the various multimorbidities of long period chronic exposure to microgravity? There is no way to know the answer to this query without one or several long-term deep space studies designed specifically to focus on these issues. And those initial studies *must* be conducted using laboratory vertebrate animals – certainly not human astronauts! This statement is so ethically fundamental that only the United States government would not only take issue with it – but would enthusiastically proceed to violate it by writing waivers to its own deeply flawed, ethically challenged, open ended policy.

ALARA on Mars

The ALARA program itself was introduced to conscientiously *reduce* exposure to ionizing radiation based on the accumulated mass of data verifying that exposures to any level of ionizing radiation comes with cellular DNA injury and is therefore accompanied by the risk for cancer and other health consequences, as defined by the CDC.[16] Once again, embedded in ALARA is the clear understanding that there is no exposure that comes without risk (including any degree of *acclimation* to cosmic or ionizing radiation), and therefore, the legal, operational guideline is one of minimizing that risk by continuously pushing exposures down *as far below the dose limits as practical.* As such, any legitimate application of ALARA operates as a cause and effect. The cause that required ALARA's implementation is that no lower risk limit or adaptation profile exists to mediate ionizing radiation. The effect is the ALARA program that intends to drive it down every year as low as reasonably achievable. Unfortunately for every NASA astronaut, they have been exempted from this protective legal requirement when their Agency strategizes the premeditated use of waivers, as evidenced by the Conclusion from the National Acadamies' Space Radiation and Astronaut Health Report, "Unless technological advancements and engineering controls provide improved radiation shielding or other protections to astronauts, for a mission to Mars to proceed, NASA would need to seek waivers to the radiation health standard both for the mission and for each astronaut."[17]

Mars is an alien planet in nearly every conceivable way, including its radiation environment. As a civilian nuclear professional for the United States Navy, I would often go to my job and pass a control point outside of a well-defined and partitioned radiation area. While there, I was continuously monitored by a personal dosimeter for my dose of ionizing radiation. Upon leaving the radiation area, I would pass through the control point again and walk through a radiation portal monitor that would scan my body and clothing for radioactive contamination. I would then be scanned once more by a radiation technician before I could walk out the door into the non-radiation area of the larger world.

On Mars, the situation is completely reversed. The entire planet is a monitored, active radiation area. Only the cosmic radiation shielded confines of the habitats are not. So, when I go outside on Mars, I enter a planetary radiation area where I am continuously monitored for my exposure to cosmic radiation. When I come back to the habitat, and once I am under its radiation shield, then I can remove my dosimeter and return to normal in the much more confining safety of the shielded environment. On Mars, ALARA *must* be in full force and not waived. Here is why it will make such a difference:

In nuclear power plants in the U.S., the Nuclear Regulatory Commission (NRC) and other monitoring and auditing organizations, such as the International Atomic Energy Agency (IAEA), monitor the exposure of workers to ionizing radiation, using the ALARA model for *reducing* exposure. The limit to exposure is used as a critical, inviolate ceiling to absolutely *avoid*, certainly not as a *goal to reach*, acknowledging that *all exposure*, regardless of how slight, *always* leads to some subliminal degree of injury. But on Mars, there will be tremendous mission pressure to reverse this logic, as is *already evolving* in the current human space exploration model.

So, what can be done? For starters, we can safely conduct uncrewed studies with vertebrate animals away from the protection of Earth and the security of the Van Allen belts. Then, once the actual limits to cosmic radiation are established based on the authentic animal model data from deep space studies, all astronauts will be legally constrained to a certain number of hours outside on the surface of Mars for each Earth calendar year – or the parameters will be modified to fit the Martian calendar based on the same data and physiological limits. As defined earlier, the goal of the ALARA model is to maintain exposures to radiation as far below the dose limits as practical; to continuously improve by driving them down – certainly not to reach the limit and then apply for another waiver to exceed it or – far worse – develop programs and processes to institutionally normalize it and thereby publicly claim ALARA adherence while actually upending it!

ALARA By the Numbers

ALARA is a process that has been characterized by continuous reduction of employee (including astronauts) exposure to various sources of ionizing radiation. There is, of course, a lower limit to that reduction beyond which the term "reasonably achievable" becomes predominant. And yet, when the exposure is so statistically low that the risk of stochastic cancers can no longer be enumerated, the ALARA process is successfully driving toward its intended result. As of 2021, the United States commercial nuclear power industry's astonishing successful implementation (with the oversight of the Nuclear Regulatory Commission) has driven worker annual exposures down to around 2 mSv – some 25 *times* below their legal annual limit. Their success has been assisted in no small part by the International Atomic Agency's Informed System on Occupational Exposure (ISOE). The ISOE is also jointly sponsored by the Organization for Economic Co-operation and Development Nuclear Energy Agency.

Authentic ALARA success is demonstrated by continuous *reductions* in worker exposure. Yet, in the reality outside NASA Headquarters, ALARA is a legally defined construct enacted by Congress specifically to protect the interests and long-term health of radiation workers and anyone exposed to ionizing radiation (regardless of its source) as a part of their occupational duties. How self-exempted NASA or private space organizations and corporations implement the cosmic radiation equivalent to ALARA on Mars or on Luna will be crucial to the future of human exploration outside the confines of low-Earth orbit. If ALARA is sincerely implemented off-planet (as it is on Earth) with the clear, unambiguous goal of maintaining exposures to cosmic radiation as far below the dose limits as practical, then it is evident that the injuries and cancers associated with living on and exploring Luna or Mars will not be crucial factors in long term mission success. But if the cowpokes ride out onto the new world deserts intent on busting the radiation exposure limits whenever possible justified by an unethical wavier process, the exploration of distant planets will be forever tainted by the tragic results.

Paraphrasing wisdom borrowed from another of my pursuits: There are bold astronauts and there are old astronauts – but there are no old bold astronauts.

In the end, galactic cosmic radiation (and its ultimate effects on soft human tissue) does not change – nor is it affected by politics, dreams of empire, corrupt federal agencies, or collective human stupidity. It is going to do what it does and will always do – permanently injure and kill biology. NASA will figure this out sooner or later. It would just be great if they could understand it before lighting the proverbial candle and respond appropriately *now* to preempt the necessity of ordering all those body bags. Getting back inside the legal limits, re-acquiescing to oversight by another agency that is legislatively responsible for that surveillance, and legitimately implementing the law would be a great first start!

Chronic Cosmic Radiation Exposure Etiology

"There was never a Jedi so wise that he could not be undone by his own assumptions."

- Claudia Gray

"When all think alike, then no one is thinking."

- Walter Lippman

WHEN TEACHING CLASSES IN BIOLOGY, CHEMISTRY, AND Anatomy and Physiology at Trident Technical College in Charleston, South Carolina, I strove to be the instructor I always wanted to be taught by in my previous academic life – tough but fair, interesting, and thorough. So, I added all the very best characteristics of my past instructors together and formed the lecturer that I would want to hear if I was one of those students facing me.

One of my habits was to let my students know when something I was about to say was fundamentally important. I stopped the lecture, paused, and let their wandering minds catch up with mine. I patiently waited for their eyes to focus on me. Then I would say something that had the not-so-hidden message that this information may well appear on a test soon. I would say, "Listen up – this is important!"

Reader, let your mind catch up with mine and...

"Listen up – this is important!"

We will now explore the physiological effects of galactic cosmic radiation (GCR) on the human body, and the body's systemic response to chronic GCR exposure in deep space, along with GCR's potential to elicit disease (etiology). Before we even begin this chapter, it is essential that we first understand the following four crucial GCR characteristics and subsequent conclusion, because together they establish the analytical pathway for the remaining pages of this chapter – and much of the remainder of this book:

Quantity – Quality – Spectra Differences – Non-linear Difference

Quantity

"An astronaut in deep space will encounter enough cosmic radiation of all forms so that every cell in their body will be traversed by a hydrogen ion particle every few days. But the same astronaut will also encounter the probability of a transit in every cell of their body by far heavier high-energy nuclei component (HZE) nuclei (^{16}Oxygen, ^{28}Silicon, and ^{56}Iron) once every month with an exponential increase in individual particle energy."[1]

Quality

"Despite their infrequency, it is the heavy ions that contribute a significant amount to the galactic cosmic radiation (GCR) dose that astronauts would incur outside of low-Earth orbit."[2]

Spectra Differences

"The character of the full spectrum of galactic cosmic radiation (GCR) differs significantly from what we have experienced in low-Earth orbit (LEO). Astronauts in LEO are not exposed to the full intensities of the GCR or its full spectrum. The GCR characteristic of deep space is significantly altered in LEO by the Earth's geomagnetic field which deflects protons and heavier ions at lower energies back into deep space. Further, significant sheltering is provided by shadow-shielding from Earth itself. Hence, particle fluence rates (the total number of particles per unit area with which a material is irradiated over time) from GCR are much lower in LEO and have a different characteristic than will be encountered in missions beyond LEO, by about a factor of three from ISS to deep space."[3]

Non-linear Difference

Because a significant percentage of HZE [high (H) nuclear charge (Z) energy (E)] particles have been selectively filtered and moderately shielded from astronauts in LEO at the International Space Station, and specifically because these particles contribute a significant portion of the cumulative radiation dose, then the expected linear increase in the full spectrum particle fluence rates of GCR by a factor of three could result in a non-linear, biologically effective dose rate absorbed in deep space, further complicating and obfuscating the predictive quality of what we have observed in LEO.[4]

Conclusion

We have very little competent radiobiology data predictive of the unique effects of this qualitatively and quantitively distinct cosmic radiation particle fluence rates and multiple spectra that we (vertebrate species) will experience in deep space – particularly that fraction that is highly biologically reactive. Specifically, our exposure will include short-term and long-term exposure to this potentially toxic GCR milieu.5 Subsequently, there is a growing body of experimental evidence that these biochemical traces will be carried back to Earth imprinted in the body of crewmembers as permanent injury of the multivariate structure and function (stoichiometry) of the astronaut's molecular biological composition.

Key GCR Spectra Differences in LEO and Deep Space

On orbit in LEO, an astronaut receives a significantly increased exposure to cosmic radiation as compared to Earth-bound observers, as is logically expected. In LEO, the atmospheric filter is absent, and the astronaut relies only on the inner and outer Van Allen geomagnetic shields that begin just above the ISS, starting at roughly 400 miles from the Earth's surface and extending out to 36,000 miles. But, because of the relative effectiveness of the shield above the crewmember, the GCR dose (quantity, defined as particle fluence rate) and its character (spectral quality) are significantly different than in deep space.

Emphasizing, again, the quantity and quality of GCR are *different in LEO* at the altitude of the International Space Station *when compared to deep space.*

The GCR spectrum in deep space is a complex combination of fast-moving ions derived from most atomic species found in the periodic table.[6] The GCR spectrum, from hydrogen (Z, or atomic number, of 1) through iron ($Z = 26$), to reiterate from the chapter "Cosmic Radiation Measured in Deep Space," is made up of about 87% hydrogen ions (protons) and 12% helium ions (α particles) mixed together with 1-2% heavier nuclei with charges from $Z = 3$ (lithium) to $Z = 28$ (nickel). Other ions heavier than nickel are also included in the GCR mix but occur more rarely. GCR ions with a charge $Z \geq 3$ are often described as HZE particles.[7,8]

While in LEO, the GCR spectrum's particle fluence rate is about one-third less than deep space – but the population of the particles and their energies have been altered by the selective filtration of the Van Allen shield. The shield is very efficient in redirecting lower energy particles away from Earth. However, the geomagnetic shield is not as effective in redirecting the same fraction of the much higher energy HZE particles. Thus, the nature of GCR in LEO packs about one-third of the energy of the deep space spectrum, but its population consists of a greater percentage of HZE particles and a smaller percentage of the lower energy particles.

Contrary to more simplistic expectations, it is not all bad news for astronauts that there is a higher relative abundance of HZE ions in the LEO spectrum below the geomagnetic shield. The mechanism of energy transfer and subsequent damage to cells by high energy cosmic particles depends entirely on the relative interactive mechanism inside the cells or tissues, or within the spacecraft walls upon penetration. If the HZE ion has sufficient energy, there is a variable probability that it may pass through the cell entirely with minimal damage.[9] However, if the heavy ion is stopped by the interaction with the spacecraft or the mass of the cellular structure, the outcome is dramatically different.

When the HZE particle, as a discrete particle of high mass and kinetic energy (the energy of motion), penetrates human tissue – regardless of the specific measure of its initial energy – it exhibits an interactive profile that can be far different than a single wave-particle of terrestrial ionizing radiation. As was discussed earlier in "A Brief Primer on Cosmic Radiation," its linear energy transfer (LET) is sufficiently high that, on occasion, the physical trace of damage that it leaves within a cell is so significant that it can be viewed with a microscope. It is an energy transfer potential so great that it is not even in the same ballpark as the energies of the highest terrestrial radiation source. In fact, the cellular damage from terrestrial radiation sources *cannot even be viewed* with any technology that currently exists. Because of the relativistic momentum of HZE radiation, when that particle enters a human cell, there is the potential that all its kinetic energy will be deposited inside the cell. It is traveling at such a velocity that, as it slows in the cellular layers, it sheds energy all along its track until the particle either exits the other side of the tissue/organism or has slowed down to a speed where it can begin to have strong interactions with orbiting intercellular electrons near its path.

At this moment in its life, the HZE particle exhibits some very interesting – and very biochemically damaging – behaviors. Once the particle begins to slow and interact with atomic structures in its path, there is a rapid loss of particle energy over a very small distance. This is the beginning of the structurally damaging LET pathway that we are able to visualize microscopically. This is described in biochemical terms as the "Bragg peak," which defines the scope of the rapid transfer of kinetic energy from the HZE particle before that particle comes to a stop in the cell itself.[10] Thus, if the HZE energy is high enough so that the particle passes through the cell without deposition of its full kinetic energy, then there is no Bragg peak, and the damage to the cellular structure is far less.

A good example of this is found in a bullet's pathway through organic tissue. Similarly, in the passage of a cosmic particle through organic tissue, many different variables will influence the "bullet's" path. In the simplest and most fortuitous of all cases, the bullet projectile will pass right through soft tissue and out the other side. (Think of your favorite cinematic hero and how they always escape the bad guy's bullets with merely a flesh wound.) While the cosmic particle's structure may remain virtually unchanged, some of its energy will be lost along a relatively small pathway in and out, but not nearly all. The stopping power of soft tissue is not significant compared to the inertia of the bullet. However, if the bullet enters more dense tissue or is deflected by bone, the projectile can easily spin, fragment, and travel in many independent pieces – and all of them completely expend their collective energies inside the organism, doing great damage. This would be an example of a Bragg peak along a bullet's pathway.

In that case, therefore, populations of high mass cosmic particles with less energy may actually be responsible for a greater potential of biological damage. This may indicate that the biological effects in LEO cannot be linearly extrapolated to the biological effects of the full GCR spectrum in deep space. In truth, we literally have not yet even begun to understand all these mechanisms with enough clarity to safely and ethically send humans into this environment.

The collective result of each interaction – defined by the sum of the energy released in the cells and tissues – ultimately defines the extent of biological injury and damage, or alteration of biomolecular function and pathways of protein synthesis. Above the optimal physiological repair threshold, over time, the radiation flux results in chronic degeneration of internal tissues and organs through a suite of biomolecular process disruptions.

Degenerative Pernicious Chronic Radiation Exposure

I personally received one illustration of the process of degenerative pernicious chronic radiation exposure as a component of a stunning announcement from my oncologist.

In the summer of 2011, I was summoned to the Lakeland Regional Cancer Center in central Florida for a consultation with my oncologist. It was for a meeting that no one ever wants to report to. I knew it before I left home. When we sat down together, I could see it in my doctor's face and hear it in his voice: I was diagnosed with prostate cancer. As he spoke, my physician gave me my treatment options along with their pros and cons. Frankly, none of them sounded like something I wanted to go through – but that was not the point. I had driven halfway across the state just to pick one.

During the discussion of therapeutic radiation treatments, my oncologist noted that if I selected radiation treatment as an option, future surgical intervention was taken off the table. If the radiation treatment was effective on the initial treatment, but the cancer subsequently returned, it could not then be dealt with by surgery. When I asked why, he pointed out that while the radiation beam is focused on the prostate gland, the beam passes through all the soft tissues between the wall of the body and the prostate, through the gland, and out the other side of the body. Since it was a focused beam, the connective tissues, blood supply, and all the other soft tissues around the prostate were permanently "adversely affected" by the radiation. In medical jargon, it is called *iatrogenesis* – or the inadvertent but inevitable result of permanent injury caused by the medical treatment itself.

In the case of brachytherapy – or internal radiation therapy with the implantation of radioactive (iodine-125 – or I-125) seeds into the prostate gland – the result was exactly the same. The seeds would induce the equivalent damage in the general area inside and outside the prostate gland, and surgery was not possible after the seeds therapy.

The complication from radiation therapy is sometimes referred to as pelvic toxicity, also labeled pelvic radiation disease (PRD), which can result from ionizing radiation-induced injury to surrounding non-cancerous tissues, often resulting in disruption of normal physiological functions.[11]

I thought about what my doctor had said at length. I also inquired if I would in any way feel or be aware of the adversely affected tissues inside my body afterward. My physician said that I would never have any sensation of it at all – it was something that only a surgeon would notice during surgery. But to a surgeon, it was a showstopper. The internal radiation-induced injury was so extensive in that area that precision surgery could not be safely and effectively performed. In this case, the radiation would significantly injure a collection of tissues inside my body that I would never be visually or sensorially aware of.

Primarily for these facts, I did not elect either form of radiation treatment. Without any reason to ponder the question prior to my diagnosis, I had labored under the simplistic assumption that radiation treatment targeted *only* the cancer. With my new awareness, I was not keen on permanently injuring other parts of my body by trying to rid myself of cancer, thereby limiting future treatment alternatives.

In deep space, the radiation particle fluence rates (or flux) to the whole body are not anywhere near the same magnitude or as extreme as medical therapeutic radiation, which are considerably lower energies per incident particle but at a far higher flux. That's the good news! The offsetting bad news is that the radiation species, quantity, quality, and spectra identification is radically distinct. galactic cosmic radiation (GCR) is a different animal altogether. Keep in mind the basketball and BB analogy from the previous chapter "A Brief Primer on Cosmic Radiation." Recall the energies of the individual particles. Remember that some of the individual particle energies are 1,000,000,000,000,000 times more energetic than the current most powerful therapeutic medical x-ray. But, if that is true, then why don't astronauts just instantly turn to ash in deep space?

Keep in mind that the radiation particle flux determines the impact on the cell and, in a larger scope, all of our tissues. A single particle of GCR traveling with that extraordinary energy will pass through a cell at a relativistic velocity (nearly light speed) and could well miss any essential organelle or the cell nucleus. The cell will then self-repair, and the cosmic drama is over, visible LET track or not. Likewise, it could also just as easily release its energy in a Bragg peak deposition event and obliterate the cell nucleus or essential organelles and kill one or many cells in a relativistic instant. The third option is profoundly important. The HZE's traverse through the cell could change the cell or tissue's molecular biology in a pathogenic manner, setting the stage for a downstream pathology, perhaps even decades later. (This phenomenon is discussed in greater detail in the next chapter, "Hitting the Physiological Wall.")

But while the HZE particles will only impact every cell in the astronaut's body *once every month,* their energy is also many magnitudes higher, resulting in their contribution to a significant amount of the full galactic cosmic radiation (GCR) dose.

It is essential that when truly understanding the etiology (cause, set of causes, or manner of causation of a disease or condition) of cosmic radiation, we are not guilty of broadly classifying the mere presence of cosmic radiation as innately deadly or crippling. It is not. All of us experience cosmic radiation on Earth each day! And therein lies the crucial point on which astronauts and their mission managers need to focus. The species, quantity, quality, and spectra identification of cosmic radiation is far different than terrestrial radiation – just as the quality and quantity of GCR encountered are different on the surface of the Earth, in low-Earth orbit, and in deep space. And all of these sources are different than therapeutic radiation options on Earth. We reach the point of concern for all radiation sources when the collective spectral energy and particle fluence rates *exceed the level of optimal repair,* and the fluence rate significantly *increases the probability of inducing downstream cancer.*

What is the limiting level of galactic cosmic radiation in deep space beyond which it becomes statistically dangerous to humans?

No one knows. Literally, *no one.*

We do not know because we have zero animal model research data on the actual cosmic radiation exposure etiology on vertebrate organisms – the consequence of our long-expired space exploration architecture. We have squandered our science, our opportunities, and our resources timidly wasting time, languishing in a ridiculous ever changing political ballet in the microgravity-induced mission irrelevance of low-Earth orbit. It is embarrassing to the human species to have squandered so much for such an insignificant return while completely missing much of what we were there to study in the first place.

In this chapter, despite the near total absence of data collected in deep space on rational limits (except for examining the long term health of the 24 members of the Apollo lunar cohort), we will focus on the bits and pieces of what we do know as we also integrate the Apollo research in some detail.

A Living Radiobiology Recording Device

The human organism is a living radiobiology recording device – a kind of organic dosimeter. When a cosmic particle passes through a given tissue or cell, it leaves a biochemical trace of many different possible downstream sequela. And if the cell survives (as a vast majority of them certainly do), the trace of its passage invariably remains – locked and recorded in the cell's molecular biology. Some are merely Bragg peak tracks left by their relativistic visitor, while others sustain intercellular organelle damage, and still others leave behind genetic imprints of the passage – some benign and others that are not. As in all exposures to terrestrial ionizing radiation, the organism responds by recording every encounter at the cellular level. If the cell subsequently programs itself to die through apoptosis because it cannot recover from the passage, then the recording is lost. But if the cumulative cells in a given tissue type store the record as protein and/or DNA damage, it has become a part of its function or dysfunction now and in later years.

Such is the case with the skin on my forearms. After spending a large majority of my life in San Diego, Hawaii, and Florida, my continually unprotected forearms have been steadily recording almost five decades of nearly constant exposure to tropical and sub-tropical solar radiation. As discussed previously, this radiation is certainly not the same species as medical ionizing radiation or cosmic radiation, but it represents a perfect illustration for both. Any high energy radiation that impacts human tissue has the potential of causing lasting damage and irreversible injury. Over the past decades, after thousands of hours of exposure to the high energy ultraviolet rays of direct sunlight, UV radiation has significantly degraded the elasticity (causing wrinkles) and depth (thickness) of the epidermal layers of my exposed skin. These maladies are the result of a pernicious and irreversible effect of solar UV radiation on my tissues.

This aberration to my integumentary system required a long period of time to evidence itself. Because of it, my increased susceptibility to contracting radiation-induced, probabilistically determined skin cancer is very high – and has, in fact, already occurred. I, and millions of other individuals, are walking examples of solar-radiation-encounter recording devices. Dr. Kristin Widick Smallwood, my knowledgeable and experienced dermatologist, reminds me at every visit, "Tanned skin is damaged skin." Initially, I ignorantly thought of her comments as akin to an overly protective, nagging nanny. Now I do not. Today, I wish I had had her conscientiously "nagging" me 50 years ago before it was too late.

In this example, the UV (ultraviolet) radiant could not penetrate more than 650 micrometers of my skin at its highest energy, and could not penetrate my thicker clothing.[12] So the damage from the Sun on my physiology was limited primarily to the easily observable exposed skin of my arms and a (fortunately) smaller amount of my upper forehead. Unfortunately, this is not at all true of cosmic or terrestrial therapeutic radiation. And yet, the human organism's internal tissues record and respond to the passage of any high energy radiation with equal effectiveness – and with the equal potential of internally imprinting irreparable damage. The best example of this has already been mentioned in the prostate cancer treatment discussion – the induction of pelvic radiation disease (PRD), which can result from internal ionizing radiation-induced injury to surrounding non-cancerous tissues.

On my forearms, I allowed unprotected rounds of skin damage and repair to occur in my body for countless cycles. On each cycle, the skin damage was evidenced by first-degree thermal burns (sunburn) followed by the skin's protective response of melatonin deposition (tanning) and healing. Just like any other exposure to any significantly powerful radiation stream, my body did not ever adapt – because it could not. It healed – but stored the biological damage of each cycle and passed it on to the new replacement cells. Cycle after cycle, my body managed the abuse and healed again and again. But the end result of such chronic injury was wholesale tissue degradation. Eventually, the cumulative cellular injury added up, and the tissues began to malfunction, disrupting the skin's normal function so that the permanent injury was evident to anyone who viewed my uncovered forearms. Ultimately, it led to skin cancer.

The same kind of cycle is occurring internally to any astronaut in space exposed to elevated GCR – the damage is just not visible, and it is not limited to the surface of the skin. Like high energy UV effects on skin, the exponentially higher and far more massive energy particle fluence rates of GCR are interacting internally with individual cells that collectively ultimately comprise the tissues of the astronaut's internal physiology. The variations are expressed by the particle fluence rates and the GCR's spectral differences. Again, in the deep space astronaut's case, as in any other exposure to any significantly powerful radiation stream, the crewmember's body does not adapt – because it cannot. It heals, but stores the biological damage of each cycle and passes the genetically recorded biochemical damage to the new replacement cells.

On Earth, we receive a flux of cosmic radiation 2000 times less than the astronaut in deep space. Just like the radiation species of a tooth x-ray is fundamentally the same as the species of a therapeutic beam of focused cancer radiation, the x-ray's particle fluence rates are many, many times less – therefore, its effect on the tissues of the patient is extraordinarily different. The average tooth x-ray may well kill a few hundred, or even thousands of cells and may be implicated in a vanishingly small (but not zero) number of head cancers – but it does not create lasting tissue injury that can be statistically documented above zero probability.

With all that in mind, as we look outward to deep space, then what is the significant difference between the effects of the cosmic radiation that each of us has always received on a daily basis (0.0009 mSv/day) on our home planet compared with the deep space astronaut's dose that is 2000 times greater (1.8 mSv/day)? Particle fluence rates and spectral quality are everything – defined as the measure of the amount of radiation received by an object from a given source and the character of that radiation when expressed in its full continuum. In this case, the object of interest here is the entire body, including all the astronaut's individual cells and internal tissues. The incident energy is far greater per unit penetration than we experience on Earth. And the spectral quality is astonishingly different, with the potential of irradiation by the heavy ions of every element in the periodic table over time. But the particle fluence rates are significantly lower than a beam of therapeutic ionizing radiation of far lower unit energy on Earth.

Obviously, a single tooth x-ray has no cumulative history of creating lasting tissue damage. But the higher particle fluence rates of a therapeutic radiation beam (and therefore the energy deposited by it) to the internal tissues and organs is an altogether different story. The therapeutic beam is not inherently evil or pernicious, but the potential to *cause* pernicious damage is a factor to be considered carefully by the patient, wholly dependent on the therapeutic conditions. Likewise, the cosmic radiation particle fluence rate that every astronaut experiences in deep space is also not inherently evil or pernicious. But, depending on the elapsed time of exposure to the relatively steady-state exposure rates and character of the GCR, it certainly does have the potential to *cause* pernicious lifelong damage to the internal tissues and organs of the astronaut.

Acute and Delayed Latent Cosmic Radiation Effects

When individual cells are penetrated by cosmic radiation, some of these cells may be directly injured and exhibit cellular toxicity caused by free radical-mediated damage to cellular DNA.[13] Other cells suffer reproductive injury when damage has been rendered to the DNA such that successful subsequent mitosis is prevented.[14] These effects are considered acute damage – or injuries that have immediate results. If they occur in a small number of dispersed cells, then the organism's tissue damage is minimized, perhaps to the point that the cells are replaced by normal replenishment, and limited lasting imprints of the damage are recorded. This is almost certainly the process that is ongoing in people exposed daily to terrestrial levels of cosmic radiation that filters down through the Van Allen shield and our atmosphere. While the cosmic ray exposures on Earth's surface also produce a steady state level of probabilistic carcinogenesis (because there is no lower exposure threshold for the development of cancer from *any* level of exposure, *however small*)[15], it is not considered statistically significant, cannot be probabilistically determined, and therefore, it is labeled as a benign background exposure.

But as the particle fluence rate of cosmic radiation is increased 2000 times in deep space and spread out over time as a chronic whole-body exposure, we have gathered considerable evidence that the effects are unmistakably different than terrestrial background radiation, even if we have not yet precisely quantified them by research on animal models in the deep space environment. As discovered recently and previously discussed here, the effect of this deep space encountered dose of cosmic radiation has unique effects on separate organs and systems of human physiology, such as the brain and the cardiovascular system. But GCR is also clearly synergistic with other negative spaceflight stimuli such as microgravity, which is then manifested as systemic multimorbidity.

One interesting contrast between cosmic radiation and terrestrial non-ionizing radiation is that cosmic radiation is particularly more damaging and permanently injurious to the cell's nuclear material than terrestrial ionizing radiation.[16] In a 2016 study with cells irradiated with terrestrial ionizing radiation when compared to a laboratory model of cosmic radiation, it was observed that there was less efficient repair when DNA damage was induced by heavy ions as compared with terrestrial sources of X-rays. The researchers noted that most of the irreparable damage was detected in a complex of DNA single strand and double strand breaks when exposed to cosmic radiation, while DNA damage induced by ionizing radiation was mostly repaired within 24 hours. This team stated that DNA damage induced by heavy ions is often complex and difficult to repair and results in persistent and irreparable DNA strand damage.[17] Again, think of the mass difference between a basketball and a BB and their kinetic energy disparity at relativistic velocities. The simple logic of such an interaction holds true, as evidenced and now documented by laboratory investigations of structures and functions affected by GCR-type irradiation inside the cell.

Also, as logically expected, another study revealed that cosmic radiation (as heavy ion species) has a high relative biological effectiveness (RBE) value because its linear energy transfer (LET) is high and can produce dense ionizing events along the particle track.[18] In contrast, the LET of X-rays is low, and X-rays only produce sparse ionizing events. Again, high LET radiation induces complex DNA damage – a unique class of DNA lesions that include two or more individual lesions within one or two helical turns of the DNA molecule. These are far more difficult to repair than individual X-ray induced lesions, and, in many instances, are irreparable – unlike the cells exposed to terrestrial ionizing radiation. When this occurs in individual cells, they "commit suicide" by activating an intracellular death program. Again, this process is referred to by cell biologists as programmed cell death, although it is more commonly called *apoptosis* (which is derived from the Greek word meaning "falling off," as leaves fall from a tree).[19] Thus, the rate of apoptosis is significantly higher in tissues exposed to cosmic radiation than in tissues exposed to terrestrial ionizing radiation. The collective results of these studies demonstrate that cosmic radiation species (high LET radiation) are responsible for a complexity of DNA damage which is difficult to repair and contributes to the persistency and latency of DNA damage after exposure to GCR.[20]

There is a pair of conspicuously missing, significantly important links we have been unable to obtain on Earth or in low-Earth orbit:
(1) Duplication of the full spectrum chronic radiation environment (which is impossible to reproduce outside of deep space), and
(2) On Earth or in low-Earth orbit, it is impossible to administer a chronic dose of full spectrum cosmic radiation.

Neither of these is possible to duplicate on Earth or in low-Earth orbit with any confidence of full fidelity. And in the instance of simulated GCR, in a majority of past cases, only a single heavy particle species or a relatively small cohort of ions is used, while the capacity to generate average HZE energies is not even possible by our most powerful accelerators. Hence, accurately replicated quality and quantity are nearly totally absent in our research data.

For example (and this will be discussed in greater detail later in this chapter), in 2007, I was part of a team that investigated the effects of a modeled *acute* cosmic ray particle fluence rate on the brains of genetically modified mice. We conducted the exposure phase of the study at Brookhaven National Laboratory's Relativistic Heavy Ion Collider in Upton, New York. There, we exposed the mice to acute doses over very short periods of time, corresponding to an average fluence of 1-3 hits of iron ions per cell. This single element, single energy particle fluence rate administered over a very short period mirrored what would have presumably been received by the mice in deep space over a period of 3-9 months. This dose was deliberately selected to mimic a deep space voyage to Mars in a *chronic* dose (a rate that was limited by our budget as beam time at Brookhaven on the Relativistic Heavy Ion Collider is expensive).

Not speaking for any other member of our team, my chief concern about the results was that they reflected the mouse's physiological response to an acute dose – not a chronic dose such as they would have experienced on an actual Mars mission over 3-9 months. In radiation, exposure *time* and *particle fluence rates* are *everything* when considering the viable physiological outcome. Again, we were limited by the project funding available, which determined our precious slice of time at the

Brookhaven Relativistic Heavy Ion Collider. And again, the collider itself at that time limited us to a single heavy ion – at a single energy – which was nowhere near the fidelity of cosmic radiation.

I may therefore summarize in a broad conclusion that as the GCR energy interacts with human cells and tissues, elevated intracellular free radicals are created as a byproduct of those interactions that initiate oxidative damage. Exposure over days to months to this level of oxidative stress begins to add up and manifest itself as radiation-induced changes that we know about at this early stage – such as the debilitating injurious effects of premature aging, cardiovascular disease, and the formation of cataracts. The extraordinary ionization capacity of GCR ions causes them to become a significant contributor to tissue damage and carcinogenesis, central nervous system (CNS) degeneration, and deleterious health outcomes, many of which are delayed and may not be expressed in the organism until decades later.[21]

Acute Space Induced Anemia

One of the acute changes we have discovered by long term (6 months or greater) functionally unprotected space exposure on the International Space Station is the development of space induced anemia, a particularly worrisome scourge that follows the astronaut back home to Earth after the end of mission.

As a *Nature* article reported in January 2022, a significant fraction of astronauts' red blood cells disappears on longer term missions in functionally unprotected space environments, such as onboard the ISS. The body's red blood cells are destroyed by a process in the body called hematolysis or hemolysis (the rupturing of the red blood cells and release of hemoglobin into the plasma) at a rate 54% higher than the normal replacement rate of the blood borne cells, leading to clinical anemia. What is unknown is why this occurs. One significant question is whether the cells are destroyed by the human body's invariably losing battle with adaptation to microgravity, by cosmic radiation, or both. Is it a process of natural hemolysis, is it a process of resorption of cells killed by cosmic radiation, or both? At this point, no one knows.[22]

What is well known is that this space induced anemia does not self-correct in every case upon return to the home planet. In 38% of returning long-term space voyagers, the rate of cell destruction had not corrected itself by a year later, and the rate of hemolysis was still 30% higher than pre-flight. According to one medical researcher, "The longer astronauts stay in space, the longer space anemia will plague them on solid land."[23] Like any other injury, the damage to the human body always returns back home with the crew.

The Game Changer of Profound Uncertainty

With all of this in mind, understand the impact of a single research paper published in *Nature*, one of the world's most prestigious journals, on 28 July 2016. The article written by the research team led by Michael D. Delp and including Jacqueline M. Charvat, Charles L. Limoli, Ruth K. Globus, and Payal Ghosh was titled "Apollo Lunar Astronauts Show Higher Cardiovascular Disease Mortality: Possible Deep Space Radiation Effects on the Vascular Endothelium."[24]

Before any discussion of its findings, it is essential to first examine what this article was and what it was not. This article certainly did *not* establish *any* level of certainty nor provide proof that these statistics are absolutely related to the physiological effects of cosmic radiation. The article pointed out this limitation clearly and stated, "it remains unknown what specific factor(s) in the space environment is responsible."[25] Furthermore, the individuals in the study numbered only 24 and therefore limited its statistical validity.[26] This article is still frequently erroneously cited in popular literature as "proof" of GCR as the definitive cause of cardiovascular injury to astronauts. It is not – and never claimed such nor intended to be interpreted as such.

However…

It is equally essential to explore the true impact of this landmark study. It was the first report to link real-world data of a critical and irrefutable medical statistic to the only human subjects ever exposed to the full spectrum of cosmic radiation in deep space. Whether the single medical factor that links them (significantly increased cardiovascular disease) was caused by cosmic radiation or by other less conspicuous lifestyle factors that they all shared will probably never be known. Science is precise for a reason.

But that precision also cuts both ways.

Why was this article's impact so profound on the future of crewed space exploration in deep space? Because its findings conclusively slipped the greatest level of uncertainty possible into the specter of future human crews flying out beyond the Van Allen belts for *any* length of time without adequate shielding. Here, a consummately researched professional paper published in the most prestigious scientific research journal on Earth clearly linked the distinct possibility of developing a significantly increased risk of long-term cardiovascular disease to *any human* who flies outside low-Earth orbit for an average of 9.5 days or longer. This uncertainty was not the result of laboratory studies but of the medical histories and biochemistry recorded and locked inside the bodies of the only 24 men who actually ventured there. While the fundamental link of cause and effect is incomplete, its implicit uncertainty is absolutely not. *Any* trip to the Moon (much less Mars) in inadequately shielded spacecraft is now at least strongly suspected of the high probability of causing serious long-term physical injury to astronaut crews from cosmic radiation – a concern that is now suggested by other documented evidence gleaned from peer-reviewed research. But this one publication alone changed the game in human deep space exploration specifically because its subjects were the only humans who had ventured out into deep space.

In rapid-fire succession, including the prior paper published by our team in 2007, careful scientific research has emerged in the past few years linking cosmic radiation with pernicious tissue damage in several key physiological systems. It is not just the *Nature* article. They are found in many respected, peer-reviewed publications. They all stand together, demonstrating the *same effect on human physiology,* as will be discussed here momentarily. Collectively, they are a clear warning that we cannot ethically allow future crews to fly into deep space without adequate protection, and that we must commit appropriate animal models to long term deep space study to determine answers before we commit humans to any future functionally unshielded deep space venture – as has been repeatedly stated by this author and is the published opinion of the United States Academy of Medicine.

Yet, clearly, this *Nature* article, and all the others, are only game changers if those who have the power to take a time-out in developing their flight manifests actually go and find out the answers without risking human lives and injury or using humans as unwitting subjects in dangerous experiments whose deleterious effects will remain locked inside their bodies, programmed into their genes throughout their lifetimes.

"Apollo Lunar Astronauts Show Higher Cardiovascular Disease Mortality: Possible Deep Space Radiation Effects on the Vascular Endothelium" was the first paper that focused on the Apollo cohort's unique statistical relationship with their astronaut peers. The study distinguished between astronauts who had flown into deep space for an average of 9.5 days, astronauts who were exposed to GCR (up to 84 days in the case of the Skylab crews) in low-Earth orbit (LEO), and those astronauts who held only ground assignments and never flew into space. The findings were eye-opening.

The study revealed that there was no statistical difference in cardiovascular disease mortality between astronauts who had never flown into space and astronauts who had only flown into low-Earth orbit. But the cardiovascular mortality rate among the Apollo deep space crews was 43% – which was a staggering 4 to 5 times *higher* than the non-flight and LEO exposed astronauts.[27]

But the study did not just lay out these statistics and stop there. As the title suggests ("Possible Deep Space Radiation Effects on the Vascular Endothelium"), it is essential to note that there is a specific physiological system in play. The study delved further by using animal models (laboratory mice) exposed to simulated cosmic irradiation at Brookhaven National Laboratory's Relativistic Heavy Ion Collider and then studied the animal's cardiovascular system after the exposure to the simulated GCR. The stated conclusion was:

"If such results translate to the human condition, then long-term dysfunction of the vascular endothelium induced by charged HZE particles could be a major contributor to the development of atherosclerotic cardiovascular disease in astronauts."[28]

As mentioned previously, nine years earlier, while participating in a similar investigation at Brookhaven National Laboratory, my team exposed genetically modified animal subjects (laboratory mice) to the simulated beam of cosmic radiation. But our interest was specifically the high energy-high mass radiation's effect on the brain. And our paper's title highlighted our conclusion: "Quiescent adult neural stem cells are exceptionally sensitive to cosmic radiation."[29]

So, what is the essential biological link between the two papers and others that have followed? That essential point is the topic of the upcoming chapter, "Hitting the Physiological Wall."

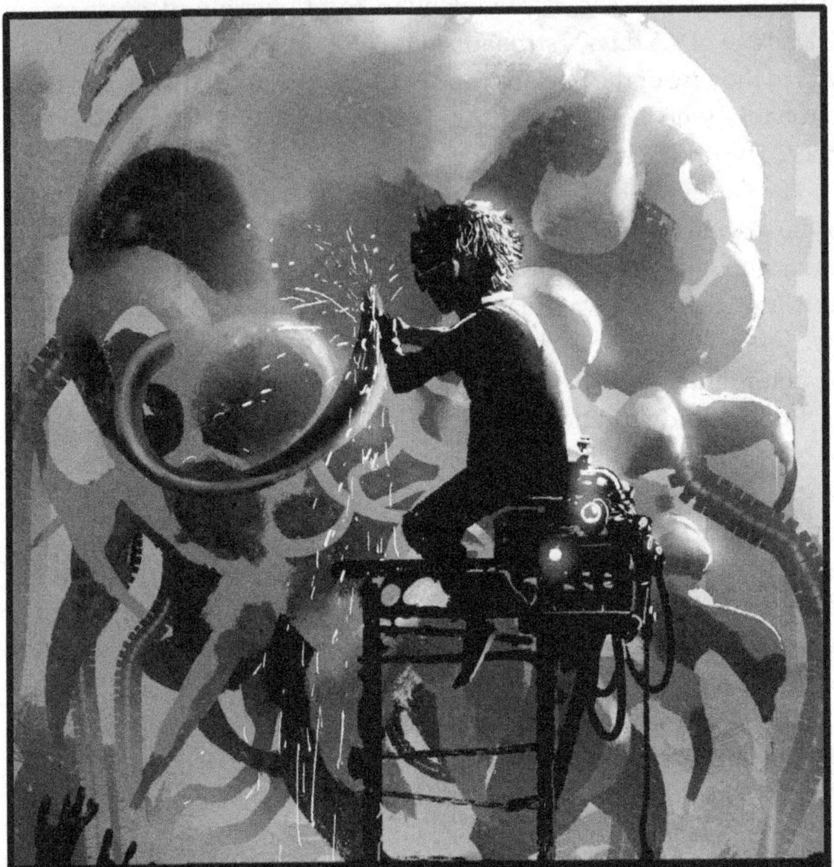

The Hiroshima Unit

"If offered a seat on a rocket ship, don't ask what seat.
Just get on."

- Christa McAuliffe

12 August 2079
The Time Travel Corporation
Announces:

The Hiroshima Experience

Using our ultra-safe TIME TRAVEL PROJECTOR A17 MODEL, you can now travel to 6 August 1945 and experience the world's first atomic attack in person! Spend 14 incredible days at just 1,500 meters from ground zero with your personal tour guide, Dr. Terufumi Sasaki, an A-bomb survivor made famous by John Hersey's classic true book, *Hiroshima*.

You will arrive in the hallway at the Hiroshima Red Cross hospital taking a walk beside Dr. Sasaki at T minus five seconds from detonation. You'll have just enough time for introductions when the blast takes out the city of 255,000 thousand people! Fortunately for you and Dr. Sasaki, you are located between windows and are just outside the overpressure zone, so that wing of the hospital will allow you and Dr. Sasaki to survive without any injury whatsoever.

You can either stick around the hospital and help Dr. Sasaki, or wander the city for two days at your leisure – keeping away from the active fires, of course. After 14 days, you will be safely transported back to the present as a survivor of the first atomic bomb! Few time travel adventures can offer such an unparalleled historical adventure. Sign up today – the opportunities for the Hiroshima Experience adventure are strictly limited!

UNFORTUNATELY FOR THE TIME TRAVEL CORPORATION the Hiroshima Experience was a total failure. Not a single individual signed up to ride to the past and experience the effects of the first atomic bomb attack in person. Six hours after it appeared on the Time Travel Corporation's social media page, the comments had to be disabled. The social media provider reported that the use of the F-word in the comments exceeded any ever recorded in the company's history. Here is the most common post (obscenities redacted):

"Are you [redacted] out of your [redacted] minds???"

Having billions of dollars invested in their nifty Time Travel Projector A17 Model, the Time Travel Corporation went bankrupt. Then, international law banned all time travel experiences as a danger to the future of humanity.

The business savvy Board of Directors for the Time Travel Corporation thereupon founded the 16 Psyche Mining Corporation to harvest the single most rich deposit of minerals in the solar system, asteroid 16 Psyche, which orbits in the Asteroid Belt between Mars and Jupiter. To assist in funding the venture, five of the mission's fifteen seats were offered for sale to tourist class crewmembers. The ad read:

12 November 2082
The 16 Psyche Mining Corporation
Announces:

The 16 Psyche Mining Experience

You can fly along on one of the most daring and important deep space adventures in human history – the 1000-day crewed mining expedition to asteroid 16 Psyche! In this incredibly exciting deep space venture, the 16 Psyche Mining crew will fly some 1.5 billion miles in solar orbit to the distant but incredibly resource-rich asteroid, strap an automated booster package on it, and then return back to Earth.

The attached booster will redirect the densely rich cosmic body to Lunar orbit where its resources – estimated at a value of $10,000 quadrillion – will be mined. This is the mission that may literally change our world from relative poverty to one of endless resources and power.

And now you can personally be a part of this celebrated crew and mission! We expect this to be the most overwhelmingly popular destination ever – so sign up today!

The 16 Psyche Mining Experience was so popular that the 5 available seats for this mission were booked just 12 minutes after the online reservations page was opened. Each of the five crew position opportunities was purchased for its advertised price of $250 million. So, what was the single most important difference between the 16 Psyche Mission and the Hiroshima Experience?

The number one stated negative opinion from potential customers of the Hiroshima Experience was overwhelmingly identified as perceived personal risk from exposure to ionizing radiation. Not a single person surveyed wanted anything whatsoever to do with walking around a fresh atomic bomb site covered by a steady rain of radioactive fallout and exposed to direct radiation and radioactive contamination from many different sources.

Yet, interestingly, in a survey of those who purchased tickets for the 16 Psyche Mining Experience, not a single individual expressed *any concern at all* for any level of perceived risk. *None at all.* But as it turns out, the radiation exposure of the Hiroshima Experience was actually multiplied by many times onboard the Psyche Mission! And yet, not a single ticket holder voiced any anxiety about exposure to radiation in deep space. To the general public, galactic cosmic radiation (GCR) is not a concern. Not in this fictional story and not in the 21st-century real world. But with the information and insight you gain from reading *Departing Earth Forever*, you are no longer a part of the "general" public.

Returning to the present, we gain some interesting insights from these fictional stories to a serious disconnect between perceived risk and two very different socially driven interpretations of scientific reality.

The Hiroshima Unit Defined

So, let us begin by instituting a benchmark for the discussion. If you suddenly materialized beside the real Dr. Terufumi Sasaki in the hallway at the Hiroshima Red Cross hospital at 08:15 on 6 August 1945 and remained by his side for two weeks, you would survive, because *he survived for another 32 years*. After an exhaustive search of the vast archives of medical data on the Hiroshima survivors, one benchmark estimates that the average dose of each of the victims in the aftermath of the bombing is around 103 millisieverts (mSv).[1] Thus, for the purposes of this discussion, we will use 103 millisieverts as our benchmark, calling it "one Hiroshima Unit" – or 1.0 HU.

Recall that the current federal limit on radiation exposure for nuclear power professionals is 50 mSv per year (so long as the average annual dose over five years does not exceed 20 mSv per year) – just half of one Hiroshima exposure unit. Also recall that the average actual exposure of every United States nuclear employee is just 4% of that or 2 mSv/year.

The origin of 50 mSv is not insignificant and was certainly not selected at random. The approximate lower limit for an increased risk of carcinogenic effects from a single exposure is around 100 mSv, so the Nuclear Regulatory Commission (NRC) established the annual limit for radiation workers at 50 mSv. The number was specifically selected for its statistically derived safety margin. Below 100 mSv, the risk of radiation induced disease is too low to be convincingly demonstrated.

Although members of the general public are limited to an exposure of 1 mSv per year (excluding ionizing radiation from medical imaging), NRC regulations limit radiation professionals to receiving 50 mSv of radiation annually for each year of a 40-year career.[2] At this dose, the risk for a United States radiation professional is low enough to be deemed comparable to that of non-radiation workers.[3] But above the 100 mSv limit – right at our 1.0 Hiroshima Unit – the statistics change. Bad things begin to appear on the radar with consistent, clearly enumerated regularity.

Therefore, the 1.0 HU dose is the start of the statistical boundary at which certain carcinogenic maladies begin to evidence themselves statistically on all probability scales. And from these statistics is precisely where NASA derives its sliding astronaut death scale they call

REID – Risk of Exposure Induced Death. As has been discussed in depth earlier in this book, NASA's current "acceptable loss" of crew from exposure to galactic cosmic radiation is three out of every 100 crewmembers.

If you did survive the real Hiroshima bombing along with Dr. Sasaki at the Red Cross Hospital and then returned home, you would have found yourself just barely above the statistically significant cancer scale and not yet on NASA's REID radar. In fact, during the two weeks living and working 1,500 meters from Hiroshima's ground zero, you would have accumulated a roughly approximated 1.0 Hiroshima Unit of radiation – exactly the same amount of radiation as an astronaut who spends five months on the International Space Station or 57 days in Lunar orbit onboard the Gateway Station in its currently planned "shielding" configuration.

Remember, the Hiroshima Unit is defined as the starting point at which statistically significant downstream cancers actually become measurable. Obviously, not all cancers are fatal, while some are far worse than others. NASA has narrowed that gap and defined it by their Risk of Exposure Induced Death statistic that is constructed to enumerate the probability of death from *fatal* cancers caused by exposure to galactic cosmic radiation.

Hence, how many Hiroshima Units are required to be included on the REID death scale? Unfortunately, the answer is not nearly as simple as that question implies. The REID estimate changes depending on the age and chromosomal makeup (sex) of each crewmember, along with factors such as smoking history and previous space radiation exposures. Therefore, let us focus on the Hiroshima Unit's usefulness in comparing terrestrial events (such as the exhaustively well documented Hiroshima survivors' massive database) to the future of space events measured in Hiroshima Units.

Earth Baseline

365 Days of Exposure to Cosmic Radiation Reaching Earth's Surface:

0.33 mSv

Equivalent to

0.0275 Whole Body CT Scans

Or

16.5 Chest X-rays

1.0 Hiroshima Unit

103 mSv

Equivalent to

9 Whole Body CT Scans

Or

5,150 Chest X-rays

The essential importance of the Hiroshima Unit is that it lies right at the statistical line between detectable disease from ionizing radiation exposure and that which is too low to convincingly demonstrate.

What About the Actual Hiroshima Experience Results?

While there was outrage and panic connected to the idea of accepting such monumental personal risks associated with the Time Travel Corporation's Hiroshima Experience, the fact is, the individual who lived the experience, Dr. Terufumi Sasaki, survived productively for another 32 years after the event. While it is true that he had a cancerous lung removed later in his life, the cause of that cancer may suspiciously be connected to the experience, but no epidemiological proof has even been sought or is forthcoming. Yet, this seminal event, measured in 1.0 Hiroshima Unit, was survivable – but with measurable risks on the order of decades – just as it all played out in reality.

Case Study 1 - The Psyche 16 Mining Mission

According to the benchmark deep space data that we have in hand, astronauts onboard a 1000-day 16 Psyche Mining Mission riding round trip on an Apollo-era design Orion spacecraft to the asteroid and then back to Earth would be exposed to 1,800 mSv of whole spectrum galactic cosmic radiation. Their REID numbers would be exceptionally high. Possibly as many as 10-20% of the crew would be killed from radiation induced cancers in their lifetimes depending on their ages and multiple other factors

Yet, in the real world, it is doubtful that any would survive the trip at all due to whole body epithelial and endothelial cell damage. This does not even account for the physically debilitating effects of their radiation exposure and its significant negative impact on work productively at the asteroid, not to mention trying to make it back to Earth as their cognitive capacities would rapidly decline en route to 16 Psyche due to the GCR. Based on these figures, these crewmembers would not survive long enough for the cancer REID probabilities to actually catch up to them.

Days in Deep Space	1000.00
Total Exposure mSv	1800.00
Hiroshima Units	17.48
Whole Body CAT Scans	150.00
Chest X-Rays	90000
Equivalent Days on ISS	2727
% Federal Limit	3600%

Analysis: Obviously, such a trip onboard our current Apollo model spacecraft which are 20th-century crude relics, dangerous, and pathetically unsuitable hardware, make this mission not even remotely survivable.

Case Study 2 - NASA Mars DRM Expedition

The NASA Mars Design Reference Missions (DRM), as advertised by the Agency, may likewise prove potentially fatal, depending entirely on several factors, including appropriate shieling on the surface of Mars, partially effective shielding en route, and the comorbidity of microgravity in deep space. Assuming NASA is going to fly on functionally unshielded ships in microgravity (as they now advertise) and improvise shielded shelters after they arrive, the outlook is not much better than the 16 Psyche Mining Mission. Using the NASA estimates of a cumulative mission of around 1000 days with the crew absorbing 1,000 mSv, the totals come out as:

Days in Deep Space + Mars	1000.00
Total Exposure mSv	1000.00
Hiroshima Units	9.71
Whole Body CAT Scans	83.33
Chest X-Rays	50000
Equivalent Days on ISS	1515
% Federal Limit	2000%

Analysis: Once again, such a trip is probably not survivable using today's crew deep space access technology, and fatal cancers would probably not claim the crew before the whole body epithelial and endothelial damage finished them off first.

Case Study 3 – The International Space Station (ISS)

YEAR	Mission Days	Total Dose	Crew Size	Avg Exposure	Avg Exp H.U.'s	Avg Exp CAT	Avg Exp Chest X-rays	High Annual Exp	Highest Exp H.U.'s	Highest Exp CAT	Highest Exp Chest X-rays	High % Over Fed Limit
2001	596	393.36	6	65.56	0.64	5.46	3278.00	111.54	1.08	9.30	5577.00	223.08
2002	599	395.34	5	79.07	0.77	6.59	3953.40	122.10	1.19	10.18	6105.00	244.20
2003	506	333.96	4	83.49	0.81	6.96	4174.50	122.10	1.19	10.18	6105.00	244.20
2004	375	247.50	3	82.50	0.80	6.88	4125.00	122.76	1.19	10.23	6138.00	245.52
2005	394	260.04	3	86.68	0.84	7.22	4334.00	118.14	1.15	9.85	5907.00	236.28
2006	386	254.76	3	84.92	0.82	7.08	4246.00	120.78	1.17	10.07	6039.00	241.56
2007	577	381.08	5	76.22	0.74	6.35	3810.80	107.84	1.05	8.99	5392.00	215.68
2008	566	373.56	6	62.26	0.60	5.19	3113.00	120.78	1.17	10.07	6039.00	241.56
2009	605	399.10	7	57.01	0.55	4.75	2850.71	131.34	1.28	10.95	6567.00	262.68
2010	730	481.80	7	68.83	0.67	5.74	3441.43	116.16	1.13	9.68	5808.00	232.32
2011	613	404.62	6	67.44	0.65	5.62	3371.83	110.22	1.07	9.19	5511.00	220.44
2012	634	418.44	6	69.74	0.68	5.81	3487.00	120.12	1.17	10.01	6006.00	240.24
2013	631	416.72	6	69.45	0.67	5.79	3472.67	109.56	1.06	9.13	5478.00	219.12
2014	671	442.86	6	73.81	0.72	6.15	3690.50	111.54	1.08	9.30	5577.00	223.08
2015	667	440.22	5	88.04	0.85	7.34	4402.20	183.48	1.78	15.29	9174.00	366.96
2016	637	420.42	6	70.07	0.68	5.84	3503.50	113.52	1.10	9.46	5676.00	227.04
2017	849	560.57	7	80.08	0.78	6.67	4004.07	161.11	1.56	13.43	8055.50	322.22
2018	888	586.08	7	83.73	0.81	6.98	4186.29	130.02	1.26	10.84	6501.00	260.04
2019	924	609.84	5	121.97	1.18	10.16	6098.40	192.06	1.86	16.01	9603.00	384.12
2020	801	528.66	10	52.87	0.51	4.41	2643.30	128.70	1.25	10.73	6435.00	257.40
2021	1317	869.22	10	86.92	0.84	7.24	4346.10	176.22	1.71	14.69	8811.00	352.44
AVG	665	438.96	6	76.70	0.74	6.39	3834.89	130.00	1.26	10.83	6500.21	260.01
MED	631.39	416.72	6	76.22	0.74	6.35	3810.80	120.78	1.17	10.07	6039.00	241.56

Estimated Cosmic Radiation Exposures Onboard the International Space Station 2001-2021 mSv and Hiroshima Units (Compiled from public data by Dennis and Claudia Chamberland)

Without repeating our extensive previous discussion of the cosmic radiation environment onboard the ISS, it is useful to be reminded here that the ISS flies below the innermost Van Allen Radiation Belt. Therefore, the astronauts receive a variable dose on orbit of roughly one-third of the unfiltered, steady state cosmic radiation stream found above 36,000 miles in deep space. The data from Case Study 3 is not linked to any individual crewmembers and was derived from publicly available data.

After compiling this data, my initial impression was that although NASA has given itself carte blanche permissions to expose its crews to any level of cosmic radiation it desires, the data from 20 years of spaceflight on the ISS describes an obviously carefully controlled, even-handed restraint by the Agency for the average ISS astronaut. In 20 years, the Agency has kept the average crew exposure down to a "reasonable" 150% of the Federal Limit for Nuclear Professionals and 25% below the statistical determination of probable downstream cancers. However, the outlying average high exposures for longer term ISS missions reach 260% of the Federal Limit. These high exposures have crossed over into statical significance for downstream cancers by 26%, or 1.26 HUs.

As was discussed previously, the long-term effects of partial exposure in low-Earth orbit to galactic cosmic radiation are not fully known – although the intensely studied astronaut cohorts in low-Earth orbit for more than half a century of human spaceflight indicate that this environment, predominantly sheltered beneath the Van Allen Shields, is "human friendly" up to a limit which, because of the ISS research, we now know.* (See "*A Crucial Reminder" at the end of this chapter.)

The ISS has proven itself to be an exceptionally significant research platform and has determined the upper limits of human exposures to GCR and microgravity. The ISS has established that we are at our limit for adaptation to the space environment and that we must change our exploration model before exposing crews to the toxic environment of deep space above 36,000 miles. In that sense, the ISS has thus fully earned its historic place in human space exploration and settlement and should be retired as soon as possible, replaced by fully shielded and spun 21st-century First Principle Solar System Class space stations.

Using the Hiroshima Unit as a Tool and Looking Ahead

Using the Hiroshima Unit as a tool to calibrate the threshold of statistical significance of future cancer pathology opens a clear window into limits of statistically significant exposure to full spectrum galactic cosmic radiation. The chart below illustrates exposure times to GCR before reaching the Hiroshima Unit threshold of statistically significant future cancer pathology. Please note that this calculated data significance does not take into account the totally unknown probability of significant disease risk defined in the next chapter as "Cosmic Radiation Induced Sensitive-Tissue Pathology Disorders" (CRISP).

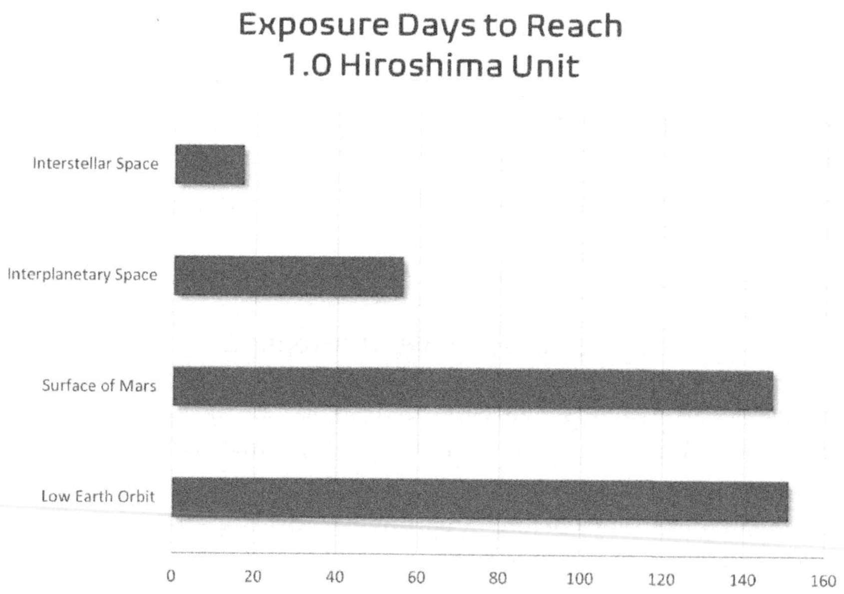

Future Planned Lunar Crewed Missions

Below are case studies of three proposed crewed lunar missions – one private and two by NASA – scheduled to occur in the mid to late 2020s. All of these are currently planned using functionally unprotected Apollo era design spacecraft and will be exposed to full energy and spectrum cosmic radiation energies as recorded by Curiosity and Lunar Reconnaissance Orbiter's deep space radiation detectors in or near cislunar space. These missions are presently baselining use of the Orion Spacecraft, SpaceX's Starship, and the Lunar orbiting Gateway Station. I have presented each mission in tabular form.

Case Study 4 - dearMoon Mission

The dearMoon Mission is intended to use a SpaceX Starship, which, in its current design, is a functionally unprotected Apollo era design spacecraft. Bankrolled by Japanese Billionaire Yusaku Maezawa, the dearMoon Mission will conduct an Apollo 8 type fly around of the Moon on a roughly 6-day mission to and from Earth. Here are its approximate stats in Hiroshima Units:

Days in Deep Space	6.00
Total Exposure mSv	10.80
Hiroshima Units	0.10
Whole Body CAT Scans	0.90
Chest X-Rays	540
Equivalent Days on ISS	16
% Federal Limit	22%

Analysis: Within known safety limits with unquantifiable risk.*

Case Study 5 - NASA Artemis 2

The NASA Artemis 2 mission is designed to fly four astronauts around the Moon and back without a surface landing. The complex mission consists of various low-Earth orbit changes and maneuvers over the first four days before departing into deep space outbound for the lunar flyby and returning to Earth.

Days in Deep Space	10.00
Total Exposure mSv	12.30
Hiroshima Units	0.12
Whole Body CAT Scans	1.03
Chest X-Rays	615
Equivalent Days on ISS	19
% Federal Limit	25%

Analysis: Within known safety limits with unquantifiable risk.*

Case Study 6 - NASA Artemis 3

The NASA Artemis 3 Mission is planned to return humans to the surface of the Moon – the first time since 1972. The mission duration is planned for 30 days in deep space, breaking the previous 12.5 days record of Apollo 17 by about 2.4 times. Of the four Artemis 3 crewmembers, only two will land on the Moon onboard a SpaceX Starship for a 6.5-day reconnoiter of the lunar surface. The other two astronauts will remain inside the Lunar orbiting Gateway Station.

Note that the Lunar surface astronauts will receive approximately 2.7 times less cosmic radiation than their Gateway Station counterparts while they are on the Lunar surface since the Moon partially shields them.

The following two Case Studies account for the crew exposures on the Lunar Gateway compared to the crew that lands on the Moon.

Case Study 6A - Artemis 3 Lunar Landing Crew

Days in Deep Space	23.50
Days on Lunar Surface	6.50
Total Exposure mSv	46.60
Hiroshima Units	0.45
Whole Body CAT Scans	3.88
Chest X-Rays	2330
Equivalent Days on ISS	71
% Federal Limit	93%

Analysis: Within known safety limits with unquantifiable risk.*

Case Study 6B - Artemis 3 Gateway Station Crew

Days in Deep Space	30.00
Total Exposure mSv	54.00
Hiroshima Units	0.52
Whole Body CAT Scans	4.50
Chest X-Rays	2700
Equivalent Days on ISS	82
% Federal Limit	108%

Analysis: Within known safety limits with unquantifiable risk.*

Unplanned NASA Missions

The following seven mission estimates are meant to convey cosmic radiation doses on currently unscheduled NASA missions in the future. This implies that the United States' Lunar presence will be maintained and expanded, as will the Lunar Gateway Station. Again, in the complete absence of any information to the contrary, current hardware design for all NASA crewed deep space missions will utilize functionally unprotected Apollo era design spacecraft. These estimates do not account for using Lunar surface regolith for shielding. If the Lunar habitats do augment with in situ shielding, the estimates below *will not* apply. These estimates are designed to communicate elevated health risks to the crews, but only those who continue to utilize functionally unprotected Apollo era crew protection engineering designs.

Case Study 7- NASA Gateway Station Mission – 45 Days

Days in Deep Space	45.00
Total Exposure mSv	81.00
Hiroshima Units	0.79
Whole Body CAT Scans	6.75
Chest X-Rays	4050
Equivalent Days on ISS	123
% Federal Limit	162%

Analysis: Within known safety limits with unquantifiable risk.*

Case Study 8 - NASA Gateway Station Mission – 60 Days

Days in Deep Space	60.00
Total Exposure mSv	108.00
Hiroshima Units	1.05
Whole Body CAT Scans	9.00
Chest X-Rays	5400
Equivalent Days on ISS	164
% Federal Limit	216%

Analysis: Outside known safety limits. This is the first mission scenario where the Hiroshima Units exceed the terrestrial radiation exposure risk to the crew, and the risk of exposure induced death becomes known in the very questionable guess-based-comparisons to terrestrial radiation energies. Because the actual data on the effects of the full spectrum of galactic cosmic radiation on vertebrate species has never been collected, the actual risk remains unquantifiable under these conditions.*

Case Study 9 - NASA Gateway Station Mission – 90 Days

Days in Deep Space	90.00
Total Exposure mSv	162.00
Hiroshima Units	1.57
Whole Body CAT Scans	13.50
Chest X-Rays	8100
Equivalent Days on ISS	245
% Federal Limit	324%

Analysis: Outside known safety limits. The Hiroshima Units exceed the terrestrial radiation exposure risk to the crew, and the risk of exposure induced death becomes known in the very questionable guess-based-comparisons to terrestrial radiation energies. Because the actual data on effects of the full spectrum of galactic cosmic radiation on vertebrate species has never been collected, the actual risk remains unquantifiable under these conditions.*

Case Study 10 - Lunar Surface Mission of 15 Days

Note that Lunar surface missions are partially shielded by the bulk of the Moon below the astronauts' feet. Therefore, the risk of over exposure is reduced to roughly one-third that of the dose received in deep space even when living and working in functionally unshielded habitats on the Lunar surface – although this is *not recommended*. (You will notice this trend in the accompanying data tables.) Further, with only modest shielding efforts using Lunar regolith, lunar habitats can be made as safe as any Earth based habitat, thereby changing the labels of these shielded habitats to First Principle Solar System class interplanetary structures.

Deep Space Transfers	6.00
Days on Lunar Surface	15.00
Total Exposure mSv	21.00
Hiroshima Units	0.20
Whole Body CAT Scans	1.75
Chest X-Rays	1050
Equivalent Days on ISS	32
% Federal Limit	42%

Analysis: Within known safety limits with unquantifiable risk.*

Case Study 11 - Lunar Surface Mission of 30 Days

Deep Space Transfers	6.00
Days on Lunar Surface	30.00
Total Exposure mSv	31.20
Hiroshima Units	0.30
Whole Body CAT Scans	2.60
Chest X-Rays	1560
Equivalent Days on ISS	47
% Federal Limit	62%

Analysis: Within known safety limits with unquantifiable risk.*

Case Study 12 - Lunar Surface Mission of 45 Days

Deep Space Transfers	6.00
Days on Lunar Surface	45.00
Total Exposure mSv	41.40
Hiroshima Units	0.40
Whole Body CAT Scans	3.45
Chest X-Rays	2070
Equivalent Days on ISS	63
% Federal Limit	83%

Analysis: Within known safety limits with unquantifiable risk.*

Case Study 13 - Lunar Surface Mission of 60 Days

Deep Space Transfers	6.00
Days on Lunar Surface	60.00
Total Exposure mSv	51.60
Hiroshima Units	0.50
Whole Body CAT Scans	4.30
Chest X-Rays	2580
Equivalent Days on ISS	78
% Federal Limit	103%

Analysis: Within known safety limits with unquantifiable risk.*

Case Study 14 - Lunar Surface Mission of 180 Days

Deep Space Transfers	6.00
Days on Lunar Surface	180.00
Total Exposure mSv	133.20
Hiroshima Units	1.29
Whole Body CAT Scans	11.10
Chest X-Rays	6660
Equivalent Days on ISS	202
% Federal Limit	266%

Analysis: Outside known safety limits. The Hiroshima Units exceed the terrestrial radiation exposure risk to the crew, and the risk of exposure induced death becomes known in the very questionable guess-based comparisons to terrestrial radiation energies. Because the actual data on effects of the full spectrum of galactic cosmic radiation on vertebrate species has never been collected, the actual risk remains unquantifiable under these conditions.*

Case Study 15 - The First Crewed One-Way Mars Mission

There is a plan to launch the first human crew to Mars in the next two or three interplanetary window cycles. Case Study 15 makes the following assumptions based on data about the mission that is available today:

(1) The average Hohmann interplanetary launch window transit time is estimated from the average over the past 60 years to be 243 days.

(2) It is assumed there may be some artificial gravity based on persistent rumors about the mission.
(3) It is assumed that upon arrival on Mars, fully shielded accommodations will be quickly available for the crew.
(4) It is assumed that the current design of a functionally unshielded Apollo era interplanetary spacecraft will be used for the journey.

Case Studies 15A&B calculate a one-way trip and a two-way trip, respectively, with a surface exposure equivalent to the surface interval in days times an estimated daily exposure rate of 4 hours surface time maximum average. This equals 0.11 mSv/day average exposure (0.67 mSv/day X 0.167 day). The average number of days on the surface equals 451 days for a total of basically 50 mSv Mars surface exposure outside the fully shielded surface habitats (451 days X 0.11 mSv/day = 49.6 mSv).

Case Study 15A - One Way Crewed Trip to Mars in the 2020-2030s

Days in Deep Space	243.00
Total Exposure mSv	437.40
Hiroshima Units	4.25
Whole Body CAT Scans	36.45
Chest X-Rays	21870
Equivalent Days on ISS	663
% Federal Limit	875%

Case Study 15B - Round Trip Crewed Trip to Mars in the 2020-2030s

Days in Deep Space	486.00
Days on Mars	451.00
Total Exposure mSv	924.00
Hiroshima Units	8.97
Whole Body CAT Scans	77.00
Chest X-Rays	46200
Equivalent Days on ISS	1400
% Federal Limit	1848%

Analysis: As much as this author would enjoy watching this grand adventure unfold, there is a probability that the crew may be somewhat incapacitated at their arrival so that, unless the craft is fully automated, a safe landing might not be possible. Please note that these case study estimates are remarkably close to NASA's Mars DRM mission estimates.

<u>**One Way Trip**</u>:

For Case Study 15A, once on the surface of Mars, even if the astronauts/colonists were fully shielded on the planet, the probability of chronic injury to their cognitive processes and epithelial systems may be severe and potentially permanent. In the total absence of any research data, it is not possible to predict anything with certainty except that this dose of very high linear energy/exceptionally high linear energy transfer radiation would probably be extremely damaging to the crew. Further, as NASA's REID death scale would predict, this mission is certainly on the targeted death scale radar for a certain percentage of the crew.

<u>**Two Way Trip**</u>:

In Case Study 15B, the crew are theoretically fully shielded upon arrival and limit their daily surface exposure to an average of four hours each day, accumulating what may be a very manageable 50 mSv exposure in their 451 days on Mars. Perhaps they will experience a time of healing of the damage they received on the inbound leg of their round trip from Earth. Unfortunately, the trip back to Earth on functionally unprotected spacecraft will double down on their accumulated radiation induced whole body physiological genetic and epithelial-endothelial trauma and will add to it. The direct injuries from CRISP disorders (see paragraph below) and indirect radiation imprinting on their genetic memory, according to NASA research, will ultimately induce fatal cancers in a significant cohort of the crew. (See next chapter.)

*A Crucial Reminder

The Hiroshima Unit was established on Earth using terrestrial ionizing radiation sources as its baseline. The disease statistics (exclusively cancer) that accompany it are determined by studies that used these solely Earth based isotopes exhibiting exponentially less energy than cosmic sources. Until vertebrate animal research is conducted in full spectrum galactic cosmic radiation in deep space, the effects of crewed exposures to this level of cosmic energy are almost totally unknown.

However, in the past two decades, we have discovered through Earth based research and approximation of cosmic energies that the far higher energy of cosmic sources causes acute and chronic damage to human tissues that are not typically exhibited by exposure to terrestrial isotopes such as in the very large cohort of Japanese Atomic Bomb survivors. That concept is covered in detail in the next chapter, "Hitting the Physiological Wall," and is defined as Cosmic Radiation Induced Sensitive-Tissue Pathology or CRISP disorders.

In the complete absence of adequate animal research on the long-term effects in deep space on vertebrate organisms, the calculable risk of any missions flown in deep space cannot be quantified. While we do not yet know the effects on vertebrate animals (including humans), because of the exponentially larger masses and higher energies involved, we are certainly assured that the effects will be significantly greater as our simulated GCR research is providing abundant evidence for this more immediate cosmic radiation induced pathology.

Hitting the Physiological Wall

"It's not what you look at that matters,
it's what you see."

- Henry David Thoreau

CRISP
Cosmic Radiation Induced Sensitive-Tissue Pathology Disorders

ASK ANY CURRENT "EXPERT" CAREER OR VETERAN astronaut or space radiologist on the planet today what the enumerated limiting factor is in human exposure to galactic cosmic radiation (GCR), and nearly all of them will give the same answer: the limiting factor of exposure to GCR is the probabilistic risk of developing downstream cancer sometime during their lifetimes.

Unfortunately, it is also almost certainly the *wrong answer*.

There is no such fantasy as a single response to that question concerning a primary constraint to the growing suite of potential human pathologies from GCR exposure in deep space.

With unfortunate end results for future space travelers above 36,000 miles enduring chronic, functionally unshielded exposure on long missions, NASA's neatly restricted and narrowly confined answer to all deep space maladies does not necessarily describe the agent that is far more likely to injure or kill them. It is the wrong answer because it does not define the agents and mechanisms that have the clear potential to catastrophically end their mission years or even decades before just one of the maladies that eventually rises up to plague them. And that cause is not simply the stochastically determined probability of death or injury by GCR-induced cancer, regardless of NASA's rough order of guesstimate – the Risk of Exposure Induced Death (REID) SWAG that the Agency applies to its deep space crews.

However, no one has yet apparently seen this handwriting on the spacecraft walls, and everyone continues to plow ahead with all due speed, committing astronaut lives to SWAGs built exclusively from atomic bomb survivors, not even one of whom ever received a single milli-Sievert of exposure to full spectrum GCR. Even the best radiobiology modeling is wholly dependent on the underlying composition of the linked variables being fed into its mathematical framework. But in the Agency's model, no one knows what the real-world deep space variables actually are – such as human exposure to the spectrum or energies that are actually out there.

The problem at NASA and every other human space exploration venture on Earth is that the obsolete Apollo-era human space exploration model is fixed on its single risk scenario – stochastic cancer. It is abundantly and unambiguously clear that the probabilistic estimates based on more than a half-century of studies of the atomic bomb survivors are quite accurate when applied to humans on Earth exposed to similar spectra and energies. But that accuracy and application are strictly limited to humans living in a one Earth-normal environment who were exposed to an acute dose of terrestrial ionizing radiation energies and spectra. Yet, exporting that Risk of Exposure Induced Death probability to chronic exposures of a deep space ionizing radiation spectrum with almost no commonality to its far distant and magnitudes weaker terrestrial relative *has no documented validity whatsoever*!

As was stated previously, NASA uses this SWAG simply because it is all they have, and no one has ever effectively challenged the approach. And, most egregiously, the world's premier space agency has never seen the need to preface crewed missions with vertebrate animal studies before committing humans to the void and the pernicious, toxic milieu of GCR – even though recent research and detailed studies have sounded loud and persistent alarms that there is another, newly discovered physiologic risk that may be more acute, and ultimately more deadly and mission threatening, than probabilistic long-term cancer. Specifically, because this risk is much more immediate than downstream cancer, for every space jockey with any intelligence at all, it should be their greatest concern before leaving the home planet. Hitting this physiological wall in deep space will not only be mission threatening, but it will also be life threatening.

I have identified this collective physiological damage to the human body as CRISP:

CRISP
Cosmic Radiation Induced Sensitive-Tissue Pathology

The Epithelial Connection

In the winter of 1971, I was pushed to my limit, wrapping up the next-to-last semester before I graduated from Oklahoma State University. It was early December, the first snow had already fallen on campus, and the Oklahoma sky had darkened and stubbornly persisted in its dreariness for days. On the OSU Stillwater campus, cutting, bitterly cold wind ripped across the prairie lands and added to the misery of just walking between classes.

And it was finals week.

Altogether, the misery could have been epic. Except, despite the myriad depressors, my mind was alight with the Yuletide joy of what lay waiting for me in the cell biology laboratory. It was almost as good as the presents that lay waiting for me under the Christmas tree as a child.

I remember this particular semester so clearly because, in my typical ADHD manner, I was imaginatively captivated by a single class: cytology – the biology of cells. Indeed, I was so transfixed that I was skipping my botany class on a regular basis to spend more time in the cytology laboratory. Quite fortunately for me, this was a forgivable sin because my botany professor had also conveniently recorded all his lectures on reel-to-reel tapes. So, on my own schedule, I sprinted into the audiovisual lab and listened to his lectures at 3X speed (an ADHD university guy's dream come true), recovering at least half an hour of time per lecture.

My laboratory presents-under-the-tree in 1971 that I was so enraptured by were a handful of flat-bottomed plastic cell culture flasks half-filled with a pink growth media. And my subjects – mouse epithelial cells isolated from mouse trachea – were layered on the bottom of the flasks. I viewed them daily with some level of schoolboy awe. There they were, with no discernable personality, just all nice and toasty, sleeping and multiplying, multiplying and sleeping, living out every college guy's dream. They were in their incubator sitting atop my carefully engineered device, which enabled the inquiry's hypothesis. I was investigating the genetic damage induced by the cell's proximity to 60-cycle low voltage current (which proved to be profound). Rodent epithelial cells are ideal for many such investigations. They mimic the same class of sensitive, fast-growing cells found in all humans.

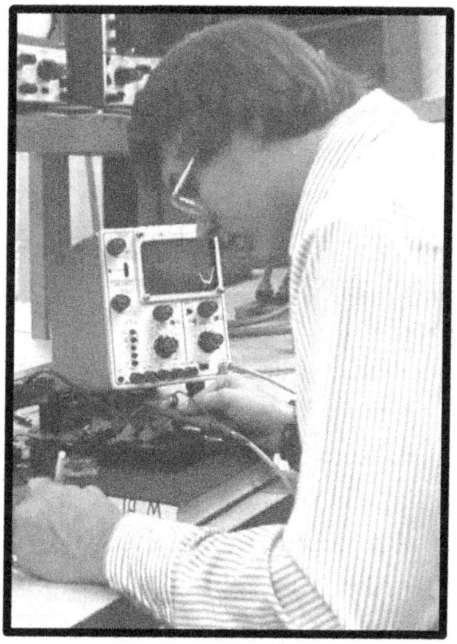

The device itself was a collaborative affair, as I had enlisted the assistance of a rather quiet but intellectually gifted fellow student by the name of Michael K. Blanchard. Mike and I created the electric coil that was designed to sit and do its 60-cycle thing while the mouse epithelial cells did their thing. (See photo) Just like humans in proximity to radiative effects of nearly any source, the rodent cells had no stimulation or impression of any kind as they were bathed in the relentless, unchanging current. Mike and I characterized the energy using a magnetic flux probe connected to an oscilloscope to map and measure its field. We then slipped it under the glass dishes in the incubator and left them to divide in peace.

I slept very little that week. But in my heart and mind, my greatest and most memorable achievement was separating the genes from those epithelial cells, then photographing, counting, and comparing them to the control cells. Of course, I made an A in cytology and an A in cytology lab. And in the end, I even passed my botany class.

Ironically, my career job would be working for a PhD botanist in a bioregenerative life support system laboratory. And, many times, I regretted that I had not invested as much passion in my botany class on that cold December in 1971 as I had with the mouse cells. But, in the end, my boss, Dr. William Knott III, was a fine instructor and certainly made up for opportunities lost!

Epithelial-class cells are not only found in mice but are also in a significant number of human systems. They are predominantly fast growing, relatively short-lived cells specifically because of their prime function. Epithelial tissue covers the outside of the human body. A specialized class of epithelial cells – called the endothelium – also lines organs, blood vessels, and the lymph system, and is found throughout the digestive system and body cavities. The endothelium is thus contiguous with the inner tissue lining of organs such as the brain, lungs, skin, and heart.[1]

Epithelial cells are fast growing and short-lived because their function requires quick replacement. They form the barrier between the body and the outside world – such as the skin. Likewise, the epithelial layers of the inside of your body cover your lips and line your nasal cavities, and act as protective membranes as well. They also create the transport layers for food in your digestive system and gas exchange membranes in your lungs. Hence, our source of mouse epithelial cells came from the trachea of the rodents, from the cells lining their upper respiratory system.

Remember the title of our 2007 genetically modified mouse experiment with cosmic radiation? It was "Quiescent adult neural stem cells are exceptionally sensitive to cosmic radiation." The adult neural stem cells of the human brain are also known as neuroepithelial cells. Also, recall the title of the *Nature* article, "Possible Deep Space Radiation Effects on the Vascular Endothelium," we reviewed in the previous chapter and in "Lies, Damned Lies, and Statistics." The connection is extraordinary because both studies focused on the harmful effects of GCR on two of the distinctly different epithelial systems of the body: the neuroepithelial (the epithelial lining of the mid-brain) and the endothelial-endocardium (the linings of the body's vascular system and heart). In both cases, the cosmic radiation was selectively injurious to fast growing, short-lived cell systems that are, by their very design function, very sensitive to adverse systemic challenges. But these examples are only two of many others, as we will review different experiments that have published their results in recent years. There are even more links to the pathological epithelial response to cosmic radiation.

But first, I will back up and describe our 2007 study in a little more depth because its findings are not only critical for astronaut crew success in deep space, but similar findings have also been repeated and verified in other studies conducted quite recently. At the end of this chapter, I will return to the epithelial connection because it is one of the primary purposes of this book to connect all those dots and, with the evidence, make a case for a radical change in our exploration model and add a game-changing new element to our predictive ensemble for deep space. These unique adjustments are the only pathway to prevent the imminent catastrophic collapse of the human exploration model and to ensure that we successfully depart Earth forever by utilizing a sane and intelligent human exploration paradigm.

Neural Stem Cells and Mars-Bound Crews

I grew up in the 1950s and 60s in a very conservative, rural mid-western home. I rarely missed a church service as a young teen, and I was very active in my youth group. It was an extremely positive and critically formative time in my maturity. As one might expect, I was raised in a conservative protestant congregation. One of the key doctrines taught was that drinking alcohol was inherently evil. Over the years, I have given considerable thought to that teaching. It was filled with factual and logical error to the point of absurdity, but in the end, its core truths almost certainly prevented much alcoholism in the local homes. Alcohol may indeed become a deadly scourge, as I witnessed in my own family, one generation removed. But as I made my way to the university environment, I finally gave in to the practice of drinking alcohol – and, ironically, my first introduction was in a Catholic church.

A friend of mine invited me one Easter season to accompany her to an observance of the Lord's Supper in her Catholic congregation. Having never attended any Catholic service, I was very intrigued. The only Lord's Supper I had ever taken part in was in my church communion service, where a quarter ounce of grape juice was paired with an unleavened bread chip in a somber and quiet moment of introspection that lasted a few minutes. That was what I expected as I made my way across campus to the Catholic church.

I was expecting to enter their sanctuary and was consumed with worry about the Catholic practice of crossing oneself on entering and on cue during the service. Except for what I had seen in movies, I was clueless, and the disquiet was building as I walked. I clearly remember thinking, "What have you gotten yourself into?" I had no idea of how or when to do this, and I did not want to embarrass her by her having to walk into service beside a protestant bozo. Instead, much to my relief, we entered their fellowship hall, where I saw rows of tables festooned with cheeses, crackers, overflowing trays of fruits, and full bottles of wine. After eyeing the feast set before us, I nervously glanced around and was quite relieved that no one was crossing themselves.

I piled my plate full of the incredible offerings, and then we all stood for a prayer of thanksgiving. I kept my eyes closed after the prayer for a second or two so as not to see any crossing behavior that would have left me embarrassed for my failure to grasp even the simplest Catholic 101 instruction. Then the fellowship of mostly college students began. After a few cheese bedecked crackers, I became thirsty. My hostess must have noticed and offered me a bottle of wine. I had never tasted wine or even held a bottle of it in my hand. I clearly remember receiving it in guilt. The idea of wine being served under the roof of a church was mind blowing in itself. Here I was sitting just feet from a crucifix and was unashamedly being offered a whole bottle of liquid evil right in front of everyone. When I left my dorm room, I honestly expected a thimble full of grape juice.

In the end, I was still a teen male and, all other motivations completely aside, I was there to impress the chick. This event and all others in my young manhood were driven by testosterone, period. I certainly did not wish to embarrass my hostess or look like a total back-woods bumpkin. With these conflicting emotions swirling in my head, with some courage, I poured a little taste of the suspicious liquid into the cup before me. I took a deep breath as I remembered all the teachings over the years that wine was nothing but rotted vegetable matter that was totally disgusting. I expected to have to suppress a gag when I lifted the glass to my lips. But, as the first drops touched my mouth, I was astonished and more than just a little taken aback.

"This is amazing stuff!" I blurted out in surprise and astonishment.

I remember so clearly the cup that sat before me. It was red plastic, about 6" tall, with those little plastic bumps all over its side. So, being the avid soda drinker that I was, I poured the red vessel full of this incredibly delicious elixir. It was then that the fellowship began in earnest.

In hindsight, it would have been far more instructive of my early teachers to have schooled me in a little reality. I had no concept that there was any difference between this glass of wine and a full glass of cola. Good is good, and it is the job of the boy to down it with gusto. So, try if you will to imagine a total ethanol virgin slamming a tall glass of wine.

I have no memory of being embarrassed or embarrassing my hostess – because I have no memory of any of it at all, save fixating on the red bumpy plastic cup. The next morning, I found myself in my bed in Cordell Hall with the most hideous headache I ever had. A note awaited me from my hostess, which read, "Call me when you are able." (Smiley face) I later discovered that she had carried me back to my dorm, up the stairs, and, with the help of my longtime friend, Freddy Hurst, they put me to bed. I was not just a lightweight – I was a virgin lightweight.

As I nursed my first hangover, I recalled the somber warnings to fresh faced and recently liberated university students that had been repeated from class to class since the founding of the initial Neanderthal campus in some cluster of African caves: alcohol was selectively dangerous to neural cells in the brain. And (added for effect in the most authoritative preface possible) it has been *scientifically proven* that neural cells do *not* reproduce. Therefore, everyone is born with the only brain cells they will ever have. Hence, if you pickle your irreplaceable neurons with alcohol, by definition, you will become increasingly stupid after each drink. As I felt my head throb, I was fairly certain that I had already crossed some boundary into catatonic cretinism, not to mention that I was also equally as convinced that my liver was likewise on strike. I had not yet heard the same rumors on liver cells, but at that moment, I was fairly sure they were included in the identical category of non-reproducing cells necessary for a long and happy life.

Interestingly, even after decades of higher education and employment as a NASA space life scientist, that fundamental belief persisted. I heard it from so many quarters that I was convinced it was actually a *"scientifically proven fact."* That is, until the day I met Dr. Dennis Steindler...

The Smartest Man in the Room

As I have mentioned previously, in 2007, our NASA team established an experimental protocol that utilized genetically modified mice as an animal subject to better understand the effects of (simulated) high energy cosmic radiation. In the initial meeting at the John F Kennedy Space Center's Space Life Sciences Laboratory, the entire team gathered. It was at that meeting that I first encountered Dennis Steindler, who was then the Director of the McKnight Brain Institute at the University of Florida in Gainesville.

Dennis is an energetic, compelling man whose face is permanently fixed with a contagious smile, his intelligent eyes framed by a set of rounded glasses. His posture and demeanor simultaneously display an expansive intellect balanced with a perceptible desire to earnestly listen *first* – a singularly unique and out-of-place characteristic on a NASA Center. I found Dennis to be one of those exceedingly rare intellects whose drive to understand is obvious enough to invite you to speak freely, even at the first meeting. I had met plenty of well-known, highly positioned individuals in my career who used their introduction as an occasion to speak about themselves as though they were applying for a job while employing conversation as a pretext to dump their resume. But with Dennis Steindler, I was thoroughly impressed that I was standing before a real human brain-man and held my steely eyed missile man's tongue.

Of course, I would have never voiced it at the time, but the only other experience I had had with a brain-man was from the movie *Young Dr. Frankenstein*. As Dennis spoke, my mind reflexively overlaid the image of Gene Wilder. The brain-man I was standing before perhaps could have told me why my own weird brain would sufficiently distract me with this spontaneous mental overlay. Fortunately, the professional Dennis Chamberland still held the keys to my mouth while speaking to the professional Dennis Steindler. (Years later, I admitted this first impression to Dennis, who received it with his trademark genuine sense of humor that lies so easily accessible just beneath his professional exterior.)

The meeting commenced and, in typical government fashion, was prefaced with the general housekeeping items. Then Dennis Steindler took the floor. Dr. Steindler does not lecture. He expertly crafts astonishing tales of science with cutting-edge real-world knowledge and personal experience. In my past, I only had a single professor who effortlessly, even breathlessly and brilliantly, taught in such a way that he had an entire assembly of a couple of hundred students sitting on the edge of their seats at the end of every lecture. The course was Organic Chemistry. And, after the final lecture, we gave him a well-deserved standing ovation.

The same feeling returned that afternoon in the KSC NASA Space Life Sciences Laboratory. I knew long before the end of his evaluation of our methodology that the smartest man in the room was certainly *not* wearing a NASA badge. And when Dennis had concluded his remarks, it was all I could do to hold back my standing-o. It was at that meeting that Steindler corrected our knowledge of the fate of alcohol-soaked neurons in the college guy's brain.

According to Dr. Steindler, there is not only an active process of neuron regeneration in the human brain, but if that process is not fully functional and always replacing certain neurons, as well as the support cells of the brain (glial cells), then the human organism will soon "run down" and finally stop altogether. Without neural sustenance and replenishment, the human could eventually lose their capacity to perform certain functions at all.

The information was gathered by the experience of directing therapeutic radiation beams to the brains of cancer victims or other people exposed to medical or environmental radiation sources, but traditionally owes its origin to a field that began with the pioneering stem cell studies of bone marrow and other tissue and organ sites done by the original radiation biologists of the 1940s and 50s. With enough radiation delivered to the forebrain limbic regions, the "run down" effect would often result. After much investigation, it was determined that the first-line victims of the radiation were the especially sensitive adult stem cells (neuroepithelial cells) lining the forebrain cerebrospinal fluid-filled cavities called ventricles. Further investigation revealed the astounding fact that in the normal functioning of the human body, through a renewal process called *differentiation*, adult stem cells lining the brain's interior limbic system selectively differentiate themselves into neural cells and travel to the brain site that needs replenishment.

Steindler related the story of one of the astonishing ways that this renewal process was verified from a prior study he was involved with at the University of Florida in collaboration with a most gifted and dedicated hematologist/oncologist, Dr. Christopher Cogle. During treatment, a female victim of a specific breast cancer required a bone marrow transplant from her brother. Red marrow material is comprised of hematopoietic stem cells (HSCs) that have the remarkable ability to differentiate into epithelial cells, or any other class of cells, as needed in any human system – including the circulatory system and neural cells. The bone marrow transplant for the woman was successful.

Unfortunately, eventually the patient succumbed to her malady. She had graciously willed her body to scientific research in order to enhance the knowledge of the disease, and Steindler's team received her brain. As they investigated, a population of neural cells was recovered from the part of her limbic brain known as the hippocampus – a seahorse shaped structure that is buried in the temporal lobe of the cerebrum and involved in lifelong memory and learning functions.

Upon a genetic evaluation of the sample, Steindler's group made the astonishing discovery that her cells had the male genetic signature from her brother! His hematopoietic stem cells had evidently migrated from her bloodstream and differentiated themselves as replenished neural cells in her body in a process called persistent *neurogenesis,* or *neuropoiesis* (just as the lifelong generation of blood cells is referred to as *hematopoiesis*).

The neuroepithelial cells that comprise the brain's limbic and periventricular adult stem cell layers are somewhat different than related epithelial cell populations located throughout the body. These cells are often described as "sensitive, fast growing, short lived" cell populations. But the neuroepithelial stem cells are described as "quiescent" or "sensitive, slow growing" populations. Comparatively, neuroepithelial cells do not have the same function as their counterparts that are at the forefront of guarding the inner body against the outer world or performing the task of separating and admitting the nutrients and gasses from the outside world into the inner body. In stark contrast, the sensitive neuroepithelial population is quiescent because its sole task is neurogenesis.

Of all the very widespread misunderstandings of human exposure to cosmic radiation in deep space (even among highly educated people), I am most surprised at the apparent inability to discern between radiation types and the consequent physiological effects on the human body. It is as though there is little or no discernment between terrestrial ionizing radiation, therapeutic ionizing radiation, the radiation that is the byproduct of nuclear weapons, radioactive contaminants, or deep space cosmic radiation. The miscomprehension is commonly expressed as, "Ionizing radiation is just ionizing radiation."

This thought process demonstrates the lack of understanding of the BB and the basketball comparison example for terrestrial vs. cosmic radiation, and the epic energies of which they are comprised. Just to get a handle on this issue and to enumerate it in a comparative sense, the nuclear profession has implemented "quality factors," "absorbed dose equivalents," and "radiation weighting factors" to even the playing field. Unfortunately, these are collectively nothing more than just another professional SWAG (scientific wild-ass guess) implemented in the darkness with a comprehensive lack of any experimental evidence. It is true that you can haul these SWAGs into a court of law. But in the end, a SWAG is just a SWAG awaiting scientific verification – of which there is yet none at all from deep space. The sad fact is, even as we design and build our obsolete deep space vehicles from plans created for low-Earth orbit that are claimed to be "on-steroids," at best, there is absolutely *zero* evidence for what the crews are going to face. It is just the SWAG factor built from an altogether different species of ionizing radiation. In my mind, I often visualize some clueless politician with little or zero science background or education making these absurd, dangerous, and truly idiotic decisions, knowing that the system will protect them from any accountability, as the Challenger and Columbia investigations clearly illustrate.

This SWAG morphs into a whole new insidious form when it is applied to living humans about to be launched into deep space. There is this Jamestown-type expectation mindset that BBs and basketballs be damned; we can just apply the SWAG and head on out into deep space legally umbrellaed by our guesses that will cover our asses because we're from NASA and we know what the hell we're doing!
Unfortunately, derrieres will ultimately be left hanging – because SWAGS and body bags always run together in the same witless crowds.

Said Elon Musk in a prescient moment of clear vision, "Everyone is trying to optimize their ass-covering."[2] That sentiment is spot-on in this case. In the Agency, this is accomplished using NASA jargon augmented with the obligatory set of PowerPoint slides and dog-and-pony-shows, marching out NASA's mission in the form of inadvertently obfuscated, somber declarations "justified" by the latest data, models, and probability assessments that only a scientist could love and understand. And somehow, they have convinced their own Inspector General that it is all by some means justified, apparently because the 21st-century NASA simply does not have the creativity or innovative spark to change their model.

Unfortunately, all this vaporware misses the point. And that point is: the next steps into deep space require, even demand, *real vertebrate data from deep space analog animal missions* – not SWAG data kludged together from Earth based analogs, and certainly not replacing the rats with living, breathing astronauts. Any time humans feel their knowledge base is so extensive that they can totally skip a vital investigative phase of gathering actual data in the field and just apply a well-used SWAG, you can go ahead and pre-order the body bags.

Imagine for just a moment that it is *your* brain stuffed in that space helmet. Astronaut-based evidence from deep space, and even from low-

Earth orbit, confirm that the radiation beyond Earth's atmosphere is radically different from what we experience on the surface of our planet. But the only way the human organism can discern its presence is when a heavy primary of GCR strikes a neuron in the visual cortex of the human brain. When it deposits its final energy load in that neural cell at the terminal end of the Bragg peak, the cell is functionally obliterated. Its last act is to send out a final signal, which the brain interprets as a flash of light. It is, in fact, the flash of death of a neuron. NASA Astronaut Jerry Linenger reported so many of these that he could not even sleep in low-Earth orbit! Meanwhile, all the other cells in the human body that are being obliterated do not have the same ability to signal their deaths to the human in charge.

So the question arises, realistically, what is the difference to your brain cells between a night of alcoholic excess and a trip to the Moon and back? In both cases, it is likely that neuron death will stack up. But, as was witnessed with the Apollo astronauts and centuries of college-boy alcohol abuse, neurogenesis takes up the load. Apparently, all is well in the end because that observation is totally backed up by 24 lunar astronauts and millions of university students, right?! Obviously, this train of thought is how rationally challenged humans walk, with eyes wide open, off 10,000 feet escarpments of logic – by applying SWAGs to disparate data. Undoubtedly, someone will publish a peer-reviewed paper on an "alcohol weighting factor" applied to cosmic radiation exposure....

The last word in science – the gold standard of approval – is, always and only, evidence from actual experimentation in the actual field. We can SWAG and compare forever, but that is only a SWAG without the data – whether the tort lawyers like it or not. And that topic is discussed in detail in the next Module of this book. The bottom line is that it is absolutely unethical to use human subjects to recover this data when it can be discovered using animal subjects – especially when there is clear, unambiguous, and repeated evidence of a high probability of long-term human injury and death.

At the end of Dennis Steindler's presentation, and after the meeting broke up, Dennis and I met in a hallway adjacent to my office at KSC's Space Life Sciences Laboratory. He looked at me and asked frankly, "What's the answer?" I knew very clearly exactly what he was asking and where he was coming from. He wanted to know NASA's official stance on this issue and how they intended to deal with it. I answered his question from the perspective of the Chairman of the Kennedy Space Center's Animal Care and Use Committee, of which, at that time, I had been the longest seated Agency ACUC Chair from all NASA Centers. I gave him a clear answer for which there were no other viable, ethical options. The answer was, quite obviously, a detailed, minimum, year-long animal cohort experiment in deep space – probably at one of the stable Lunar Lagrange points between the Earth and Luna.

To the inevitable, quite logical, very intelligent follow-up question of how to make such an experiment happen, I had no choice but to default to an incredibly embarrassing answer. It was not that there was not any funding available to answer the single most important question in human space exploration. Rather, there was *no political will to do so* inside NASA or out – and there were very few advocates for the necessary research at NASA Headquarters. At my level, the political waters around the Agency always seemed fouled by interagency intrigue, invisible power brokers, and substantial Congressional and administrative meddling in the ever-morphing, never clearly defined NASA mission.

While the most important answers to the radiation and altered gravity issues should logically have been obtained by animal models in deep space no later than the mid to late 1970s, NASA funding was controlled very little by science or logic as much as it was used as a tool to accommodate social engineering by hijacking NASA's high-visibility platform. The Agency was literally unable to keep its eye on the ball because the ball was constantly being switched, relabeled, and communally reshuffled in mid-flight.

Now, 15 years later, nothing has changed. The single most important answer before sending humans into deep space still has zero resolution. And, most incredibly, it has not occurred to anyone to actually go gather the data – except by sending living humans out into the void armed and protected only by the ephemeral, sacrosanct SWAG.

But I clearly sensed even *that* was not the deepest meaning of the query posed by Dennis Steindler. He was asking the classic question voiced so clearly by Harry Stamper (Bruce Willis) in the movie *Armageddon*. Steindler was asking me, a NASA life sciences engineer and also bureaucrat, what the NASA answer was – the ingenious answer from the most successful science and engineering organization in human history. The Agency that was required and fully expected by society and history to have the right answers all the time.

Harry Stamper looked at NASA manager Dan Truman (Billy Bob Thornton) and asked the same question with unhidden exasperation and wholly unfulfilled expectation:

"And this is the best that you ca... that the...the government, the U.S. government can come up with? I mean, you... you're NASA, for cryin' out loud; you put a man on the Moon; you're geniuses! You... you're the guys that think this shit up! I'm sure you got a team of men sitting around somewhere right now just thinking shit up, and somebody backing them up! You're telling me you don't have a backup plan, ...that's what you're telling me?" (Whoever advised the *Armageddon* scriptwriters must have warmed a seat at NASA Headquarters for a season.)

As I write this book many years later, I still have no answer for Dr. Dennis Steindler to the pair of single most important questions in human space exploration. But, in the upcoming chapter in *Departing Earth Forever* – Book Two, "It's the Biology, Stupid," I will address it at length.

Returning to our 2007 investigation, the resultant paper was titled "Quiescent adult neural stem cells are exceptionally sensitive to cosmic radiation" and was published in the *Journal of Experimental Neurology* in November 2007. Regarding the acute exposure to the simulated beam of GCR (as taken from the paper's Abstract), "We found that quiescent neural stem cells, rather than their rapidly dividing progeny, are most sensitive to radiation. Since these stem cells are responsible for adult neurogenesis, their death would have a profound impact on the production of new neurons in the irradiated adult brain. Our finding raises an important concern about cognitive and emotional risks associated with radiation exposure."

Our paper would not be the only one to clearly sound this specific alarm. Nine years later, another study would enumerate the concerns we expressed about cognitive and emotional risks associated with radiation exposure.

Long Term Mid-Brain Injury

Two years prior to our study, another paper published in the *Journal of Radiation Research* announced the finding titled, "High-LET radiation induces inflammation and persistent changes in markers of hippocampal neurogenesis." Like our study, this research also focused on the core brain hippocampus region and also reported clear evidence of gross inflammation and injury to the neuroepithelial lining of the mid-brain. Further, the research also reported that the layers of these sensitive neuroepithelial cells responsible for neurogenesis that received the acute injury did not return to normal functioning after a time of recovery. Instead, they continued to deteriorate in the months after the exposure to the GCR simulant.

"Our current data suggest that these changes are not only persistent but may worsen with time. Loss of precursor cells was also associated with altered neurogenesis and a robust inflammatory response. These results indicate that high-LET radiation has a significant and long-lasting effect on the neurogenic population in the hippocampus that involves cell loss and changes in the microenvironment."[3] This finding was verified three years later in a 2008 paper published in the *Journal of Radiation Research* titled, "^{56}Fe-particle radiation reduces neuronal output and attenuates lipopolysaccharide-induced inhibition of long-term potentiation in the mouse hippocampus."[4] The authors also state, "Our data show that ^{56}Fe-particle radiation affects the response of the hippocampus to an immunological stressor and that the alterations progress over time." Once again, even the relatively weak, Earth-based simulated GCR induced injury to the mid-brain region populated by sensitive neuroepithelial cells.[5]

GCR Induces Alzheimer Like Effects

In 2012, in a peer-reviewed paper published by *PLOS One* titled, "Galactic Cosmic Radiation Leads to Cognitive Impairment and Increased Aβ Plaque Accumulation in a Mouse Model of Alzheimer's Disease,"[6] the risk of long-term neural injury was discovered for the first time. The study also utilized simulated cosmic radiation at Brookhaven National Laboratory's Relativistic Heavy Ion Collider. It stated that their results demonstrated "that HZE particle radiation can increase Aβ plaque pathology in a... mouse model of Alzheimer's disease." In addition, the study also reported that "6 months *after* (acute) exposure to 10 and 100 cGy ^{56}Fe radiation at 1 GeV/μ, that mice show decreased cognitive abilities."[7] As in our study, once again investigated was the effect of simulated GCR on the neuroepithelial cells in the inner-brain region of the hippocampus.[8] The study also demonstrated that the effects may not rebound upon return to Earth.

Cognitive Dysfunction and Neural System Anatomical Injury

In 2016, a paper was published in the journal *Nature* titled "Cosmic radiation exposure and persistent cognitive dysfunction."[9] The study also utilized simulated acute GCR particle exposures (^{16}O and ^{48}Ti at 600 MeV) to vertebrate models of mice and rats at Brookhaven National Laboratory's Relativistic Heavy Ion Collider. These animals were then tested for cognitive behaviors after exposure and compared to a non-exposed cohort of control animals.

The findings of the paper were startling and encompassed not only cognitive, behavioral, and higher brain function abnormalities, but evidenced neural anatomical injuries as well. The study found that the GCR simulant evoked "spatial, episodic and recognition memory (that) were temporally coincident with deficits in executive function and... elevated anxiety." The study discovered that the artificial GCR also caused injury to the anatomical structure of the vertebrate animal model's nervous system. "Irradiation caused significant reductions in dendritic complexity, spine density and altered spine morphology along medial prefrontal cortical neurons known to mediate neurotransmission interrogated by our behavioral tasks. Cosmic radiation also disrupted synaptic integrity and increased neuroinflammation that persisted more than 6 months after exposure."[10] And, once more, the study specifically pointed to the sensitive region of the core brain populated by the neuroepithelial cells and stated, "Rodents exposed to cosmic radiation exhibit persistent hippocampal and cortical based performance decrements...."[11]

Refined Dose Simulation – A New Clear Warning

A paper published 5 August 2019 in the journal *eNeuro* was titled, "New Concerns for Neurocognitive Function during Deep Space Exposures to Chronic, Low Dose-Rate, Neutron Radiation." This truly landmark study was one of the first to lend a more realistic view to the effects of a simulated *daily dose* of GCR, and was the first to expose vertebrate animal models to a simulated GCR stream that mimicked the daily dose of particle radiation found in deep space with a significantly higher fidelity than an acute dose. Because of the prohibitive expense of dominating an entire massive facility such as Brookhaven's Collider-Accelerator for half a year, previous studies in simulated space radiation had to be conducted by giving the animal subjects an acute, very short dose of what they would accumulate for months in deep space. As I related previously, this did not allow the species to repair and recover as they would attempt to do in a much lower dose accumulated over an extended period of time.

The study's remaining chief limitation was, obviously, the deep space spectrum represented by the periodic table of the elements or the much greater incident energies. In this experiment, the research team utilized an encapsulated single source of Californium-252, a neutron and gamma emitter. The vertebrate animals (mice) received a total dose rate of ~1 mGy/day. This represented a mixed dose with a single source neutron contribution of 80% of the total dose per day, and photons contributing 20%, at an approximate distance of 1.8 m from the source.

This was the first time animal subjects were exposed to a more realistic daily, albeit lower energy, single source dose over a six month period of time which simulated an estimated one-way voyage to Mars – 150,000 times lower fluence than the exposure to the Brookhaven simulated GCR beam. After the half-year exposure period, the animals were subjected to behavioral studies and extensive physiological system and tissue evaluations.

The Abstract summarized the findings by stating that they discovered "diminished hippocampal neuronal excitability and disrupted hippocampal and cortical long-term potentiation. Furthermore, mice displayed severe impairments in learning and memory, and the emergence of distress behaviors. Behavioral analyses showed an alarming increase in risk associated with these realistic simulations, revealing for the first time some unexpected potential problems associated with deep space travel on all levels of neurological function.... we have uncovered striking neurobehavioral and electrophysiological defects."[12]

Again, the study reported reduced hippocampal neuronal excitability. The hippocampus is, once more, that region of the mid-brain that is lined with sensitive neuroepithelial cells that play not only an essential role in neurogenesis, but is also the active site of the regulation of the vertebrate organism's motivation, emotion, learning,

Countermeasure Madness

"Many of the cataclysmic leadership failures were failures of rationality. The pendulum of leadership development needs to swing back toward the rational: strategy, creativity, foresight, decision-making, and analytics."

- Paul Gibbons

"Science is what we have learned about how to keep from fooling ourselves."

- Richard Feynman

The Penalty for Exploring Outside the Range of Human Adaptation

I READ SCOTT KELLY'S BOOK, *ENDURANCE*, WITH RAPT fascination. I was also quite impressed with his candid, unvarnished, detailed revelation of what he endured coming home after nearly a year in low-Earth orbital space. I never met Scott Kelly in my 30 years with the Agency, which is not at all surprising. But as the words flowed off the page, I could sense many things because I personally knew the organization he functioned under quite well. And I found a person I could relate to – a straight talking individual, fully dedicated to human space exploration, with absolutely nothing to hide, and the courage to tell a very difficult story. I instantly liked and very much respected Scott Kelly on a personal level. And it was obvious that he was completely transparent.

I was also quite surprised NASA did not discover a way to stop his book's publication, heavily redact it, or somehow find a way to discourage him from publishing it. His description of the outer limits of human adaptation to spaceflight was so brutally detailed and accurate that it revealed that Scott Kelly was articulating that boundary at which we can no longer push beyond by exploring space utilizing a 50-year-old, flawed exploration paradigm. As soon as I read it, I put the book down and said to Claudia, "Now we know where the microgravity limit is." The problem with knowing where the limit is requires that something radical happens. We must ethically either stop exploring beyond that point or change something dramatically. Unfortunately, neither NASA nor any private or international space agency is ready for either, still pressing forward at all speed to the next phase of human exploration in a system that has officially exceeded its safe physiological limits. The system within NASA that allows this process to go forward against all odds and sanity with virtually no endpoints is called "countermeasures."

Yet, discovering limits is a legitimate, natural, and essential part of all human exploration. These define the boundary of change. These limits demand necessary change or the goal of the exploration itself is irrevocably stalled. Kelly described the limit we have reached so that we must now wake up and change our paths specifically so that human exploration of space can continue unimpeded under a new model without deliberately and knowingly injuring our astronaut crews. But change does not come easy, as Kelly also clearly described. When forced with two untenable paths – stop at the limit or radically change models in mid-stream – most behemoth organizations will default to doing more of the same. This time NASA is doing more of the same "on steroids."

For example, what was Scott Kelly's own solution upon reaching the documented limit? More countermeasures. He courageously admitted the limit had been reached and then suggested the solutions in a single sentence, as he said, "Our space agencies won't be able to push out farther into space, to a destination like Mars, until we can learn more about how to strengthen the weakest links in the chain that makes spaceflight possible: the human body and mind."[1]

But embedded in his summary sentence is the solution as NASA sees it: to "strengthen the weakest links... the human body and mind." Please note this not even subtle reminder to everyone – the human body *must adapt.* And, when we reach the limits, we must then *strengthen it* until it *adapts some more* so that the mission can go on. And what is the process by which that is accomplished? In NASA parlance, again, it is called countermeasures.

Countermeasures We Employ Each Day

As long as the human species exists, we will sensibly employ countermeasures every day of our lives so that we can continue to exist and thrive on Earth. As Desmond Morris so famously described us, we are but naked apes walking about on an unforgiving planet that is fully capable of inflicting injury and harm on our person. The first countermeasure we employ each day is clothing. Then we adjust our home and office temperatures and humidity. We don sunglasses to protect our eyes against solar radiation. We rub on sunblock to protect our skin from ultraviolet rays. We spray insect repellant on our exposed skin to ward off disease. We climb into our cars and fasten our seatbelts. We employ glass windshields to keep the bugs out of our face and prevent frostbite. During global pandemics, we strap on masks, avoid crowds, and slather up with hand sanitizers. All of these are countermeasures, and all of them are absolutely necessary to survive on planet Earth without injury or running into the grim reaper before we are ready.

NASA and all the other private and government space programs have merely employed versions of exactly the same countermeasures concept so that we can also live in space using their collective ideas of space exploration. Unfortunately for everyone, their idea of space exploration, when extended beyond the limits they have discovered, has been shown to be hazardous to our health. We have now encountered the absolute unprotected adaptability threshold and the effective limit to unprotected countermeasures. But it appears that the guy that delivers these memos around NASA Headquarters has been fired, and everyone in every office on all nine floors is apparently still clueless.

Instead of any admissions of that limit and its most profound meaning for the future of human exploration of deep space, the world's space agencies plod on using an antiquated, obsolete exploration model as effective as a screen door on a submarine. And while that illustration may sound outrageous, it is quite applicable to NASA, the ISS, and their future plans for deep space exploration.

It is very possible that, just like the models from the early days of human habitation of the oceans, all submarines could, in fact, use a screen door on the underside of their craft. Interestingly enough, nearly all modern military submarines employ a similar open ballasting system anyway. If they did install such a screen door opening on their undersides for the crew, every time they submerged, they would have to pump compressed air inside the submarine to counter the pressure outside as they traveled deeper in the ocean to maintain the inside at ambient pressure, and the air would bubble out as they decreased their depth on the way to the surface. In an engineering sense, this could actually be accomplished. And if engineered properly, the crew could (theoretically) perhaps "adapt" to the changing pressures. If they operated this way, the United States Navy could save hundreds of millions of dollars by using thin aluminum hulls instead of thick, expensive hulls of exotic classified alloys to keep the ambient pressure out. So why do they not do this?

Because it is both completely absurd and ridiculous, of course, and would be harmful to the crew. Such an engineering design would be just as insane as launching a trillion-dollar orbiting space station asset into low-Earth orbit with inadequate radiation shielding in unrelenting microgravity. In both cases, the crew will soon exhibit injuries for many distinct reasons, and, ultimately, the mission objectives would become eclipsed by the fatally flawed and wholly unworkable exploration model itself. Both models are equally absurd and ridiculous. And yet, here we are....

What is the weak link in both cases? Human adaptability.

What is the obvious solution in both cases? Protecting the human asset as the first priority.

What is the root cause of this ongoing engineering debacle soon to culminate in a dead-end catastrophe? Stubborn human intransigence and an innate limited capacity of our species to rapidly change and correct.

And what is NASA's ultimate go-to solution? More countermeasures. More screen doors.

So, is NASA seriously flying the space version of a submarine in space with a screen door?

Unfortunately, yes, it certainly is.

But just like the Navy wisely skipped that phase with an easy fix by protecting the human asset from exposure to the alien environment outside as a going-in design priority, the fix for NASA is likewise equally straightforward and may begin today. But regrettably, it is not even a topic up for discussion anywhere in the Agency. Unfortunately, NASA is apparently quite fond of its screen door approach to exploring space and shows no sign of throwing in the towel on its fatally defective exploration model just yet. The Space Agency is apparently now all-in and committed to the body bag phase, which seems to be inevitable for NASA management, just as the Apollo 1, Challenger, and Columbia catastrophe investigations all unambiguously revealed.

NASA, unlike the Navy, has decades and half a trillion dollars invested in its model. And, just like all other juggernaut bureaucracies, it is not likely to admit its limits, its false paths, and its manifold errors and then retool. How can they when they have tattooed "Failure is Not an Option" on their foreheads? Sadly, because human intellect is powered by emotional drivers, Hollywood fantasies, and knuckle dragging politics, we are typically wired such that a total human catastrophe is required before intelligent, thoughtful change is demanded by the sheer magnitude of the tragedy. To say, "But it doesn't have to be that way," is like whistling in the dark in the real world of immense organizations and entrenched political ideology, as each of NASA's post-disaster investigations have all clearly described.

Countermeasure Madness

Scott Kelly said we need to "learn more about how to strengthen the weakest links in the chain that makes spaceflight possible: the human body and mind."[2] He distinctly did not say that we need to change our exploration model and do business in a totally different way. He did not say that we need to forget about living with the effects of the screen door and build a space station that protects the humans inside and build better space vehicles so that *no astronaut ever must go through what he went through ever again*!

As Kelly implied while dutifully echoing the company line, NASA's approach has always been to employ more and better countermeasures – interminable, never-ending countermeasures layered upon other countermeasures. Why so many? Because in order to protect the idea that humans can endlessly adapt – that we can adapt in microgravity while at the same time enduring the awful energies of cosmic radiation – all we need is more layers of clothing or more sunblock or more insect repellant, a better N95 mask, and so on, and so forth, etcetera, etcetera, etcetera. Then, things will soon be just fine.

But beyond reasonable limits, good ideas about fixing fatally flawed models can quickly morph into madness, because there are never enough good ideas. There are never enough countermeasures to repair arrogance and overt, stubbornly unrecognized stupidity by propagated vast government committees. The stacking of the band-aids plan is always fully engaged when the system is about to utterly fail, taking many down with it.

The absolute limit of the countermeasure idea is always the magic bullet of genetic engineering. If we can just genetically engineer humans to become "space creatures" so that we can adapt to microgravity and the effects of cosmic radiation (good luck with that one), then we can fly about in space endlessly with our thin aluminum walls and screen doors. Make it so![3] If that fails (as all completely ludicrous ideas are wont to do), there is always the drug approach. If we can just gin up a drug that will somehow counter space induced osteoporosis, cardiac atrophy, brain squeeze, blindness, and projectile vomiting, then we can float around freely in space for generations to come! This forgets that the very drugs themselves have their own set of problems and disregards the contraindications that may morph the inside of the crewmember's bones to a Frankensteinian labyrinth of malformed calcium deposits.

But NASA always needs just one more study of microgravity. If the billions of dollars already squandered on microgravity research over the decades were not enough, there is always the specter of one more key question and one more experiment that will somehow miraculously transform into the ultimate countermeasure – the breakthrough NASA needs to allow us to continue forward, literally floating about in this absurd model. Congress is sure to fund yet another round because they will apparently believe and fund just about any wet dream from NASA. The scientifically clueless lawyers on Capitol Hill nearly always seem to bite whenever microgravity research model is dangled before them.

At this point, it would be useful to pause the discussion for a moment so that we can write a letter that will clearly illustrate the degree of madness we have actually lapsed into. Let us compose a letter to the Department of the Navy and suggest that they save billions of dollars by building thin hulled submarines with a screen door on the bottom and develop all the engineering countermeasures required to vary their inside pressure to ambient. They can then spend billions to develop other countermeasures to repair the resulting human injuries created by this model simply by genetically engineering new sailors who can withstand rapid pressure changes and will not need decompression. Further, they can invest billions more to develop new countermeasure drugs to offset the resultant injuries. We can even assure them that it is so expensive to build pressure resistant hulls that if we do not do this, we may have to stop using submarines at all.

If the Navy responded in any manner, I would be truly shocked. This letter would be placed in the crackpot file and would be fodder for years of jokes around the office. But if they did respond, the reply would go something like this (paraphrasing the actual meaning from the professional reply): "We strongly suggest that you seek professional help."

And yet, NASA actually somehow continues this nearly identical narrative with a straight face and without meaningful challenge – which, in the end, is not at all a good sign for the human species.

But when it all comes down to everyday life in the miserable grip-less void of microgravity, absolutely nothing brings it all home like using the toilet in NASA's space. The space toilet is a countermeasures engineer's wet dream machine in every sense.

You begin this process by strapping your bare rear onto a space toilet that more resembles a Rube Goldberg throne costing tens of millions of dollars. Whether male or female, you have to insert private members into or onto cold devices and turn on the suction machine – before doing your business (God forbid an emergency bathroom call). Then the control panel is activated in the proper order. If the astronaut gets any of this out of order or wrong, the cleanup can be traumatic. You never actually "sit" anywhere in microgravity – not even on a toilet. There is no gravity to allow the process of sitting to actually take place. Instead, you grab hold of the handles, pull down, and strap in, especially when it is important that things coming out of your body actually end up in the right place at the right time.

Of course, training for this probable misadventure consists of loading a camera in the bottom of the training toilet on Earth so that, after strapping in, you can aim your privates so that they are aligned in the right direction before letting go. I related this entire story to a passing, approximately 10-year-old boy accompanied by his parents at a NASA Smithsonian sponsored exhibit on the Mall in Washington DC in 2008. He had just queried me with the single most frequently asked question by members of the public to NASA officials, "How do you go to the bathroom in space?" When I finished giving him the same spiel I had given to thousands of other somber faced, more mature people in hundreds of gatherings before, he looked at me for a protracted sober moment like I was having him on. When he saw I was most serious, he exploded into a fit of convulsive, red-faced laughter that he literally could not stop. His spontaneous burst of laughter was so infective that the tables around us turned to enjoy his hilarity with him. Finally, as his mother led him away, still laughing, she turned back to me and said, "You have certainly made one little boy's day!"

In retrospect, the innocent young boy's uncontrollable laughter is wholly appropriate for all of us. As I remember that day, if I had been standing beside him looking on at my own serious, academic face describing in my professorial tone this engineering malefaction from hell as though it was a perfectly normal process, I probably would have joined him in his uncontrollable mirth. What the peerless cadre of scientists and engineers have come to believe is perfectly normal is actually just as preposterous as not declaring the King is, in fact, butt naked – and ultimately, just as hilarious to children, as it well should be to every sane adult.

Another bathroom countermeasure involves controlling surface tension. On Earth, the powerful effects of surface tension are all regulated and held at bay by gravity. Not so in NASA's space model. Surface tension is a formidable force that literally commands the flow of fluids along surfaces – all surfaces, including human skin. For example (being delicate here), if you stick your thumb in a can of soup in microgravity, the soup will flow along the thumb and eventually around the wrist. With no gravity to hold fluids in place, they will run along surfaces until something stops them. When using the bathroom in space, surface tension must be controlled by something other than gravity. In the case of the space toilet, rapidly moving air is the countermeasure. The air is sucked along the tubes that privates are inserted into or the other devices that orifices are covered with. The air "captures" the bodily fluid wastes and pulls them safely inside. As far as solid wastes are concerned, another air stream confines them and "safely" captures them for disposition.

Now multiply the complexity of the space toilet by every engineering system in NASA's flight hardware. At this point, we are racking up hundreds of countermeasure systems that alone cost tens of billions of dollars and add tons of mass to the model. So, what are the cumulative financial burden and the mass penalty for all the ever-increasing mass load of countermeasure madness? NASA does not advertise that number, probably because NASA itself has absolutely no idea.

The Penalty for Exploring Outside the Range of Human Adaptation

In an ideal space model that employs just one of the two essential future primary countermeasures (artificial gravity), a hypothetical male astronaut would walk (not float) into a toilet, stand (not be strapped) in front of a plastic urinal that costs $500 (ridiculously expensive, I know, but "space equipment" is absurdly pricy because the government will actually pay the sticker price), unzip his pants, and urinate. He would then flush, zip things back, and wash his hands in a sink with running water that flows over his hands and runs down the drain. (Yet another grossly overpriced unit that costs another $500.) He would then dry his hands on a towel without worrying about the wash water climbing up his arms, and go back to work. This flight model employs the single countermeasure of artificial gravity, and all these systems work perfectly in a one Earth-normal environment. This model costs the taxpayer literally 260,000 times less than the single absurd microgravity model and requires no specialized training to use. Once again, the lowly space toilet is a truly marvelous example of just how far down the rabbit hole the world space agencies have descended. Indeed, the space toilet is the countermeasure poster child for an absolutely end-of-the-road, expired, and obsolete human exploration model. Unfortunately, the space toilet also has the average human IQ written all over it.

Yet, I can clearly and unmistakably hear the supercilious clearing of the academic throat in the inevitable ramp up of the defense for the status quo. They will argue that "We just need a little more time to find that Holy Grail discovery in microgravity research that has nothing at all to do with the scientific questions about surviving and staying healthy trying to actually live in it, but that Nobel prize winning discovery in physics or metallurgy or medicine that will make it all worthwhile!" A sober and professional response would logically be, "We will gladly build your detached microgravity laboratory 100,000 meters outside our space station, and you can spend as many centuries as you wish out there investigating your atrophied hearts out and racking up damage to your musculoskeletal and cardiovascular systems as you whistle while you float freely around and work." And under my breath, I would add very unprofessionally, "investigating your single discipline questions without making the rest of us live in its injurious misery and dramatically detracting from every other single, far more significant scientific question in space exploration and settlement."

It is true that any adequately protected and shielded human can venture to any environment that we are aware of – even to the surface of Venus where the temperature is 900° Fahrenheit, the pressure is 90 atmospheres, the clouds are comprised of sulfuric acid droplets, and windows must be made of diamond just to survive. Yes, with the right equipment, we can go there and explore its surface, boot to searing ground! But it will also cost us – a lot. And the risks will be astonishingly high. But we can do it. Likewise, we have demonstrated that we can also build a massive human space station in low-Earth orbit with little adequate radiation shielding and no artificial gravity. It will cost us financially. And as we approach the limit of human adaptability, the inevitable price will be in the form of human injury. If we exceed the limits, those injuries will be permanent.

Go park your car on a railroad track and then try and work out adequate "countermeasures" to protect yourself from the injury you will receive inside the car when the train inevitably hits. Meanwhile you have completely disregarded the best of all countermeasures – *don't park on railroad tracks*. In deep space exploration that basic wisdom equates to beginning your long journey by leaving home with adequate shielding and artificial gravity to begin with!

We have built that massive space station and, with it, we have learned our limits. We are now at the crossroads dictated by our most essential findings. Will we continue to respond with even more countermeasure madness? Or will we finally admit, as was spoken before, that:

The countermeasure burden itself defines the end of this dead-end human space exploration paradigm. It is the penalty for exploring outside the range of human adaptation.

The ISS was a brilliant platform to learn what we needed to learn in low-Earth orbit below the Van Allen shields. But it has finally outlived its usefulness. It has provided us with the key data points that we needed to understand what the next generation of human space exploration system will look like – but its time has expired. Today, it is time to retool, shred the ever-lengthening list of countermeasures, and reverently and with appropriate ceremony, dump the ISS into Pont Nemo. This releases us to rebuild new stations based on our lessons learned in low-Earth orbit by courageous astronauts like Scott Kelly and Mikhail Kornienko, who took great personal chances to acquire all the essential data so vital to our future beyond Earth. And the best news of all is that the new generation of astronauts that replace them should never have to endure their injuries or live with the permanent traces of future disease because they have bravely paved the way by gathering the data so that everything can be fixed and the rest of us who follow do not have to be injured.

Why Do
We
Persist?

"Leaders need to correct for cognitive biases the way a sharpshooter corrects for wind velocity or a yachtsman corrects for the tide."

- Paul Gibbons

The Exploration Model Fueled by Biased Irrationality

AS YOU VIEW THE CURRENT REALITY AND BREAKDOWN of the human space exploration model of the 21st century, after clearing way the cognitive biases – the insistence to crew injury to preserve the prevailing exploration model appears irrational. Indeed, the real irrationality is, in fact, real. In more frank terms, Justin E.H. Smith, professor of philosophy in the Department of History and Philosophy of Science at the University of Paris, has stated that it is the seed of irrationality that is part and parcel of human nature itself. States Professor Smith in his 2019 book *Irrationality - A History of the Dark Side of Reason*, "The problem is that the rational gives birth to the irrational and vice versa in an endless cycle, and any effort to permanently set things in order sooner or later ends in an *explosion of unreason*."[1]

In science, this irrationality is admitted, but it is described more sympathetically as a "cognitive bias" – a powerful capacity of the human brain to exclude the flow of logical information that would eventually suggest more appropriate corrections in behavior or course. Indeed, the genesis Apollo Program of the 1960s was *rationality* on speed – inventive genius of the truly extraordinary kind. But its management and nurturing were given over to the most irrational of all human institutions – presidents, congresses, and their subservient administrative bureaucracies. That wholesale irrationality should result ought to be a surprise to absolutely no one.

The Adaptability Psychosis

Why do we persist with our current disastrous model of human exploration? We persist because the *full expectation of* an ever-irrational patchwork of countermeasure enforced human adaptation in space has become our determined practice – our *modus operandi*. It is how we do things in the Agency. We do it because we have always done it this way. It is locked into our exploration mind and is thus expressed as a chronic engineering practice. It is doubtful that few people anywhere in the Agency have ever even thought about it, much less challenged it, because its solution is just one more countermeasure breakthrough away. Indeed, in the latest round of speculative countermeasures, it has been unabashedly suggested that we should redesign the human form by genetic engineering to create in our laboratories a new astronaut species.[2,3,4] No one even blinks an eye at the manifest evolving irrationality of such a suggestion. That is just how deeply the adaptation psychosis has embedded itself in our psyche. It has thus bloomed into a ludicrous madness, and few appreciate that it represents the singularity interface separating us from the ultimate rabbit hole. Restating University of Paris Professor Justin E. H. Smith, "any effort to permanently set things in order sooner or later ends in an *explosion of unreason*."

I am not entirely certain if the idea of limitless human adaptability originally stemmed from the fact that near "perfect" specimens were selected for spaceflight in the form of highly intelligent, experienced, and extremely healthy young test pilots. Yet, fairly, in the 1950s, NASA inherited such expectations from the culture. Certainly, this theme of limitless adaptability was the grand premise of nearly all 20th-century exploration paradigms. From wherever it originally came, this altogether mythical idea ultimately culminated in the wholesale failure of the first thrusts into the human colonization of the oceans because of the inability of humans to adapt long-term to hyperbaric environments.[5] If we continue down this path, we will likewise ensure the failure of the first thrusts to permanently colonize off our home planet, and ruin countless lives in the process. Limitless adaptability was such a tenacious and ultimately malevolent philosophy that humans seemed altogether incapable of looking beyond its error to correct it with relatively simple engineering solutions.

Nothing has changed, either in human exploration or the human capacity to correct this wholly irrational cognitive bias, in the opening decades of the 21st century. The philosophy of unlimited adaptability was, and remains, a human philosophic construct of monstrously destructive potential that even the most intelligent humans seemingly cannot circumvent. It leaves the distinct impression that we persist in clinging to it to perhaps protect some deeply embedded emotional hubris of which we just will not – or cannot – let go. We would rather spend a billion dollars on 1,000 adaptation countermeasures than give it up and spend ten million dollars on two engineering solutions that make adaptation completely unnecessary and reveal all those engineering countermeasures as simply preposterous. It is so persistent and so thoroughly embedded in our exploration zeitgeist that it has become a psycho-social artifact representing a true human intellectual limitation – or even a kind of mental fixation that we somehow cannot seem to get around or over. Simply expressed, it appears that more than anything else, we humans demand screen doors on our submarines – and we will do anything or sacrifice anyone to keep them!

These are the reasons why we persist – until its tragic collapse clears our thinking with what should have been obvious truth all along. It is the way of humans.

A Strategic Necessity

"Greater than any army with banners is an Idea whose time has come."

- Victor Hugo

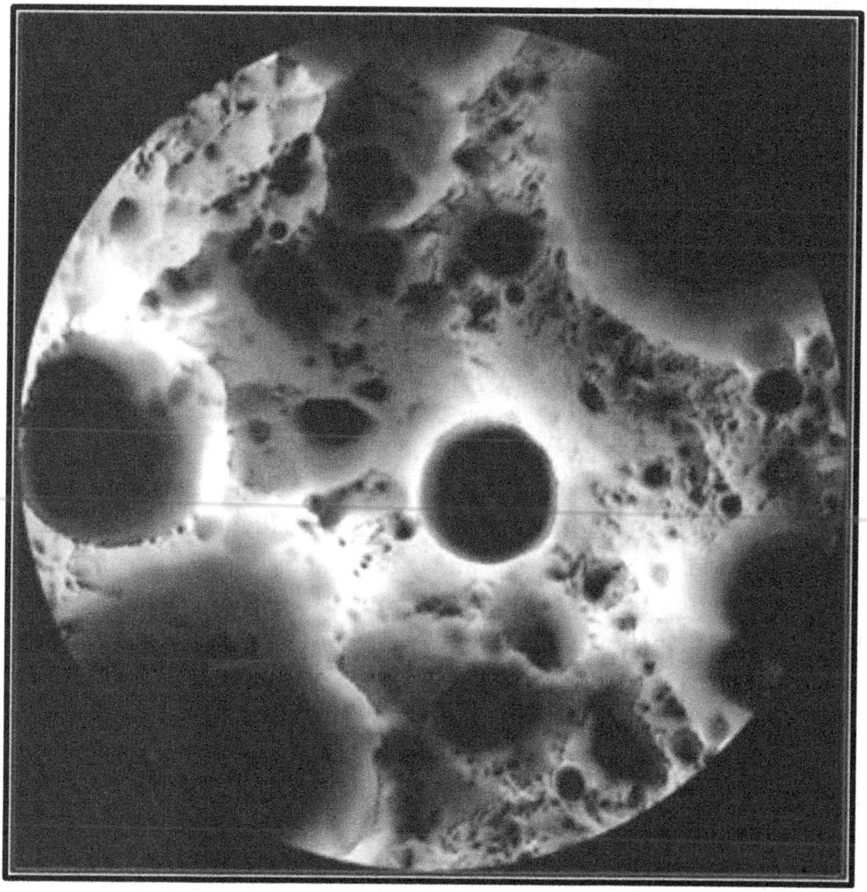

A Photo from Lunar Orbit of the Proposed Landing Site for the Next Human Mission to the Southern Pole of Luna

ONLY THE MOST NAÏVE INDIVIDUAL WOULD CLING TO the Luddite notion that humanity's future is fixed to the surface of Earth. It is certainly not. Humanity's future is not found on Earth, and it is certainly not in low-Earth orbit. We are already a space faring species – at least at this moment – even while we are shackled by irrational leadership and management of space exploration planning and the current wholly deficient exploration philosophy limited to excursions beneath the Van Allen shield. But once our exploration model is corrected and repaired (as is inevitable) and implements the First Principle of Human Exploration, we will harvest a vast majority of our natural resources and energy from outside Earth's sphere. Likewise, the greatest and most powerful of all human societies will be built far from our global shores. Embedded in this simple paragraph, dear reader, is a not-so-subtle hint of a certain future:

The first nation that corrects the exploration paradigm will rule space
– and everything else.

You see, space exploration is and will always be all about strategic necessity – because that *a priori* necessity assures our culture's survival. It was the not-so-well-kept secret of Apollo. Humankind ventured to another word for the very first time out of strategic necessity. In reality, our space exploration program is not primarily about science and never was. While the fact that strategic necessity is actually the foundation for our space endeavors may annoy and concern some people, it is the truth nonetheless. Science is secondary to the strategic necessity. Why? Because if any nation cannot first seize the strategic imperative, a scientific opportunity will not be available to them.

How can I state that with such absolute certainty? Because it is profoundly basic. If I may be permitted to return to the screen door on a submarine analogy, it would be the same as declaring that the first nation that replaces the screen doors on their submarines with a sealed pressure hull will rule the seas. Fundamentally repairing our exploration model is likewise so outrageously simple that it should embarrass the human species – as the manner in which we now fly in space is so intellectually challenged, it would shame our Neanderthal ancestors (to put it as kindly as possible). But the supremely humiliating part of it all is that we not only persist, but we stubbornly insist on it, and then aggressively defend the wholly irrational and inevitably doomed false path. We are the rubes in our galactic neighborhood who insist on either exploring space on the turnip truck, or we do not explore at all!

Assuming, however, that we will somehow snap out of our knuckle-dragging stupor, there are immediate strategic benefits for the nation or the private enterprise that first changes the status quo and overhauls the paradigm, creating a Solar System Class First Principle human space exploration system.

Unlimited Resources

Figure 1

As mentioned earlier, there is a single asteroid, 16 Psyche, that has an estimated value of $10,000 quadrillion. That is about 100,000 times the current annual gross world product (GWP)[1] Today, 16 Psyche is not within our reach because of our obsolete space exploration system, the deleterious impact of prolonged exposure to cosmic radiation, and the toxic effects of long-term immersion in microgravity. Yet this amazing resource is simply two relatively easy fixes away. The first nation or private enterprise that makes those fixes will seize that prize. That single asteroid alone is a strategic game-changer – and it is only one of perhaps millions of other such celestial objects available for harvest. Due to 16 Psyche's inestimable value and equally immeasurable immediate importance, NASA is already preparing a spacecraft (shown in Figure 1 above) to examine it up close. The Arizona State University-led NASA Psyche Mission is slated for launch in August of 2022 and scheduled to arrive at the asteroid in early 2026 after a 2023 slingshot gravitational assist from Mars. In orbit for 21 months around 16 Psyche, it will collect data about her true nature and may be able to determine whether she is the motherload that she is now suspected to be.[2]

Can you imagine expecting any crew to fly out to that object and begin to harvest it while fighting off the crippling effects of microgravity and chronic exposure to cosmic radiation? No advanced space faring species would even consider that level of insanity as a going-in plan. And the nation or private entity who actually plans to bring that prize home will certainly not!

While governments tend to be idiotically stubborn to the point of madness, we are now witnessing a new breed of private enterprise that focuses on profits and enterprise building. And, if success is on the line, they will protect their greatest asset to assure the victory is won. No one in their right mind will waste a cent of corporate cash or a second of company time on an exploration model that has proven that the crew/workforce will be sick or injured when they reach their objective – much less be capable of effectively working in that environment and maintain any hope of one day returning to Earth! Imagine trying to recruit a competent crew based on the certainty of debilitating, permanent injury. And then envision the litigation disaster upon return (shades of Ripley and her grueling 3 ½-hour testimony before the Company Board – only the Company will be the one being grilled this time).

16 Psyche, again, is only one of millions of potential space resources. Another asteroid we have already visited and examined up close, 433 Eros, contains more gold than has ever been extracted on Earth.[3] But harvesting both of these asteroids requires long space voyages and establishing mining facilities far from Earth. In his landmark book *The High Frontier*, the late American physicist and space activist Gerard K. O'Neill advocated mining our own Moon and using mass drivers on its surface to launch unprocessed lunar regolith into lunar orbit. There the regolith would be processed by big, mostly automated systems and would churn out sheets of metals for use in building space colonies, space vessels, and, most importantly, for cosmic radiation shielding without having to launch all that mass from the bottom of Earth's

gravitational well. These machines would be serviced by nearby teams of astronaut crews living in orbiting colonies that are spun to create artificial gravity and are fully shielded from GCR.[4]

The First Vital Extraterrestrial Strategic Resource Claim

The new, profoundly serious human space race was officially announced on 26 March 2020 by former Vice President Mike Pence at the fifth meeting of the National Space Council held at the Marshall Space Flight Center in Huntsville, Alabama. Pence called for NASA to return humans to the Moon by 2024. He stated that "NASA already knows that the lunar south pole holds great scientific, economic, and strategic value, but now it is time to commit to go there. Today, the National Space Council will recommend that when the first astronauts return to the lunar surface that they will take their first steps on the Moon's south pole." Pence also declared, "Make no mistake about it, we're in a space race today, just as we were in the 1960s. The first woman and the next man on the moon will both be American astronauts, launched by American rockets, from American soil."[5]

Apparently, no reading between the lines is necessary. The race is on.

As was discussed in the previous chapter in this volume titled "SLS-Orion – NASA's Dysfunctional Stepchild on Steroids," the obsolete although conditionally functional Apollo-on-steroids architecture of SLS-Orion is fully capable of making such a journey to Luna. Under the current mission parameters, SLS-Orion is slated for a very short exposure to microgravity and GCR during the initial crewed Artemis Exploration Mission III as they follow in the footprints of the Apollo crews of half a century earlier.

Again, when the Vice President announced the Lunar landing site, he was speaking of the most strategically critical point on the entire celestial world – the southern Lunar pole. Why? It is the only place on the body that has what appears from orbit to be relatively easily accessible water-ice deposits at the surface, protectively concealed in the eternal darkness of craters, hidden by the permanent Lunar polar shadows. Yet its crater walls are illuminated by the relatively easy harvest of eternal sunlight. Only on the Lunar southern pole are these specific conditions available.[6]

The Unlimited Reach of the High Ground

As the best example from history available, the primary reason that initiated the original 1960s panic-driven space race between the United States and the Soviet Union was specifically to preserve the militarily strategic high ground. From the very first days of warfare, gaining the high ground ensured success in any battle. The worst possible position in any conflict was to be fired down upon from a high place. It is an almost indefensible position. When the Russians launched Sputnik in 1957 (the first satellite to orbit Earth), in a single stroke, they gained the strategic high ground, and the race was on, for the prevailing public fear was that if they could launch a satellite, they could also orbit a nuclear bomb and drop it whenever they wished anywhere on Earth. While we may have developed a combination of historical amnesia or succumbed to a precarious historic revisionist naivety, we may certainly and very confidently recall that the Apollo program in all of its historic glory was, in fact, a quasi-military effort to preserve the ultimate strategic high ground at any cost – a presence on the surface of Luna. As the Soviets failed to achieve that high ground and ceased their attempts, Apollo was canceled because the program's strategic necessity was made irrelevant.

"High Ground is not only more agreement and salubrious, but more convenient from a military point of view; low ground is not only damp and unhealthy, but also disadvantageous for fighting."

- Sun Tzu *The Art of War*

But the idea that space or our Moon can be used as a military platform to achieve the strategic high ground was not invented in the Soviet Union. The high ground was a central theme of the magnum opus of human warfare, Sun Tzu's *The Art of War*, penned in the 5th century BC.[7] And the idea was expanded brilliantly to space in science fiction writer Robert Heinlein's 1965 award winning novel *The Moon is a Harsh Mistress*. In Heinlein's premise, Lunar colonists broke free from the government of Earth by strategically catapulting moon rocks (which landed with more kinetic energy than an exploding atomic weapon) down onto the home planet.

While there have always been space treaties and international declarations pontificating to the contrary, the nation that commands deep space also simultaneously commands the strategic high ground. And it is simplistic, wishful thinking to actually trust in treaties and international declarations, which are almost universally broken and disregarded when push comes to shove. The only assurance of peace has historically been through parity of strength and strategic equality.

Unlimited Buffer – Unlimited Control

There is a raft of space treaties floating about that consist of lofty, legal language (best read aloud with a pronounced nasal twang and said appendage held slightly higher than average) which all proclaim the resources and domains of space are the exclusive property of all mankind and are not only to be equally distributed to all upon their return to Earth, but that beyond the orbit of the home planet, there will be no ownership or sovereignty. Interesting. Look around you and see how that works out day-to-day here on the home planet. Such treaties are not just nonsensical academic jurisprudence, they also repudiate basic human nature and the entire sweep of human history.

This reality was not lost on a handful of United States capitalists who convinced their representatives to pass the Space Act of 2015 – a wholesale revision of the original Space Act of 1958, which established NASA. In the Space Act of 2015, the law allows U.S. space exploration and mining companies full legal rights to own and sell all the harvested natural resources that they can mine and return to Earth from asteroids and any other body in solar orbit.[8]

Some scholars and legal experts assert that this U.S. law violates space treaties already in place that claim that these resources belong to all mankind.[9] But if history is any precedence, the law opens the door for U.S. mining concerns to have the financial incentive to go out, mine the resources, and sell them in the USA to cover their costs and make a profit, as all capitalist interests are designed to do. This 2015 law is the perfect illustration that pie-in-the-sky treaties do not hold water when it actually comes time to seize the high ground and the astonishing resources available there for real.

The fundamental, irrefutable expression of all human power is this: Treaties and agreements between nations are *only as valid as the capacity to enforce them.*

Enforcing treaties that are projected through legal imposition on unsigned sovereign nations – such as future planetary, Moon, or space colonies – will initially require sanctions and, ultimately, military intervention if the sanctions are ineffective and Earth desires or requires their resource streams. But the stark reality is that such colonies will be hundreds of millions of miles distant, on moving planets that can easily calculate launch windows, and just as easily track incoming space vessels who cannot zig-zag to avoid offensively launched assaults.

The United States developed into a world power today for a single, primary, overriding geographic reality: the distance between North America and Europe provided a strategic buffer between the U.S. and European antagonist states. It was just too expensive for nation states in the 18th and 19th centuries to launch attacks across a wide ocean. This strategic buffer allowed the United States to expand both geographically and in the accumulation of natural recourses. Such a buffer gave the U.S. virtually unrestrained control of its destiny. Even the semi-intelligent among us who may or may not be really paying any attention will clearly understand that this reality defines the future of those who will conquer the high grounds of deep space with their First Principle ships, habitats, and hardware.

Unlimited Potential

The very moment the exploration model is corrected, interplanetary space will become the fully exploited domain of humankind. The corporations and nations who are the first to build and launch advanced, First Principle Solar System Class ships and architectures will rule and reign. The possibilities are staggering indeed – from inconceivable resources and new expansive territories to settle to unique and manifestly powerful interplanetary global and nation-states yet unborn. The potential of the space frontier is not just great, it is, in fact, greater than we have *ever imagined*. And it opens to us the very moment the lightbulb finally flickers on above the head of the astonishingly stubborn human and the new exploration model is adopted.

It cannot be overstated – the nation or private corporation who corrects the exploration model first will literally seize command of the space frontier and the ripe, first low-hanging fruits of the available resources on out to infinity.

What We Risk

"The Devil is in the details, but so is salvation."

- Hyman Rickover

"It is useless to attempt to reason a man out of a thing he was never reasoned into."

- Jonathan Swift

Saturday, 1 February 2003
Merritt Island, Florida

IT WAS MY DAY OFF, AND I TOOK MY SATURDAYS VERY seriously. As my first NASA mentor, Ed Pottmeyer, would say with such characteristic eloquence, "When I drive off the rocket ranch on Friday afternoon, I stop thinking about everything and leave it at my desk. If I ever forget something, it would be useless to try and return because I wouldn't remember how to find my way back!" Ed was an old-school pad-rat pioneer from Gemini, Apollo, Skylab, and Shuttle. He knew each of the pre-shuttle astronauts on a first name basis and met them as they arrived on what he called "his pad." He was the undisputed expert on how to live the NASA life off center and on. And once Ed reported back to work, he owned his professional sphere. After 30 years of reporting to duty on NASA soil, I found no one who lived the NASA life better than Ed Pottmeyer – no one.

On that Saturday morning, I was in my Ed Pottmeyer mode for real, driving on the Merritt Island Causeway, when my cell phone rang just after 9:00 a.m. It was my daughter Katy. She blurted breathlessly, "Daddy, they lost the space shuttle!"

I was very aware of the STS-107 mission, being an active participant in the payload design and processing efforts, and knew the crew was returning to Earth that morning. My Institutional Animal Care and Use Committee had reviewed all of the mission life sciences payloads onboard – and there were more than a handful. Each of these payloads was outfitted for launch from our suite of laboratories at Hanger L on the Cape Canaveral Air Force Station, and a couple of the STS-107 astronauts had been to the Hangar in the weeks before launch to be briefed on their payload responsibilities. I knew the ship and her crew were scheduled to land at the KSC runway just about eight miles north of where I was then located, and that sometime today, the life sciences payloads would be transported to our waiting teams at the Hangar. While I was the Animal Care and Use Committee Chairman, I did not have any offloading and payload processing responsibilities that day.

But Katy was most insistent that they had lost the Space Shuttle Columbia. "Katy, sweetheart," I assured her, "you can't just *lose* a space shuttle! Where did you hear this? Did you read this on the internet?"

"It's on every channel!" she replied, and then her voice broke. "Daddy, they're showing it coming down over Texas in pieces!"

It is not possible to describe what I felt at that moment. My first thought was of the Columbia astronaut who had dropped by the hangar for a briefing just a couple of weeks prior. My hair was literally standing up on my neck. Then I thought of the teams who were at that moment waiting to receive their payloads at Hangar L.

"Katy," I said, "I'm sorry, I have to go now, sweetie. Thank you for letting me know. I love you."

I looked over to Claudia. "I have to get out to the Hangar right now!" She nodded and lay her hand reassuringly atop my leg. But as I drove her back home, my mind was shrouded in disbelief. Losing a space shuttle and her crew was something so unthinkable to those of us who supported crewed spaceflight that it was an event that was so far off the radar that it lay squarely in the "not-even-possible" category.

I had spent countless hours inside and around Columbia, the flagship of the United States Space Shuttle Fleet. She was the first shuttle – the old girl with all the first-to-come-out debutant glory and all the stories of been-there-done-that-first. She was also the first of the shuttle fleet I had climbed inside and sat on her seats. To me, her inside bulkheads seemed to carry more rubs, miscellaneous patchworks of Velcro, and general signs of wear than the other shuttles. I always felt that each shuttle had its own distinct odors. At one point, I believed that if I were led to the OVs (orbital vehicle) blindfolded, I could still tell you which shuttle I was in.

My first assignment as a NASA Space Vehicle Safety Specialist was at the newly opened Orbiter Maintenance and Repair Facility (OMRF) which later became Orbiter Processing Facility (OPF) 3. For its very first use (a couple of years before they installed work stands that wrapped around and serviced the orbiters), they rolled Columbia inside the OMRF. I will never forget the day that I opened the door to the bay and saw Space Shuttle Columbia positioned in the middle of the cavernous bay with no work stands ruining the view. I could see her standing there alone, bathed in brilliant white light and in her literal glory. As a space nerd from childhood, it was truly breathtaking. I remember reminding myself that this was my job. They were actually paying me for doing this. Seriously, I would have paid them just to stand there like that! Like a kid staring at the Christmas tree, I could happily stand there all day. No one was going to challenge the NASA Safety Specialist assigned to this facility and, now for a couple of weeks, assigned to Columbia!

As I drove Claudia back home, the idea that this beautiful ship and her crew were gone – lost at the very edge of space – seemed impossible. The waking nightmare was slowly settling in, and I struggled with the emotions that were about to overwhelm me.

The moment we arrived home, I switched on our television while I dressed for the office. On the screen were the horrendous views of Columbia's return to Earth as a cluster of flaming meteors pointed due east over southeast Texas, right along her flight path back to the Kennedy Space Center. Claudia must have witnessed my struggle and came to the rescue with a sweet embrace, pulling my eyes away from the screen.

The reporter then announced the words spoken by NASA Flight Director Leroy Cain to the Mission Control Room in Houston the moment communications were lost from the orbiter. "Lock the doors," he commanded soberly.

I clung to Claudia for a long time, trying to fathom the feelings the families of the Columbia crew must have been enduring at that very moment. NASA is a family in every sense. And the worst possible tragedy for the family is the loss of a crew.

Claudia was a child living in Cocoa, Florida in 1967, her father a NASA engineer, when the Apollo 1 crew was lost to a fire on the launch pad at Complex 34. She could still describe the tears and feelings of anguish in their home decades before. Then the Challenger was lost in 1986. And now the tragedy was being repeated for the third time. The

NASA family had lost another crew. I knew that I had to get out to the Hangar as soon as I could. But, at that moment, I had no idea of the complicated tangle that lay just ahead of me.

"I should be back by lunch," I stated optimistically and kissed her goodbye. I was seriously wrong.

I drove with all due diligence (speeding) to my office some 18 miles east. Somehow missing a talk with the Brevard County Sherriff Department's finest deputies in their green and white squad cars, I pulled inside the fence at Hanger L on the Cape Canaveral Air Force Station, just across the fence from the Kennedy Space Center. The lot was jammed. That was no surprise considering the number of life sciences payloads that were to be delivered and, obviously, because of the unfolding disaster.

As I walked into the crowded confines of the adjoining Hanger, "Little-L," I ran directly into my boss, Charley Quincy. "Thank you for coming, Dennis," he said with a tight smile. "I'm going to need an update on the animal payload status whenever you can pull it together."

Charley Quincy was a fantastic boss in an unbroken line of such world class managers I was privileged to be assigned to at the Kennedy Space Center. My first boss, Dr. Bill Knott, was incredible, first class, and amazing. Yet even after Bill, they just kept getting better.

Charley was a trim, intelligent, and engaged supervisor who listened well and made decisions easily. He involved himself in his employees' lives as much as they wished but was not prone to any degree of overbearing behavior. He was quick to smile and even smiled in anticipation of a pleasant conversation. He was trusting, defended his people at every opportunity, and was deeply involved in our career building.

I nodded and turned to go find Ramona Bober, Kennedy Space Center's Chief Animal Care Technician. I quickly walked outside over to Hangar "Big-L" just across the parking lot, around the massive Biomass Production Chamber, and into the KSC Animal Care Facility. Ramona was standing in the hall, staring into a closed laboratory full of control animals for the Life Sciences payloads aboard Columbia.

Beginning years prior to this event, I had claimed that Ramona Bober was a "single point failure" in the United States Space Life Sciences program. She was a contract employee and as valuable and irreplaceable as any NASA team member. If we lost Ramona for any reason, I was certain that we could never fully recover. She was not just excellent – she was exceptional, world-class superb. No one on the planet knew the workings of an animal care facility better or more comprehensively than Ramona. She was a good-natured individual with a quick smile and a focused intellect. She left no doubt that she was an all-in team member but that the animal care facility was her baby. And that made me incredibly happy.

Ramona looked at me with her signature kind, calm expression then moved her eyes over to the laboratory in front of us. "See anything unusual?" she asked.

I looked into the laboratory and saw nothing out of place. "No..." I replied quizzically.

She then pointed to the small, round hasp clips in the upper right quadrant of the door. And there, looped though the hasp, was a wire sealed with a lead disk. The door was restrained by an official seal.

I was incredulous. "Who did this?"

"Security," she said. "I came in after it was already done."

"We need to get these animals out of these laboratories," I stated uselessly.

"Well," she replied with a flat smile, "*you're* the NASA guy..."

As I looked down the long hallway, every door was sealed with wire and lead. Behind many of them were live animals, but it was absolutely illegal under federal law to touch, remove, or tamper with any of the seals. When Flight Director Leroy Cain said, "Lock the doors" at Shuttle Mission Control in Houston, he was speaking to the entire Agency. "Lock everything." I was witnessing the results of that directive firsthand. It certainly had not taken very long.

Anything that flew onboard Columbia was now officially suspect until it could be cleared. All emails, written notes, or lab reports were now locked and protected against deletions or changes. At that moment, everything pertaining to the final mission of the Space Shuttle Columbia was now embargoed. The formal investigation into Columbia's loss began the moment the communications were lost.

Whoever put the seals on the doors was either unaware that there were live animals inside, or they did not care. The standing order Agency wide was to lock everything down hard. And they did. Even the laboratories with live animals inside of them!

I thought about the situation for a few seconds and then spoke the obvious, "I'm going to have to go directly to the Center Director... like... now..."

"I guess that's why they pay you the big bucks," Ramona replied wryly, obviously not at all envying my imminent trek up to the mountaintop on an extraordinarily problematic day.

I was not at all exaggerating the gravity of the moment. At the Kennedy Space Center, the Animal Care and Use Committee was not founded by NASA policy and directives but by Federal Law. The Animal Welfare Act of 1966 established strict governmental controls over the use of laboratory animals in government sponsored research activities. From it evolved the Health Research Extension Act of 1985. That Act engendered the *Guide for the Care and Use of Laboratory Animals* which sets the standards for care and housing that must be provided for animals in government funded activities. In this public law, it directed that the head of each institution was solely responsible for the adherence to the law, and that the ultimate responsibility could not be delegated. It did allow for delegation of duties, and, as such, I was delegated by the KSC Center Director the responsibility of Chairman of the KSC Institutional Animal Care and Use Committee and for carrying out and coordinating its activities. But the ultimate legal responsibility for the welfare of all research animals fell squarely on the shoulders of the Center Director – alone.

On one occasion prior to this, Charley Quincy's supervisor asked him what it was that Dennis Chamberland did every day. Charley replied in a single sentence, "He keeps the Center Director out of jail."

It was apparently a satisfactory answer.

I quickly returned to Little-L from the Animal Care laboratories and found Charley. I gave him "the look" as I attempted to formulate the requirement into as few words as possible. He patiently waited.

I informed him that the laboratories had all been sealed with the animals still inside and then waded through the torturous legal requirements as quickly as I could before finally laying it out. "We have to go see the Center Director, now." I must have presented my declaration with all the emotional impact of, "Help me, Obi-Wan Kenobi. You're my only hope!"

Charley obviously immediately understood the impact of a demand to see the CD on this of all days in the history of Kennedy Space Center. But he also clearly grasped our dilemma and the compromised legal position of the CD in all of it. He merely nodded, picked up his cell phone, and dialed his supervisor. I may have had a legitimate, legally protected straight-line access to the CD by virtue of my delegation, but

I would never have used it without notifying the customary chain of command – and neither would Charley. We always dutifully checked the boxes on the way up. We received our clearance and made the drive to the KSC Headquarters Building across the Banana River.

The Kennedy Space Center's Headquarters was full on this Saturday morning. To no one's surprise, it appeared far more crowded than on a normal working day. Outside, there were at least half a dozen news vans with raised antennas sitting helter-skelter everywhere. We boarded a crowded elevator up to the fourth-floor offices of the Center Director's senior staff and made our way along the packed hallways to the CD's office, which was also jammed. We hesitantly stepped up to the desk, and his Executive Assistant, Mrs. Jane S. Kleinschmidt (whom I also knew from my civilian life), greeted me in her consummate professionally calm and controlled way and said, "What can I do for you, Dennis?"

"We have to speak to Mr. Bridges about an urgent laboratory situation."

"Can it wait until next week?" she asked dutifully.

"No," Charley replied frankly. "Someone sealed live laboratory animals in the laboratories, and we are not permitted to break the seals."

"Oh, dear," she sighed. "I can get you inside in a few minutes, but promise that you'll be as brief as possible."

"Of course."

We nervously waited in front of her desk until the door opened, and out walked United States Senator Bill Nelson. "You're next," she said to us with a weary smile.

Inside we were surprised that Center Director Roy Bridges was not there. Instead, sitting in his place was one of the KSC senior managers. "Sorry, but Mr. Bridges is scheduled somewhere else right now," he explained. We then reviewed the laboratory situation for him as quickly as we could, placing particular emphasis on the CD's ultimate legal responsibility. He thereupon picked up the phone and called NASA Headquarters in Washington DC. When he hung up the phone, he stated, "They're working it. We can't do anything from here. Removing seals is above anyone's paygrade here." Charley gave him his cell phone number, and we left the towering madhouse for the relative sanity of the cape-side Hangar.

As soon as I returned, I briefed Ramona and then took stock of the life science payload situation. As far as laboratory animals, we had three cohorts onboard Columbia. But there were at least four other life science teams with their research loaded onboard also. Some of the research team members looked dazed. Others had tears in their eyes. Not only did we lose a crew of astronauts and Columbia, but some individuals lost years of research preparation for this rare flight opportunity. The disaster had many facets, and there were countless disparate reasons to grieve on that tragic Saturday.

After sunset, I was still at Hanger L waiting for clearance to release the animals. About 6:30 p.m., the call finally came. But we were not to touch the seals. A security team was scheduled to arrive, remove the seals, witness the removal of only the animal enclosures, and then reseal the laboratories directly under their watchful eyes.

As I waited for the arrival of the Security team, memories came to me woven together in my mind like a heartbreaking tapestry of ill-fitted, disjointed pieces. The Columbia astronaut I had seen from a distance in Hanger L what seemed to me like just days before. The recollection of sitting so many hours onboard the Columbia shuttle herself, admiring and memorizing every inch of her otherworldly interior. The tears of the research teams who had just lost years of their life's work. The jammed office of the Center Director. The sense of heartbreak, loss, and catastrophe that settled like a fog everywhere. It was monumentally depressing.

I walked back to the Animal Care Laboratory alone and looked at the animals in their enclosures through the windows of the sealed doors. In a few days, I would be involved in the surgical removal of the tracking chips between their scapula as they were finally dispositioned. I could not help but reflexively think about their matched sets that flew into space and wound up ejected from Columbia into the near vacuum at the edge of space at the end of her final orbit. There was no other possible avenue leading from that thought but to the fate of the Columbia crew flying back home after nearly 16 days on orbit.

I stood outside the main surgical suite and looked inside at the gleaming stainless-steel tables as I considered these things. Then I recalled that it was in this very laboratory that the remains of the Challenger astronauts were brought for their government autopsies. That last thought was way too much for my already overloaded emotions. I finally found a quiet, private place and openly wept for the tragedy that had swept upon us all out of the Saturday morning twilight.

The Lessons that Must Not Be Re-Learned

I recall the words from the designer of what easily could have been the most dangerous vehicle ever built – the nuclear submarine. Admiral Hyman Rickover revealed his mindset in his design constraints before the initial nuclear submarine's first bolt was turned. He demanded stringent safety requirements and a powerful focus on quality standards to protect the crews. When once asked why, he responded, "I have a son. I love my son. I want everything that I do to be so safe that I would be happy to have my son operating it."

Notice that Admiral Rickover did not give a tinker's damn about the heroics of narcissistic submariners who wished to star aboard his submarines. As a matter of fact, he individually interviewed every officer and made a personal decision about their fitness just to weed out these kinds of narcissistic aces. He cared solely about his chief design criteria: building a vessel that was so safe that he could openly trust the most precious thing in his world to its safety and reliability – his own son. Thus, Rickover unknowingly designed with the First Principle philosophical code:

"In every design activity, we will protect the human as a primary objective."

Yet today Rickover's common-sense design philosophy has been replaced outside the U.S. Navy by the wholly ridiculous tough-guy-ready-to-adapt-or-die creed. Today we are holding hands and skipping merrily down the path toward the imminent loss of yet another human space crew. Arrogantly, we even openly advertise our collective, quite astonishing level of engineering incompetence. And based on that incompetence, we have given future crews over in advance to injury and death, expecting them to adapt or die per our cultural messaging. Here are some examples from published works of that unfortunate message which seems indelibly fixed in the human space exploration ether:

Said one NASA astronaut about the short voyages to the Moon, "It's going to be hard. There's a good chance of death. Going in a little can through deep space, it's a very harsh environment. We think you can come back, but nothing is certain."[1]

One article is titled "Boiling Blood and Radiation: 5 Ways Mars Can Kill."[2]

And another, "Elon Musk's plan for sending people to Mars is probably a suicide mission."[3]

Or, how about this one: "Mars mission astronauts must 'prepare for non-survivable illnesses or injuries'"?[4]

And then there is this one that effectively summarizes it all: "NASA Can't Ethically Send Astronauts on One-Way Missions to Deep Space."[5]

Remember when someone finally asked the Academy of Medicine (AOM) for their opinion? AOM took a careful look at the deep space programs then replied, "The committee finds relaxing (or liberalizing) current health standards to allow for specific long duration and exploration missions to be ethically unacceptable."[6]

I will use the context of this book to go on record. Anyone who has not personally lost a family member, a friend, or a colleague in a tragedy on a crewed space mission – anyone who has not seen and felt the tears and lived with the tragedy – should not dare to lecture any of us who actually lived these things and worked with the heroes about risk and courage. They should not use their platform to advocate for more tragedy because they do not have the simple intellect and fundamental human courage to take the time to correct the exploration model so that tragedy and injury are not integral parts. While it is true that science is often changed one funeral at a time, then let it be *their* funeral, and do not enthusiastically advocate the learning of the lessons in deep space burning up other people's lives.

Obviously, NASA is a federal agency and has already demonstrated that they can and will ultimately do what they damn well please, simply writing themselves reams of waivers as they go. Private space ventures have even fewer restrictions and can fly just about anybody to their deaths or long-term injury just for the thrills and a paid ticket. The only restriction to such private ventures is the inevitable after-the-catastrophe avalanche of survivor lawsuits and the unavoidable new laws that will protect subsequent waves of travelers off-planet. Based on the incredible volumes of information available on the certainty of the risks today, they will not be able to convince any jury that they proceeded with integrity and without prior knowledge of impending mission disaster. A single unfortunate mission could well bankrupt any company. And no hold-harmless waiver will ever be airtight enough, as history has clearly shown in American jurisprudence. With the kind and volume of evidence available today, in the end, the civil suits will inevitably cascade into well-deserved wrongful death criminal penalties as well.

Here the handwringing begins in earnest, with wails of, "Now what do we do? It's all over. We are trapped on Earth forever!"

Seriously? Is that how dumbed down our immediate gratification addicted species has become? Let us take a deep breath, close our eyes, and just *get a grip*.

What are all these people fretting about? The number one worry for long duration exploration missions is exposure to cosmic radiation. The second worry is prolonged exposure to microgravity. The third worry is how these toxic environments synergize to injure and kill crews with greater efficiency. The fourth worry is about a rocket explosion or crash. The final worry is packing a crew with clueless, self-centered dweebs who will all kill each other before making it to Mars into a relatively small space from which no one can escape.

Five worries – four solutions. Unfortunately, even with the best laid plans of mice and men, there is no solution to the occasional explosion or crash. That is truly an act of God, save the infrequent engineering misstep or incompetence. But that particular worry is perhaps less than the probability of death on the nation's highways. The other four concerns we have complete and total control over – even packing the can with personality challenged idiots.

Somehow, we have convinced ourselves that the old Apollo human exploration model is the only one that will ever work, having God's and Neil Armstrong's certified stamp of approval. In the next two constructs of this book, we will begin to unwind our error in detail and re-evolve our human space exploration model... for the genius of Admiral Hyman Rickover. For the mind of Dr. Richard Feynman. For the 21 United States astronauts and Russian cosmonauts who have died strapped to the seats of their spacecraft trying to figure all this out.

What do we owe these of our species who have done so much to leave us with their genius? What do we give to future generations so that we can move forward on the trail of human history, the way illuminated by the brilliant legacy of our scientific lineage? We owe them a short introspective breather in that history. We owe them a pause to get our bearings right this time – before, not after, tragedy. We are obligated to live out the courage to stop for just a moment and apply our deepest genius to correct our model. Is it really going to be our legacy that we left for deep space with the philosophy "Take chances or stay home" as we willingly committed so many individuals to spacecraft that we already knew were inadequate and obsolete?

Or will we leave the legacy of brilliant human foresight prefaced by an equally brilliant analysis of the need to apply a significant course correction on the way out to the permanent human conquest of our solar system and beyond? That is what we risk. The integrity of the journey, and the history by which we shall be remembered, and our generation of explorers are judged.

We get to decide. Will it be idiocy, tears, regret, and typical stupid-human-disgrace that forces the inevitable change? Or will we earn all the intelligence we have already convinced ourselves we possess by exploring with brilliance, courage, and the ability to learn from our mistakes and recognize the more than obvious blind alleys?

We get to decide.

But the clock is ticking.

Why We Will Depart Earth Forever

Reason 4

"It is strange to be known so universally and yet to be so lonely."

— Albert Einstein

ONE MORNING IN 2001, I RECEIVED A CALL FROM MY GOOD friend and fellow aquanaut, Mark E. Ward. Mark was many things – a man who wore many hats. One of them was being the publicist for the late Mercury Astronaut Scott Carpenter – the first astronaut-aquanaut in history. Mark asked if I would be able to pick Scott and his wife Patty up at the Orlando International Airport. Of course, I replied, it would not be any problem at all.

I had met Scott 16 years earlier during an interview I conducted for an article I was writing for the U.S. Naval Institute *Proceedings* about the splashdowns of American spacecraft returning to Earth. We had become friends, and I visited with Scott and Patty at their Aspen, Colorado, home. We collaborated on the *Scott Carpenter Space Analog Station*, I had the privilege of teaming with him on a professional paper, and I was honored when Scott wrote the Foreword for my Mars colony sci-fi/action-adventure book *Abyss of Elysium* – revised and republished as *Mars Base One*. Claudia and I often lunched with Scott on his visits to the Kennedy Space Center. I never met any astronaut that had Scott Carpenter's professional bearing and kindness toward his fans. And, as for me, Scott literally never turned me down when I needed experienced advice. I believe him to be one of the major contributors to the success of my NASA career.

Scott and Patty arrived in Orlando on schedule, and we headed due east on the Beachline toward Cocoa Beach, where Scott was meeting with the surviving Mercury Astronauts for their annual get-together. The remaining four of the seven original astronauts – Scott, John Glenn, Wally Schirra, and Gordon Cooper – were in attendance. Gus Grissom was lost in the Apollo 1 fire, Deke Slayton passed in 1993, and Alan Shephard joined them in 1998.

On the way to their hotel, I asked Scott if he could please do me a favor, to which he agreed. I asked if he could kindly get the autographs of his fellow original astronauts in my copy of the book *We Seven*, one of my prized possessions. The book was published in 1962 and was written in part by each of the original Mercury 7 astronauts. I had read it in 1962 at age 11, the very moment it hit our library shelves. Then I had checked it out and read it again. And then I read it *again*. Eventually, I owned my own copy.

"I'll take care of that, Dennis!" Patty replied immediately as Claudia handed her the book. A couple of weeks later, my book arrived in the mail from Aspen, signed by Scott, John Glenn, Wally Schirra, and Gordon Cooper. "Gordo" Cooper passed away just three years later in 2004, and Wally Schirra in 2007. My dear friend Scott died in October 2013, and John Glenn joined them in 2016.

They are all gone now.

Just after Scott passed, I got up from my desk across from the towering Vehicle Assembly Building (VAB) at Launch Complex 39 and drove out to Launch Complex 14 on the Cape from which all but Shepherd and Grissom had launched. The steel had long since been torn down. The salt air from the ocean adjacent to the launch pad eats steel away like so much liquid cancer. Thus, the pad was dismantled for safety reasons.

Leading into the pad complex is a 13-foot-tall stainless steel monument on which the Latin phrase "*Si monumentum requiris circumspice*" is engraved, which, when translated, reads, "If you would seek my monument, look around you." The Project Mercury symbol (a 7 outlined by the symbol for the planet Mercury) stands over the plaque. As the inscription states with such eloquence, the eternal monument to the Mercury 7 astronauts is defined not by plaques and symbols, but by their courage to be among the first humans in space, as is evidenced by the place from which they departed Earth at Complex 14.

As I reverently stood there alone at Complex 14, I thought long and hard about the one astronaut that I had personally known from among them and all the grace and kindness he had shown me. I had always understood that he had absolutely no obligation to do so. And I knew that he had done so just because he was that kind of a man. He was an authentic explorer, a man of true honor. His great personal courage was matched only by his gentle compassion for everyone else – extraordinary traits with which so few humans have been blessed. It was just who Scott Carpenter was. So, as I stood there, I prayed that Patty would have peace, and I gave thanks to Scott's Creator that He would make such a man at all – and, in His infinite grace, would allow our paths to cross.

But in the early 60s, these seven were not just dauntless explorers, they were like modern-day movie stars, as famous as anyone had ever been. With their selection as the first American astronauts, they inherited something.

The fourth reason why many will depart Earth forever is:

Fame and Fortune

The only astronauts that have exceeded the fame of the original seven Mercury astronauts were Apollo 11 astronauts Neil Armstrong, Buzz Aldrin, and Michael Collins. Some fame was granted, to a lesser extent, to the other 21 "Lunar Crew" Apollo astronauts. From the hundreds of astronauts since Apollo, none of them ever came close to any level of fame and fortune except as conferred by the rather nameless label of "astronaut." And, with the advent of tourist flights which will bestow that label on thousands in the next decade, even that distinction will be forever lost.

However…

There is a new breed of astronaut that is just now emerging in our culture – the Interplanetary Astronaut. Another window of historical fame and fortune is about to open for the first crews bound for Mars – especially that extraordinarily fortunate ONE who will be the first human on another planet.

At some top level intellectual thought process, true explorers who are far too focused on what they are doing often complain that they have no need to be distracted by aspects of fame. They know very well that the media circus is ultimately constructed for a cash flow advantage for the great media empires, who contract to share a slice of it with the explorer, on occasion. Every astronaut and explorer I have ever met steadfastly denies that fame has ever been a factor.

Of course, all these explorers and astronauts are such liars. (LOL)

These folks are human, and nearly all of them instinctively crave attention and cash flow. It is just too simple to miss – or to deny. So, there it is – the mice and the cheese. Fame and fortune are still indelible, undeniable parts of the early 21st-century astronaut experience.

Further, astronauts are also generally held in the instinctive awe of "famous" people and are socially bestowed with an added allure because of their fame. If that may seem outrageous, you clearly do not know many astronauts very well. Hence, regardless of whether one is a nameless, anonymous astronaut to the world at large, in their own private circles, they very much remain rock stars with several layers of additional attraction to the culture. Astronauts are generally considered to be of significantly above-average intelligence and are thought to have extraordinary abilities and equally amazing life adventures. In reality, much of this is actually overblown hype, often on the edge of absurdity, which *almost* every astronaut will be quick to assure you. But they also appreciate that this persona is relatively fixed in the cultural mind.

Fame and fortune. It was and will always be one of the most powerful motivators to the human imagination and is fixated as a major component of every culture on Earth. Now it is about to go off-planet…

Module
Three
Ch.01
Ch.02
Ch.06
Ch.07
Ch.08
Ch.10
Ch.11
Ch.12
Coopting an
Alien Mind

The Power of Knowing Ourselves

"And this is the best that you, the government, the U.S. government could come up with? I mean, you're NASA for crying out loud. You put a man on the moon. You're geniuses! You're the guys that're thinking shit up! I'm sure you got a team of men sitting around somewhere right now just thinking shit up and somebody backing them up! You're telling me you don't have a backup plan?
That's what you're telling me?"

- Harry Stamper
(Bruce Willis – *Armageddon*)

"To know thyself is the beginning of wisdom."
- Socrates

CORRECTING OUR EXPLORATION MODEL REQUIRES THAT we first recognize that it needs correcting. Understanding the necessity to correct it and precisely why entails a further knowledge of how we engineer that change by an unencumbered decision-making process. And finally, knowing how to direct and forge change requires a philosophical foundation to pin it to. Therefore, what we primarily require as a species that dares to expand into the cosmos is a fundamentally new way of thinking and decision making. We need to co-opt a far different, comparatively alien mind and its attendant thought processes. Fortunately, we, and our civilization, already have this full capacity embedded in our genes. Our task today is to recognize that our chief limitation is self – our ignorance of the viable process and how we have exchanged other faulty practices that obscure the powerful ones that actually accomplish meaningful advances.

As one example, let us examine the core subject of this book: taking even one additional step into interplanetary space at this point without a major review and alteration of the exploration model is not just dangerous, it would reveal the greatest human frailty – the inability to know the difference between essential, immediate change and a frivolous reorganizing activity that merely covers weakness and confusion. Often attributed to the Roman voluptuary Petronius Arbiter from the reign of Nero is the phrase, "We trained hard – but it seemed that every time we were beginning to form up into teams we were reorganized. I was to learn later in life that we tend to meet any new situation by reorganizing, and what a wonderful method it can be for creating the illusion of progress while actually producing confusion, inefficiency, and demoralization."

Petronius Arbiter described a common human attribute expressed in established groups of people loosely defined as organizations. Any organization will do. Here, Arbiter describes the process of human organizational breakdown that results from an abject loss of any discernable philosophical foundation which ultimately elicits this behavior. As humans, we have great difficulty understanding the wide difference between appropriate change and illusory change because of a central reflexive failure within the human psyche – the innate difficulty to achieve *nosce te ipsum.* That is, simply to authentically know ourselves and why we behave the way we do – individually or in groups. Without fully understanding ourselves, our collective behavior will nearly always dissolve into U.S. Congress-like actions – conflicted, disorganized, monumentally wasteful, and, ultimately, dangerous and reckless.

This is so because human beings are sentient, homeostasis seeking biochemical machines that fundamentally maneuver between a pure, albeit simplistic, sentience and an existential reality animated by a rather strange, convoluted process of reasoning, wholly energized by pseudological organically derived constructs that sometimes resemble logic. But through even the most simplistic of observations, it can never be said that human beings have a Spock-like ability to wield true, emotionless logic. Ours is actually a process of pseudological decision-making – a process that even we have difficulty partially understanding. That is because it is layered, more often than not, with an inconsistent, often contradictory confusion of logic laced through and through with emotional irrelevance. And none of it is anchored by a reasoned philosophy – personal or corporate. We have developed atomic weapons and deep space travel while retaining all the organizational awareness of our distant Neanderthal cave dwelling ancestors.

As if to probe the depth of our intellectual shortcomings, we have convinced ourselves in glowing terms how gifted our species is. Indeed, we have difficulty understanding why, if it exists, the local galactic group of intelligent societies has not yet invited us to join their league of civilizations. Said one internet meme, "I'm sure the aliens lock their doors when flying past Earth."

At the same time, our capacity to justify truly insane, illogical choices across the globe is utterly self-evident to the point of being legendary. This perfectly illustrates the intransigent, blinding pseudologic we employ in our everyday decision making. We have a self-image that does not at all match our behavior as a species. And yet we have deliberately blinded ourselves to that more than evident truth, which explains our truly bizarre patterns of self-destructive space exploration that, if flown in space hardware developed under an Apollo legacy exploration philosophy, can only lead to failure, mass injury, and death. And yet, as a species, we are enthusiastically lining up by the millions to volunteer as meat for this great cosmic cook-out.

Without a shadow of a doubt, the most essential academic course that can be taught to any genuinely educated people is the study of human philosophy, which gives our self-aware species the tools to actually know ourselves and understand our processes of individual and common thought. And yet philosophy has become mostly an elective academic course of study and has, in fact, fallen out of favor in many scientific disciplines.[1] Indeed, many scientists hold a degree as a Doctor of Philosophy (PhD) having never completed a single course in their core academic degree field – philosophy! Literally, we have unleashed a virtual limitless tsunami of knowledge on the world and have given ourselves not a single tool to understand or to guide its rational use. Thus, when the exploration model breaks down so spectacularly (as it has), such a level of degreed philosophical ignorance becomes strikingly apparent and should be alarming for many reasons totally apart from its immediate and inevitable downstream effects on human space exploration and settlement.

The truth is, we all act out our philosophical constructs every hour of each day, whether we are consciously aware of them or not. That is why the study of philosophy (the organized, disciplined evaluation of human reason) is so essential. It teaches us how to use the tools of thinking and reasoning, and to understand the basis of our actions and belief constructs rather than just to react and respond. And when that understanding breaks down or (as it is today) is entirely absent, we begin to act in inconsistent, contradictory, and truly irrational ways. Without a clear and unambiguous grasp of our patterns of thinking, and without the comprehensive tools to manage our decision-making processes, we devolve into the reasoning of momentarily expedient, politically biased thoughts and other human reflexive patterns of thinking. Literally, we respond to the immediate environment and stimuli without an application of a more comprehensive allocation of logic based on our philosophical foundation. If the immediate stimuli are more complex than our philosophical tools, then our reasoning will inevitably break down. Such is the power of the silent philosophy. Such is the potential power of truly knowing ourselves. And such is the root cause of the breakdown of our human space exploration model.

The Deadly Results of the Invisible Philosophy

In 30 years of professional employment in the life sciences and aerospace engineering, I never once sat in any meeting that discussed the fundamental philosophic reasons for doing what we were engaged in.

Not one time in 30 years.

While there were absurdly superficial requirements to box-check mission statements and phrase-lift mission goals for PowerPoint chart making, it was generally assumed that somewhere deeper than these government boiler-plate aphorisms, there *must* be an animating philosophy that fundamentally energized it all. It was certainly there, of course, embedded invisibly in the background, almost surreptitiously in the undercurrent of what we were doing, but I never ever heard it discussed even once in three decades. And yet, the underlying, transparent philosophy formed the basis of all that we did and determined the placement of each nut and bolt on every crewed spacecraft.

And thus, in the absence of a known, clearly communicated, and thorough understanding of the actual underlying philosophy of crewed space exploration, humans reflexively substitute something in its place, because the human chain of intellectual logic – even pseudologic – demands it. I have heard this, or pieces of this, voiced hundreds of times in my career – on the hangar floor beneath the space shuttles; inside the space shuttles as a Spaceflight Vehicle Safety Specialist; and even during an occasional discussion with an astronaut. It was spoken in many voices, but I always heard the voice of William Shatner asserting the guiding ethos:

"Space: the final frontier. Our continuing mission: to explore strange new worlds, to seek out new life and new civilizations, to boldly go where no one has gone before."

And thus it is that our fundamental philosophy was taken from a 1960s-era television show and became a part of the NASA psyche. It resonated with us because it was all we had and because it was a part of my generation's psychological pedigree. It literally defined our philosophy of human space exploration. But as you reflect on that reality, there was a serious disconnect between *Star Trek* space exploration and another core philosophy of the 1960s.

Human pseudologic thrives in complexity – even two-dimensional complexity. While the *Star Trek* crews flew about the galaxy in one gravity and never even spoke of or had any concern about the space radiation environment outside the walls of the Starship Enterprise, the humans back on the real Earth had inherited, and then expanded on, an accompanying exploration philosophy based wholly on an 18th-19th century exploration paradigm: adaptation to extreme environments. And so it was that we were going to explore strange new worlds and boldly go wherever we wished, but we were going to adapt or die in the process because that is what rugged explorers have always done! We routinely send pilots up in untested aircraft to see if the vehicles are going to fall apart (as they frequently do), and everyone knows the risks. Thus, when the occasional pilot or crew does not come home, it is part of the underlying acceptable social philosophical bargain forged in generations of world wars.

As if to seal this unspoken but motivating system of beliefs, test pilots were selected for the first classes of astronauts specifically because of their capacity to adapt to extreme environments and situations. And they were selected because the suite of contracted risks was acceptable to them, and the nation would not grind to a halt when a test pilot was lost doing their duty... that is until they became astronauts. And then, the application of the social philosophical contract changed. No one knew why then, and no one knows why today. It just changed.

This selection criterion made perfect sense in April of 1959 when NASA chose the original seven astronauts for Project Mercury. But the underlying, unspoken paradigm (the implicit strategic mission to enhance national security) also drove the design of the spacecraft. In the 1960s, the expectation of human adaptation was also the cheapest, fastest way to the Moon – and it was indeed a hurry-up-and-get-to-the-Moon-first race laid at the feet of brave but expendable test pilots. The Moon mission morphed into the perfect storm of circumstances, and its brilliant, historic success sealed its exploration model for generations to follow. Indeed, even to this day, the Apollo model of human space exploration (and all its expectations) is the only one in existence. And it is driven by the underlying expectation of its undeclared genesis: unlimited adaptation and extreme risk taking as we boldly go where no one has gone before.

That was also the high-risk, full adaptation paradigm evidenced by the early 20th-century barn storming aviators who flew in open cockpits in their sketchy flying machines. But we quickly changed that standard to make aviation a commercial venture modeled as low risk, with full adaptation to one Earth-normal. Can you imagine flying cross country in an open cockpit today? Picking bugs out of your teeth and enduring extreme temperatures would certainly not be positive criteria for an acceptable business plan for a 21st-century airline. Grandmother is going to be very unhappy when and if they can get her out of the cockpit.

But NASA is openly advertising that there are "no appropriate workarounds" to the 1960s Mercury-Gemini-Apollo model; therefore such a standard cannot be accomplished in the space environment in the 21st century.[2] It is fascinating (and a bit tragic) that the Agency that flew Americans first to the Moon can no longer manage the innovations necessary to improve on their original barn storming model for space exploration and has openly advertised the next step as "impossible."

Unfortunately, NASA's philosophical model has reached its endpoint even while we ramp up for a dramatic expansion of our deep space exploration push out and away from Earth for good. But our goal in no way matches our failing paradigm. Human adaptation to the space environment has reached its absolute limit. And yet, even as we continue on with the formerly brilliantly successful Apollo model, we apparently do not possess the philosophic tools to identify our own illogical and dangerous behaviors. We still have no clues to understanding our pseudological processes.

But now, just before NASA orders the body bags (or, as on Everest, decides to allow the corpses to remain on Mars as too expensive to pick up and return home), we must stop the multi-billion dollar train so that we may come to an understanding of the philosophy that actually drives us, and therein create a new one that will fit the real-time data. It will be exceedingly difficult, to be sure. But only then can we carry on, taking a considerably more enlightened path – one that is actually openly articulated and clearly understood by all. A new path that does not have to be forged after the catastrophic collapse of the old. Envision it as a well understood philosophy that is so perfectly disseminated that it is used as a part of the engineering and mission planning process itself. Imagine, if you will, a future where spacecraft and exploration systems are designed by a well-known and clearly articulated energizing exploration philosophy.

As each second ticks away, as each dollar is now wasted on the antiquated, deadly Apollo model, humanity comes ever closer to disaster in deep space and risks yet another very human pseudological conclusion – that exploring deep space was never practical in the first palace. We are on the very doorstep of determining that the next step is not possible, as did Jacques Cousteau and the United States Navy, who abandoned the possibility of achieving a permanent undersea presence in the 1970s. Why did they reach that conclusion? Because they were stuck inside their blind alley of exploration and stubbornly refused to back out, admit error, and redesign. If we do not correct our pathway now, we may end up stubbornly responding as a culture to the inevitable tragedy that lies just ahead by declaring that human exploration and the settlement of deep space is not possible at all just as they did.

It is way past time that we finally achieve *nosce te ipsum* – simply, to come to know ourselves. For, in truth, nothing less than the destiny of the human race and all our future empires is at stake.

Familiarizing the Unknown

"Human beings are curious creatures, driven along and ahead along their life streams by deep psychological forces that they are often blissfully unaware of. Humans and their organizations are quite often thus manipulated by their own subconscious biases that, for the most part, they will vehemently deny even exist. This primal reality alone is responsible for a vast cohort of human individual, cultural, and organized systemic failures."

- Dennis Chamberland

UNDERSTANDING SELF IS ONE OF THE MOST BASIC OF all human survival instincts, and yet it always seems powerfully and conspicuously elusive. Thus, the idea has been recognized since antiquity as a fundamental human philosophy. As we briefly discussed in the previous chapter, the term "know thyself" (*nosce te ipsum*) was one of the Delphic maxims and, according to the Greek writer Pausanias, was inscribed in the forecourt of the Temple of Apollo at Delphi.[1]

Most people indeed struggle to truly know their individual selves. But as we knit our lives together into our broad and nearly infinitely complex social tapestries, knowing ourselves as a part of a social construct is daunting for anyone. Giambattista Vico (1668-1744), an 18th-century Italian philosopher, is well remembered for his truly seminal work, first published in 1725 as *Scienza Nuova* (*New Science*), in which he establishes some central principles intrinsic to human culture and, in his timeless thesis, reveals an account of the stages common to the development of all societies defined by the pure science of their complex philosophic frameworks.[2] Although the full title of the work is staggering in itself (*Principles of a New Science Concerning the Nature of Nations, through Which Are Recovered the Principles of Another System of the Natural Law of the People*), Vico was able in just two fundamental steps to summarize the vast power in first knowing our psychological construct, which *must* be antecedent to actually knowing ourselves.

Thus, Vico reveals a critical weakness in the organization of human awareness which arises from the fact that "man makes himself the measure of all things" (Axiom I, §120, p.60) and that "whenever men can form no idea of distant and unknown things, they judge them by what is familiar and at hand" (Axiom II, §122, p.60). In two brilliant steps, Vico nails down the Jamestown Effect that cost so many lives in one of the earliest American colonies, and the innate capacity of humans to project the familiar upon the alien shores of every new world they encounter.

The philosopher clearly defines such a projection as not an asset at all. Instead, he describes it as a powerful human *limitation*. But Vico teaches the antidote to such human weakness by clarifying that any observation of the unfamiliar can only be known by the evidence of their origins, or *per caussas* (through causes). He states, "Doctrines must take their beginning from that of the matters of which they treat" (Axiom CVI, §314, p.92), not by attempting to familiarize them.

Further recognizing this process of subconscious corruption of truth (already endemic in 18th-century science), Vico called for a continuous purging of the body of scientific truth by the pure application of the scientific process itself. Notice that Vico's early cure of an already corrupt scientific establishment leveraged its remedy via a thorough understanding of its primal philosophic construct. Thus, the false insights that had crept into the scientific method that were based on the human proclivity to project a full understanding by a process of human biases were exposed so that they could never gain a foothold. Said Vico, "It is one great labor of...Science to recover...[the] grounds of truth – truth which, with the passage of years and the changes in language and customs, has come down to us enveloped in falsehood" (Axiom XVI, §150, pp.64–5).

According to the philosopher, stripping away and revealing falsehoods and preconceived biases is the *only* available path to true scientific wisdom. And perpetuating falsehood by rote tradition and overlay of the new onto the old systems of thought is not science, but superstition in the name of science. I have seen the process of rote tradition play out its ugly hand in the laboratory and in the spacecraft processing bays year after year. Such human weaknesses and systemic failures were ultimately blamed for the loss of both the Space Shuttles

Challenger and Columbia, along with their 14 precious, irreplaceable souls.

Embodiment of The Jamestown Effect

As we look backward in time over four centuries, even before the birth of Giambattista Vico, such a process of "whenever men can form no idea of distant and unknown things, they judge them by what is familiar and at hand" was a major part of the planning and development of the distant, ill fated Virginia colony.

As I stated in a video that I created about the disaster for the Wings of Eagles Discovery Center Mars Base Eagle exhibit, "Humans have this unfortunate tendency to project unrealistic expectations on alien places – like the early settlers at Jamestown, Virginia, did in the 17th century. The British settlers moved to the new colony of Virginia, located thousands of miles across the Atlantic Ocean, fully expecting that they were moving a part of familiar Britain along with them. Unfortunately, they weren't just wrong, but tragically so. At the end of the first few years, more than 80% of the colonists had died of disease, starvation, and freezing to death. Ultimately, the colony descended into cannibalism."

What caused the settlers to project civilized Britain on Virginia and then prepare accordingly? As I stated in the video, "The colonists made a lot of poor decisions even before they departed England – and then continued to do things wrong based on their misunderstanding of the alien place called Virginia once they arrived. They failed because of their arrogance, which fueled their incompetence. Because they arrived at the New World expecting England, they failed to close their life support system, and famine was the inevitable result. But their most serious failure was their total lack of understanding and awareness of the limits of their adaptation to an alien place."

The colony planners presciently played out Vico's insight 61 years before his birth by their actions: "whenever men can form no idea of distant and unknown things, they judge them by what is familiar and at hand." I call it the **Jamestown Effect.**

The Jamestown Effect on Luna and Mars

It is astonishing to watch the Jamestown Effect unraveling itself once more in real time today as we make plans and build hardware to colonize Luna (the Moon) and Mars. I recently spoke with an experienced and well-known NASA astronaut who stated that they often communicate with people "how much Mars is like the Earth." That astronaut is fully encouraged by the Agency to speak this line to the public. For example, from a NASA website, students are taught:

"Mars is similar to Earth in many ways, having many of the same 'systems' that characterize our home world. Like Earth, Mars has an atmosphere, a hydrosphere, a cryosphere, and a lithosphere. In other words, Mars has systems of air, water, ice, and geology that all interact to produce the Martian environment."[3]

While all of those statements are factual, the fundamental presentation is utterly misleading and, for future colonization, it is dangerous. The rest of the online article is fixed on "finding past signs of life on Mars." This search implies that Mars is or has been so much like Earth that the probability is high that traces of past life will inevitably be discovered on the Red Planet. Not a single sentence was given to the absolutely alien-like characteristics of Mars or the difficulty in establishing a human colony specifically because of these characteristics. And not a solitary paragraph was devoted to the severe environment of Mars that renders its surface continuously sterile through many mechanisms. Not one phrase was devoted to the fact that the alien characteristics far, far outweigh the Earth-like nature of Mars. Such facts do not at all make it impossible to settle human colonies on Mars – but the philosophy of the presentation is skewed, clearly illustrating Vico's warning to the 21st century:

"Man makes himself the measure of all things" (Axiom I, §120, p.60) and that "whenever men can form no idea of distant and unknown things, they judge them by what is familiar and at hand" (Axiom II, §122, p.60).

Currently, the SpaceX private enterprise is actively planning human missions to Mars and shows so much promise that it (if its success continues) will probably land the first humans on the Red Planet. Unfortunately, at this date, SpaceX is following NASA's example by using the Apollo exploration model. Yet, any exploration venture to Mars that continues on this course and does not alter its exploration philosophy and model is likely to either permanently injure most of its participants or fail at the cost of every crew member. The model is independent of nation or agency, so this is a critique of the failure of humanity to advance and perpetuate disaster, not a critique or criticism of any entity.

Why is the model destined for failure? There are four overriding reasons:

(1) Those who employ it are thumbing their noses in the face of an active cosmic radiation field consisting of the most formidable, unrelenting power stream in the universe in the bizarre and quite illogical expectation that humans can "adapt and take it,"

(2) They are absolutely ignoring the data returned from the International Space Station that six to nine months of exposure to microgravity is destructive to the human body,

(3) None of them have even considered the effects of multimorbidity when combining extended exposure to galactic cosmic radiation and microgravity, and finally,

(4) Neither the government nor private enterprise literally has any clue of the fundamental requirements of advanced life support systems with limited resupply. They have not worked out the reliability of complex carbon dioxide removal systems, in situ oxygen production systems, or how to provide the energy required to run everything necessary. If they are to survive, they must change their exploration model. At this point, the probability of success without significant permanent injury and fatalities is so low it approaches the odds of the straight pick-four lotto.

Each of the four reasons mentioned above has viable solutions. The fact is no one needs to die on the way to settle Mars. Not a single astronaut needs to be crippled or inevitably face cancer or cardiovascular disease later in their life. But it starts with rejecting the concept that we, our biases, our prejudices, and our pathetically limited knowledge base are the actual measure of all things. With this knowledge of self, with a sure cognitive purpose and deliberation, we can thus utterly reject the idea that Earth can be successfully overlaid on Mars.

That is where we must begin. But starting down this road will require that we first back up and change our path – after we discover the philosophical basis for our most elementary patterns of reasoning.

The question is, do we possess the intellectual wherewithal to make that happen without relying on the dysfunctional tragedy-before-correction model we have always favored before?

The Mind of Hyman Rickover

"Optimism and stupidity are nearly synonymous."

"As a guide to engineering ethics, I should like to commend to you a liberal adaptation of the injunction contained in the oath of Hippocrates that the professional man do nothing that will harm his client."

- Hyman Rickover

WHEN SEEKING TO ACHIEVE THE ADVANTAGES OF AN alien mind in order to escape the tragic and often self-destructive effects of the social hivemind of the past, it is always instructive to look to individuals who have exhibited evidence of such uniquely alien and highly effective thinking. One such individual is United States Navy Admiral Hyman G. Rickover.

In the spring of 1984, I was a fresh-faced, new-hire United States Navy Civilian Nuclear Engineer assigned to the Mare Island Naval Shipyard in Vallejo, California, located on the edge of San Francisco Bay. I had spent six years in the U.S. Navy, and now, after completing my graduate studies in Bioenvironmental Engineering at Oklahoma State University, I found myself once more a part of the U.S. Navy fold.

Having the background of an experienced naval officer, I immediately sensed that working as a civil servant in the Navy was in many ways much like working as a shore based active-duty officer. In some respects, it was almost indistinguishable. Thus, I felt very much at home from the very first day at Mare Island.

Our task was to maintain radiological controls during the overhaul and repair of the U.S. Navy Pacific Fleet's nuclear submarines. In a nutshell, our job was to ensure that all nuclear materials and ionizing radiation, no matter how small, were kept securely inside the controlled radiation zones. We were overseen by the U.S. Navy Naval Sea Systems Command, simply called NAVSEA. And NAVSEA's Radiological Controls was an office created by Admiral Hyman G. Rickover.

Admiral Hyman G. Rickover is known as the father of the nuclear submarine. He was its creator, chief architect, father, mother, and conscience. And not a single individual in the United States Navy, in Congress, or in the White House would dare argue with that fact. Said President Richard Nixon at the ceremony awarding Admiral Rickover his fourth star, "the greatness of the American military service, and particularly the greatness of the Navy, is symbolized in this ceremony today, because this man, who is controversial, this man, who comes up with unorthodox ideas, did not become submerged by the bureaucracy, because once genius is submerged by bureaucracy, a nation is doomed to mediocrity."[1] And President Jimmy Carter was so personally influenced by the Admiral early in his Navy career that he titled his 1976 book *Why Not the Best?* after a pointed question personally poised to him by Rickover in a job interview onboard one of the Admiral's submarines.

I never encountered a single Naval officer in my years with the Navy that did not know about the legendary Admiral. It did not escape a solitary individual with any knowledge of naval history that this one extraordinary leader singlehandedly masterminded and personally directed the construction of one of the most complex and important inventions in human history – the nuclear submarine. That this man is a certified legend in U.S. history escapes no one. He served for 63 years as a Naval Officer only because Congress intervened and allowed him an exception beyond the typical retirement age. He served under 13 Presidents, from Woodrow Wilson to Ronald Reagan. His *magnum opus* was the creation of the first Nuclear Submarine in history, the USS *Nautilus* – SSN-571. Its keel was laid in June 1952, and it was launched in a mere 19 months. 30 years later, almost to the day, my supervisor at Mare Island Naval Shipyard called me into his office.

The Keys to the Kingdom

I was no stranger to the occasional odd assignment. In the active duty Navy, as in the civil service, they were called "Duties as Assigned." This term was mentioned in the last line of the professional job description as a catch-all so that no matter what your superiors ordered you to do, it was covered in the Duties as Assigned caveat. As I walked into my boss's office, I sensed one was coming.

It was fast and unceremonious. He stood up, reached into his desk drawer, pulled out a silver key on a ring, and handed it across his desk to me saying, "You are being assigned to the 571."

I was more than somewhat confused at being assigned to a sub. First, I had not yet been trained for a sea trials assignment. And this duty could certainly not be an assignment to a submarine crew, which civil servants are never assigned. And I had never seen a set of orders presented in the form of a key. He must have read the clear fog of confusion and bewilderment on my face.

"I'm sorry, Dennis," he said. "You are the new Radiological Control (radcon) Engineer for the refurb of the *Nautilus*. She's your baby now. I thought you might like that assignment." He added a sly wink because he owed me a huge career favor, and now he was in the payback mode.

"The *Nautilus*? SSN-571… *THE Nautilus*?" I stammered. "I thought she was decommissioned long ago…"

"Yes, three or four years ago."

"And she's here? At Mare Island?" I exclaimed, struggling to keep from actually drooling. I gripped the key tightly like he was going to change his mind and take it back. My geek genes kicked in, and I started to actually tremble. "May I go see her… now?"

"She's being converted to a museum. They've created a standup entryway on her deck with stairs leading belowdecks. That key opens the door on her deck. Maybe you can wait until tomorrow when I can arrange for a tour."

"What's my job, sir?"

"Well, whenever they call for radcon controls, you'll be the guy. At this point, they've almost wrapped it up. But somebody has to be the POC (point of contact), and now that's you!"

I stared at him like he had just handed me the Medal of Honor. "Oh my God," was all I could manage to breathlessly stammer.

"That's what I thought you would say," he beamed. "She's all yours, Dennis. Have at it," he added, sitting back down and resuming his work.

I did not return to my desk. Instead, I walked out the front door in a daze, repeatedly mumbling my boss' key words, "She's all yours, Dennis… she's all yours, Dennis…" He had just handed me the keys to the very first nuclear submarine – the USS *Nautilus*!

I walked directly over to the duty office, where I slyly looked at the status board to discover the sub's position. And there, marked by a magnet, was the location of SSN-571. I moved quickly back down to the row of piers and found her alone at the end of a relatively isolated dock, sitting unceremoniously tied up with thick lines and an open gangway. On her deck, I could see an oddly out of place glassed-in enclosure outfitted with a standard glass entry door. I could hear the boss' words in my head: "You might want to wait until tomorrow when I can arrange for a tour." I guess he did not realize he was speaking to the Nuclear Engineer ADHD poster child.

But surely, it would be okay if I just walked up on her deck.... After all, I was now officially the USS *Nautilus'* newly assigned Radiological Control Engineer. I paced up to her gangway and faced a sign which read in bold letters, "No Admittance. Controlled Area." I immediately reasoned that this sign was for anyone who did not actually have a key to the boat. In a surreal moment, I inserted the key into the completely out of place aluminum framed door on the deck of the nuclear sub and entered. Locking the door behind me, I turned and stared down a set of OSHA compliant, yellow and black striped steps leading belowdecks.

It was just weird. No submarine sailor had *ever* accessed *any* submarine by a flight of standard stairs complete with a handrail. But the *Nautilus* was being converted into a tour boat – and rightly so. It was one of the most important and historic artifacts in the United States Navy. A true treasure and heritage of the American people and their genius – in this case, as breathed into reality by the mind and formidable will of Hyman G. Rickover.

I had been onboard operational submarines before: a WWII-class diesel boat (SS), nuclear fast attack (SSN) boats, and a nuclear ballistic missile (SSBN) submarine. At one point in my naval career, I had orders to join the crew of the nuclear-powered USS *Puffer* (SSN-652) before another assignment intervened. But, of course, none of those subs were boarded by OSHA approved stairs. And none of them were completely empty when I came onboard. I was acutely aware of my solitude as I descended into the darkness of the *Nautilus* for the very first time.

I easily moved into the adequately lighted Officer's Wardroom. After soaking in the view, I assumed a seat at the wardroom table. As a former Naval Officer, I had sat at similar tables for the years I was an assigned officer onboard three surface ships. While this was a 1950s vintage submarine, not much had changed in terms of the government furnishings of U.S. naval vessels. And yet, I was immediately lost in the surroundings of the *Nautilus*. It was the first of its kind in history – an atomic powered vessel capable of incredible things. It split seawater to create breathable oxygen so that it could remain submerged for weeks or even months. Plus, on 3 August 1958, this boat traveled undersea beneath the Arctic ice shield to the geographic north pole. And all of it was created from the mind and the intense genius of Hyman Rickover, who accomplished what most said was not even possible.

The retrofitting crew had been assigned to wall off the various shipboard compartments in transparent plexiglass so that a visitor could walk down the passageways and clearly see inside them, but they were not able to enter. As I arrived, that level of interior work had barely begun, so most of the ship (save the after compartments, crew quarters, and engine room) had not been blocked by the plastic walls. Thus, as I found the newly installed banks of light switches, I could see that I had the majority of the empty ship open and ready for inspection, all to myself.

I was most curious about the reactor compartment, of course. But, not surprisingly, it was locked and chained shut. While the fuel was long gone, the reactor components themselves were present, and their radioactive half-lives still had decades of activity remaining, although they were more than adequately shielded from the rest of the boat. I made my way over to the attack center, settled into a portable chair with pulldown arms resting against one bulkhead, and relaxed, eyes drinking in the magnificent sight. I sat there among the dark faux wood panels and green walled submarine, stunned to my deepest core at having the privilege of being here at all. I was all alone inside the very first of history's most stunning technological achievements. Inevitably, I recalled the memory and image of the one who designed and built her: Hyman Godalia Rickover.

I sat there utterly awestruck and looked about Admiral Rickover's life's triumph. For the first time in history, this very submarine navigated some of the most inaccessible and treacherous of the world's regions, remaining submerged for weeks at a time – a feat that was impossible by all other submarines before the *Nautilus*.

But soon, I could feel Rickover's cold breath breathing down my neck. He was not one to tolerate sloth, and certainly not by a former U.S. Naval Officer or Civil Servant onboard his greatest achievement. And so, I stood, smiled, and patted my key confidently. This was not going to be my last private visit to the USS *Nautilus*! Subsequently, on very frequent visits, I let myself in and enjoyed my bagged lunch alone and perpetually astonished in the *Nautilus* wardroom. I felt very, very much at home to the core of my being, and quite comfortable with the stern spirit of Admiral Hyman Rickover reminding me when my half-hour was over and it was time to go back to work. And yet, at the same time, his memory was prompting my mind to digest where I was and what I was seeing. Even more importantly, I was prompted time and again to reflect on its mission capacities, how it operated, and how it actually functioned – foundations that coalesced into what I would later document as a First Principle exploration system.

200 Years of Tradition Unchallenged by Progress

On 13 October 1975, I celebrated the bicentennial birthday of the United States Navy on duty as an enlisted quartermaster/navigation assistant at the Pearl Harbor Naval Station. Less than a year later, on July 4, 1976, I would celebrate the nation's bicentennial birthday as a freshly minted commissioned officer onboard a ship at Naval Base San Diego. The running joke in 1975 (that ostensibly originated from the U.S. Naval Academy) was that the Navy was proudly celebrating 200 years of tradition untainted by progress. No one was more aware of this than Hyman Rickover.

By any account, Rickover was a certified genius. Therefore, by definition, he had to operate well outside the bubble of longstanding Navy tradition since his passion was radical progress and technological advance. In his multiple layers of genius, Rickover's first enabling talent was creating his own sphere of operations where he was the isolated single point of contact, thereby bypassing the crippling effects of the hivemind, groupthink committee – a standard of all government operations. In many ways, therefore, Rickover was a walking, personal skunkworks for the U.S. Navy. A skunkworks is defined as a project developed by a small, informally structured group that research and develop a project using truly innovative concepts and radical innovation. The term originated with Lockheed's World War II Skunk Works project.

As a contemporary genius to Rickover, Albert Einstein mused, "Great spirits have always encountered violent opposition from mediocre minds." Hyman G. Rickover was certainly no exception. The formal reason the selection boards gave for their deliberate snub of Rickover and passing him over for promotion twice (in 1951 and again in 1952), attempting to force his retirement even before his historic achievement was complete, was that "he was too specialized to meet the qualifications of a flag office."[2] But, the actual reason was more akin to whether or not he was going to be invited to join the very clubbish upper echelons of self-selecting Naval leadership.

However, Washington DC soon discovered that Rickover was just as adept at playing the Beltway system as he was at inventing undersea nuclear propulsion. After a few well-placed telephone calls (one presumably to the White House) and intervention by the U.S. Congress and the Secretary of the Navy, Rickover was ultimately promoted to Admiral in 1953.[3] Rickover seriously gave not one tinker's damn about jumping the chain of command to preserve Navy nuclear propulsion and his singular leadership over it.

Anyone who knew Admiral Rickover knew that the traditional Navy's actual reasons for repeatedly passing him over for promotion to flag officer had little or nothing to do with his degree of specialization. This is how one Rickover biographer described it:

"Many believed that Rickover's caustic personality was the problem, plus the fact that he didn't act like a Navy man. He rarely wore his uniform, and he once bragged that he'd never had a book of Navy regulations in his office. And he suffered fools badly. As he said to Diane Sawyer on *60 Minutes* in 1984, 'I never thought I was smart. I thought the people I dealt with were dumb. Including you.'"[4]

When it came to Hyman G. Rickover, it mattered not one iota that the Navy he worked in and around represented 200 years of tradition untainted by progress – because he applied his genius to work above it, under it, around it, and through it, however it suited him. The Navy was going to get progress, even if he had to jam it down their throats, which he did. This isolated Rickover from the organization that he depended on to achieve his historic result – the same one that he in many ways despised. But in other respects, that self-imposed isolation was part of his plan.

As I sat in the *Nautilus* wardroom on one of my many visits there, I reflected on these thoughts and what I was experiencing in the dim silence of her final makeover. At that moment, Admiral Hyman G. Rickover was still alive and retired, living in Virginia. I would never have the privilege of meeting him before he died three years later. But the genius and the fire of the irascible tyrant always lived close aboard. In some ways, he has been my inspiration – not just in his genius, but also by his extraordinary capacity just to get things done despite the system and its often-crippling traditions.

Recalling the special privilege of my assignment to the USS *Nautilus*, I would later reassemble the lessons learned from it in my next assignment. I left the profession of a U.S. Navy Civilian Nuclear Engineer to join NASA and the United States Space Shuttle Fleet at the John F. Kennedy Space Center just three years later.

What I had learned during those magic hours spent inside the *Nautilus* alone with Admiral Rickover's mystical presence was this: when comparing the USS *Nautilus* with any space shuttle orbiter, the space shuttle was decidedly inferior technologically and nowhere near as advanced as even the first nuclear submarine. Indeed, the shuttle's life support system was antiquated from the beginning and wholly inadequate to protect its crews. Unfortunately, much of the technology derived from the shuttles' Apollo-vintage life support system subsequently morphed into the life support system for the International Space Station. There were advances, to be sure. But the advances were wholly inadequate to even begin to address the U.S. space fleet's most serious and dangerous shortcomings that would soon start to systematically injure our astronaut crews.

In the next few years, I would find myself assigned as a NASA Safety Specialist at the Kennedy Space Center in Orbiter Processing Facility Bay 2. Whatever orbiter happened to pull into my Bay, I was the onsite NASA Safety Point of Contact (POC). When I arrived at KSC, Space Shuttle Challenger had been lost two years earlier, leaving Columbia, Discovery, and Atlantis. At that time, Endeavour was still under construction.

With the new rules following the Challenger disaster, safety professionals had been granted wide authority. Thus, as the NASA Safety Officer for the facility, I was able to enter any orbiter in my bay and inspect it internally and externally. I had all the KSC access "area numbers" on my badge and enjoyed full access to all the spaceflight hardware areas on the Center. As I stood inside each of these magnificent machines, I could not help on each and every occasion but compare them to the USS *Nautilus*. While the shuttles' aerodynamic capabilities made them truly incredible flying machines, in stark contrast, their life support systems were absolutely primitive.

And though I did not have the freedom or luxury to hang out inside the spaceships as much as I had in the *Nautilus*, I did have the opportunity to spend many hours in each of them performing my NASA duties. I also had the occasion to speak at length to many of their astronaut flight crews. Thus, by 1989 when I transferred to the KSC Biomedical Office as an Advanced Life Support Systems Engineer, I had amassed a particularly good background in the operational characteristics and capabilities of the space shuttle orbiters.

The Best, the Brightest, and The Clueless

Upon arrival in my new office at the Life Sciences Support Facility Hangar L on the Cape Canaveral Air Force Station in 1989 (literally just across the fence from the Kennedy Space Center), my new boss, Dr. William "Bill" Knott III, immediately let me know that he ran his office like a university campus and respected each of us as though we were professional faculty. We were in the business of developing the most advanced life support system ever devised – a bioregenerative system that integrated plants that grew food for the crew, filtered water, absorbed the humans' waste gas (carbon dioxide), and produced life-giving oxygen in return. It was monumentally complex but brilliant in its design and execution. It is without question that Dr. Knott had staffed his laboratory with the very finest NASA and contractor support scientists and engineers in the world. Had it been allowed to continue, his program would have ultimately produced support system technology that would have maintained human crews requiring minimal Earth-based resupply anywhere off the planet, in Earth orbital space, around the Sun, or on any planetary body such as the Moon, on Mars, or even a dwarf planet like Ceres. Such a system clearly had the unmistakable potential to leap generations ahead of the space shuttles, and even Rickover's marvels.

Unfortunately, The Kennedy Space Center's Bioregenerative Advanced Life Support System effort ultimately fell victim to Washington DC's presidential administration shuffles and subsequent serial agency reorganizations. By 2009, it had totally disappeared from the landscape of NASA work. Yet, as I arrived at the KSC Life Sciences Facility in 1989, the United States embarked on pouring billions of dollars into the design of the permanent human facility in space to be called Space Station Freedom.

The orbital space project was the first in a series of post-Apollo era Presidential and Congressional multi-billion-dollar legacy building blunders in human space exploration that would stretch out over decades. Said young Lieutenant Junior Grade John F. Kennedy in a private correspondence to a friend just before his PT-109 sinking, "The Navy has a supernatural capacity to screw up everything it touches."[5] That same future President, who would set into motion NASA's thrust of humankind into the new frontier by a simple vow, may have been surprised to learn that his promise would also unleash his successors and subsequent Congresses on a series of post-Apollo conflicted, stumbling, ultimately self-canceling space exploration quests that would make the United States Navy with its capacity to screw up everything it touched look like a rank amateur. Indeed, the "cancel culture" of 21st-century fame had its elemental roots in the Presidential and Congressionally designed U.S. Human Space Program.

Space Station Freedom had been announced some three years prior to my arrival at KSC by President Ronald Reagan at the State of the Union address in 1984. The incoming administration of George H.W. Bush did not interfere with its planning and execution. But after the election of President Bill Clinton, Freedom was canceled. And, in 1993, he asked NASA to create a new space station with different priorities while he invited the Russians and other nations to participate, naming it the International Space Station.

The International Space Station arose out of the ashes of Freedom with a new design and purpose – what the Clinton Administration called an international Mission to Planet Earth. While one of Freedom's major purposes was to provide a way station and assembly point for Mars expeditions,[6] the International Space Station (ISS) had no such mission. Again, the shift of Presidential Administrations ended one space direction, radically reversed itself, and more precious resources and ultimately decades of research and development were wasted on personal visions of empire. In the Clinton years, the look was back to Earth, and any discussion of crewed Mars missions was forbidden.

After the tensions between the United States and Russia began to rise inside the ISS and seriously unravel in 2021, the *Washington Post* looked backward and recalled the ISS's Frankensteinian political legacy. "The ISS was 'the product of a fairly unique moment in time when the U.S. government was looking to change the relationship it had with Russia following the collapse of the Soviet Union,' said Brian Weeden, the director of program planning at the Secure World Foundation, a think tank. The station was born 'for foreign policy reasons and to keep Soviet scientists and engineers working on space instead of selling their services to the highest bidder. It's clear those conditions have changed.'"[7] Ultimately, the Russian space agency, Roscosmos, announced on 26 July 2022 that they were pulling out of the ISS partnership altogether after 2024.

So much for an intelligently conceived space platform initially designed from the ground up for coherent science objectives. Political machinations inside both the DC Beltway and NASA headquarters effectively nuked any possibility that the ISS would become a useful platform for investigating long term human exploration and settlement of space. The fog of the death throes of the cold war had obviously also clouded the ISS's scientific mission from the outset, as would become painfully obvious as soon as humankind began to machine the ships for the inevitable expansion of our species to other worlds. The Clintonian vision was parked and securely tethered back in low-Earth orbit, and it would remain so substantially until it was finally reversed in October of 2016 by yet another political paroxysm ignited by Barak Obama in the waning days of his Administration.

After he had canceled the Shuttle's first successor, Constellation, President Obama declared, "We have set a clear goal vital to the next chapter of America's story in space: sending humans to Mars by the 2030s and returning them safely to Earth, with the ultimate ambition to one day remain there for an extended time."[8] President Trump also expressed support for NASA to explore Mars and to first focus on creating a base on the Moon. And, as a totally refreshing note, upon taking office, the new Administration of President Joseph Biden did not cancel or appreciably change any of the previous administration's human space exploration objectives – a truly unexpected outcome after the uninhibited vitriol of the election campaigns.[9] Yet, the previous decades of canceled and shifting priorities and mission goals, and constant retooling of the schizophrenic American human space vision have taken its toll at an extraordinary cost of resources, time, and squandered opportunity.

Back in 1990, at the KSC Life Sciences Facility, Bill Knott scheduled regular weekly seminars to maintain the group's academic excellence. He invited experts from around the Agency and the world to share their individual life support expertise. One afternoon, a group of life support engineers arrived from the Johnson Space Center to brief us on their design for a system that was destined for installation onboard Space Station Freedom. In the meeting, they unveiled a long, multipage list of literally more than a hundred individual air contaminants that would be continuously monitored for onboard the space station. Then they described the multimillion-dollar suite of instruments that would be required to keep track of these gasses in the parts per million and parts per billion range.

The last time I was on a nuclear submarine, to the best of my recollection, the only gasses that were continuously monitored for were oxygen, carbon dioxide, carbon monoxide, and methane. As for the rest, every submarine's air was famously saturated with the redolent blend of hydraulic fluid, food preparation, oils, various body odors, and (until 2010) heavy cigarette smoke – and that is just the short list. After a long cruise, it was impossible to get the unique smell out of your clothing regardless of how many times you washed them. With that personal experience, I looked at this team of bright-eyed experts and could not resist raising my hand.

"*Why*?" I asked.

I did not realize it at that moment, but I was about to hear the single answer that covered not just a multitude of sins, but every single sin in the United States crewed space program. It was the ultimate absolution for every crewed space exploration dilemma, uncertainty, suspicion, or funding need.

"*Microgravity,*" was the single word response to my single word question.

"*Excuse me*?" I replied.

I suppose my purposefully evocative follow-up question stemmed from the more reptilian Rickover mind embedded in my USS *Nautilus* themed mind, and thus I fully anticipated the non-verbal cues that I ultimately received. I was not disappointed. I had just revealed myself to them as the one they expected all along – the unwashed, ignorant beggar of knowledge from the learned and the keepers of all pertinent information and all judicious wisdom. In other words, I just poked the Johnson Space Center hornet's nest filled with irritable critters who never had any patience for Kennedy Space Center engineers or scientists or their inevitable dumb-ass questions.

With a rapid fire answer filled with a jumble of acronyms and "of course as you must know" interludes, the basic response was that the effects of long term microgravity were unknown on astronaut health, and therefore many resources were absolutely required to safeguard the astronaut's long term well-being. Then the holy grail of all excuses was verbalized: "This system will be required for assurance of astronaut health and *safety*."

Unfortunately, leveraging the "S" word in conversation with a professional NASA Safety Nazi is a bad idea. My respect and tolerance needle dropped all the way to zero.

Yet there is it was, and it would be repeated again and again over my tenure with NASA. Elevated above heaven, Earth, and the solar system, and now linked to health and safety, was the sacrosanct specter of microgravity – the great unknown. Cost and resources be dammed, all wisdom and knowledge would have to flow through its single portal. No one could argue this, of course, because it comprised its own *ipso facto* justification and, it goes without saying, the very reason Freedom was being built at all. And how dare I, a mere lowly former naval officer, publicly question the astronaut's very health and wellbeing? Comparing NASA with the U.S. Navy in an academic setting was apparently just fundamentally loutish.

Unfortunately, I was already wound up and sufficiently annoyed. I simply could not resist the opportunity to add, "Well, many thousands of crewmembers on nuclear submarines around the world are routinely sealed up inside for days, weeks, or months smoking cigarettes and packing along any old off-gassing item that will fit down the hatch inside their seabags. Their crews are a lot larger than the crews projected for Freedom, they remain down with sealed hatches for months at a time without surfacing, and they never reported any problems."

I knew even before I opened my mouth that the observation would be wasted. The design teams were so isolated, so incredibly well-funded and steeped in the Agency's narrative, that they were, in effect, in the end, essentially clueless to the world outside their labs. Their great space station was a hideously overpriced albatross whose design was driven by serial Presidents and politicians. It was a product of fundamentally conflicting social ideologies that had nothing whatsoever to do with its ultimate function – which shifted almost aimlessly from administration to administration. After nine years since its inception, without a single piece of space-rated hardware to show for its life, Space Station Freedom (Reagan Administration) was ultimately placed on hold after squandering billions of tax dollars. Eventually, Freedom was renamed the International Space Station (Clinton Administration). The entire debacle was completely the fault of revolving door administrations, and congressional ambivalence of waste and abuse of taxpayer resources as NASA truly did its best to keep up with ever evolving mission properties.

What the well-meaning but quite lost JSC assembly and their Agency managers distinctly lacked was the mind of Hyman Rickover. Said one of Rickover's closest aides, "He had little tolerance for mediocrity and no tolerance for stupidity. A friend of his from Chicago once stated, 'If a man is dumb, Rickover thinks he ought to be dead.'"[11] These bright, enthusiastic engineers were totally optimistic that their expensive system would work exactly as advertised. But as Rickover has said, "Optimism and stupidity are nearly synonymous."[12] Yet, if a defense may be mounted on their behalf, he is also quoted as saying, "In Greek mythology, Antaeus was a giant who was strong as long as he had contact with the earth. When he was lifted from the earth, he lost strength. So it is with engineers. They must not become isolated from the real world..."[13]

The mythological Greek goddess Aite (the deity of mistakes) provides another example of any organization that is philosophically unanchored. It follows that the manifestation of Aite's catastrophic process is triggered by a nebulous or altogether missing mission philosophy. But with such a philosophical anchor, each step of the design and mission execution playbook becomes clearly evident, exorcising mythical Aite's ruinous dynamic.

Such a systemic failure of this underlying philosophic underpinning in those responsible for carrying out the nation's future in space was illustrated by the visit of a shuttle astronaut to our bioregenerative life support systems laboratories in the mid-1990s. He listened very carefully to our description of how the crops were planted, automatically cared for, and harvested while the atmosphere was recycled, stored, buffered, and released by computer control.

At the end of our presentation, he bluntly promised, "Well, I would never do any of that."

When we asked him why, he replied that he was an astrophysicist astronaut, not a space farmer. He said, "Give me one of those potatoes and a microwave and just let me do my job."

And he was not joking. Over the decades since, I have often reminisced on his commentary in the face of the most advanced space life support system ever developed or conceived. Instead of responding in awe, he was personally annoyed. He wanted his damn potato and solitude so that he could just do his own thing – the coherent duties of each crewmember in space be totally damned. It seemed his astronaut gig was all about him. *His* mind and *his* concept of *his* job were arrogantly centered on *his* needs and *his* interests. He harbored no thought for the vast organism of the advanced life support system nor its conjoined mission that kept him alive and allowed him to do his job at the same time everyone else was doing theirs.

In the United States Navy submarine service, to be qualified as a submariner and "wear the dolphins," it is rigidly required that every sailor onboard learn something about everyone else's job and become qualified in a broad spectrum of all of them. It is all about the difference between the Navy's philosophy of teamwork and NASA's *laissez-faire* culture of personality and individual greatness. Admiral Rickover said he created his nuclear submarines to be "sailor proof" as his bottom-line design principle. Perhaps NASA should retool their astronaut program to make it "astronaut proof" as well. Mission single point failure is defined by philosophical vacuum and division of purpose.

Somewhere between the Florida launch pads and orbit, highly intelligent NASA engineers became truly isolated from the real world. Instead of building life support systems that would fully protect their human crews as the single most fundamental, going-in priority (acting therefore as the principal design philosophy), they routinely invest billions of dollars into equipment that will detect the extent of injury their life support systems will inflict on the crews, and closely monitor and record the exact moment when it happens. They are fixated on documenting injury and then going about the task of building more countermeasures to offset injury. But not a single cent is leveraged on designing, building, and launching a life support system that would actually protect the crews and thereby prevent the injuries and the need for an avalanche of countermeasures in the first place. What is missing? The philosophical foundation, which is discussed in depth in the next section of this book.

The mind of Hyman G. Rickover had a much different approach. He was going to build his exploration device (the nuclear submarine) based on a rigid exploration philosophy that protected the lives of his crew *first* as a going-in priority. His device was packed with the most highly concentrated radioactive substance ever developed, and his vessels traveled in permanently dark and bitterly cold places under extremely high pressures. So, Rickover applied the First Principle of

Exploration, not demanding that his crew "adapt" to any level of radiation exposure or pressure changes at depth. Again, he based his design philosophy on this very personal premise: "I have a son. I love my son. I want everything that I do to be so safe that I would be happy to have my son operating it. That's my fundamental rule."[14] Would Hyman Rickover have sent his son into space to face 10 times the legal limit of radiation for nuclear workers? His response probably would not be printable in this book. As for the person who asked him the question, Rickover would certainly deem them too stupid to live.

If we project that rule to NASA, it begs the same question. Would every NASA manager responsible for jacking up the deep space radiation limits continue to do so if their own children were flying? Or would they suddenly insist that ALARA be applied as it was designed and is actually used internationally – as leverage to continuously *lower* radiation dose expectations, as they have carefully documented now for well over a decade?

NASA desperately needs the mind and ethics of Hyman Rickover. NASA even more desperately needs a space vehicle version of the *Nautilus*. NASA must create a spacecraft designed to protect the crew first and foremost and not require the humans to adapt to the space environment. NASA must apply Rickover's fundamental rule – design spacecraft and exploration systems that would be so safe that any NASA manager would be happy to have their own children operating it.

If it functioned just like a *Nautilus* type craft, what would that space vehicle look like? It would shield the crew from cosmic radiation, and it would utilize artificial gravity. Therefore, just like a submarine crew, they would not be exposed long term to the space environment. And, even after a career of decade after decade of continuous spaceflight, they would return to their families as healthy as any other career professional on Earth.

Said Rickover of his profession and his approach, "It is important also to recognize that morals and ethics are not relative; they do not depend on the situation. This may be the hardest principle to follow in working to achieve goals. The ends, no matter how worthy they appear, cannot justify just any means... As a guide to engineering ethics, I should like to commend to you a liberal adaptation of the injunction contained in the oath of Hippocrates that the professional man does nothing that will harm his client. Since engineering is a profession that affects the material basis of everyone's life, there is almost always an un-consulted third party involved in any contact between the engineer and those who employ him, engineering ethics ought therefore to safeguard their interests most carefully."[15] Unfortunately, that Rickoverian ethic certainly does not apply in NASA.

Hyman Rickover and Albert Einstein on Mars

After sitting alone for hours on many occasions inside the *Nautilus*, and later spending even more time in the United States space shuttle orbiters as well as in crewed undersea habitats, I have become well accustomed to appreciating the human genius invested in them all. I have never sat in one of these steel cocoons designed for alien places and not been totally awestruck. Never, on any occasion, have I ever lost the primal emotion of wonder and reverence for what each of them enabled for our species. And at no time did I ever lose sight of the unique minds of the individual humans that made them happen regardless of the obstacles.

Now, as the Moon and Mars beckon as places for permanent human habitation that will ultimately become our greatest empires, we need to seize the genius of Hyman G. Rickover. And such genius is not to be invested in a single individual – but in many.

As someone who has invested much creative time in both NASA and the United States Navy, I will say that they have much in common. Indeed, in many ways, they are each separate tentacles of a single, vast, monolithic bureaucratic cephalopod (speaking as an Aquanaut, of course). After 42 years of combined service in the military and government, I have witnessed and been a part of several completely different versions of the massive, often byzantine enterprise. And if asked how I relate to it all in a single phrase, I would simply conjure up the image of another civil servant: Swiss Patent Clerk, Albert Einstein.

While working as a Swiss Civil Servant from 1902 to 1909, Albert Einstein wrote what is arguably the most important seminal document of modern physics: *The Special Theory of Relativity*. The question that shall always remain a mystery, particularly perhaps to the Swiss government, is how much of Einstein's civil service office time was spent developing the Special Theory? Was Einstein working on his own personal career-making boondoggle at his desk when he should have been reviewing patents, all the while on the Swiss clock? If he was ethically innocent and left his notepad at home during working hours, was Einstein's mind otherwise floating about in relativistic space during his 9-5 job? I suspect that such mind experiments, as he described his deepest thoughts, were unavoidable. Einstein himself hinted strongly at the connection. He even referred to his office there as "that worldly cloister where I hatched my most beautiful ideas."

Einstein's comment opens up a wider and most essential discussion as we envision allocating Einstein and Rickover desk space, complete with a pair of Position Descriptions, on the new worlds.

In my career with the United States government, I have occupied a pair of completely different work environments. In one of them, my boss let me know that I was a part of his "faculty" and that my task was to innovate and create new technologies that would allow mankind to be able to live on Luna and Mars bases indefinitely. It was truly one of the most exhilarating and fulfilling times of creativity in my life. He let me know many times that my job was to create the elements of a new civilization. His only caveat was to keep him out of trouble in the process.

But in another assignment years later, I was appointed to an office where I was required to sign in and out when going to the bathroom or lunch. In this job, I was obligated to produce long, tedious bureaucratic reports and contract reviews which were exposed to multiple layers of Senior Executive Service (SES) level edits where literally hours were spent in what one supervisor called "the death spiral" over the use or misuse of conjunctions and adverbs and whether or not the word under the microscope on any given page was "behaving as a conjunctive adverb or, instead, was it actually acting as an independent adverb being used to modify two independent clauses and join them together, somewhat like coordinating conjunctions? Seriously, can't you see this?" This was, of course, of monumental consequence between the two SES-ers (Senior Executive Service managers). But after all its drama, the end product – a monotonous, piteously tiresome tome, which was in the end only lightly skimmed by their intended audiences – was eventually filed away forever next to the Holy Grail in some obscure cavern guarded by the ghosts of seriously-bored-to-death Civil Servants. These few years were the most unproductive and miserable time in my government service. If it were not for my very compassionate and professional lead (who for some reason awarded me the highest performance rating in my career up to that time), I most certainly would have taken the leap out the window after the first six months. Said Hyman Rickover, quite familiar with this process, "A system under which it takes three people to check what one is doing is not control; it is systematic strangulation."

Amen. Been there and saved the noose as a souvenir.

Ultimately, thank God, the cosmic stars aligned, and I was eventually rescued by another engineering lead. She moved me to her office and overnight completely restored my career independence, professional dignity, and ultimately my full creative productivity. Often, such assignments are the luck of the draw – a result of sheer time and place probability. Asked one discouraged fellow Ensign on my first Navy ship, "What if the Navy dispatchers put all the commissioned assholes in the Navy on the same ship?" To which I instantly replied, "Yeah, *and what if we're on it*?" The double entendre was just too outrageously hilarious to forget. Later, after a couple of beers in the O'Club, we discovered that it was even funnier.

Looking back, the contrast between all the assignments in my government career was profound and extreme. And so was the useful product to the people of the United States and, ultimately, to the people of those distant worlds. As we look forward to creating that new civilization on the Red Planet, humankind can be quite certain that on the transport ships, there will be many Einsteins and Rickovers. How do we deal with the new colonists so that the products of such geniuses do not get crushed and strangled by the antiquated expectations of the 9-5 human treadmill and its attendant projected personal morality drama?

First, we must allow these geniuses room to think, to work, and to create independently. I personally worked, thrived, and created in this kind of environment – for, and inside, the United States government, no less. It is entirely possible. I do not know how Einstein generated his creative space. I always envisioned him sitting at his desk in the Swiss Patent Office and, between reviews, staring out in the aether (which he conclusively disproved sitting at that very desk, by the way) and imaging the universe as it really was before committing that vision in his head to mathematics so that the rest of us could understand and use it in our engineering systems. But that required uninterrupted time to formulate, a quiet space to pull it all together in his mind, and then to commit it to paper. Somehow, Einstein working at his desk at his office or at his home somehow made it happen – sans disruptions such as television, telephone calls, mindless, trivial assignments, or frequent drop-by visitors. All of these are as poisonous to the art of radical creativity as dumping boron into a nuclear reactor.

I am very familiar with how Rickover generated his creative space. As a U.S. Naval Officer, I was quite acquainted with the work ethic of other officers. Rickover created his space by sheer brute force. Rickover's personality was such that the Navy's often useless traditions, and even its very presence, were entirely voided out by his laser like, close-to-supernatural focus. Rickover worked legendary hours. He said, "A man, by working 24 hours a day, could multiply himself 3 times. To multiply himself more than 3 times the only recourse is to train others to take over some of his work."

But, in NASA, that approach was not so easy to come by. One colleague shared with me a story from his research office that was much like our own. He, just like we at KSC, was working strange hours. He and several of his colleagues would arrive at their desks at 5 a.m. while other peers would come in mid-morning and stay past midnight. Literally, no one knew when they came and went, except that someone was always there. When they filled out their timecards, they reported 40-hour weeks – because if they reported their actual hours (often double that), the bean counters would go off scale critical (speaking in reactor terms). They were all, to an individual, creatures who loathed the bureaucratic side of things. And since they were dutifully and unfailingly reporting all their required hours, they deemed it completely ethical and were giving the government double free labor, and often more, as a bonus!

As any seasoned career government employee might expect, one day, their blissfully innocent and quite benevolent relationship with their timecards came to an abrupt halt. It is often said that the worst thing you can do when in government employ is to even whisper you are content in any way because you will end up reassigned the very next day. As such, running an office with a rational, working plan is not allowed once it is uncovered in any government operation.

So, they were ordered to complete their timecards by writing down every minute they worked, to the minute, or they would be guilty of the federal crime of timecard fraud. With that morbid threat hanging over their heads, they did precisely what they were asked – for a single pay period. When the head office added up the overtime bill, they panicked again. Another visit from the bean counters ensued. The next order was to stop working over 40 hours per week. That resulted in a virtual collapse of productivity. They were often forced to leave their desks sometime during the day on Wednesday and not return until the following Monday as they had already completed their 40 hours. Often, they would be forced to work their weekend tests by only working Friday through Sunday. Soon, they were ordered to work a strict 9-5 schedule, five days per week. That, of course, interfered with their most efficient biological clocks and completely wrecked the ability to conduct long scientific tests and investigations, many of which were scheduled on the weekends, but they were told to suck it up.

By this time, the productivity of their laboratories was so critically off kilter that another secret and very personal meeting was held, and things silently and without fanfare went back to the original work plan. The government once again received at least double labor – 40 hours for the bean counters and the next 40 and more for the new civilizations on Mars and the Moon. They loved their jobs, and it was a certainty that no one from the head office was going to drive out to the remote laboratories at 5 a.m. – or midnight – to check up on the hopelessly driven scientists and engineers. The government got the 40 hours they paid for, and the civil servants freely donated the rest of their personal time to the future of mankind.

I relate this story for a specific reason. Embedded in it is the exceedingly powerful, pervasive morality drama I spoke of earlier in this chapter. Our culture is still quite locked into the mechanical, gadget production assembly line work ethic of the industrial revolution. But that work ethic absolutely cannot apply to the radically creative professional. That is why companies that must show a profit hire these kinds of individuals on salary.

Here is where it gets difficult for those who are still locked into the work ethic morality of the industrial revolution. Employment of a truly creative mind requires four essential elements from the employer. It begins with explicit trust, then requires space, authority, and resources.

May I repeat this?

Employment of a truly creative mind requires four essential elements from the employer. It begins with explicit trust, then requires space, authority, and resources.

This may be received as wholly and unreserved sacrilege to the old school supervisor, but the timecard process is a useless appendage and is actually an impediment to the creative mind. It is profoundly ridiculous to imagine Albert Einstein, Hyman Rickover, or Nikola Tesla tied to a timecard or a 40-hour work week. Indeed, Rickover said, "More than ambition, more than ability, it is rules that limit contribution; rules are the lowest common denominator of human behavior. They are a substitute for rational thought."

The fact is, those with truly creative minds truly never stop thinking and rarely sleep. When they do sleep, their mind is constantly waking them to surprise them with yet another new layer of unique thought. Their brain is constantly engaged regardless of where the salary comes from or even where they are. Attempting to chain this mind to a desk and a timecard is not even possible.

On Luna and on Mars, we must prepare the work environment in advance for the geniuses that will arrive with all the attendant trust, space, authority, and resources they will need. Before then, we must pray fervently for these unexpected, often hidden figures to appear out of the mists of "reality" and prepare the colonies to harvest unanticipated flashes and surges of genius from very unexpected people. Without them, without their alien minds, we may as well stay at home. For without them, the cause on other planets will be entirely lost.

The mind of Hyman G. Rickover delivered to our home planet an unexpected technological device of historic importance. Understanding his mind and preparing to develop that genius in others will unlock even more historical creativity in many Rickover's to come – in this century and beyond – on this planet and countless other words as yet unexplored.

Alien Mind

"When the writing is on the wall, closing and hiding the book will have very little impact on the outcome, neither will be walking away or shrugging the responsibility that the message is directed to others, it always takes efforts and determination to clean the wall."

- Shahenshah Hafeez Khan

"The question is always far more important than the answer."

- Elon Musk

WILLIAM S. BURROUGHS IS QUOTED AS SAYING, "MAN is an artifact designed for space travel."[1] Perhaps. But only if the human from Earth departs their home planet as something other than what we know as an Earthling. That sentient physical entity must quickly evolve to a new Homo sapiens *before* launching away from home. It may come as a shock to many readers, but the human we are familiar with does not actually belong in space – at least not before that species undergoes a dynamic metamorphosis. We cannot survive there unless we submit to a significant transformation – and that is not physically, not mentally, and not as a social unit. Something is going to have to change. A *lot* is going to have to change. But we are fully equipped with the intellectual acumen and resources to make the necessary transformation. The culturally heroic question we are now facing is this: Do we have the *will* necessary to turn the corner and make those modifications of our own intelligent volition – or will we have change forced upon us through historic scarring and epic tragedy, as is always the seemingly insurmountable human obsession?

The artifact of which Mr. Burroughs speaks is not solely the physical being of humanity, but the sum total of a radically changed being – one possessing a new mind integrated into the one Earth-standard body.

Any individual from Earth is no more suited for space travel than a fish out of water in a very literal sense. But anyone from Earth who has developed adequate engineering preparations and who has been sufficiently changed psychologically can morph into a literal new creation of their own making and become an individual exceptionally well suited for what an Earth-bound observer would reckon as an alien lifestyle. Specifically, because of the advanced capacity of at least a few of us for intelligence and intellectual adaptation, we are, as Burroughs has clearly stated, an artifact designed for space travel. But there is not merely "one small step for a human" just ahead of us. It is rather a "giant leap" that is far more than just a slogan.

Intellectually and psychologically, we are well designed to acclimatize under the right circumstances. But physiologically, we are exceedingly limited. The two adaptation processes are not at all alike, even though we have convinced ourselves that they are one and the same. This misunderstanding (or total absence of understanding) of the coordination of the two branches of the adaptation process has led to many blind alleys in exploration and to human disasters because the processes are often in open conflict. How we manage these thought processes is defined by our fundamental human space exploration philosophy. You may be surprised to discover that NASA *does* have an exploration philosophy – it is just that the Agency has no idea how to articulate it. Yet, the Agency quite blindly writes it large on its spacecraft models and in its mission designs.

As one of my fellow explorers has stated (and quite accurately so), "My mind keeps writing checks my body can't cash." This little nine-word cliché sums up the difference between the two adaptation processes nicely. Yet another powerful explorer's aphorism blends nicely into the same idea: the body cannot go where the mind has never been. These two platitudes together perfectly illustrate the great divide of human adaptation to any new and alien environment – particularly the space environment. In order to come to terms with achievable adaptation, body and mind must somehow meet before the whole organism can be changed to adequately take on the challenge. If they are not changed together, catastrophe becomes the only inevitable consequence.

If we look at these two sayings, something interesting emerges as a conjoined philosophical whole. Examine them side by side in a single expression:

My mind keeps writing checks my body can't cash – the body cannot go where the mind has never been.

Note that the controlling entity is the mind, and that the body is being required to follow. But also note that while the body is limited by the mind's leading, there is the real potential of a more than conspicuous limiting conflict. The mind, as the controlling entity, can well (and quite often does) force the body into a situation where it cannot acclimate.

This clearly and simply illustrates the state of the current great human space exploration collapse. We have made a key *psychological assumption* expressed as a *philosophical motivation* consisting of an expectation of out-of-range adaptation. This philosophically driven error forces the body to go where the human physiology breaks down, driven powerfully forward by a nonnegotiable philosophical construct of which few people even have any understanding. Our oblivious but typical human response to that philosophical error? We use layered countermeasures to correct the human physiology to literally force the environmental imbalance into our philosophic construct instead of fundamentally correcting the engineering systems that support human physiology. Thus, the most simple, logical, and straightforward solution completely escapes the limited human intellect.

Why do we do this? Unconsciously, we are protecting our relatively subliminal philosophical base, which demands out-of-range adaptation. Instead of logically changing our base philosophy to one that protects humans first, we are fully willing to injure the humans in order to protect the engineering systems because the humans are required by the underlying energizing philosophy to adapt to the system and the environment. For those who are having trouble believing this absurd and tragic state of affairs, review once more our cosmic radiation exposure guidelines and what we are planning to do to our very expendable astronauts in the future.

As an engineer, I clearly understand where this discussion has led because "*I are one,*" as most engineers would confess with a wry grin. Lapsing into a serious conversation of one's philosophy with a typical engineer had better wait for the second or third beer. There are rare exceptions, of course, but very rare. Most engineers have had zero exposure to philosophy because of the breakdown in our educational process and therefore have no idea whatsoever that our human spaceflight philosophical base, in fact, is the same foundation that drives the whole train of all human endeavors. Most engineers I know, if honest, would admit, "I don't have a philosophy." While that admission is in itself tragically false, the fact that a highly educated, otherwise intelligent, professional individual would make this admission represents a wholesale failure of our cultural learning system. We are harvesting the result of that systemic malfunction as a culture as we stand at the brink of a catastrophic breakdown of our deep space human exploration paradigm.

Of course, the statement, "I don't have a philosophy," is false. Not only does every person have a philosophy, but it is also their underlying philosophy that powers their thoughts, ideas, and responses to the intellectual and technical world around them. The individual who does not first recognize this and understand their own base philosophy is prone to error and confusion at every level of life, particularly when exercising their professional responsibilities.

What we discover as we look up the NASA corporate ladder is that extremely capable, very successful engineers climb to positions of extraordinary power in the Agency sans any knowledge or understanding of how their personal philosophy feathers into the Agency's human spaceflight thinking or even precisely what the Agency's human spaceflight philosophy actually is. I have reviewed NASA's statements of mission, vision, and goals for decades past and in preparation for this chapter, and they are absolutely devoid of any discernable philosophy of human spaceflight. One can almost hear NASA screaming between the lines of their documentation, "We don't have a philosophy!"

But the truth is, yes, they do! It is incredibly powerful, and it is integrated into every design characteristic of each spacecraft and every mission planning objective. And it literally has not changed one iota since 5 May 1961 when Alan Shepherd became the first American to depart Earth. If NASA had published an Agency human spaceflight philosophy in 1961, not a single word would have changed in sixty years.

Yet, from our coherent understanding of the First Principle of Human Exploration philosophy, and by integrating it into our engineering systems, our new destiny will be created from intelligent minds that have evolved along with our engineering systems. We will construct our new worlds logically by knowing our strengths and by integrating our body and mind together to formulate a single, clearheaded system of rational exploration. By application of specific engineering paradigms to physically shield our weaknesses, we will develop brilliantly conceived advanced exploration systems that become not just new strengths, but the very tools that will allow us to command the cosmos as we expand infinitely outward.

Yet, before we can achieve this philosophic and cultural pinnacle to become authentic human explorers of the cosmos, we must have the elemental intelligence to understand and then apply the new philosophy of human space exploration. And, in the process of doing so, we will evolve a new, alien mind. When we reach this juncture in our collective intellect, only then will we become an advanced race worthy of escaping the bonds of Earth. As we individually discover this enhanced fusion of an intelligently developed and clearly understood philosophy that matches human strengths, weaknesses, and engineering systems, we become new individuals with new minds – alien minds that are suited for living off-planet.

Likewise, if we fundamentally misunderstand our adaptation limitations, we will depart our atmosphere into deep space as bungling human fools, intent on permanently injuring or even killing ourselves in the process of trying to look like brave outer space heroes that our current primitive philosophy literally demands. We deny this vociferously, of course. Denial is always a part of the human contest to pile on self-injury in many areas in our inclination to inflict personal harm on ourselves and others.

History needs to be our guide. If it is not, we are certainly not fit to expand beyond Earth's atmosphere. In history, there has never been a single case of honor extended to ill-conceived exploration ventures that wound up killing or maiming their crew. If such endeavors are our destiny, the universe on the other side of the Van Allen shields would be far better off with our self-imposed quarantine to the surface of the Earth.

The great explorer, Ernest Shackleton, is erroneously reported to have advertised these words when looking for a crew: "Men wanted for hazardous journey. Low wages, bitter cold, long hours of complete darkness. Safe return doubtful. Honour and recognition in event of success." These words are, astonishingly, even etched on the walls of the Mars exhibit at the Kennedy Space Center. But this saying is a complete fabrication. Ernest Shackleton never wrote it or used it in advertising, and certainly would never have utilized it in any form at all.[2] A reward has even been offered for any proof that Shackleton ever said this.[3] To date, no one has been able to come forward and claim that reward.

Ernest Shackleton was a professional, world-class explorer who valued the lives and well-being of his team. Such a caviler attitude toward the lives and safety of his people would have been unthinkable for Shackleton. He never had the going-in expectation of loss and failure. That is absurd. In a careful reading of Shackleton's story in *Endurance,* the author reveals clearly the opposite – a rigid and meticulous planning for all eventualities. Tellingly, in the end, even after monumental struggles and mishaps in the world's most inhospitable and dangerous region, Shackleton did not lose even a single man of his 28-member team.[4]

It is, of course, unthinkable that NASA would even consider posting these words in a public human space exploration venue. But it does clearly illustrate their philosophical cluelessness and mixed, incoherent signals. Does the Agency really expect to lose their crews through maladaptation to their limited, poorly designed, and functionally obsolete engineering systems? Does NASA actually plan to launch crews on extended missions on SLS-Orion into the full flux of galactic cosmic radiation with inadequate shielding?

Sadly, *yes, they do.*

And by posting the fictitious words falsely attributed to a world class explorer, the Agency is also apparently clearly advertising their intent to lose their crews as a bizarre badge of heroism and of very dubious honor in what history will inevitably judge as an ethical and moral catastrophe.

But, in the near future – either through calamity or, hopefully, from intelligent foresight – there may be a different outcome altogether. The explorer who has applied all the available data on any new frontier and, in the process, adapted their equipment as well as their minds, becomes a truly formidable tool and a brilliant representative of our race who will eventually and inevitably overtake the cosmos. Such a declaration is not wishful thinking in any sense. It is and has always been our manifest destiny. It is who we are. We are the intelligent seed of the people who will eventually populate our entire solar system and beyond. The only question is, how we will make this happen? Are we smart enough this time to skip the cataclysm and tragedy phase?

Therein lays the chief caveat fully enabling that vision of our human destiny: *only the intelligent seed.* In order to graduate to this level of possibilities, we must literally come to our senses and recognize where we are, how we got here, the engineering monstrosities that are currently on our drawing boards and launch pads, and why we have even allowed them to exist for so many decades. The train is fully up to speed now. Who will have the courage to stand up and pull the brake line?

Any species that would deliberately expose themselves to a significant level of galactic cosmic radiation (more than a week in deep space) on purpose with the idea that their genetic complement can adapt and handle it is absolutely not an illustration of our intelligent seed. Indeed, such individuals are great examples of a genetic blueprint that will phase itself out of the pool – and very quickly – specifically because of the lack of any discernible actual intellect. Prolonged exposure to microgravity, once again, demonstrates a clear absence of sufficient brainpower to leave the home planet. All these lessons that we have learned and do not immediately respond to correct clearly place us in the "less than desirable species to expand" category in our cosmic neighborhood.

I cannot help but instinctively laugh when I think of whole armies of humans from several countries who are at this very moment scurrying about frantically gearing up to explore deep space with unshielded human space systems, planning to float around in prolonged microgravity. It is actually so red-faced embarrassing for humanity that it is unavoidably and outrageously hilarious.

Our so called best and brightest engineers and scientists are spending a cool trillion dollars each to line up their people to become the meat for a celestial cookout driven quite literally by the energetic fires of countless supernovae and particle accelerators called PeVatrons that are the size of our solar system. They have even seized the legal process from the hands of the intelligent regulators to ensure that no one stands in their way. In the scene that is playing out today before the cameras and recorders of history, *they actually know this,* and yet they (the human system of politicians both inside the agencies and outside who are driving the train) are too intellectually challenged to stop the train, fix it, and avoid the inevitable Jamestown class die-off of the first crews.

And how do they audaciously describe their heroic, historical efforts at their public exhibits? With a mythical phase attributed as truth on the walls of a government supported facility: "Safe return doubtful. Honour and recognition in event of success."

What is the problem in NASA and the other space exploration activities all competing with one another to become the first meat for the slow cosmic roast? No one has yet understood that the alien mind (the organic human mind that is antithetically alien to the standard human intellect – the truly advanced intellectual mind) must drive the process. Not the unprepared mind. Certainly *not* the political mind. And absolutely not the minds that are intellectually inadequate and unqualified, who often wind up in charge. Unfortunately, we have to work with less than desirable intellects at the controls because of the political process and educationally deficient system that has created and enables this ludicrous human inanity and ultimate failure to succeed in the near future.

But there is at least a glimmer of hope that the new alien mind will *somehow* come to control the process. The former Earthling who has evolved into a new species of alien Homo sapiens will depart their planet and create a niche of protection for themselves on their new home that very closely mimics their planet of origin *specifically* so that they will remain sufficiently healthy to *actually arrive* at their destination and remain healthy enough to *build their new empire*. These individuals will literally engineer their weaknesses into their strengths. How? By understanding the philosophy that underlies their every decision, thought, and design process for equipment and the model of the mission itself.

This new human will leverage their exceptional intelligence to radically change their thinking and their expectations to become an alien as they adapt within clearly defined biological constraints. The reason these individuals are so intellectually advanced and capable of this radical transformation is that they understand that in the perpetual internal battle between human emotion, reason, and ambition, the core biological limits and processes must be nourished and protected first and primarily before their ambitions of empire are engaged. There can be no building, or even thoughts of building, on distant worlds before this is well understood and integrated into the human experience.

A clear understanding of, and intelligent response to, our actual hard limitations is the primary intellectual tool that will ultimately drive authentic, real productivity in the new empire. Limitations can be overcome, but not by human organic adaptation. These limitations are fully overcome by engaging engineering system adaptations. It should have been abundantly obvious decades ago to a species of even average intelligence. But not everyone is genetically qualified to actually wield the title of "average intelligence."

Recalling the words of Robert Heinlein:

"Space travel can't ease the pressure on a planet grown too crowded not even with today's ships and probably not with any future ships – because stupid people won't leave the slopes of their home volcano even when it starts to smoke and rumble. What space travel does do is drain off the best brains: those smart enough to see a catastrophe before it happens, and with the guts to pay the price – abandon home, wealth, friends, relatives, everything – and go. That's a tiny fraction of one percent. But that's enough."

How It Begins – The Advanced Spacecraft

This new space exploration paradigm *will* come to pass. It will either evolve from the intelligent realization that we are about to send many people out to their deaths or to grievous injury, or it will come about *after* those serial tragedies. Either way, mankind *cannot* explore space with the current exploration architecture based on its fatally flawed philosophy. It is, after all, a self-limiting madness that we are wasting countless trillions of dollars to design, develop, and build. How do we as a species get this correct and earn our right to become a legitimate member of the cosmos? We can start the process today by designing and building a truly advanced spacecraft.

An advanced human spacecraft has never been built before. But it has been conceived in fiction many times over. The advanced human spacecraft is one that is engineered to solve three problems simultaneously, *in order of priority*:

(1) Its first and most essential duty is to protect its human occupants from the harmful environment of space: temperature extremes, vacuum, galactic cosmic radiation, plus the long- term effects of exposure to microgravity and the multimorbidity associated with its toxic mix.

(2) To safely enable reliable transport of its crew from one location to another.

(3) To effectively conduct its designed mission without deliberate exposure to potentially lethal environments.

This short list should seem like a no-brainer. But in our culture, it is *actually resisted* with rage and anger.

An advanced human spacecraft has never been built before because the relatively rudimentary craft we have constructed and launched have all failed to meet their first and most essential priority of protecting their human crews from injury. An advanced human spacecraft will do all the things the Apollo, Shuttle, and Orion spacecraft were designed to do, but will incorporate artificial gravity and adequate cosmic radiation shielding as well as *sustainable* advanced life support system components. Once artificial gravity, adequate shielding, and sustainable advanced life support systems are added, they will then become truly advanced spacecraft. Until that time, the craft designed to depart low-Earth orbit (Orion) will be little more than monstrously expensive long-term death and disease traps.

This rule does not apply for short-term flights into suborbital space or short-term spaceflights under the Van Allen Belts, and perhaps for single flights to the Moon and back with very little time spent on the surface, outside Earth's protective magnetic shield. This statement is qualified specifically because of the research data presented in the Chapter "Lies, Damned Lies, and Statistics," which discussed the statistical significance of cataracts and cardiovascular disease experienced by the cohort of Apollo astronauts. They all departed Earth on relatively short-term expeditions to the Moon, either to orbit or land on its surface, then returned to Earth. Given this limited research data, there is some discernable chance that *any* exposure outside the Van Allen belts (deep space) may incur a significant probability of long-term physiological effects.

As has been previously discussed in depth, it is important to make this clear distinction. There is a class of spacecraft currently in production that are designed for the tourist market with merely incidental exposure that is clearly *not* injurious to their passengers. I would certainly fly on any of them *and bring along my grandchildren*! It is essential to understand that short flights into suborbital space and flights to low-Earth orbit for brief exposures have clearly and unambiguously shown (by in-depth research on the International Space Station) that there are very limited risks to human health under these conditions. For *all* these flights, advanced spacecraft are *not* required, and humans may freely frolic (and vomit) in microgravity for a "fun" flight, or even a weekend, or a week or so in an orbital space hotel.

Yet, *any* space voyage outside the Van Allen Belts into deep space for more than a week *must* be conducted in an advanced spacecraft. Examples of this include the planned Lunar Orbital Platform – Gateway, a Lunar Base, or any Mars expedition. There is currently no announced spacecraft on the drawing board that is advanced enough to safely permit exploration outside the Van Allen shield. Today, there is zero level of discernment between suborbital, low-Earth orbit, or deep space expeditions at all. The going in assumption in every space exploration paradigm from sub-orbit to Lunar and Mars is all managed by human adaptation and unlimited, stacked, and mostly ineffectual countermeasures – the model that has existed from the flight of Alan Shephard.

With no existing advanced spacecraft and none on the drawing boards for missions now being planned for deep space, we are not just witnessing a Jamestown moment, we are willing participants in it. If the alien mind inside each of us does not make its intellectual arguments known very soon, many individuals will never return to Earth. And the ones who do may not be considered the lucky ones in the long term.

So, how do we turn the corner before it is too late? There is not much time left. And, frankly, if we do not stop the speeding train that has gained so much inertia – and rethink, retool, and rebuild – our first long term experiences into deep space will not end well. Then we will be forced by the tragic results of our own limited human intellect to rethink, retool, and rebuild.

And yet, there is an amazing new view of the evolution of that alien mind ongoing in our own culture, for those who have eyes to see it rapidly advancing. In the next chapter, we will unwind the full and powerful expressions of that innovative mind brought to you by a new media that has literally appeared from the mists of the world wide web.

Watching
Humanity
Departing
Earth Forever

"Believe nothing until it has been officially denied."

— Claud Cockburn

"In old days men had the rack.
Now they have the Press."

—Oscar Wilde

THE MEDIA, ACTING AS THE POTENT FOURTH ESTATE IN the United States, has long been considered a *tour de force* in American life. But with the advent of electronic communications and its eventual morphing into an Internet driven content, the expression of media power has quickly and radically shifted into a vast new, unique media empire. Quite unlike the media's identical, conjoined triplets of the "big three" in the mid-20th century (CBS, NBC, and ABC), today's American broadcasting zeitgeist has fractured from a single orchestrated hivemind into many thousands of media outlets, each speaking their own wholly disparate worldviews as dictated by their individual prevailing narratives. While the old predominant singular narrative previously decided what Americans should or should not think, today's media is not so exclusively controlled. As the social media giants attempt to define and enforce a singular 21st-century narrative, creative independent journalists are stepping in to effectively redirect the narrative into a powerful and influential media archipelago consisting of a more sane and even-handed reality.

Except for the social media empires, the vast majority of the media islands are all autonomous voices, each defending and advocating their individual narratives. One of these islands is the media outlets actively promoting human space exploration and permanent settlement off planet. They have generally started from scratch with young men and women just out of college who uniquely, creatively, and brilliantly from their home computers cover the new human expansion outward. Being internet savvy, they have captured their audiences (some small and some very large) by sheer individual ingenuity and often astonishing talent on budgets so modest they often resemble a typical university student's allowance or paycheck from a weekend job. And yet, nearly all of them produce podcasts or cover live events such as launches and other space related proceedings with more skill and creativity than the media giants. The broadcasting corporations spend tens of millions of dollars attempting (mostly unsuccessfully) to compete with these new-media prodigies who are funded through viewer support.

In this chapter, I highlight the work of three remarkable, driven, and incredibly creative young individual geniuses, and two networks of space advocates/enthusiasts, none of which have any peers at all in the anachronistic old media empires.

MARCUS HOUSE

This extremely talented young Australian broadcasts the day-to-day state of humanity's reach into space to the world each week from his native Tasmania. That statement alone strongly reflects the profound power of the new media. Broadcasting from what was formerly considered one of the most distant and remote parts of the planet is now absolutely no different than one of the past's broadcasting greats speaking to the world from their news desk in New York City. Watching one of Marcus' weekly Saturday pre-recorded broadcasts, it is impossible to know from where it is being transmitted. In the 21st century, whether Tasmania or New York City – it has become totally irrelevant. In a few years, such broadcasts will be viewed on Earth from Luna, Mars, or interplanetary destinations.

With my experience as an advanced space life sciences design engineer and Principal Investigator of several scientific investigations, I have a very keen sense of who knows what they are talking about when it comes to space exploration, and Marcus House is literally at the top of his game explaining human space exploration science and engineering in his dramatically interesting and exceptionally talented, easy-going approach. And every Saturday morning, as unerringly regular as any scheduled network broadcast, Marcus gives the review of progress and events that occurred during the past seven days.

I have to point out that this degree of up-close-and-personal access has never been allowed in human history to any historical event of this magnitude or any other! The credit is shared between Marcus and his peers doing all the boots on the ground work, but it is also equally shared by Elon Musk and SpaceX, who have gone to astonishing and extraordinary lengths to keep opportunities and the doors of their facilities open to these independent journalists. They have even allowed frequent (almost daily) access to the skies over SpaceX facilities at Starbase for aerial photographers such as RGV Aerial Photography to record each movement of every small and large piece of flight hardware every week that Marcus documents and shares with his viewers. Literally, no other private company, or even NASA, has ever allowed this kind and degree of access so that the rest of humanity can enjoy the ride to our destiny as a multi-planetary species on a day-to-day basis.

In a standard weekly broadcast, Marcus masterfully utilizes media from many different content providers to document our sprint toward human exploration and permanent occupation of other worlds. He will typically document launches of key space vehicles. Then Marcus begins to describe the events ongoing at South Texas Starbase and Launch Complex 39A at Florida's Kennedy Space Center, leading to the construction and launching of SpaceX Starships and other crewed Lunar and Mars hardware and activity. Extremely detailed high definition videos are presented that show how the assembled hardware will be used in future flights and activities on Earth, Luna, Mars, and in interplanetary space. Everything from cataloguing each major piece of hardware produced, delivered, and utilized (including rocket engines) is laid bare every week. NASA, never in its wildest imagination, ever dreamed of allowing such access to their day to day activities.

The bottom line is that no one provides a better, more comprehensive, more accurate, more reliable presentation of humanity's greatest adventure than Marcus House! But he is joined by some incredibly talented 21st-century media prodigies also reporting this adventure in their own very unique, talented, and interesting ways.

TIM DODD THE @VERYDAY ASTRONAUT

Tim Dodd, aka the Everyday Astronaut, has very wide-ranging talent and ample energy to match. He produces periodic live streams and has become a live host for private launch providers and launch streaming services. Tim is perhaps best known for individual tours of Starbase grated to him personally by Elon Musk. The synergy between Dodd and Musk is sincere, easy going, and very informative. Tim Dodd has amassed the highest subscriber base in this social media niche in the world for many good reasons.

FELIX SCHLANG WHAT ABOUT IT!?

Felix Schlang, formerly of Germany, is now a resident of Florida's Space Coast. Felix, self-described as a "Husband, Father, SpaceTuber, Educator," has hosted the widely popular YouTube channel "What About It!?" (WAI – pronounced "why") since 2019 that offers a weekly review of the push outward to settle the space frontier. His broadcast is quite informative, very accurate, and well developed, utilizing a wide range of content providers. Felix has an innate talent for this kind of instructive production and keeps it easy and lighthearted (occasionally even sporting blooper outtakes).

After spotlighting these three individuals, I would be remiss if I did not also mention two other teams that likewise produce some of the best crewed spaceflight coverage in the world.

NASA SPACEFLIGHT AND LAB PADRE

NASA Spaceflight (not affiliated with the National Aeronautics and Space Administration) was perhaps one of the very first private groups to fill the void between government and network space media and the degree of coverage that many private individuals were looking for. It describes itself as "a news website and forum which launched in 2005, covering crewed and uncrewed spaceflight and aerospace engineering crews."

NASA Spaceflight began by covering interesting and highly anticipated space launches (both crewed and uncrewed) on a private basis. By the 2020s and beyond, NASA Spaceflight is internationally known for its un-attended web cams that run 24/7 as well as its online livestream moderated events that literally span hours. NASA Spaceflight has no set start time or end time and may run as long as 8-10 hours with non-stop discussion between 1-3 moderators addressing a seemingly endless stream of web-chat questions. Tuning into NASA Spaceflight stream is literally like being there with no one waiting for you in the car to go home even after 6-8 hours!

NASA Spaceflight maintains its own suite of 24/7 cameras both at Starbase in Texas and on NASA (the government NASA) property.

Lab Padre is another distinct group of youthful, incredibly talented, like-minded spaceflight media crews who descend on Starbase or Kennedy Space Center to cover individual launches. Lab Padre also maintains a set of capable 24/7 cameras at Starbase as well as hosting their own moderated pre-launch (as well as most every pre-test) events at Starbase. It is truly a dream for those of us who have an intense interest in these experiences that NASA Spaceflight and Lab Padre offer parallel coverage of these events that provide moderated broadcasts for hours and unmoderated webcams 24/7. If you tire of a specific conversation on one of these engaging YouTube channels, just switch over to the other.

I am dismayed when I discover that many enthusiasts are unaware of these media offerings. And when I pass this information to those who share this interest, the same comment inevitably is offered, "Why doesn't anyone know about this?" It is truly one of the most astonishing miracles of our technology: free 24/7 quality access to the most significant historic human enterprise in history – the days, months, and years of our migration from the cradle of humanity into the wider universe. And we can watch it live for as long as we wish from any computer or phone.

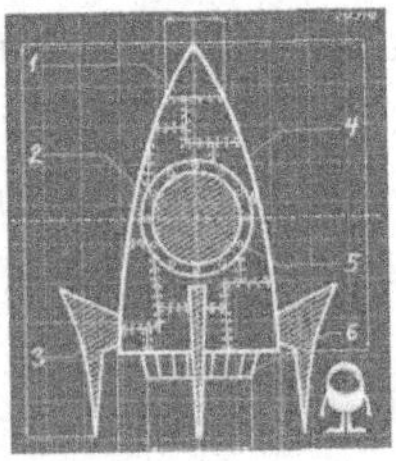

CONTENT PROVIDERS

All of these private individuals/groups utilize truly world class graphics that are developed by other exclusive, self-supported people. The big networks keep their own staff of content providers in house. But these 21st-century media entrepreneurs look outside for assistance from the private content providers working as independent entities. These individuals are supported by their fan base and YouTube channel subscribers by one-time or recurring donations through platforms like Patreon and PayPal. But unlike providers that restrict access to their products solely to paying members, these world class contributors offer their products to all for the benefit of humanity's advancement in space.

Their work is truly top tier professional consisting of rendered videos of future interplanetary space hardware that are practically indistinguishable from authentic video shot on actual launches. following is a short list of some providers, each of whom has Twitter accounts and YouTube channels so that anyone can follow their products. These are just a few of the true 21st-century masters of the new media, illustrating humanity's push to become a multiplanetary species. The content provider's name and Twitter account are provided. Many can be subscribed to on YouTube as well.

Cosmic Perspective / Considercosmos

Starship Gazer / StarshipGazer

Greg Scott / GregScott (Photography)

RGV Aerial Photography / RGVaerialphotos (Photography)

Tony Bela / InfographicTony (Infographics)

Ryan Hansen Space / RyanHansenSpace (Exploration Renders)

Eric X / ErcXspace (Exploration Renders)

Together, the broadcasters and content providers (as well as many others not mentioned here), each working independently and self-supported, have completely reimagined and redesigned access to the greatest human venture in our history. It is these young, ingenious, driven individuals who have teamed up to open the window to the new worlds so that every human on planet Earth can witness it all happen in exhilarating detail and in real time!

Module
Four
Ch.01
Ch.02
Ch.03
Ch.04
Ch.05
Ch.06
Ch.07
Ch.08
Ch.09
Ch.10
Ch.11
Ch.12
Assured
Success

Getting It Right

The Bad News:

"Man is so intelligent that he feels impelled to invent theories to account for what happens in the world. Unfortunately, he is not quite intelligent enough, in most cases, to find correct explanations. So that when he acts on his theories, he behaves very often like a lunatic."

— Aldous Huxley

The Good News:

"Our greatest weaknesses and propensity to fail are more than offset by humankind's brilliant, far more powerful strengths. These small creatures inhabit a tiny blue world orbiting an average star riding along the edge of their average galaxy's spiral arm. But they also have an astonishing potential that wholly offset their limitations. And in that potential lies an equally astounding power – the power to correct – the power to right their wrongs – the power to fix what they have broken - the power to understand their failures and thus defines their power to create a civilization that will reach far beyond their relatively insignificant sphere of influence. For small beings such as these – their capacity to intelligently adjust their course fully illustrates the true nature and power of their intellect and thus will define their course to the stars."

— Dennis Chamberland

A Summary of Where We are Today and How We Got Here

AS I HAVE STATED REPEATEDLY, HUMANKIND *WILL* correct its flawed exploration model and make things right specifically by constructing First Principle Solar System Class human deep space exploration ships and habitation structures to explore and settle space successfully. And humankind will do so whether by intelligent foresight and appropriate application of data or through tragedy. We *will* change, or we will not ever depart low-Earth orbit. Regardless of our determination, stubbornness, and will, we are rigorously limited by our exquisitely balanced and delicate physiological frame, which is not altered or strengthened by determination, stubbornness, or willpower – however epic we think ourselves to be.

Such a revolution in human spaceflight will be accomplished by a process of overcoming error with a complete understanding of reality and by constantly correcting our course. Reality is defined only by hard, incontrovertible data, which we already have in hand to a significant degree. What we do not have are those data sets associated with the vertebrate adaptation profile in partial gravity and a complete vertebrate injury profile in deep space to chronic exposure to full spectrum galactic cosmic radiation. Progressing beyond low-Earth orbit on long term missions without these two data sets will almost certainly lead to disaster and a loss of mission.

As discussed previously, human adaptation to chronic GCR exposure is not possible as defined by the already well established and understood international conventions that state, per the United States Nuclear Regulatory Commission, "There is no threshold below which radiation doses are safe."[1] NASA has legally exempted itself from this reality by a manipulated loophole that has nothing whatsoever to do with the effect of a cosmic primary traveling at relativistic velocity on any human cell. Again, currently, the limit of this shamefully unethical behavior is set and will never yield to anything but injury and death in deep space.

Ultimately, either through epic, completely avoidable disaster or through applied intellect, we will correct the current errors in our human spaceflight models by adding adequate radiation shielding to obtain exposure doses lower than U.S. federal limits, and by provisions for artificial gravity in spacecraft and human habitats that fully meet First Principle advanced human space system designs. That is the only viable, clear, and unambiguous process whereby our errors are corrected, our crews fully protected, and mission success is completely assured. All other approaches will ultimately fail at the cost of human lives, permanent injury, and loss of mission. We get to decide.

But we can realize that moment of clarity right now, today, by just a single second of institutional courage and a redesign to First Principle protected engineering. If not, there will be a historic struggle followed by the perfect hellstorm of disaster and the requisite, cyclic national contrition. Yet, for the remainder of this book, as the author, I have deliberately chosen to assume that humankind can do this, and that we will avert that hellstorm and inevitable tragedy. The warning has been clearly and unambiguously articulated. It is time to move to correction, solutions, and detailed revelations of the realities of alien worlds.

A Win-Win Plan for SLS-Orion

It is incontrovertibly true that today NASA's SLS-Orion human space exploration architecture is hopelessly obsolete, dangerous on many levels, and will inevitably injure astronaut crews on long term space missions in deep space as it assiduously drains down the national treasury. These are facts that are simply beyond any doubt, argument, or debate. However, there may still be a few viable missions involving the SLS-Orion hardware – crewed and uncrewed. While we regroup and redesign our human space exploration systems as First Principle ships and interplanetary habitats, SLS-Orion system components that have already been built may have a few coherent, logical, and relatively (but very borderline) safe applications for short crewed missions on the order of fewer than two weeks beyond LEO – such as the first crewed missions that will return humans to the Lunar surface.

Wasting these astonishingly expensive components would not be wise as there are potential missions that relate to the development of a First Principle human space exploration system architecture that can use this existing hardware. To date, some $50 billion has been spent on SLS-Orion, with another $20 billion allocated for more hardware in the next several years, for a total investment of $70 billion – with three launches scheduled during that period.[2,3,4] These *three* NASA launches cost the taxpayer the same as the expense for 778 Falcon Heavy launches, less spacecraft.[5] It would be immoral to squander those kinds of resources and the incredible suite of possible human exploration opportunities. But instead of using $70 billion as a perfect excuse to injure astronaut crews, the SLS-Orion hardware can be used to help engineer a First Principle space exploration system.

The virtually unshielded Orion capsule is an ideal containment for a long-term flight to a stable Lagrange orbit in cislunar space to study the prolonged effects of full spectrum cosmic radiation on vertebrate animal cohorts. The Orion capsule could be stripped out to bare walls and outfitted with six animal containment units in an arc around the center of the capsule. Each containment unit would be individually connected to the Orion life support system. At two-month intervals, each of the animal containment units would be jettisoned back to the ISS, from where they would be transported back to Earth for evaluation of the vertebrate organisms. In this manner, the vertebrate animal cohorts will have experienced from 2 to 12 months or more of unprotected exposure to GCR and can be evaluated extensively for short- and long-term effects.

In another experiment that would help resolve the next most essential unanswered question in human space exploration, the Orion capsule could again be outfitted in a different configuration. The capsule would be adequately shielded to protect the vertebrate organisms to legitimate current Federal guidelines and also configured as before with the six cohorts of vertebrate animals, each scheduled for two to 12-month exposures, but this time to variable gravity in a spun Orion capsule. As before, the animal containment units would be jettisoned back to the ISS, from where they would be transported back to Earth for evaluation. Several other Orion capsules could thus be outfitted at their respective stable Lagrange orbits in cislunar space to study varying levels of partial gravity that simulate Mars, Luna, and other partial gravity gradients of interest.

Other Lunar-class missions can also utilize the SLS-Orion system for very short term missions outside LEO with the understanding that the short term exposure risk to the cardiovascular system, lens of the eye, and other epithelial systems is poorly understood and may result in long term physiological damage to crews with sequalae not manifesting until decades post mission. Ideally, lunar habitation structures will all be robotically shielded *before* any human landings are attempted.

In these ways, the SLS hardware can be used responsibly, ethically, and for the express purpose of employing the hardware to build First Principle human space exploration architecture – rather than be dumped into the ocean, as was its predecessor, to save waste processing costs. Leave it to the U.S. government to expertly seek and implement the lose-lose opportunity… Every. Single. Time.

A Win-Win Plan for Humanity

If you stand back and objectively look at where we are today in human space exploration, it is clear that we are at a critical crossroads. And the reasons that we are here are distinctly understandable and can be unwound, as has been done in this volume of *Departing Earth Forever*. You would think – or you would *hope* – that humanity is not so rationality-challenged that we actually sit around and purposefully plan for disasters of our own making. Disasters such as the ones that we face today are typically hidden by our vast, labyrinthine, systemic processes of government, human emotional constructs, and even our grossly overinflated perceptions of self. Combined, they tend to hide the obvious and distort reality so that disaster eventually morphs into a nearly inevitable byproduct, as the investigations of the loss of the spaceships Challenger and Columbia and their crews so readily attested. In many ways, thus humanity often struggles mightily in order to guarantee spectacular failure. Unfortunately for our often intellectually challenged species, the "ah-ha" moment more often than not comes *after* the catastrophe instead of before.

Hopefully, in this case, a NASA insider (such as your author) comes forward with a very public critique, a clear enumeration of the details of the future disaster to come, its root causes, and even an offsetting set of solutions to sidestep the inevitable calamity. Here, the clear warning has been sounded before the disaster. The question is this: will the messenger be attacked? Of course he will! That result is inevitable in our easily triggered culture. It is human nature. (Your author is way too far down the road to care about the attacks and only reads pertinent emails.) But the real question is this –will there be anyone in a position of authority who will actually listen and make the changes necessary to protect future astronaut crews? Remember – your favorite self may be one of those future space farers!

And, to be clear, please understand that this is not only a NASA issue. It is one that belongs to every private group, every international group, and every other nation besides the United States planning to fly humans in deep space. All of them who are preparing for human missions above low-Earth orbit are facing exactly the same limitations – and exactly the same opportunity. As I stated before very clearly, **the first company or nation that gets this right literally commands access to the stars and every world – and every resource on them**. If the United States fails, it will lose to the nation that does get it right. But if the United States, China, Russia, and every private company stop right now and retool to build First Principle exploration systems, the cosmos opens to all. Until that moment, humankind is running full speed down the wrong track toward a brick wall at the other end.

As time is of the essence, then let us use the time to correct things now before the competition does – because every nation and corporate entity faces the inevitable redesign challenge. The grand prize winner of first access to the solar system and all of its inestimable riches is the nation or entity that accomplishes this first. **Literally, the one who shifts to the correct track first will take home the gold for all of humanity – and will harvest the most astonishing wealth and power in human history**! Unfortunately for their competitors who try to harvest those resources in obsolete Apollo-era ships – even if those systems are "on steroids" – they will probably never return home. And when they arrive at their destinations in deep space, their crews will be too sick and injured to work and harvest the available treasures. In fact, we will truly pity the ones who are actually able to return.

Right now, we are struggling mightily to guarantee spectacular failure just ahead. But we do not have to. As you have clearly heard the warning, now, just as clearly, hear the promise. Our greatest weaknesses and propensity to fail are more than offset by our brilliant, far more powerful strengths. We, as such small creatures, inhabit a tiny blue world orbiting a typical star riding along the edge of our average galaxy's spiral arm. But we also have astonishing potential that wholly offsets our limitations. And in that potential lies an equally astounding power – the power to correct, the power to right our wrongs, and the power to understand our failures and thus define our power to create a civilization that will reach far beyond our relatively insignificant sphere of influence. For small beings such as us, our capacity to intelligently adjust our course fully illustrates the true nature and power of our intellect – and thus will define our course to the stars.

We can get this right, fellow colonist... ***and we will!***

DEPARTING EARTH FOREVER
Book Two
Alien Worlds
The Manual for Today's Colonists Preparing to Launch to Mars and Our Moon
"A brilliant expose...
...the bible for human space exploration beyond our moon."
Lt. Col. Robert S. Miller
United States Air Force, Ret.
DENNIS CHAMBERLAND
RETIRED NASA SPACE LIFE SCIENCES ENGINEER
Preparing for Book Two

"Give me six hours to chop down a tree and
I will spend the first four sharpening the axe."

- Abraham Lincoln

YOU HAVE ALMOST COMPLETED BOOK ONE OF *DEPARTING Earth Forever*: *Warning and Promise*. In Book Two, we shall explore *Departing Earth Forever*: *Alien Worlds*.

You may be quite surprised to learn that in the constructs to come, the alien world that is nearest you, and the one that must be conquered first, is between your ears. Before a legitimate explorer can competently step out the front hatch, make their way to *any* new frontier, and expect to remain alive to tell the tale, there must first be a change in their mind and significant conditioning in their inner being. Each explorer must discover the alien mind that lies ready to come forward, seize all these golden opportunities and adventures, and build all the human empires that lie just ahead. As Book One has exhaustively championed, humanity cannot even begin to change its exploration model until there is an understanding – and then an intelligent execution – of the philosophical transformation that must first take place. We are exclusively creatures of thought before we are beings of exploration and change. One cannot come before the other without seeding disaster and tragedy on a historical level, as our unfortunate history has repeatedly demonstrated.

Literally, in every sense, *the body cannot go where the mind has never been.*

Thus, it is the purpose of Book Two to initially address, correct, and strengthen access to the inner frontier before addressing the engineering and voyages to the alien worlds to which we shall all travel together. We cannot exercise the intellectual muscles of human exploration before gathering and assimilating a complete, detailed scientific understanding of the worlds to which we aspire

A wonderful illustration of this is found in the very early ramp-up to the Apollo Moon landings, humanity's first expeditions to another world. Before any thought of departing Earth for another celestial body, there had to be a reckoning between the mind and the hardware itself. The human mind knew very well where it wanted to go – but it had to instruct its machine masters precisely how to do that. That was accomplished through the most complex mathematical equations in our early 1960s inventory. And, before computers, all of that had to be worked out in the human mind employing only slide rules, huge clunky calculators, and number two pencils.

You have heard of the incredible story that was aptly titled "Hidden Figures." And what was the core drama of that tale? Before NASA could get to another world, it had to first conquer human stupidity, human bigotry, human hate, and human intolerance of other humans. And when it finally overcame those hurdles, it released the rare intellectual power of those few who made it happen through the intellectual achievements of Katherine Johnson, Dorothy Vaughan, and Mary Jackson, and the simultaneous sea change of our cultural philosophy.

Before the brains of those equipped to solve the problem could even start their work, and before humanity could realize its ancient dream, it first required corrections of the mind. That is precisely where we are today – faced with a battle of the minds before those who are equipped to lead us forward can be released to do what they do so very well! Integrating a new philosophical approach in any human endeavor is always a challenge. But it is even more so if the tools to make that happen have never been taught. It will become evident that these powerful implements of reasoning are a just as important part of all human exploration as the nuts-and-bolts engineering. Even without the awareness of the builders, deciding upon rockets, habitats, and life support systems all require a complete understanding of our subliminal reasoning and expectations. This is so because humans do not explore by machines but by the power of their exploration philosophies that guide their ultimate destinies.

Departing Earth Forever Book One represented those discussions. But Book One only embodied the beginning of the discussion of expanding our human empires into the Solar System. Nearly all of *Book Two* fleshes out that exploration philosophy into nuts-and-bolts hardware and processes whereby we will not just get to these other worlds, but what we have to do to actually survive and thrive when we get there.

All human exploration is always wholly energized by a thought process so powerful and effective it may surprise you, even though so many do not even know it is that very energy that is driving them ahead. As we have discussed, and as we shall see clearly in *Book Two,* failing to understand these principles almost always eventually leads to failure and eventual death of many of the explorers. But the process encouraged by these two books is the opposite. Using the tempered, well considered data harvested from more than 50 years of space exploration, we shall create the new machines that lead to assured success, life, and new generations of people who will certainly one day populate other worlds. Book Two lets you in on the processes that will actually make that happen.

So, fellow space traveler, fellow citizen of other worlds, let us learn all those things here on the home planet under the safety of our beautiful and carefully shielded sky!

But just before we together turn the corner to a new age of human exploration and discovery by turning the very first pages of Book Two, let us close out *Book One* by bunking out together on the surface of another world and see what it is like to wake up on another planet. This will give you something of a flavor of what is in store in Book Two.

Waking on Another World

"There is life on Mars, and it is us. We are the Martians. We give ourselves a gift of us. We are more than water, we are more than earth, we are more than sun. We are the Life Force giving itself a reason for being."

- Ray Bradbury

FOR 31 MORNINGS OF MY LIFE, I HAVE HAD THE extraordinary privilege of waking in a very different world within our world – in a habitat anchored to the bottom of the sea. Outside, there was no air to breathe, the pressure was much higher than anywhere on the surface, and, in order to stay alive, the machine in which I slept had to shield me and nourish me from the harsh and very unforgiving world surrounding us.

Obviously, waking in a habitat on Mars is different in many ways – but not so very different in many others. In that as I write these words there are only four humans who are still alive who have ever actually awakened on another world besides Earth, let me share with you the feelings of one who has slept in a NASA space analog habitat on the seafloor, one that I specifically designed and had constructed to simulate what it would be like to live and thrive in space or perhaps on another planet.

On one of those assignments, I was acting as the Mission Commander for a pair of seafloor missions occurring simultaneously in two undersea habitats located about 100 feet from one another. In the evening, before lights out, I swam from my habitat over to the other. Already inside and preparing for bed were two famous individuals who are very well known to just about every one of you, but whose names I will keep private. One of them was a younger man who was already in his bunk, covered with a blanket. I walked over to him and bid him a good night, but in his face and eyes, I could read a certain level of discomfort and unease. I understood the feeling perfectly. After the excitement of entering the habitat for the first time, dining on the ocean floor, and then settling down for the night as the nighttime blackness encompassed the seafloor, lights out tended to focus the explorer on not where they were at the moment, but – more importantly – what lay just outside the thin hull of the habitat, literally inches away in the darkness.

I confessed to him that I knew exactly how he felt, remembering my own first night there a handful of years before. I clearly recalled the heightened sensations, the sounds from outside, and especially the machine noises that surrounded me. And then, after the lights went out, everything seemed magnified. It was not fear, but it was instead a deeper sense of appreciation and attachment to the safety of the metal cocoon that surrounded me and the machine noises that meant that the structure's heartbeat and lungs were working together as advertised. Unlike any shelter on the surface of Earth, this one was not easily escapable in case of an emergency. It drove home the realization that this habitat, its shelter, and all of its machines were very special and all somehow very personal. After a couple of nights, that heightened sensation all goes away and is replaced by the opposite feeling – a sense of intensified security.

I have never slept so deeply, nor have I ever felt as secure as in an undersea habitat on the ocean floor – never. But the real treat when sleeping in a habitat off the surface of the Earth is the awakening. And I am certain that that extraordinary feeling will also accompany each Earthling, including you and me, who dares travel the hundred million miles to Mars or any other world.

Your First Morning on Mars

As you, the explorer, awaken from deep sleep ensconced in your warm, soft bed at Mars Base One, the first sensation before even opening your eyes is the quiet, white noise of the habitat's machines working ceaselessly and tirelessly to keep you alive. This white noise surrounds you and, like your warm blanket, ceaselessly but in a very powerful way, also caresses you. Depending on the habitat, you may also feel the almost imperceptible, nearly subliminal vibrations of those machines purring all around you. As your mind begins to focus on awakening itself, your body wants to continue sleeping. But your brain begins to command the process and opens the new day with a single overriding thought – where am I?

The answer is so profound that your body begins to lose the struggle to fall back to sleep. Because – your mind reminds itself that you are awakening on another world.

Your body struggles to rest some more, as your mind focuses on the soft white noise all around. The cool, fresh filtered, conditioned air wafts gently across your face and one exposed shoulder. Your mind begins to relax at the secure sound of it all – of the monotonous, barely perceptible hiss of the ventilators and the purr of the metal beast all around. It is nearly enough to lull you back to sleep. Until your eyes slowly open.

The space is still darkened. Your eyes slowly focus on a series of very small red, green, and amber lights alternately blinking and remaining stationary on the desk opposite the bed. And there is an exceptionally dim safety light over the threshold to the bathroom on the far side of the stateroom. Yet, in the nearly complete darkness, these tiny lights have combined just enough of a fog of photons to lend a festive, soft glow to the space, increasing the feeling of security here in your very own Martian cocoon. This muted light and the soft noises are powerful attractants to close your eyes and drift off to sleep once more, wrapped in a layer of soft bedclothes, with your head resting on a simulated down pillow, and protected by the habitat walls from all that lies outside, literally inches away.

Then you recall what it is that lies so close to you and your soft, warm, comfy bed. The outside temperature at this season on the northern hemisphere of Mars, just 12 degrees north of the equator, is minus 102 degrees Fahrenheit – just above the freezing point of carbon dioxide – and the atmospheric pressure outside is equivalent to about 30 miles elevation back on Earth, or almost five and a half times the elevation of Mount Everest. Your life expectancy, if those conditions invaded your space, would be measured in just a few horrible seconds. And if it were not for six feet of Martian regolith (surface sand) piled atop your habitat, the cosmic radiation dose you received while resting in your bed would be enough to expose you to the Earthly equivalent of 2.5 chest x-rays of radiation every night while you slept. But because your habitat is designed as a First Principle Solar System Class accommodation, you received no more radiation than your cousin in Denver!

But those thoughts are forgotten immediately because you never would have been stupid enough to fly to Mars or live in a colony designed around anything less than First Principle hardware!

You sigh with the deepest contentment, close your eyes, and give in to the sleep that calls you – but not for very long. Once again, the reminder is repeated in your thoughts. You are awakening for your very first morning on another world!

Both eyes open wide now. They focus on the clock on the desk. 05:15 – some 45 minutes before the alarm. Perfect! Nearly an hour to lay and rest just a bit longer after the incredibly long voyage of 243 days across the solar system. While the spacecraft that transported you from the home planet to Mars was a First Principle Solar System class ship outfitted with artificial gravity and fully shielded so that the flashes of your dying neurons did not keep you awake every night, the moment-arm effects of rolling over at night made every night's sleep a continually interrupted mess. But here on Mars in 1/3 Earth gravity, no more "Coriolis sleep." It was your best night's sleep in more than 9 months. In fact, it is incredible how amazingly comfortable you are on Mars in your own room and in your very own bed.

Your eyes are now focusing on the individual lights and clock on your desk. Your brain, fully shielded from space radiation and undamaged during the voyage, is totally awake now. Your musculature and bones are as good as the day you departed Earth thanks to the shielding and artificial gravity on your First Principle ship. And your mind now shifts to your assignment and those crew members with whom you became such good friends in transit. Your excitement builds as you consider that in a few days, you will be allowed to venture outside the new outpost.

You stretch, sit up on your elbow and peer out your window at the Marscape just outside. It is beginning to awaken from the darkness of the Red Planet's very unique night. On Mars, sunrises and sunsets are blue, while its daytime skies have a ruddy mustard or reddish hue depending on many atmospheric circumstances. This morning, the Sun is still below the horizon, but there is a distinct blue glow where Sol's bright blue disk will soon be exposed over the distant mountain range.

While it is still too dark to make out many details, you arise from your bed and step over to the transparent pane of radiation and impact resistant polycarbonate and peer outside. It is difficult to believe that the temperature outside is really minus 102 degrees Fahrenheit. Your room, with its temperature at 68 degrees and maintaining a cabin pressure of 14.7 pounds per square inch, is luxurious! You have also been informed that the Martian winter is coming, and the relatively mild minus 102 degrees outside will occasionally drop at Mars Base One to a level that will freeze the gaseous carbon dioxide to a layer of wintertime frost each morning before sunrise. Of course, carbon dioxide makes up 95% of the relatively thin atmosphere, so there is always an abundant supply of CO_2 for frost, or even carbon dioxide blizzards, on the planet's polar caps.

As these things occur to you, Sol's disk finally peeks over the distant ranges. The first thing you notice is that the Sun's disk is considerably smaller than on Earth. Indeed, its diameter appears just over half in angular size as you remember it from an Earth perspective. And it will take some adjustment to get used to blue sunrises and sunsets.

As the sunlight begins to illuminate the Marscape, you notice that your room has a great and mostly uninterrupted view of the vast Elysium Plain that stretches forth from Mars Base One southward across the Red Planet's equator. As it turns out, the Elysium Plain may be a dry, desiccated desert on its surface – but just beneath its drifting, fine sands lies a virtual hidden sea of perpetually frozen water in the form of extremely dense and ancient permafrost. This fact was first discovered by the European Mars Express orbiting spacecraft in 2005. But the subsurface frozen sea under the Elysium desert was actually first described and published in science fiction writer Dennis Chamberland's 1994 book *Mars Base One - Abyss of Elysium*! The presence of relatively vast amounts of that critical resource is precisely why Mars Base One was located here. Mining the water is the chief product of this colony – and with it, life and the human presence on another world are assured.

You move back toward your bed but, unaccustomed to the 0.376 g gravity here on the Red Planet, you almost stumble. Apparently, it will take some getting used to. Even though the artificial gravity on the First Principle Solar System class interplanetary transfer ship had been gradually reduced during the flight outbound, living on a spaceship or in colony habitation is different in many ways.

Safely seated on your bed, you sigh with the greatest contentment of your life. You made it – against many odds. And now, here you are on Mars; on the very first morning of your new life. Many things are about to change on the Red Planet. It is the fastest growing human city in the solar system. The 3D printers outside are busy completing five new private human rated habitats every day, exclusively using Mars materials to build them. New water factories and rocket fuel centers are already under construction. Food production facilities, resource recovery hubs, and liquid carbon dioxide refineries are being built. The human footprint is enlarging itself daily as new colonists are streaming in by the shiploads. And you are a part of it all!

You arise to engage the unique water conserving Mars shower that recycles your own personal shower water. And then you will be off to breakfast at your residential hub's cafeteria, where you will enjoy a manufactured foods breakfast of sausage, scrambled eggs, toast, strawberry yogurt, and milk – all grown locally by the food facilities that make it all from stem cell and unicellular organisms genetically designed for flavors and textures, and plant materials. Most of the solid food on Mars is printed with 3D food-grade printers to guarantee texture to match the incredible flavors. All their elemental carbon is ultimately derived from human resources (the term waste is never permitted on Mars) and carbon dioxide from the Martian atmosphere.

You step before your slightly red tinted bathroom mirror (mined from Martian sand) and shake your head in wonder. After hundreds of millions of miles in the complete comfort and safety of your orbital Mars transfer ships, you are here and ready to go to work – building the first human presence on another world.

Then you remember the first Mars rumor you heard even before making it to your residential hub. The colonists are about to vote on a new name. They want to change the name from Mare Base One to Musktown – which apparently will be pronounced "Muskton."

You smile. You are definitely going to love it here!

The next chapter of your other world adventure begins with page one of *Departing Earth Forever: Alien Worlds*. In that volume, you will open the first book ever written specifically for Lunar and Mars colonists, which will lay out the landscape of your new alien worlds in graphic and extensive detail. *Departing Earth Forever: Alien Worlds* is the definitive second volume for the first citizens of a vast human solar system empire.

Dennis Steindler is the former Executive Director of the McKnight Brain Institute at the University of Florida and recently the former Senior Scientist and Director of the Neuroscience and Aging Lab at Tufts University.

HUMANKIND TOTALLY LUCKED OUT THIS TIME, HAVING A Dennis Chamberland on this small but beautiful blue planet, to help with an ambitious but seemingly predestined cause of exploring deep space and inhabiting planets including Mars, for the good of us all and civilizations to follow. At times, I feel like I am much the same guy as Dennis Chamberland, having grown up during the beginning of the space race, being totally taken with the space program as far back as I can remember, and feeling my pulse race every time I heard or read about rockets, space travels, Martians, NASA, astronauts, and future planetary colonization.

Dennis describes in his new book how we met, and began our "stem cells in space" collaboration, while I was in the midst of running a wonderful institution for brain research at the University of Florida, at the same time working with great teams of scientists to discover and apply the fruits of stem cell biology and regenerative medicine toward improving treatments for some of our most debilitating neurological disorders. Our discoveries of both the normal human brain stem cell that resides in our cerebrum throughout life, and its over ambitious "cousin" who makes too much brain tissue, the brain cancer stem cell, were the foundation of studying how the continual generation of brain cells through life, so-called "neurogenesis" or "neuropoiesis" (as in blood cell generation that is referred to as "hematopoiesis") is involved in higher brain functions for life and at the same time possessing cell and molecular interactions that are at-risk for environmental insults, including radiation. I assumed that Dennis and NASA contacting us was a result of the Kennedy Space Center's close proximity to our brain institute, but not knowing that what lay before our budding collaboration was an issue actually driven by the astronauts themselves. As Dennis describes in the book, our astronauts were experiencing certain physiological responses during space travel that had to involve something that might not be so favorable to our brains and other organs - e.g., cosmic radiation – that poses serious challenges to conventional forms of shielding during long distances or protracted periods of space travel.

The field of human regenerative medicine also has seen its share of controversy, with our work and the work of other investigators generating lively debate as to whether or not there is truly a significant production of new neurons in the adult human brain, as seen in other mammalian and vertebrate species, in spite of an evolutionary notion that nothing exists or persists in the cell and molecular architecture of our bodies unless it functions and has a purpose. The field nonetheless had to design and implement experimental studies to document the utility or even necessity of maintaining a stem cell niche in the brain throughout life, when one considers that the dogma at the time was no new neurons are generated in the adult brain because all they would do is complicate already well-established circuitry that underlies all of our behaviors and functions. To make a real long story very short, we did a study of humans and their brain stem cells under the most respectful and stringent of experimental conditions and found, like others who followed, that these cells do appear to be involved in repair and function of the adult brain, including a potentially major role in lifelong memory and learning functions. The work Dennis and our group did together, for the cause of studying the potential vulnerability of these cells during long time exposures to cosmic radiation as seen during deep space travel, definitely added credence to a notion that we must understand the negative consequences of such travel, and prevent and mitigate them using all means possible.

Orson Welles created an over-the-top attempt at a realistic, apocalyptic radio production called *War of the Worlds*, which had a huge impact on the original listeners, and the feelings and thinking of generations to follow when it came to the notion of space travel and an alien mystique that was represented by bloodthirsty Martian invaders. I also remember the truly-scary-for-me *Mars Attacks* trading cards that I collected in the early 1960s, and that also dealt with a Martian invasion with extremely futuristic space vehicles, weapons, and unusual, to say the least, beings coming to commandeer our Earth in a far from friendly way. I vividly remember the Martian's excessively large and totally exposed brains that exhibited gross anatomical features not so different than ours, and those folds of gray matter that must have had even more of an impact on me, since the human version has been the focus of a great deal of my attention for almost fifty years.

Unfortunately, I often bear witness to respected colleagues and all who choose such a career path in science having to tolerate way too many portrayals in the mainstream media of their line of work being associated with monsters or motivated by power and glory rather than true workings for and total dedication to the cause of discovery for the betterment of humankind. Other than perhaps wanting to sell more newspapers, magazines, books, or movie tickets with dirty laundry or evil but memorable characters like Dr. Jekyll and Mr. Hyde, God only knows why such negative portrayals must outweigh the depiction of most scientists as devout in a mission dedicated to end human suffering via searching their microscopes, test tubes, and experimental devises and data for better treatments and cures for our most frightening and deadly life and health challenges, including the challenges associated with human space travel. Of course, I am biased about such issues, but the too common depiction of science and the search for truths in life that we face as something dominated by the malevolent and ego-driven is simply not true. The science and search for truth in preparing for and executing safe and successful deep space travel, that drives the likes of Dennis Chamberland, is simply driven by the human spirit and instinctual curiosity that compel us to want to explore the heavens and translate related discoveries to improving life for all of us and for all who will follow.

So, many previous efforts at mainstream attention and marketing of anything to do with space travel often include evil geniuses or scary alien creatures with motives that do not always appear so virtuous. But not this time. And who would have ever guessed that decades after my collecting Martian trading cards, and carrying outside in the middle of the darkest nights a difficult to use reflecting telescope gift from my loving parents, and getting so extremely excited every time I saw real moon craters up close, or the barely visible white spots at the Northern and Southern poles of Mars, that in the so-called autumnal years of my life I would be writing a Foreword for such a most important book like this, from such a gifted and dedicated engineer, scientist, space guy, human being, author, my friend, and collaborator? Dennis, along with his wife, editor, and frequent writing collaborator, Claudia, have totally dedicated themselves to the cause of improving our lives through exploration and discovery that includes the most distant heavens.

At this point in time, it is we who will be the Martian "invaders," since we have already explored the Red Planet via uncrewed spaceships that have sent back to Earth fundamental information on the outer and inner workings of this new world that has been the fantasy and folly of so many astronomers and lay advocates of Earth-based cosmic studies over the course of history, and who early on spotted the polar ice caps that even in the most rudimentary of telescopes had obvious similarities to ours. Who also would have ever guessed that as a case in point, I read a previous Foreword for another one of Dennis's works by one of his collaborators and admirers, and definitely one of my heroes, Scott Carpenter, who was one of our first astronauts and a member of America's elite and pioneering Mercury Seven group. The company that I keep in this current venture of extolling the present work is truly humbling, and it is also an incredible honor for me to have been asked by Dennis Chamberland to do this and to help introduce what I see as an extraordinary, brilliant, complete, and up to date compendium of all you need to know about deep space travel, functioning then as a Complete Handbook for the Earth-Based Mars Colonist who will face extreme and dangerous conditions that require knowledge and understanding of everything from cosmic radiation and the required shielding of such, to adult human brain cell generation that is at risk during long distance human space travel.

Dennis Chamberland writes brilliantly, authoritatively, yet with what he often refers to as a casual style of prose in general that successfully conveys the most valuable knowledge and insights from working in this most demanding arena of engineering, founded on his roots in engineering, underwater and space habitats, and the physics and medicine of deep space exposure to the most dangerous-for-life elements that constitute the cosmos. As an example of his gripping, truthful, insightful, and spellbinding composition, in reference to a consistent and predictable history of us humans taking on challenge and the unknown with often historically reckless abandon, he cites examples of our innate curiosity- and technology-driven risk behavior, as with the goal of traveling to and inhabiting Mars and "nearby" dwarf planets like Ceres,"...Daring human enterprise always seems inevitably plagued by its genomic heritage colored with simultaneous measures of sheer brilliance, hopeful confidence, a sharp intellect that is often energized by the negative energies of stubborn ignorance, and arrogance. Unfortunately, this genetically driven package is also quite dangerous when it regularly morphs into an emboldened and wholly irrelevant sense of entitlement that invariably leads to self-deception and then folds into disaster. It is the cyclic curse of the true brilliance of humans habitually entangled in their inevitable sense of misplaced awesomeness." This is but a sample of the brilliance of prose in this book, a book and a story that elegantly and colorfully lay out the history, at the same time including a practical prospectus, from a scientist and engineer who truly knows his stuff, on how to best travel to deep space destinations.

Dennis Chamberland has written this book for all the right reasons. Despite our unbridled enthusiasm for planetary exploration, originally founded on the bravery and risking of life and limb that so many test pilots have offered and endured over the last many decades of flight testing and space travel. Deep space travel poses a great many obstacles and challenges that require the most sincere and informed debate possible, relying on the expertise of the most knowledgeable of scientists and engineers, when exposing astronauts such as the Mars colonists to potentially debilitating and life-threatening conditions. Dennis is religious, spiritual, and believes in the greater good of humankind. He has dedicated his life to the study and caring for brave men and women who have also dedicated themselves to the mission of space exploration for the good of humankind, and he has written many books on his religious beliefs with an underlying awe of the beauty of life here on Earth and in the cosmos. Dennis has relied on his lifelong work and mission – the exploration of the heavens – to write both fiction and non-fiction on space travel and, in particular, travel to and colonization of Mars. The current book tells the truth and consequences of deep space travel that is not founded on both efficacy and safety for the explorers. Regardless of any controversy associated with this subject, "stubborn ignorance and arrogance" should not have a seat at the table of discussion for when and how is the best way for us humans to explore and inhabit distant planets and stars. With Dennis Chamberland's profound respect for such a mission, along with well-engrained awe and veneration for humankind, life, and the amazing beauty of our planet and the cosmos, we all should read this book with an eye toward also acknowledging that one's reverence and spirituality celebrates the wondrous achievement of human physiological design, a design that is resilient yet still so very vulnerable to extreme physical insults, including deep-space radiation and other currently unfavorable conditions for human life that, as Dennis writes about, should be adapted to suit us rather than visa versa. At the same time, such a dedication and effort acknowledges the beauty and wonder of the creation of humankind, our planet, and the universe, that all together epitomize something that just seems to have required an original divine intervention.

Dennis A. Steindler - Boston, MA

Half a
Million
Minutes

"Invention, it must be humbly admitted, does not consist in creating out of void but out of chaos."

- Mary Shelley

EVERY BREATHING HUMAN WAKES EACH MORNING with the same incomparable gift: one thousand four hundred and forty minutes to paint their heart's desire across the canvas of their life. Subtracting quality sleep, life maintenance, and mandatory family or social time, every human winds up with about 700 minutes to spend however they wish. And if you are working diligently on a massive project for two years, those 700 minutes add up to more than half a million sweeps of the big hand across the face of the clock!

During the past two years plus, my prime focus for those more than half a million minutes has been spent on what you hold in your hands. It has weathered every obstacle and challenge – from death in our family and home to a global pandemic. And through it all, I never for one second took my eye off this ball. Because even though visited by the most unwelcome of all life's tragedies, and while changing our lives in an unprecedented way, the message of this book was preeminent over all other considerations.

Most days began before sunrise with writing and editing the pages you hold in your hand, as well as Book Two. There were also days of in depth discussion and deep research. And, in many cases, significant rewriting to ensure that every detail was as right, up to date, and as accurate as we could make it.

The "we" I refer to here is myself and the project's Editor-in-Chief, Claudia Chamberland. We make quite the team – blessed by God with the strongest of marriages spanning three decades, a commitment to one another, and an equally strong commitment to the thesis of these books, which is sharing the detailed and deep truth that will ensure the success of the future of human space exploration and settlement of other worlds. We have been privileged to explore the undersea regions of Earth together as crewmates living on the seafloor for more than 30 days combined in habitats not much bigger than a family SUV.

But successfully managing all the complex interconnected details of a large family of six grown children and seven grandchildren, as well as an expansive extended family, and keeping so many balls in the air at once, is not easy under the best of circumstances. Toss in a double-book project of this scope, and chronic stress always seemed to be part of our lives.

So, as the author, let me be the first to fully acknowledge that the quality, the pinpoint accuracy, and the sharp focus of this work are as much or more the result of Claudia's tireless efforts and unrelenting quality control from cover to cover.

Going all the way back to the book's genesis in 2018, the influence of Dr. Corinne Rutzke and her selfless dedication to the future of human spaceflight and colonization initially made this work possible, although there was no way I could have imagined that the effort would morph into a nearly three-year endeavor!

Certainly, this book would be far too premature without the ongoing work and inspiration of Elon Musk. The settlement of the worlds of the solar system is being fundamentally underwritten by his startling innovative genius alone. Musk has literally left every other nation, company, and private venture in the dusty regolith of several worlds.

Much appreciation is given to retired United States Air Force Lieutenant Colonel Bob Miller for his tireless critiques from the beginning of the process of book development throughout development and into marketing strategies.

I cannot say enough for the ongoing encouragement of Dr. Dennis Steindler that resulted in his writing of the book's Afterword even though he wrestled with many of life's issues – including moving to his retirement home and other profound impacts in his life. Dennis is one of those rare individuals gifted with the capacity to encourage and energize, even in the worst of times.

And Claudia and I both very much appreciate the contribution of extraordinarily talented Steve Goddard in his audiobook expertise and reading of the Afterword. Steve and I share personal stories as boyhood friends in a small Oklahoma town. The friendship has only grown stronger with time.

And I must extend long deserved recognition to Keith Cowing. Keith is the founder and editor of *NASA Watch* – an online forum built and edited by a real former NASA guy designed to allow everyone else (including a very wide NASA readership) to understand the "real NASA" and what's really going on inside the hallowed halls. For decades, Keith's work has routinely frightened the living daylights out of a literal army of NASA Senior Executive Service members and Administrators running the Agency – and all for a good purpose. He is one of the few devotees passionately driven to shining bright lights on places that do not want bright lights. For example, if it were not for Keith and *NASA Watch*, the evolution of the soon to explode partial gravity scandal described earlier in this book would probably have been lost to history. Bravo Keith! *Nil desperandum carborundum illegitami.*

We finally extend our acknowledgements to all future colonists. Without you, there would not be any reason to tell this story. Therefore, because of you, there is about to be a revolution in how we humans explore space! And explore space we will, just as we will depart Earth forever, never to return to our home planet. As Ayn Rand stated about those of us who stand at the shorelines of our planet about to debark in our advanced, newly designed, outbound First Principle Solar System Class Ships:

"The question isn't who is going to let us – it's who is going to stop us?"

Deus est;
Deus est omnia…

God is; God is everything.

About the Author

"I am a scientist. I am an engineer. I am an explorer. I am also a man of faith. They are entirely consistent. Indeed, one cannot exist with full integrity without the other. They build upon one another and make each stronger. Without my faith - the rest would be empty, futile and; in the end, terribly meaningless. And without my science, I would understand nothing of the world I was created to change..."

- Dennis Chamberland

DENNIS CHAMBERLAND IS AN EXPLORER, AUTHOR, bioengineer, award winning nuclear engineer, space life scientist, and aquanaut. Following his role as a United States Naval Officer and after completing graduate studies, he again worked with the Navy as a civilian U.S. Government Nuclear Engineer. Dennis was a radiological control professional for Navy Nuclear submarines and was involved in a Nuclear Emergency Planning study at the Harvard School of Public Health.

He is the chief architect, design engineer, and builder of several undersea habitats, including NASA's *Scott Carpenter Space Analog Station*. Dennis served as the NASA Mission Commander for 14 undersea missions and was the Principal Investigator for the planting and harvesting of the first agricultural crop ever grown in a habitat on the seafloor. Dr. Chamberland is the author of the seminal and visionary work *Undersea Colonies* and is widely considered the world's leading professional in permanent undersea colonization. He was featured in National Geographic's documentary series *Naked Science* "City Under the Sea" episode, and Motherboard's "The Aquatic Life of Dennis Chamberland."

Dennis Chamberland was a design engineer at NASA's John F. Kennedy Space Center for Advanced Space Life Support Systems being considered for Moon and Mars bases, where he conceived the phrase and developed the philosophy of Resource Recovery in lieu of Waste

Processing, which was adopted across the Space Agency. During this time in his career, he was an active annual participant in the presentation of his work at the Princeton Space Studies Institute, chairing multiple sessions there.

During his 30-year distinguished career at NASA, Dennis was also privileged to apply his skills to the NASA Safety, Quality, and Reliability Office as an Operational Safety Specialist for the Space Shuttle orbiter fleet. While at KSC, Dennis additionally managed NASA scientific contracts, and sat as the Chairman of the NASA Kennedy Space Center Institutional Animal Care and Use Committee (IACUC) for 14 years, reviewing during his tenure all vertebrate animal life science payloads on Shuttle and those flown by Space Shuttle to the Russian Mir Space Station and later to the International Space Station. He was also an active member of the NASA KSC Speakers Bureau for over 20 years and spoke to groups and students across the U.S.

Dennis published the 1986 groundbreaking cover story on the bioethics of genetic engineering in *Christianity Today,* quoted as the conclusion of the President's Commission for the Study of Ethical Problems in Medicine and Behavioral Research. During his tenure as the Chairman of the KSC IACUC, he acted as a NASA representative on several Agency biomedical ethics review boards.

He was a Principal Investigator for a landmark scientific study conducted by a team from the John F. Kennedy Space Center, Brookhaven National Laboratory, McKnight Brain Institute at the University of Florida, and Cold Spring Harbor Laboratory that inquired into the chronic neurological effects of galactic cosmic radiation related to a crewed mission to Mars utilizing the Brookhaven Collider-Accelerator.

Dr. Chamberland is also the 2018 recipient of the prestigious Nuclear Professional of the Year Award from the International Atomic Energy Agency – ISOE North American Technical Center.

On the surface of Mars, all signals from Earth cease without warning and with no apparent cause. One minute the Earth is the brightest radio source in the known universe - the next minute it is as if the third planet from the sun had never existed at all. To the relatively new, first human colony on the Red planet, self-sufficiency is a distant dream. If Earth ceases to exist for whatever reason - their own days are likewise numbered. For the human race to survive, a lot of people are going to have to die on the bitterly cold and rarified air on Mars. Mars Base One is filled with the full sweep of human emotions from the bittersweet romance between men and women whose days are clearly numbered, to the mystery of losing contact with all they have ever known on earth, and the fear of a silent and cunning enemy intent on destruction and domination. The pages are filled with interplanetary intrigue, the race against an emotionless clock, a spectacular, savage wilderness and the specter of a war of insanity waged to survive just one more day. This is the 2nd edition - published and updated in 2022.

Available on Amazon.com in paperback, hard cover and Kindle.

THE ABYSS OF SPACE TRILOGY

The parallel novel trilogy to
Mars Base One - Abyss of Elysium

The desperate journey of a small band of survivors begins with an impossible escape from Earth in an antique ICBM to the hijacking of the International Space Station after the home planet is destroyed by nuclear winter. And that is only the beginning of their trials.

Join Aaron Seven as he leads the remnant of humanity on a thrilling, deadly adventure full of suspense, catastrophe, romance, and intrigue to their final destination at Mars Base One.

But not everyone will make it....

AN AARON SEVEN ADVENTURE

THE NOVEL TRILOGY BY DENNIS CHAMBERLAND

BOOKS BY DENNIS CHAMBERLAND

Aaron Seven Adventure Trilogies

Quantum Storms – Quantum Genesis

Quantum Storms – Quantum Devastation

Quantum Storms – Quantum Defiance

Abyss of Space – Armageddon

Abyss of Space – Singleton

Abyss of Space – Betrayal

Apocalypse Moon – Silent Death

Apocalypse Moon – Refuge Luna

Apocalypse Moon – Regeneration

Aquatica – Skyfall

Aquatica – Empire

Aquatica – Deliverance

Fiction by Dennis Chamberland

Mars Base One – Abyss of Elysium

Alyete – Dogs of Eros Damned

Science and Technology

Undersea Colonies

Departing Earth Forever: Warning and Promise

Departing Earth Forever: Alien Worlds

Thriving in the Days of COVID-19

Science Fact with Claudia Chamberland

The Proxima Manual of Space Exploration

Children of God Series

Consuming Fire

Proverbs for My Children

The Way Back Home

Eternity's Child

Eventide

Psalms of Ascent at Noon

The Spiritual Knight of the 21st Century

The Power of the 21st Century Knight

Don't Miss Your
Online
Companion!
DepartingEarthForever.com

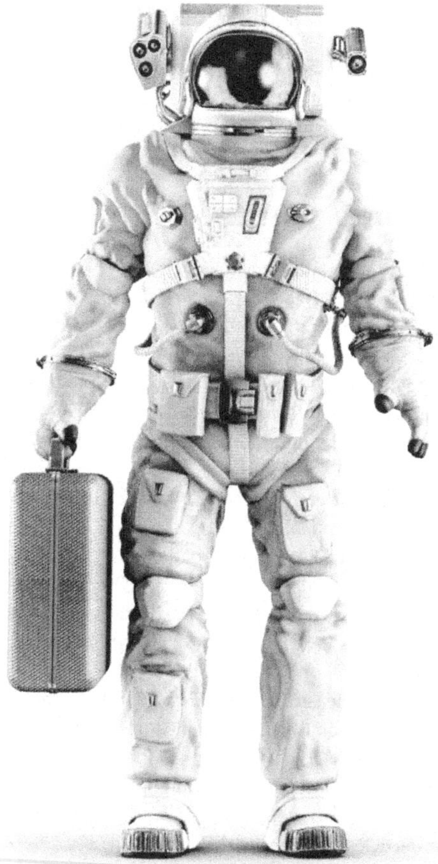

Chapter References Book One

Skies Are Clear

1. https://www.nationalgeographic.com/science/article/where-is-the-edge-of-space-and-what-is-the-karman-line
2. https://www.worldatlas.com/articles/what-is-the-karman-line.html
3. https://en.wikipedia.org/wiki/Kármán_line
4. Gangale, Thomas (2017). "The Non Kármán Line: An Urban Legend of the Space Age" (PDF). *Journal of Space Law*. 41 (2): 155.
5. https://www.nasa.gov/astronauts
6. https://www.nationalgeographic.com/science/article/where-is-the-edge-of-space-and-what-is-the-karman-line
7. https://www.epa.gov/radiation/radiation-terms-and-units

Why We Will Depart Earth Forever – Reason 1

1. https://www.forbes.com/global/2001/1029/060.html#48cc06f22080

Jamestown, Virginia

1..www.slate.com/articles/health_and_science/science/2014/06/pocahontas_wedding_re_enactment_john_rolfe_john_smith_and_native_americans.html)
2. John Thompson, The Journals of Captain John Smith: A Jamestown Biography, ISBN 1426200552, 2007, p. 139
3. http://www.pbs.org/wnet/secrets/death-jamestown-background/1428/
4. https://historicjamestowne.org/archaeology/jane/
5. https://www.britannica.com/topic/Virginia-Company

Let the Revolution Begin

1. The NASA Cosmic Radiation Research Team – 2006-2007: Juan M. Encinas, Marcelo E. Vazquez, Robert C. Switzer, Dennis W. Chamberland, Harry Nick, Howard G. Levine, Philip J. Scarpa, Grigori Enikolopov, and Dennis A. Steindler
2. https://www.sciencedaily.com/releases/2013/05/130530145930.htm
3. explorecuriocity.org/Explore/ArticleId/6421/carbon-dioxide-on-earth-and-on-the-iss.aspx
4. https://www.co2.earth/daily-co2
5..https://ec.europa.eu/environment/integration/research/newsalert/pdf/very_high_co2_levels_decrease_antioxidant_content_green_vegetables_452na5_en.pdf
6. https://www.nature.com/articles/npjmgrav20151
7. Office of the Inspector General. *NASA's efforts to maximize research on the international space station*. National Aeronautics and Space Administration, Report no.: IG-13-019 (8 July 2013).
8. http://theconversation.com/five-key-findings-from-15-years-of-the-international-space-station-51540
9. https://www.independent.co.uk/news/science/five-key-scientific-findings-from-15-years-of-the-international-space-station-a6796211.html

The Limits of Human Adaptation

1. https://www.livescience.com/34128-limits-human-survival.html
2. https://www.mayoclinic.org/first-aid/first-aid-fever/basics/art-20056685
3. https://www.psychologistworld.com/stress/general-adaptation-syndrome
4. Chamberland, Dennis, Undersea Colonies, Quantum Editions, 2007, p. 97.
5. Madsen, A., 1986, Cousteau – An Unauthorized Biography, p. 137
6. Miller, James W; Koblick, Ian G (1984). *Living and working in the sea*. Best Publishing p. 432.
7. Chamberland, Dennis, *Undersea Colonies*, Quantum Editions, 2007, p. 100-111
8. Ibid, p. 5.

A Brief Primer on Microgravity

1. physics.stackexchange.com/questions/29929/gravity-on-the-international-space-station
2. http://physicsbuzz.physicscentral.com/2015/03/the-happiest-thought-of-my-life-100.html
3. http://www.startribune.com/space-travel-tips-from-one-who-went-there/419430034/

4. Shuttle Mir – The United States and Russia Share History's Highest Stage, 2001, NASA SP-2001-2445, pp. 80
5. Martina Heer, William Paloski, Space motion sickness: incidence, etiology, and countermeasures, *Autonomic Neuroscience* Vol.129, no.1-2, 2006. Pp. 77-79
6. Shuttle Mir – The United States and Russia Share History's Highest Stage, 2001, NASA SP-2001-2445, pp. 80
7. www.ge.com/reports/post/74545065351/ge-researchers-to-investigate-link-between/

A Brief Primer on Cosmic Radiation

1. http://www.aasc.ucla.edu/cab/200708230001.html
2. worldnuclear.org/uploadedFiles/org/Features/Radiation/4_Background_Radiation%281%2
3. https://en.wikipedia.org/wiki/Sievert
4. https://www.nuclear-power.net/nuclear-engineering/radiation-protection/equivalent-dose/sievert-unit-of-equivalent-dose/
5. Princeton Univ.Open Source Radiation Safety Training, Module 3: Biological Effects
6. https://earth.gsfc.nasa.gov/geo/research/geodynamo
7. https://www.nasa.gov/vision/earth/lookingatearth/29dec_magneticfield.html
8. https://www.sciencedirect.com/topics/earth-and-planetary-sciences/solar-cosmic-rays
9. https://three.jsc.nasa.gov/articles/hu-spes.pdf
10. https://www.nasa.gov/mission_pages/sunearth/news/curiosity-daredevil.html
11. Kim, M.H.; George, K.A.; Cucinotta, F.A. Evaluation of skin cancer risks from lunar and mars missions. Adv. Space Res. 2006, 37, 1798–1803.
12. Hakura, Yukio (1976). "Interdisciplinary summary of solar/interplanetary events during August 1972." Space Sci. Rev. 19 (4–5): 411–457.
13. Carter, Brett (November 7, 2018). "Blasts from the Past: How massive solar eruptions probably detonated dozens of US sea mines." *The Conversation*.
14. https://www.space.com/spacex-starlink-satellites-lost-geomagnetic-storm
15. https://www.weather.gov/news/190504-sun-activity-in-solar-cycle
16. spaceweatherarchive.com/2019/07/16/cosmic-ray-update-new-results-from-the-moon/
17. https://www.space.com/heliosphere-map-solar-wind
18. https://iopscience.iop.org/article/10.3847/1538-4365/abf658
19. https://www.livescience.com/voyager-2-detects-heliopause-plasma-shield.html
20. https://www.cdc.gov/nceh/radiation/cosmic.html
21. Kim, M.H.; George, K.A.; Cucinotta, F.A. Evaluation of skin cancer risks from lunar and mars missions. Adv. Space Res. 2006, 37, 1798–180318.
22. sciencemag.org/news/2016/03/milky-way-s-black-hole-may-be-spewing-out-cosmic-rays
23. Chancellor, Jeffery C., Scott, Graham B.I., Sutton, Jeffrey P., Space Radiation: The Number One Risk to Astronaut Health beyond Low Earth Orbit. *Journal Life*, 2014, 4,491-510.
24. Simpson, J.A. Elemental and isotopic composition of the galactic cosmic rays. Ann. Rev. Nucl. Part. Sci. 1983, 33, 323–381.
25. Schimmerling, Walter (5 February 2011). "The Space Radiation Environment: An Introduction" (PDF). *The Health Risks of Extraterrestrial Environments*. Universities Space Research Association Division of Space Life Sciences.
26. Can People go to Mars? Science@NASA, February 17, 2004.
27. Schimmerling, W (5 Feb 2011). "The Space Radiation Environment: An Introduction". *The Health Risks of Extraterrestrial Environments*. Universities Space Research Association Division of Space Life Sciences.
28. Contribution of High Charge and Energy (HZE) Ions During Solar-Particle Event of September 29, 1989 Kim, Myung-Hee Y.; Wilson, John W.; Cucinotta, Francis A.; et al., NASA Johnson Space Center; Langley Research Center, May 1999
29. http://www.citycollegiate.com/atomic_structureXIg.htm
30. https://en.wikipedia.org/wiki/Fermi–Dirac_statistics

31. Cucinotta FA, Nikjoo H, Goodhead DT. The effects of delta rays on the number of particle-track transversals per cell in laboratory and space exposures. Radiat Res 1998; 150
32. Chancellor, Jeffery C., Scott, Graham B.I., Sutton, Jeffrey P., Space Radiation: The Number One Risk to Astronaut Health beyond Low Earth Orbit. *Journal Life*, 2014, 4,491-510.
33. https://cosmic.lbl.gov/SKliewer/Cosmic_Rays/Primary.htm
34. https://www.nature.com/articles/s41586-021-03498-z
35. https://phys.org/news/2021-05-observatory-dozen-pevatrons-photons-exceeding.html
36. National Council on Radiation Protection and Measurements (NCRP). Information Needed to Make Radiation Protection Recommendations for Space Missions Beyond Low-Earth Orbit; NCRP: Bethesda, MD, USA, 2006.
37. Linenger, Jerry M. (13 January 2000). *Off The Planet: Surviving Five Perilous Months Aboard The Space Station MIR.*
38. https://curiosity.com/topics/astronauts-can-actually-see-cosmic-rays-curiosity/
39. National Council on Radiation Protection and Measurements (NCRP). Information Needed to Make Radiation Protection Recommendations for Space Missions Beyond Low-Earth Orbit; NCRP: Bethesda, MD, USA, 2006.
40. Wang H, Wang Y. Heavier ions with a different linear energy transfer spectrum kill more cells due to similar interference with the Ku-dependent DNA repair pathway. *Radiation Research*. 2014;182(4):458-461. doi:10.1667/RR13857.1
41. https://www.universetoday.com/140217/plans-for-a-modular-martian-base-on-that-would-provide-its-own-radiation-shielding/)
42. nuclearsafety.gc.ca/eng/pdfs/Fact_Sheets/Fact-Sheet-Background-Radiation-eng.pdf
43. sci-news.com/space/science-mars-radiation-measurements-surface-01629.html
44. https://www.newscientist.com/article/2167224-amazing-gif-shows-dust-and-cosmic-rays-raining-down-on-comet-67p/
45. https://www.smithsonianmag.com/smart-news/video-comet-snowstorm-likely-shows-dust-particles-180968889/
46. ibtimes.com/nasa-photo-shows-mysterious-cosmic-ray-light-mars-horizon-2811489
47. www.nasa.gov/sites/default/files/atoms/files/sf_radiation_stu_bob.pdf
48. Schaefer H, Benton E, Henke R, Sullivan, J. Nuclear track recordings of the astronauts' radiation exposure on the first lunar landing mission Apollo XI. Radiat Res 1972; 49:245–71.
49. https://observer.com/2016/07/space-radiation-devastated-the-lives-of-apollo-astronauts/
50. ntrs.nasa.gov/api/citations/20200001314/downloads/20200001314.pdf
51. https://www.nationalacademies.org/our-work/assessment-of-strategies-for-managing-cancer-risks-associated-with-radiation-exposure-during-crewed-space-missions
52. https://www.nasa.gov/pdf/284273main_Radiation_HS_Mod1.pdf
53. https://www.sciencefriday.com/segments/nasa-to-the-moon-and-definitely-beyond/
54. https://www.space.com/22252-women-astronauts-radiation-risk.html
55. https://www.nasa.gov/pdf/284273main_Radiation_HS_Mod1.pdf
56. newscientist.com/article/dn2956-space-station-radiation-shields-disappointing/
57. Kennedy, Ann R., Biological effects of space radiation and development of effective countermeasures. *Life Sciences in Space Research* (2014),
58. 60 Minutes, "Who Killed Alexander Litvinenko?" 7 January 2007.
59. scienceblog.cancerresearchuk.org/2008/08/29/radioactive-polonium-in-cigarette-smoke/
60. https://us.macmillan.com/books/9780374534714
61. https://www.mdpi.com/2075-1729/4/3/491/htm
62. Chancellor, Jeffery C., Scott, Graham B.I., Sutton, Jeffrey P., Space Radiation: The Number One Risk to Astronaut Health beyond Low Earth Orbit. *Journal Life*, 2014, 4,491-510.

Cosmic Radiation Measured in Deep Space

1. https://ntrs.nasa.gov/archive/nasa/casi.ntrs.nasa.gov/19730010172.pdf

2..https://www.ndeed.org/EducationResources/CommunityCollege/RadiationSafety/radiation_safety_equipment/thermoluminescent.htm
3. https://iopscience.iop.org/article/10.1088/0031-9155/40/2/009
4. https://www.ncbi.nlm.nih.gov/pubmed/17578870
5. https://radiopaedia.org/articles/thermoluminescent-dosimeter?lang=us
6. https://ntrs.nasa.gov/archive/nasa/casi.ntrs.nasa.gov/19730010172.pdf
7. https://mars.nasa.gov/msl/spacecraft/instruments/rad/
8. https://www.hq.nasa.gov/alsj/tnD7080RadProtect.pdf
9. https://ntrs.nasa.gov/archive/nasa/casi.ntrs.nasa.gov/19730010172.pdf
10. https://www.hq.nasa.gov/alsj/tnD7080RadProtect.pdf
11. sites.nationalacademies.org/cs/groups/depssite/documents/webpage/deps_179047.pdf
12. https://www.nasa.gov/pdf/284277main_Radiation_MS.pdf
13. Reitz, Guunter. Berger, Thomas. Matthiae, "Radiation exposure in the moon environment", *Planetary and Space Science,* V 74, Issue. 1, December 2012, 78-83
14. https://spaceflightnow.com/2019/06/18/10-years-since-its-launch-nasa-lunar-orbiter-remains-crucial-for-planned-moon-landings/
15. spaceweatherarchive.com/2019/07/16/cosmic-ray-update-new-results-from-the-moon/
16. Reitz, Guunter. Berger, Thomas. Matthiae, "Radiation exposure in the moon environment", *Planetary and Space Science,* V 74, Is. 1, December 2012, 78-83
17. https://agupubs.onlinelibrary.wiley.com/doi/pdf/10.1002/2013SW000995
18. United Nations Scientific Committee on the Effects of Atomic Radiation (2008). Sources and effects of ionizing radiation. New York: United Nations (published 2010). p. 4.
19. https://phys.org/news/2019-06-showstopper-mars-exploration.html

Lies, Damned Lies, and Statistics

1. https://www.cdc.gov/niosh/ocas/pdfs/42cfr81/42cfr81.pdf
2. http://hps.org/publicinformation/radterms/radfact142.html
3. cancer.gov/about-cancer/causes-prevention/risk/substances/radon/radon-fact-sheet
4. http://mathscinotes.com/2015/01/cat-litter-and-radioactivity/
5. Energy Employees Occupational Illness Compensation Program Act of 2000
6. Guidelines for Determining the Probability of Causation 42 CFR 81
7. https://radiationcalculators.cancer.gov/irep/
8. "Space Radiation and Cataracts in Astronauts", *Radiation Research* 156(5 Pt 1):460-6·Nov 2001
9. "Apollo Lunar Astronauts Show Higher Cardiovascular Disease Mortality: Possible Deep Space Radiation Effects on the Vascular Endothelium", *Nature,* Scientific Reports volume 6, Article number: 29901 (2016)
10. https://glycanspacexr-agency.com/docs/cataracts.pdf
11. https://three.jsc.nasa.gov/articles/AstronautRadLimitsFC.pdf
12. "Apollo Lunar Astronauts Show Higher Cardiovascular Disease Mortality: Possible Deep Space Radiation Effects on the Vascular Endothelium", *Nature,* Scientific Reports volume 6, Article number: 29901 (2016)
13. https://www.nasa.gov/hhp/treat-act
14. https://www.cdc.gov/niosh/topics/aircrew/cosmicionizingradiation.html
15. ntrs.nasa.gov/api/citations/20190001111/downloads/20190001111.pdf

A Pretense of Knowledge

1. https://cafehayek.com/2011/12/hayek-on-scientistic-hubris.html
2. https://www.nature.com/articles/srep29901

The Malpractice of Failure is Not an Option

1. https://www.thetimes.co.uk/article/the-dead-planet-you-may-die-on-my-spaceship-says-musk-7fz3jqsqx (March 12 2018)
2. https://en.wikipedia.org/wiki/Failure_Is_Not_an_Option

3. https://www.nasa.gov/johnson/HWHAP/apollo-8-part-1
4. https://www.nasaspaceflight.com/2017/07/falcon-heavy-prepares-debut-musk-urges-caution-expectations/
5. https://www.aip.org/fyi/2021/fy22-budget-request-nasa
6. https://employeeorientation.nasa.gov/contractors/default.htm
7. https://www.scribd.com/document/497259653/SpaceX-Response-DOJ-Hiring-Subpoena#from_embed?campaign=SkimbitLtd&ad_group=35871X943606X1605a55d6345a14763edac7ee3c18c8b&keyword=660149026&source=hp_affiliate&medium=affiliate
8. https://www.foxbusiness.com/features/elon-musk-on-mars-flights-good-chance-first-passengers-will-die

SLS-Orion

1. "How much does it cost to launch a Space Shuttle?". NASA. Retrieved June 28, 2011.
2. https://www.space.com/1567-nasa-moon-plans-apollo-steroids.html
3. Ibid.
4. https://archive.triblive.com/news/nasas-new-rocket-to-fly-twice-cost-38-billion/
5. www.cnn.com/2016/10/11/opinions/america-will-take-giant-leap-to-mars-barack-obama/
6. https://www.nasa.gov/content/what-is-nasa-s-asteroid-redirect-mission
7. astronautix.com/v/vonbraunstation.html
8..https://www.nasa.gov/sites/default/files/atoms/files/19690804_manned_mars_landing_presentation_to_the_space_task_group_by_dr._wernher_von_braun.pdf
9. https://www.theverge.com/2017/5/10/14886570/nasa-space-launch-system-rocket-ula-blue-origin-spacex
10. https://www.space.com/nasa-artemis-moon-program-93-billion-2025
11. cbsnews.com/news/artemis-moon-program-cost-delays-nasa-inspector-general/)
12. https://www.space.com/spacex-starship-super-heavy.html
13. https://arstechnica.com/science/2020/05/nasa-will-pay-a-staggering-146-million-for-each-sls-rocket-engine/)
14. spacenews.com/nasa-audit-reveals-massive-overruns-in-sls-mobile-launch-platform/
15. www.bloomberg.com/opinion/articles/2021-02-18/scrap-nasa-s-space-launch-system
16. https://arstechnica.com/science/2020/05/nasa-will-pay-a-staggering-146-million-for-each-sls-rocket-engine/
17. https://sam.gov/opp/d20ca9e5714b4fa08b75b0a3612f38f9/view
18. https://arstechnica.com/science/2021/10/nasa-wants-to-buy-sls-rockets-at-half-price-fly-them-into-the-2050s/
19. https://spacenews.com/bridenstine-hints-artemis-3-could-land-near-apollo-site/
20. https://www.nasa.gov/sites/default/files/atoms/files/artemis_plan-20200921.pdf
21. https://www.whitehouse.gov/briefings-statements/remarks-vice-president-pence-fifth-meeting-national-space-council-huntsville-al/

Blind Alleys in Exploration

1. https://www.chemistryworld.com/news/science-really-does-advance-one-funeral-at-a-time-study-suggests/3010961.article
2. Madsen, A., *Cousteau - An Unauthorized Biography*, Beaufort Books, New York, 1986, P137
3. sciencing.com/important-thomas-edison-invention-light-bulb-6305.html
4. https://www.mattmcwilliams.com/winston-churchill-success-quote/
5. Kelly, Scott (2017). *Endurance: A Year in Space, a Lifetime of Discovery*. With Margaret Lazarus Dean. Alfred A. Knopf, division of Penguin Random House.

Why We Will Depart Earth Forever – Reason 3

1. https://www.businessinsider.com/space-travel-per-seat-cost-soyuz-2016-9
2. https://posey.house.gov/news/documentsingle.aspx?DocumentID=394240
3. https://wccftech.com/spacex-could-earn-30-billion-annually-from-starlink-10x-of-sending-iss-supplies-elon-musk/

4. www.the-sun.com/lifestyle/tech/1456612/elon-musk-plans-for-mars-timeline-spacex/)
5. https://www.independent.co.uk/life-style/gadgets-and-tech/news/elon-musk-ticket-mars-spacex-journey-price-cost-date-a8775096.html
6. https://www.usatoday.com/story/tech/nation-now/2017/01/18/nasa-planning-mission-asteroid-worth-10000-quadrillion/96709250/
7. https://news.ycombinator.com/item?id=358351

Radically Divergent

1. https://www.washingtonpost.com/news/the-switch/wp/2018/02/26/after-2016-rocket-explosion-elon-musks-spacex-looked-seriously-at-sabotage/
2. http://aviationweek.com/blog/nasa-cnes-warn-spacex-challenges-flying-reusable-falcon-9-rocket?sfvc4enews=42&cl=article_4
3. http://nasawatch.com/archives/2019/11/sls-sticker-sho.html
4. https://arstechnica.com/science/2016/09/nasa-chief-says-hes-not-a-big-fan-of-private-investment-in-large-rockets/
5. https://arstechnica.com/science/2020/09/former-nasa-administrator-says-sls-rocket-will-go-away/
6. http://fortune.com/2017/07/22/elon-musk-falcon-heavy-test-failure/
7. www.washingtonpost.com/technology/2020/02/28/boeing-admits-starliner-testing-flaws/
8. https://www.cnbc.com/2020/08/13/elon-musk-spacex-competitor-ula-a-complete-waste-of-taxpayer-money.html
9. https://medium.com/illumination/spacexs-starship-will-make-all-other-rockets-obsolete-71240a143260
10. https://arstechnica.com/science/2019/11/nasa-does-not-deny-the-over-2-billion-cost-of-a-single-sls-launch/
11. https://capitalresearch.org/article/nasa/
12. Tarpley, Ashley F.; Starr, Brett R.; Tartabini, Paul V.; et al. *"Ares I-X Range Safety Trajectory Analyses Overview and Independent Validation and Verification"*

One Earth-Normal

1. Chamberland, Dennis. *Undersea Colonies*, Quantum Editions, 2007. Pg 5.
2. http://www.wiseoldsayings.com/excuses-quotes/
3. https://sayingspoint.com/excuses-sayings/
4. https://sayingspoint.com/excuses-sayings/
5. https://www.cnet.com/news/elon-musk-hints-spacex-running-track-will-be-2001-a-space-odyssey-inspired/
6. https://www.therichest.com/rich-list/the-19-coldest-and-hottest-inhabited-places-on-earth/
7. Aitken, et al., International Journal of Andrology, April 23(2) P 116-120

Ethics of Neverland

1. "Tuskegee Study - Timeline". *NCHHSTP. CDC. June 25, 2008*. Retrieved December 4, 2008
2. Kim, Oliver J.; Magner, Lois N. (2018). A History of Medicine. Taylor & Francis
3. "These Atomic Bomb Tests Used US Troops as Guinea Pigs", Smithsonian Society Video, August 17, 2019
4. https://www.militarytimes.com/veterans/2016/06/02/we-were-guinea-pigs-documentary-puts-atomic-veterans-in-limelight/
5. https://www.militarytimes.com/opinion/commentary/2021/12/02/veterans-of-us-atomic-age-deserve-compensation/
6. http://www.pbs.org/wgbh/nova/next/space/nasa-mars-radiation-rule/
7. https://oig.nasa.gov/docs/IG-16-003.pdf
8. https://phys.org/news/2014-04-health-mars-mission-nasa-limits.html
9. Ibid
10..https://science.house.gov/imo/media/doc/Williams%20Astronaut%20HealthCare%20Testimony%20FINAL.PDF

11. nap.edu/resource/26155/Space-Radiation-Study_Recommendations-and-Conclusions.pdf
12. https://www.nrc.gov/reading-rm/doc-collections/nuregs/staff/sr0713/v41/index.html
13. https://ncrponline.org/publications/reports/ncrp-reports-116/
14. nap.edu/resource/26155/Space-Radiation-Study_Recommendations-and-Conclusions.pdf
15. Ibid.
16. https://www.cdc.gov/niosh/topics/aircrew/cosmicionizingradiation.html
17. https://www.space.com/nasa-changing-astronaut-radiation-exposure-limit
18. Ibid.
19. https://ncrponline.org/publications/reports/ncrp-report-98/
20. https://www.nasa.gov/pdf/284277main_Radiation_MS.pdf
21. https://www.nrc.gov/about-nrc/radiation/health-effects/rad-exposure-cancer.html
22. Schweitzer, Albert, *Civilization and Ethics* (Sydney: Allen & Unwin,1961) Preface

Expendable Astronauts

1. https://www.hq.nasa.gov/alsj/MelAverner.html
2. Professional Notes – Dennis Chamberland
3. http://www.world-nuclear.org/uploadedFiles/org/WNA/Publications/Nuclear _Information/pocket_guide_radiation.pdf
4. Ibid
5. https://nuclear.duke-energy.com/2012/08/21/radiation-protection-for-nuclear-employees
6. https://www.nrc.gov/about-nrc/radiation/health-effects/rad-exposure-cancer.html
7. https://www.ncbi.nlm.nih.gov/pmc/articles/PMC3997516/#idm140417238913968title
8. Chancellor, Jeffery C., Scott, Graham B.I., Sutton, Jeffrey P., Space Radiation: The Number One Risk to Astronaut Health beyond Low Earth Orbit. *Journal Life*, 2014, 4,491-510.
9. Fry RJM, Powers-Risius P, Alpen EL, Ainsworth EJ (1985) High LET radiation carcinogenesis. Radiat Res 104: S188–S195.
10. Alpen EL, Powers-Risius P, Curtis SB, DeGuzman R (1993) Tumorigenic potential of high-Z, high-LET charged particle radiations. Radiat Res 88: 132–143.
11. Dicello JF, Christian A, Cucinotta FA, Gridley DS, Kathirithhamby R, et al. (2004) In vivo mammary tumorigenesis in the sprague-dawley rat and microdosimetric correlates. Phys Med Biol 49: 3817–3830.
12. Weil MM, Bedford JS, Bielefeldt-Ohmann H, Ray AF, Gernik PC, et al. (2009) Incidence of acute myeloid leukemia and hepatocellular carcinoma in mice irradiated with 1 GeV/nucleon 56Fe ions. Radiat Res 172: 213–219.
13. Imaoka T, Nishimura, Kakinuma S, Hatano Y, Ohmachi Y, et al. (2007) High relative biological effectiveness of carbon ion irradiation on induction of rat mammary carcinoma and its lack of H-ras and Tp53 mutations. Inter J Radiat Oncol Biophys 69: 194–203.
14. Trani D, Datta K, Doiron K, Kallakury B, Fornace Jr AJ (2010) Enhanced intestinal tumor multiplicity and grade in vivo after HZE exposure: mouse models for space radiation risk estimates. Radiat Environ Biophys 49: 389–396.
15. Datta K, Suman S, Kallakury BV, Fornace Jr AJ (2013) Heavy ion radiation exposure triggered higher intestinal tumor frequency and greater b-catenin activation than c radiation in APC Min/+ mice. PLoS One 8 e59295.
16. Grahn D, Lombard LS, Carnes BA (1992) The comparative tumorigenic effects of fission neutrons and Cobalt-60 c rays in B6CF1 mouse. Radiat Res 129: 19-36
17. Wolfe C, Lafuma J, Masse R, Morin M, Kellerer A (2000) Neutron RBE for induction of tumors with high lethality in spraque-dawley rats. Radiat Res 154: 412–420.
18. National Council on Radiation Protection and Measurements (1989) *Guidance on radiation received in space activities.* NCRP Report 98: Bethesda, MD.
19. ncbi.nlm.nih.gov/pmc/articles/PMC3997516/#idm140417238913968title
20. National Safety Council, Injury facts 2011 edition. National Safety Council: Itasca IL.
21. https://oig.nasa.gov/audits/reports/FY16/IG-16-003.pdf

22. Institute of Medicine. 2001. *Safe Passage: Astronaut Care for Exploration Missions.* Washington, DC: The National Academies Press. https://doi.org/10.17226/10218.
23. Ibid.
24. https://phys.org/news/2014-04-health-mars-mission-nasa-limits.html
25. https://www.centralflalaw.com/culpable-negligence.html
26. https://scitechdaily.com/nasa-releases-moon-to-mars-deep-space-exploration-objectives-seeks-public-input/

Multimorbidity

1. https://en.m.wikipedia.org/wiki/List_of_space_travelers_by_name
2. http://planetary-science.org/planetary-science-3/exploration-2/the-physiological-and-psychological-aspects-of-sending-humans-to-mars/
3. http://clowder.net/hop/railroad/sched.html
4. Roach, James M.; Schoene, Robert B. (2002). "High-Altitude Pulmonary Edema". In Pandolf, Kent B.; Burr, Robert E. *Medical Aspects of Harsh Environments* (PDF).
5. Bärtsch, Peter; Swenson, Erik (2013). "Acute High-Altitude Illnesses". *The New England Journal of Medicine.*
6. https://news.nationalgeographic.com/2018/03/scott-kelly-astronaut-space-station-dna-health-science/
7. https://www.livescience.com/62000-scott-kelly-dna-twin-study.html
8. Li X, Fang P, et al. (March 2017). "Mitochondrial ROS, uncoupled from ATP synthesis, determine endothelial activation for both physiological recruitment of patrolling cells and pathological recruitment of inflammatory cells". Can J Physiol Pharmacol. 95 (3): 247–252.
9. Li X, Fang P, Li Y, Kuo YM, et al. (April 2016). "Mitochondrial Reactive Oxygen Species Mediate Lysophosphatidylcholine-Induced Endothelial Cell Activation". *Arteriosclerosis, Thrombosis, and Vascular Biology.* 36 (6): 1090–100
10. https://www.ncbi.nlm.nih.gov/pmc/articles/PMC3412069/
11. https://news.nationalgeographic.com/2018/03/scott-kelly-astronaut-space-station-dna-health-science/
12. https://www.sciencedirect.com/science/article/pii/S1084952117303968
13. Shanmugarajan S, Zhang Y, et al. Combined Effects of Simulated Microgravity and Radiation Exposure on Osteoclast Cell Fusion. Int J Mol Sci. 2017 Nov 18;18(11):2443. doi: 10.3390/ijms18112443. PMID: 29156538; PMCID: PMC5713410.
14. María Moreno-Villanueva, Michael Wong, Tao Lu1, Ye Zhang and Honglu Wu, "Interplay of space radiation and microgravity in DNA damage and DNA damage response", *Nature Partner Journal of Microgravity*, 10 May 2017
15. Chancellor, Jeffery, Scott, Graham B.I., Sutton, Jeffrey, "Space Radiation: The Number One Risk to Astronaut Health beyond Low-Earth orbit." *Journal Life*, 2014, 4,491-510.
16. https://www.ncbi.nlm.nih.gov/pubmed/27421249
17. https://www.nrc.gov/about-nrc/radiation/health-effects/rad-exposure-cancer.html
18. https://www.ncbi.nlm.nih.gov/pubmed/27421249
19. https://www.nature.com/articles/srep29901
20. https://www.jstor.org/stable/40305968
21. Maks, C.J.; Wan, X.S.; Ware, J.H; et al. "Analysis of white blood cell counts in mice after gamma- or proton-radiation exposure." Radiat. Res. 2011, 176, 170–176.
22. Sanzari, J.K.; Wambi, C.; Lewis-Wambi, J.S.; Kennedy, A.R. *Antioxidant dietary supplementation in mice exposed to proton radiation attenuates expression of programmed cell death-associated genes.* Radiat. Res. 2011, 175, 650–656.
23. Wilson, J.M.; Krigsfeld, G.S.; Sanzari, J.K.; Wagner, E.S.; Mick, R.; Kennedy, A.R. Comparison of hindlimb unloading and partial weight suspension models for spaceflight-type condition induced effects on white blood cells. Adv. Space Res. 2012, 49, 237–248.

24. Chancellor, Jeffery C., Scott, Graham B.I., Sutton, Jeffrey P., Space Radiation: The Number One Risk to Astronaut Health beyond Low-Earth orbit. Journal Life, 2014, 4,491-510.
25. Kelly, Scott. "*Endurance: A Year in Space, A Lifetime of Discovery*". With Margaret Lazarus Dean. Penguin Random House (2017) Prologue.
26. https://www.space.com/1567-nasa-moon-plans-apollo-steroids.html

Chronic Microgravity Exposure Etiology

1. Dennis Chamberland, Professional Career Notes, "Personal Discussions with NASA Flight Surgeons."
2. Kelly, Scott (2017). *Endurance: A Year in Space, a Lifetime of Discovery*. With Margaret Lazarus Dean. Alfred A. Knopf, division of Penguin Random House.
3. Stavnichuk, M., Mikolajewicz, N., Corlett, T. et al. A systematic review and meta-analysis of bone loss in space travelers. npj *Microgravity* **6,** 13 (2020). https://doi.org/10.1038/s41526-020-0103-2
4. www.space.com/1732-martian-dust-major-risk-manned-mission.html
5. parabolicarc.com/2021/02/07/what-happens-to-the-human-body-in-space/
6. Ulbrich C, Wehland M, Pietsch J, Aleshcheva G, Wise P, van Loon J, Magnusson N, Infanger M, Grosse J, Eilles C, Sundaresan A, Grimm D. The impact of simulated and real microgravity on bone cells and mesenchymal stem cells. Biomed Res Int. 2014;2014:928507. doi: 10.1155/2014/928507. Epub 2014 Jul 10. PMID: 25110709; PMCID: PMC4119729.
7. "*Space Bones*". NASA. October 1, 2001
8. Cavanagh PR, Licata AA, Rice AJ. Exercise and pharmacological countermeasures for bone loss during long-duration space flight. Gravit Space Biol Bull. 2005 Jun;18(2):39-58. PMID: 16038092.
9. Stavnichuk, M., Mikolajewicz, N., Corlett, T. et al. A systematic review and meta-analysis of bone loss in space travelers. npj *Microgravity* **6,** 13 (2020). https://doi.org/10.1038/s41526-020-0103-2
10. Shanmugarajan S, Zhang Y, et al. *Combined Effects of Simulated Microgravity and Radiation Exposure on Osteoclast Cell Fusion*. Int J Mol Sci. 2017 Nov 18;18(11):2443. doi: 10.3390/ijms18112443. PMID: 29156538; PMCID: PMC5713410.
11. Jeffrey Willey, "Effects of the Spaceflight Environment on Bone Health", Radiation and Flight – A Down to Earth Look at Risks, National Council on Radiation Protection and Measurements (NCRP) 57th Annual Mtg, April 2021
12. https://www.tasnimnews.com/en/news/2020/09/08/2344155/injection-could-help-astronauts-stop-loss-of-muscle-bone-during-spaceflight
13. https://pubmed.ncbi.nlm.nih.gov/10908405/
14. https://www.thelancet.com/journals/lancet/article/PIIS0140-6736(00)02217-0/fulltext
15. www.hughesfulfordlab.com/PublicationPDFs/physiological%20effects.pdf
16. Jeffrey Willey, "Effects of the Spaceflight Environment on Bone Health", Radiation and Flight – A Down to Earth Look at Risks, National Council on Radiation Protection and Measurements (NCRP) 57th Annual Mtg, April 2021
17. Ibid.
18. Ibid.
19. Ibid.
20. Stavnichuk, M., Mikolajewicz, N., Corlett, T. et al. A systematic review and meta-analysis of bone loss in space travelers. npj *Microgravity* 6, 13 (2020). https://doi.org/10.1038/s41526-020-0103-2
21. Ibid.
22. Ibid.
23. Joanna B Tyrovola, The "Mechanostat Theory" of Frost and the OPG/RANKL/RANK System, National Library of Medicine, February 2016

24. Stavnichuk, M., Mikolajewicz, N., Corlett, T. et al. A systematic review and meta-analysis of bone loss in space travelers. npj *Microgravity* 6, 13 (2020). https://doi.org/10.1038/s41526-020-0103-2
25. https://www.nasa.gov/pdf/64249main_ffs_factsheets_hbp_atrophy.pdf
26. physoc.onlinelibrary.wiley.com/doi/full/10.1113/jphysiol.2010.188508
27. https://www.wired.com/2010/08/astronaut-muscle-waste/
28. https://www.newscientist.com/article/2223705-low-gravity-in-space-made-some-astronauts-blood-flow-backwards/#ixzz6600a7oyn
29. https://jamanetwork.com/journals/jamanetworkopen/fullarticle/2755307
30. https://www.cnn.com/2019/11/15/health/astronaut-blood-flow-clot-scn-trnd/index.html
31. Ibid.
32. https://www.nejm.org/doi/full/10.1056/NEJMc1905875
33. https://academic.oup.com/eurheartj/advance-article-abstract/doi/10.1093/eurheartj/ehaa359/5840773?redirectedFrom=fulltext
34. www.nasa.gov/mission_pages/station/research/experiments/1038.html
35. https://www.washingtonpost.com/national/health-science/the-mysterious-syndrome-impairing-astronauts-eyesight/2016/07/09
36. Tain RW, Bagci AM, Lam BL, Sklar EM, Ertl-Wagner B, Alperin N. Determination of cranio-spinal canal compliance distribution by MRI: Methodology and early application in idiopathic intracranial hypertension. J Magn Reson Imaging. 2011 Dec;34(6):1397-404. doi: 10.1002/jmri.22799. Epub 2011 Oct 3. PMID: 21972076; PMCID: PMC3221868.
37. Brandon Macias, et al., Intracranial Pressure And Its Effect On Vision In Space And On Earth, World Scientific Publications, June 2017
38. https://www.dovemed.com/diseases-conditions/choroidal-folds/
39. https://www.ncbi.nlm.nih.gov/pubmed/29091569
40. https://www.sciencealert.com/brain-structures-change-size-long-term-space-travel-2017
41. https://www.space.com/26903-astronaut-immune-systems-in-space.html
42. Ibid.
43. https://science.nasa.gov/science-news/news-articles/your-immune-system-in-space
44. https://www.ncbi.nlm.nih.gov/pmc/articles/PMC2804794/

Partial Gravity Scandal

1. "Space Bones". *NASA. October 1, 2001*
2. Ulbrich C, Wehland M, Pietsch J, Aleshcheva G, Wise P, van Loon J, Magnusson N, Infanger M, Grosse J, Eilles C, Sundaresan A, Grimm D. The impact of simulated and real microgravity on bone cells and mesenchymal stem cells. Biomed Res Int. 2014;2014:928507. doi: 10.1155/2014/928507. Epub 2014 Jul 10. PMID: 25110709; PMCID: PMC4119729.
3. Jeffrey Willey, "Effects of the Spaceflight Environment on Bone Health", Radiation and Flight – A Down to Earth Look at Risks, National Council on Radiation Protection and Measurements (NCRP) 57th Annual Meeting, April 2021
4. https://www.nasa.gov/pdf/64249main_ffs_factsheets_hbp_atrophy.pdf
5. nature.com/scitable/forums/science-in-africa/plants-defying-gravity-22738096/
6. https://www.mdpi.com/2075-1729/11/2/109/htm#B84-life-11-00109
7. REMOVED
8. https://forum.nasaspaceflight.com/index.php?topic=14987.0
9. https://spacenews.com/nasa-halts-work-space-station-centrifuge-research-hardware/
10. Dennis Chamberland, Apocalypse Moon, Quantum Editions, 2018.
11. Ibid.
12. https://www.astrobio.net/moon-to-mars/shades-of-gravity/
13. https://www.ncbi.nlm.nih.gov/pubmed/28860998
14. http://lis.arc.nasa.gov/lis/Programs/Cosmos/overview/Cosmos_Biosat.html
15. http://lis.arc.nasa.gov/lis/Programs/Cosmos/Cosmos_782/Cosmos_782.html

16. https://lsda.jsc.nasa.gov/Mission/miss_payload/85
17. https://www.astrobio.net/moon-to-mars/shades-of-gravity

ALARA Range of Cosmic Radiation

1. https://www.nrc.gov/reading-rm/basic-ref/glossary/alara.html
2. http://www.oecd-nea.org/rp/pubs/2009/6386-isoe.pdf
3. https://pubs.rsna.org/doi/10.1148/radiol.2020200200
4. https://srag.jsc.nasa.gov/Publications/TM104782/techmemo.htm
5. https://www.nasa.gov/pdf/284273main_Radiation_HS_Mod1.pdf
6. https://www.nasa.gov/centers/johnson/pdf/514218main_FC-01_Overview_Integrated_SRP_Program.pdf
7. L. Walsh, U. Schneider, A. Fogtman, et al., Research plans in Europe for radiation health hazard assessment in exploratory space missions, *Life Sciences in Space Research*, Volume 21,2019, Pages 73-82, ISSN 2214-5524, https://doi.org/10.1016/j.lssr.2019.04.002.
8. https://www.nationalacademies.org/our-work/assessment-of-strategies-for-managing-cancer-risks-associated-with-radiation-exposure-during-crewed-space-missions
9. www.nationalacademies.org/assessment-of-strategies-for-managing-cancer-risks-associated-with-radiation-exposure-during-crewed-space-missions-public-webinar
10. https://www.nationalacademies.org/our-work/assessment-of-strategies-for-managing-cancer-risks-associated-with-radiation-exposure-during-crewed-space-missions
11. https://www.nap.edu/resource/26155/Space-Radiation-Study_Report-Highlights.pdf
12. https://www.cdc.gov/niosh/topics/aircrew/cosmicionizingradiation.html
13. Ibid.
14. https://ntrs.nasa.gov/archive/nasa/casi.ntrs.nasa.gov/20070010704.pdf
15. https://www.nasa.gov/pdf/284273main_Radiation_HS_Mod1.pdf
16. https://www.cancer.gov/about-cancer/understanding/statistics
17. nap.edu/resource/26155/Space-Radiation-Study_Recommendations-and-Conclusions.pdf

Chronic Cosmic Radiation Exposure Etiology

1. Townsend, L.; Cucinotta, F.; Wilson, J.; Bagga, R. Estimates of hze particle contributions to space radiation exposures on interplanetary missions. Adv. Space Res. 1994, 14, 671–674.
2. Schimmerling, Walter 5 February 2011. "The Space Radiation Environment: An Introduction" *PDF*. The Health Risks of Extraterrestrial Environments. Universities Space Research Association Division of Space Life Sciences
3. ncrponline.org/wp-content/themes/ncrp/PDFs/ExecSumm-NCRP-Report-No-153.pdf
4. https://www.ncbi.nlm.nih.gov/pmc/articles/PMC3473681/
5. Chancellor, Jeffery C., Scott, Graham B.I., Sutton, Jeffrey P., Space Radiation: The Number One Risk to Astronaut Health beyond Low Earth Orbit. *Journal Life*, 2014, 4,491-510.
6. Simpson, J. A. Elemental and isotopic composition of the galactic cosmic rays. Annu. Rev. Nucl. Part. Sci. 33, 323–382 (1983).
7. Ibid.
8. Badhwar, G. D. et al. Radiation environment on the Mir orbital station during solar minimum. Adv. Space Res. 22, 501–510 (1998).
9. Chancellor, Jeffery C., Scott, Graham B.I., Sutton, Jeffrey P., Space Radiation: The Number One Risk to Astronaut Health beyond Low Earth Orbit. *Journal Life*, 2014, 4,491-510.
10. Ibid.
11. https://www.ncbi.nlm.nih.gov/pmc/articles/PMC5705828/
12. spiedigitallibrary.org/journals/journal-of-biomedical-optics/volume-13/issue-4/044030/Wavelength-dependent-penetration-depths-of-ultraviolet-radiation-in-human-skin/10.1117/1.2957970.full?SSO=1
13. https://www.uhms.org/11-delayed-radiation-injury-soft-tissue-and-bony-necrosis.html
14. Ibid.

15. https://www.cancer.org/cancer/cancer-causes/radiation-exposure/x-rays-gamma-rays/do-xrays-and-gamma-rays-cause-cancer.html
16. Author's Notes from NASA conference. (August 17, 2010)
17. Ibid.
18. Bailey. O.T.; et al. (1973). "26. Biocore Experiment". Apollo 17 Preliminary Science Report (NASA SPP-330).
19. https://www.ncbi.nlm.nih.gov/books/NBK26873/
20. Haymaker, Webb; Look, Bonne C.; Benton, Eugene V.; Simmonds, Richard C. (January 1, 1975). *"The Apollo 17 Pocket Mouse Experiment (Biocore)"*. In Johnston, Richard S.; Berry, Charles A.; Dietlein, Lawrence F. SP-368 Biomedical Results of Apollo (SP-368). Lyndon B. Johnson Space Center. OCLC 1906749.
21. Walker, S. A., Townsend, L. W. & Norbury, J. W. Heavy ion contributions to organ dose equivalent for the 1977 galactic cosmic ray spectrum. Adv. Space Res. 51, 1792–1799 (2013).
22. https://www.nature.com/articles/s41591-021-01637-7
23. Ibid.
24. Michael D. Delp, J. Charvat,C. et al., "Apollo Lunar Astronauts Show Higher Cardiovascular Disease Mortality: Possible Deep Space Radiation Effects on the Vascular Endothelium", *NATURE*, Scientific Reports volume 6, Article number: 29901 (2016)
25. Ibid.
26. Ibid.
27. Michael D. Delp, J. Charvat,C. et al., "Apollo Lunar Astronauts Show Higher Cardiovascular Disease Mortality: Possible Deep Space Radiation Effects on the Vascular Endothelium", *NATURE*, Scientific Reports volume 6, Article number: 29901 (2016)
28. Ibid.
29. https://www.ncbi.nlm.nih.gov/pubmed/18076878

Hiroshima Unit

1. Wakeford, Richard, "Japanese atomic bomb survivors", *Journal of Radiological Protection*, Volume 20, Number 3
2. www.seekhealthz.com/health/recommended-annual-dose-limits-for-radiation-exposure/
3. https://www.mayoclinicproceedings.org/article/S0025-6196(15)00591-1/addons

Hitting the Physiological Wall

1. https://www.thoughtco.com/animal-anatomy-epithelial-tissue-373206
2. https://www.inc.com/bill-murphy-jr/in-6-words-elon-musk-explained-why-so-many-people-are-afraid-to-take-risks-achieve-greatness.html
3. https://www.ncbi.nlm.nih.gov/pubmed/16187787
4. https://www.ncbi.nlm.nih.gov/pubmed/18439042
5. Ibid.
6. https://pubmed.ncbi.nlm.nih.gov/23300905/
7. https://www.ncbi.nlm.nih.gov/pmc/articles/PMC3534034/
8. Ibid.
9. https://www.nature.com/articles/srep34774
10. Ibid.
11. Ibid.
12. Munjal M. Acharya, Janet E. Baulch, Peter M. Klein, et al., "New Concerns for Neurocognitive Function during Deep Space Exposures to Chronic, Low Dose-Rate, Neutron Radiation.", ENEURO.0094-19.2019
13. https://www.ncbi.nlm.nih.gov/pmc/articles/PMC3548359/
14. https://link.springer.com/article/10.1007/BF01221035
15. Santosh Kumar, Shubhankar Suman, A. Fornace Jr., Kamal Datta, "Space radiation triggers persistent stress response, increases senescent signaling, and decreases cell

migration in mouse intestine." Proceedings of the National Academy of Sciences, pnas.1807522115
16. https://www.eurekalert.org/pub_releases/2018-10/gumc-ass092618.php
17. https://gumc.georgetown.edu/news-release/animal-study-suggests-deep-space-travel-may-significantly-damage-gi-function-in-astronauts/#
18. https://www.vox.com/science-and-health/2019/5/14/18306893/apollo-50-nasa-spaceflight-human-body-twin-study
19. Ibid.
20. https://morgridge.org/news/publications/fearless-science/pagliarini/
21. Cucinotta F, Nikjoo H, Goodhead D. The effects of delta rays on the number of particle-track transversals per cell in laboratory and space exposures. Radiat Res 1998; 150: 115–19.
22. Alston, C., Rocha, M., Lax, N., Turnbull, D., and Taylor, R. (2017). The genetics and pathology of mitochondrial disease. *J. Pathol.* 241, 236–250. doi: 10.1002/path.4809
23. https://www.frontiersin.org/articles/10.3389/fphys.2020.00955/full
24. Nissanka, N., and Moraes, C. T. (2018). Mitochondrial DNA damage and reactive oxygen species in neurodegenerative disease. *CT FEBS Lett.* 592, 728–742. doi: 10.1002/1873-3468.12956
25. Ibid.
26. Harman, D. (1972). The biologic clock: the mitochondria? *J. Am. Geriatr. Soc.* 20, 145–147. doi: 10.1111/j.1532-5415.1972.tb00787.
27. Alwood, J, Ronca, A, Mains, R, et al. (2017). From the bench to exploration medicine: NASA life sciences translational research for human exploration and habitation missions. *NPJ Microgravity* 3:5. doi: 10.1038/s41526-016-0002-8
28..nasa.gov/sites/default/files/atoms/files/radiation_protection_technical_brief_ochmo_021420.pdf
29. "Cancer Risk Assessments for Astronauts Participating in Deep Space Missions", Tony Slaba, NASA Langley Research Center, Radiation and Flight – A Down to Earth Look at Risks, National Council on Radiation Protection and Measurements (NCRP) 57th Annual Mtg, April 2021

Countermeasure Madness

1. Kelly, Scott (2017). Endurance: A Year in Space, a Lifetime of Discovery. With Margaret Lazarus Dean. Alfred A. Knopf, division of Penguin Random House.
2. Ibid.
3. https://www.wired.com/story/ideas-jason-pontin-genetic-engineering-for-mars/

Why Do We Persist?

1. https://press.princeton.edu/books/audio/9780691193472/irrationality
2. https://www.technologyreview.com/s/604142/engineering-the-perfect-astronaut/
3. www.bbc.com/future/story/20171123-will-we-ever-have-genetically-modified-astronauts
4. https://www.space.com/9439-biologist-space-travelers-benefit-genetic-engineering.html
5. Chamberland, Dennis; Undersea Colonies, Quantum Editions, 2007

Strategic Necessity

1. https://www.usatoday.com/story/tech/nation-now/2017/01/18/nasa-planning-mission-asteroid-worth-10000-quadrillion/96709250/
2. https://solarsystem.nasa.gov/asteroids-comets-and-meteors/asteroids/16-psyche/in-depth/
3. https://news.ycombinator.com/item?id=358351
4. O'Neill, Gerard K., The High Frontier, 1977
5. https://www.space.com/us-astronauts-moon-return-by-2024.html
6. Ibid.
7. https://suntzusaid.com/book/9
8. https://foreignpolicy.com/2016/04/28/the-asteroid-miners-guide-to-the-galaxy-space-race-mining-asteroids-planetary-research-deep-space-industries/

9. https://www.iflscience.com/space/who-owns-space-us-asteroid-mining-act-dangerous-and-potentially-illegal/

What We Risk

1. https://www.ibtimes.com/nasa-mars-mission-astronauts-admit-good-chance-death-due-harsh-environment-watch-2850121
2. https://www.space.com/36800-five-ways-to-die-on-mars.html
3. https://thenextweb.com/science/2019/10/02/elon-musks-plan-for-sending-people-to-mars-is-probably-a-suicide-mission/
4. https://www.wired.co.uk/article/mars-mission-astronauts-must-prepare-for-non-survivable-illnesses-or-injuries
5. https://www.vice.com/en_us/article/4x3wbd/nasa-cant-ethically-send-astronauts-on-one-way-missions-to-deep-space
6. https://www.ncbi.nlm.nih.gov/books/NBK222144/

Power of Knowing Ourselves

1. https://qz.com/1132948/why-philosophy-is-so-important-in-science-education/
2. https://oig.nasa.gov/docs/IG-16-003.pdf

Familiarizing the Unknown

1. Pausanias, Description of Greece, 10.24
2. https://plato.stanford.edu/entries/vico/
3. https://mars.nasa.gov/programmissions/science/

Mind of Rickover

1. https://www.presidency.ucsb.edu/documents/remarks-promotion-ceremony-for-admiral-hyman-g-rickover
2..https://www.dixonvalve.com/sites/default/files/publications/articles/From_Steerage_to_Four_Stars_summer_2007_DIXBOA.pdf
3. Rockwell, Theodore (2002). *The Rickover Effect*. Lincoln, NE: IUniverse. p. 155. *ISBN* 0-595-25270-2.
4..https://www.dixonvalve.com/sites/default/files/publications/articles/From_Steerage_to_Four_Stars_summer_2007_DIXBOA.pdf
5. *PT-109 - An American Epic of War, Survival, and the Destiny of John F. Kennedy*, William Doyle, Harper Collins, October 2015
6. http://www.astronautix.com/craft/spaeedom.htm
7. https://www.washingtonpost.com/technology/2021/12/21/us-russia-space-station-tension/
8. https://en.wikipedia.org/wiki/Space_Station_Freedom
9. https://www.cnn.com/2016/10/11/opinions/america-will-take-giant-leap-to-mars-barack-obama/index.html
10. https://www.nasa.gov/topics/moon-to-mars
11. https://althistory.fandom.com/wiki/Admiral_Rickover_(Space_Race_Didn%27t_End)
12. https://www.azquotes.com/quote/730278
13. Rockwell, Theodore (2002)*The Rickover Effect*. Lincoln, NE: IUniverse. ISBN 0-595-25270-2
14. "Thoughts on Man's Purpose in Life", at San Diego Rotary Club (10 February 1977
15. https://www.cee.org/sites/default/files/media-pdfs/Admiral_Rickover_brought_the_nuclear_age_to_the_US_Navy.pdf

Alien Mind

1. http://www.seasky.org/quotes/space-quotes-space-travel.html
2. http://discerninghistory.com/2013/05/shackletons-ad-men-wanted-for-hazerdous-journey/
3. Ibid.
4. Endurance: Shackleton's Incredible Voyage, Lansing, Alfred; 1959

Getting It Right

1. https://www.nrc.gov/reading-rm/doc-collections/fact-sheets/bio-effects-radiation.html

2. https://arstechnica.com/science/2019/02/nasa-nears-50-billion-for-deep-space-plans-yet-human-flights-still-distant/
3. https://www.nextbigfuture.com/2019/07/28-billion-into-sls-through-2019-and-59-69-billion-total-cost-sls-by-2024.html
4. https://www.rocketlaunch.live/?tag=series-nasa-artemis
5. "Fiche Technique: Falcon Heavy" [Technical data sheet: Falcon Heavy]. Espace & Exploration. No. 51. June 2019. pp. 62–63.

Made in United States
Orlando, FL
04 November 2026

90201494R00364